EOS Annual Meeting (EOSAM 2021)

EPJ Web of Conferences Volume 255 (2021)

Rome, Italy
13 - 17 September 2021

Editors:

Concita Sibilia
Alessandro Belardini
Gilles Pauliat

ISBN: 978-1-7138-3916-3

Printed from e-media with permission by:

Curran Associates, Inc.
57 Morehouse Lane
Red Hook, NY 12571

Some format issues inherent in the e-media version may also appear in this print version.

No changes have been made to the content of these proceedings. There may be changes to pagination and minor adjustments for aesthetics.

Printed with permission by Curran Associates, Inc. (2021)

For additional information, please contact EDP Sciences – Web of Conferences at the address below.

EDP Sciences – Web of Conferences
17, Avenue du Hoggar
Parc d'Activité de Courtabœuf
BP 112
F-91944 Les Ulis Cedex A
France

Phone: +33 (0) 1 69 18 75 75
Fax: +33 (0) 1 69 28 84 91

contact-edps@webofconferences.org

Additional copies of this publication are available from:

Curran Associates, Inc.
57 Morehouse Lane
Red Hook, NY 12571 USA
Phone: 845-758-0400
Fax: 845-758-2633
Email: curran@proceedings.com
Web: www.proceedings.com

TABLE OF CONTENTS

Preface

The 2021 annual congress of the European Optical Society (EOS), EOSAM, has been celebrated hybrid, both online and onsite at the same time for the first time in its history. This has been a bold yet necessary movement, forced by the exceptional circumstances of the global pandemic still faced world-wide. The hybrid format has been a great success and the response from the European optical community has risen to the occasion as is shown by the number and quality of the communications presented that are published in this volume.

The conference was held in September 13-17 in Rome Italy, bringing back the highly awaited international in-person events. EOS organized the conference in close co-operation with the Società Italiana di Ottica e Fotonica, SIOF, the Italian optical society Branch member of EOS, and the Università di Roma La Sapienza.

The hybrid conference has been extremely successful in its contribution to the main mission of the EOS: bring together and encourage the cooperation in Europe of all with an interest in optics, photonics and related scientific fields, to make practical use of research results, and to support the industrial exploitation of optics and photonics.

In total, 445 talks and posters were presented at the conference The five-day onsite and online conference included several plenary talks, tutorials and other invited presentations given by outstanding scientists working in the most important areas of optics and photonics. In addition, submitted contributions were accepted on all topics, as oral and poster presentations.

Each session was held onsite in Rome, and streamed to attendees online. Discussions were held onsite together with online participants and speakers. In addition, a call was made for submission of selected papers for publication in JEOS:RP, the journal of the European Optical Society, covering the latest advances in both theory and applications of optics and photonics grouped on the following TOMs (Topical Meetings):

TOM1 Silicon Photonics and Guided-Wave Optics

TOM2 Computational, Adaptive and Freeform Optics

TOM3 Optical System Design, Tolerancing and Manufacturing

TOM4 Bio-Medical Optics

TOM5 Resonant Nanophotonics

TOM6 Optical Materials: crystals, thin films, organic molecules and polymers, syntheses, characterization and devices

TOM7 Thermal radiation and energy management

TOM8 Nonlinear and Quantum Optics

TOM9 Optics at Nanoscale (ONS)

TOM10 Optical Microsystems (OMS)

TOM11 Waves in Complex Photonic Media

TOM12 Optofluidics

TOM13 Ultrafast Optical Technologies and Applications

TOM14 Advances and Applications of Optics and Photonics

EU Project Session

Early Stage Researcher Session organised by SIOF

Grand Challenges of Photonics Session

We would like to express our gratitude to the European Optical Society, the Italian Optics and Photonics Society (SIOF), Università di Roma La Sapienza, and the chairs of the different TOMs for their support and help in the organization of the Conference. We finally want to acknowledge all the authors that have submitted their work to be included in this volume and very specially thank all those collaborating persons who acted as reviewers of the papers included in this volume for their excellent work.

The editors:

Concita Sibilia, Sapienza Università di Roma, Italy, Local Chair
Alessandro Belardini, Sapienza Università di Roma, Italy, Local Chair
Gilles Pauliat, Université Paris-Saclay, Institut d'Optique Graduate School, CNRS, Palaiseau, France, President of EOS

Statement of Peer review

In submitting conference proceedings to *Web of Conferences*, the editors of the proceedings certify to the Publisher that

1. They adhere to its **Policy on Publishing Integrity** in order to safeguard good scientific practice in publishing.
2. All articles have been subjected to peer review administered by the proceedings editors.
3. Reviews have been conducted by expert referees, who have been requested to provide unbiased and constructive comments aimed, whenever possible, at improving the work.
4. Proceedings editors have taken all reasonable steps to ensure the quality of the materials they publish and their decision to accept or reject a paper for publication has been based only on the merits of the work and the relevance to the journal.

Title, date and place of the conference EOSAM 2021, 13-17 September 2021, Rome, Italy

Proceedings editor(s):

Alessandro Belardini

Date and editor's signature

02/11/2021

17, avenue du Hoggar - PA de Courtabœuf – BP 112 - 91944 Les Ulis Cedex A (France)
Tél. : 33 (0)1 69 18 75 75 - Fax : 33(0)1 69 07 45 17 – www.edpsciences.org

edp sciences

Web of Conferences

Statement of Peer review

In submitting conference proceedings to *Web of Conferences*, the editors of the proceedings certify to the Publisher that

1. They adhere to its **Policy on Publishing Integrity** in order to safeguard good scientific practice in publishing.
2. All articles have been subjected to peer review administered by the proceedings editors.
3. Reviews have been conducted by expert referees, who have been requested to provide unbiased and constructive comments aimed, whenever possible, at improving the work.
4. Proceedings editors have taken all reasonable steps to ensure the quality of the materials they publish and their decision to accept or reject a paper for publication has been based only on the merits of the work and the relevance to the journal.

Title, date and place of the conference

Proceedings editor(s):

Concita Sibilia

Date and editor's signature November 5th, 2021

17, avenue du Hoggar PA de Courtabœuf BP 112 - 91944 Les Ulis Cedex A (France)
Tél 33 (0)1 69 18 75 75 Fax 33(0)1 69 07 45 17 – www.edpsciences.org

edp sciences

Web of Conferences

Statement of Peer review

In submitting conference proceedings to *Web of Conferences*, the editors of the proceedings certify to the Publisher that

1. They adhere to its **Policy on Publishing Integrity** in order to safeguard good scientific practice in publishing.
2. All articles have been subjected to peer review administered by the proceedings editors.
3. Reviews have been conducted by expert referees, who have been requested to provide unbiased and constructive comments aimed, whenever possible, at improving the work.
4. Proceedings editors have taken all reasonable steps to ensure the quality of the materials they publish and their decision to accept or reject a paper for publication has been based only on the merits of the work and the relevance to the journal.

Title, date and place of the conference EOSAM 2021, Rome, Sept 13,17, 2021

Proceedings editor(s): Gilles PAULIAT

Date and editor's signature Nov. 2nd 2021

17, avenue du Hoggar - PA de Courtabœuf – BP 112 - 91944 Les Ulis Cedex A (France)
Tél. : 33 (0)1 69 18 75 75 - Fax : 33(0)1 69 07 45 17 – www.edpsciences.org

 https://doi.org/10.1051/epjconf/202125501001

Building high-performance integrated optical devices using subwavelength grating metamaterials -INVITED

Alejandro Sánchez-Postigo[1,*], *Pablo* Ginel-Moreno[1], *Alejandro* Ortega-Moñux[1], *J. Gonzalo* Wangüemert-Pérez[1], *Robert* Halir[1], *Daniel* Pereira-Martín[1], *Abdelfettah* Hadij-ElHouati[1], *Jens H.* Schmid[2], *Shurui* Wang[2], *Martin* Vachon[2], *Dan-Xia* Xu[2], *Winnie N.* Ye[3], *Jordi* Soler Penadés[4,5], *Milos* Nedeljkovic[4], *Goran Z.* Mashanovich[4], *Pavel* Cheben[2], and *Íñigo* Molina-Fernández[1]

[1]Telecommunication Research Institute (TELMA), Universidad de Málaga, CEI Andalucía TECH, Louis Pasteur 35, 29010 Málaga, Spain

[2]National Research Council Canada, 1200 Montreal Road, Bldg. M50, Ottawa K1A 0R6, Canada

[3]Department of Electronics, Carleton University, 1125 Colonel By Drive, Ottawa, Ontario K1S5B6, Canada

[4]Optoelectronics Research Centre, University of Southampton, Southampton, SO17 1BJ, United Kingdom

[5]Currently at VLC Photonics S.L., Valencia, Spain

Abstract. The use of subwavelength grating structures in silicon waveguides have fuelled the development of integrated optical components with superior performance. By a judicious lithographic patterning of the grating, the optical properties of the synthesized metamaterial can be accurately tailored. In this work, we review our latest advances in subwavelength-grating-engineered silicon and germanium planar devices.

1 Introduction

Subwavelength grating (SWG) metamaterials are becoming established as an advantageous tool to manipulate light in integrated photonic devices [1–3]. SWG nanophotonic waveguides comprise an arrangement of structures with a period (Λ) that is smaller than half the effective wavelength in the waveguide (i.e., $\Lambda < 0.5\lambda/n_B$, where n_B is the effective index of the fundamental Bloch–Floquet mode that is guided). Under this condition, the SWG structure behaves as an anisotropic homogeneous artificial material. The optical properties of the SWG waveguide, including the refractive index contrast and the wavelength dispersion, can be controlled by choosing the geometrical parameters (e.g., period and duty cycle) of the lithographically defined patterns. In this work, we leverage the refractive-index engineering possibilities of SWG metamaterials to implement high-performance waveguide-based devices for the near- and mid-infrared wavelength ranges.

2 Broadband surface grating couplers with sub-decibel coupling efficiency

Efficient and broadband fibre-to-chip coupling is a major challenge in silicon photonics [4]. To address this problem, surface grating couplers are often used, as they allow wafer-scale testing and provide robust misalignment tolerances. These devices are diffractive, periodic waveguide structures that satisfy the phase matching condition between a guided Bloch–Floquet mode and a free-propagating off-chip plane wave, as described by [5]

$$n_a \sin(\theta) = n_B + m\frac{\lambda}{\Lambda}, \qquad (1)$$

where n_a is the refractive index of the medium to which the leaky Bloch–Floquet mode is radiated, θ is the radiation angle, n_B is the effective index of the mode, m is the radiation order, λ is the operating wavelength, and Λ is the grating period. To collect most radiated power, an optical fibre is conveniently positioned, with a given tilt angle, above the surface grating coupler.

The two important parameters in surface grating couplers are the coupling efficiency (CE) and the 1-dB bandwidth (BW). For a given fibre mode field diameter, there is a trade-off between these two parameters and the coupling efficiency–bandwidth product (CE·BW) is an important figure of merit. We have recently shown how to break the coupling efficiency–bandwidth product limit of surface grating couplers by using SWG metamaterials and a silicon prism to radiate in the zeroth diffraction order [6].

Our proposed zero-order grating coupler, illustrated in Fig. 1(a), consists of an SWG waveguide of pitch Λ and duty cycle DC = a/Λ. By operating in the subwavelength regime, all diffraction orders are frustrated. However, by positioning a silicon prism ($n_a \sim 3.45 > n_B$) on top of the subwavelength grating, radiation is enabled for m = 0. Therefore, in a ZGC, the phase matching condition becomes

$$n_a \sin(\theta) = n_B, \qquad (2)$$

* Corresponding author: asp@ic.uma.es

 https://doi.org/10.1051/epjconf/202125501001

where the dependency between the radiation angle and the wavelength only arises from the silicon refractive index [$n_a = n_a(\lambda)$] and the grating effective index [$n_B = n_B(\lambda)$], thus increasing the bandwidth with respect to conventional grating couplers. The radiation strength, and hence the coupling efficiency, depends on the prism angle ϕ_{prism}, the tilt angle ϕ_{clad}, and the initial thickness t_0 of the wedge-shaped air gap between the prism and the grating.

The light injected into the grating from the input waveguide is evanescently coupled to the prism and then into an SMF-28 optical fibre via an anti-reflection coated facet. As shown in Fig. 1(b), a two-stage transition at the beginning of the grating adapts the fundamental mode of the input waveguide (width W = 500 nm) to the Bloch–Floquet mode of the grating waveguide (width W_g = 15 μm). This transition is composed of a 200-μm-long linear taper and a 3-μm-long fork-like structure formed by a transverse series of trapezoids (W_{in} = 500 nm, W_{out} = 300 nm).

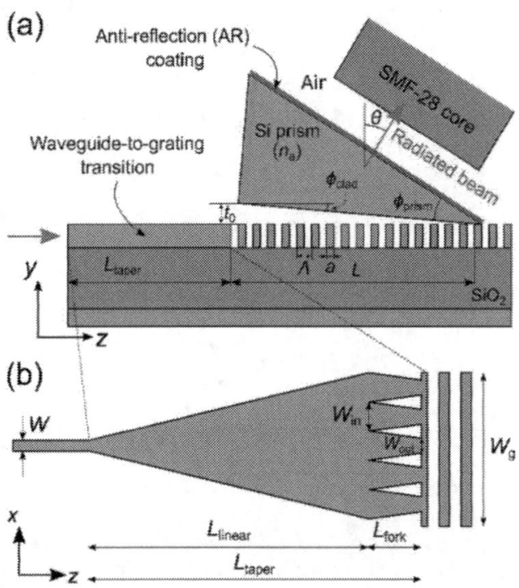

Fig. 1. (a) Schematic of a zero-order fiber-chip coupler (side view). (b) Schematic of the waveguide-to-grating transition (top view) used to adapt the guided mode of the input homogeneous waveguide to the Bloch–Floquet mode supported by the zero-order grating coupler.

The metamaterial refractive index is engineered to minimize the effective index dispersion of the grating waveguide and therefore maximize the bandwidth. Specifically, we choose a pitch of 200 nm and a duty cycle of 50%, which lead to a minimum feature size of 100 nm compatible with standard deep-ultra-violet (DUV) lithography processes. For this SWG waveguide, a theoretical coupling efficiency >90% can be achieved by using a prism with ϕ_{prism} = 35.7° and length L = 20.3 μm, tilted ϕ_{clad} = 1° (t_0 = 355 nm).

Using the zero-order grating coupler, we achieve experimentally a coupling efficiency < 1 dB and a 1-dB bandwidth of 94 nm, yielding a CE-BW product exceeding 75 nm, the highest performance yet reported for a silicon photonics surface grating coupler.

3 Millimetre-long optical antennas with DUV-compatible feature sizes

The rising need for compact, lightweight, and inexpensive light detection and ranging (LiDAR) systems and free-space communication transceivers has fuelled research in silicon photonics optical phased arrays (OPAs) [7,8]. Integrated optical antennas are the key components of OPAs and can be designed to enable rapid, precise, and non-mechanical steering of optical beams [9–12].

To shape high-resolution, ultra-narrow beams in the far field, grating antennas with very low radiation strengths and lengths of several millimetres are required. Achieving such long antennas in high-index-contrast platforms such as silicon on insulator makes necessary the lithographic patterning of very small perturbations (corrugations below 10 nm). This requirement hinders fabrication using typical 248-nm and 193-nm DUV lithography, with fabricable minimum feature sizes between 70 nm and 100 nm [13].

We have recently reported a millimetre-long silicon optical antenna composed of an SWG waveguide core that is loaded with evanescently coupled radiative elements (see Fig. 2) [14]. The SWG metamaterial allows us to delocalize the propagating mode to accurately control the radiation strength. We choose an SWG core W of 650 nm, a period Λ_{SWG} of 170 nm, and a duty cycle (a/Λ_{SWG}) of ~53%. The loading segments, of width W_s = 200 nm and length L_s = 200 nm, are separated a distance g of 300 nm from the SWG core. Radiation is enabled by using a period Λ_g = 680 nm. Our experiments confirm a record beamwidth of ~0.1° for a minimum feature size of 80 nm and a wavelength sensitivity of ~0.13°/nm.

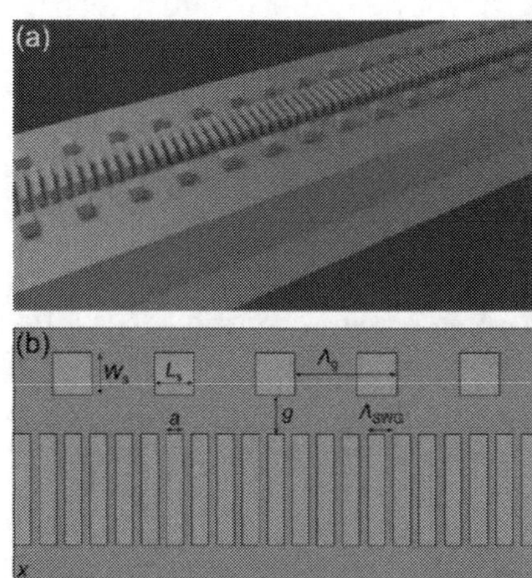

Fig. 2. Schematic (a) 3D view and (b) top view of a millimetre-long optical antenna with SWG waveguide core and evanescently coupled loading segments.

4 Suspended germanium waveguide platform for the mid-infrared band

The mid-infrared band has attracted great attention for important emerging applications such as biochemical sensing and free-space optical communications [15]. The high material losses of silicon dioxide and silicon above wavelengths of ~4 μm and ~8 μm, respectively, prevent the use of the SOI platform in the longwave range. Germanium, a CMOS-compatible semiconductor that is transparent up to ~15-μm wavelength, can be used to develop waveguide-based devices at long wavelengths [16].

We have recently demonstrated a suspended germanium platform, based on Ge-on-SOI, operating at a wavelength of 7.7 μm [17]. The waveguide is defined by etching an SWG cladding that anchors the waveguide core to lateral slabs. This cladding synthesizes a metamaterial that is judiciously engineered to provide the lateral index contrast required for light guiding. The buried silicon and silicon-dioxide layers of the Ge-on-SOI wafer are removed by a hydrofluoric (HF) acid solution, which flows through the venting holes of the lateral SWG cladding. The designed suspended waveguide has a germanium guiding layer with a nominal thickness of 1 μm. The core and cladding widths are 2.7 μm and 3.8 μm, respectively, while the SWG lateral cladding has a period of 1 μm and a duty cycle of 20%. Figure 3 shows a scanning-electron-microscope (SEM) image of one of the fabricated suspended waveguides. At a wavelength of 7.7 μm, we measured a propagation loss of 5.3 ± 1.0 dB/cm. Furthermore, by simulation we have confirmed that the proposed platform can be adapted to operate with low loss up to a wavelength of 15 μm.

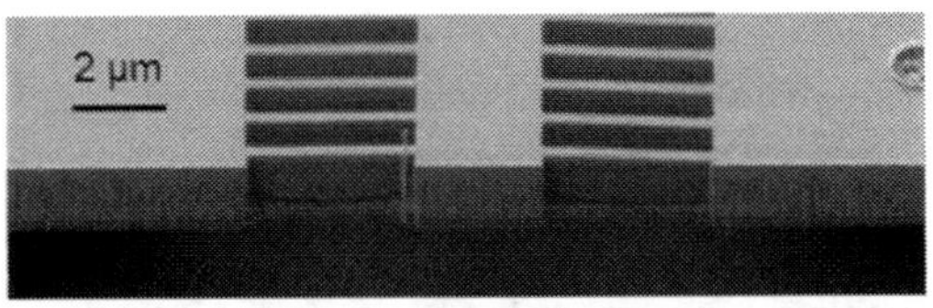

Fig. 3. SEM image of a suspended germanium waveguide designed to operate at a wavelength of 7.7 μm.

To couple light into the suspended germanium waveguides from a commercially available, high-numerical-aperture chalcogenide optical fibre, we utilized suspended micro-antennas [18]. These couplers exhibit a respectable coupling efficiency of ~40%, a broad 1-dB spectral bandwidth in excess of 430 nm, and a 1-dB angular bandwidth of ~20°.

These results demonstrate the practical use of subwavelength-grating metamaterials in the longwave infrared regime.

We acknowledge funding from Universidad de Málaga, Ministerio de Ciencia, Innovación y Universidades (MCIU) (PID2019-106747RB-I00), Consejería de Economía, Conocimiento, Empresas y Universidad (CECEU) (UMA18-FEDERJA-219, P18-RT-1453, P18-RT-793), and National Research Council of Canada (NRC) Collaborative Science, Technology and Innovation Program (CSTIP) (HTSN 209).

References

1. P. Cheben, R. Halir, J. H. Schmid, H. A. Atwater, and D. R. Smith, Nature **560**, 565–572 (2018).
2. R. Halir, A. Ortega-Moñux, D. Benedikovic, G. Z. Mashanovich, J. G. Wangüemert-Pérez, J. H. Schmid, Í. Molina-Fernández, and P. Cheben, Proc. IEEE **106**, 2144–2157 (2018).
3. J. M. Luque-González, A. Sánchez-Postigo, A. Hadij-ElHouati, A. Ortega-Moñux, J. G. Wangüemert-Pérez, J. H. Schmid, P. Cheben, Í. Molina-Fernández, and R. Halir, Nanophotonics **10**, 2765–2797 (2021).
4. R. Marchetti, C. Lacava, L. Carroll, K. Gradkowski, and P. Minzioni, Photonics Res. **7**, 201–239 (2019).
5. T. Tamir and S. T. Peng, Appl. Phys. **14**, 235–254 (1977).
6. A. Sánchez-Postigo, R. Halir, J. G. Wangüemert-Pérez, A. Ortega-Moñux, S. Wang, M. Vachon, J. H. Schmid, D. Xu, P. Cheben, and Í. Molina-Fernández, Laser Photon. Rev. 2000542 (2021).
7. M. J. R. Heck, Nanophotonics **6**, 93–107 (2017).
8. J. He, T. Dong, and Y. Xu, IEEE Access **8**, 188284–188298 (2020).
9. T. Kim, T. Ngai, Y. Timalsina, M. R. Watts, V. Stojanovic, P. Bhargava, C. V. Poulton, J. Notaros, A. Yaacobi, E. Timurdogan, C. Baiocco, N. Fahrenkopf, and S. Kruger, IEEE J. Solid-State Circuits **54**, 3061–3074 (2019).
10. C. V. Poulton, A. Yaacobi, D. B. Cole, M. J. Byrd, M. Raval, D. Vermeulen, and M. R. Watts, Opt. Lett. **42**, 4091–4094 (2017).
11. W. Ma, S. Tan, K. Wang, W. Guo, Y. Liu, L. Liao, L. Zhou, J. Zhou, X. Li, L. Liang, and W. Li, Appl. Opt. **59**, 9985–9994 (2020).
12. Q. Wang, S. Wang, L. Jia, Y. Cai, W. Yue, and M. Yu, Opt. Express **29**, 10509–10517 (2021).
13. M. Kamandar Dezfouli, Y. Grinberg, D. Melati, P. Cheben, J. H. Schmid, A. Sánchez-Postigo, A. Ortega-Moñux, G. Wangüemert-Pérez, R. Cheriton, S. Janz, and D.-X. Xu, Opt. Lett. **45**, 3701–3704 (2020).
14. P. Ginel-Moreno, A. Sánchez-Postigo, J. De-Oliva-Rubio, A. Hadij-ElHouati, W. N. Ye, J. G. Wangüemert-Pérez, Í. Molina-Fernández, J. H. Schmid, P. Cheben, and A. Ortega-Moñux, Opt. Lett. **46**, 3733–3736 (2021).
15. R. A. Soref, S. J. Emelett, and W. R. Buchwald, J. Opt. A Pure Appl. Opt. **8**, 840–848 (2006).
16. R. Soref, Nat. Photonics **4**, 495–497 (2010).
17. A. Sánchez-Postigo, A. Ortega-Moñux, J. Soler Penades, A. Osman, M. Nedeljkovic, Z. Qu, Y. Wu, I. Molina-Fernández, P. Cheben, G. Mashanovich, and J. G. Wangüemert-Pérez, Opt. Express **29**, 16867–16878 (2021).
18. A. Sánchez-Postigo, A. Ortega-Moñux, D. Pereira-Martín, Í. Molina-Fernández, R. Halir, P. Cheben, J. Soler Penadés, M. Nedeljkovic, G. Z. Mashanovich, and J. G. Wangüemert-Pérez, Opt. Express **27**, 22302–22315 (2019).

Heterostructured silicon-germanium-silicon p-i-n avalanche photodetectors for chip-integrated optoelectronics -INVITED

Daniel Benedikovic[1,2,*], *Leopold* Virot[3], *Guy* Aubin[2], *Jean-Michel* Hartmann[3], *Farah* Amar[2], *Xavier* Le Roux[2], *Carlos* Alonso-Ramos[2], *Milan* Dado[1], *Eric* Cassan[2], *Delphine* Marris-Morini[2], *Jean-Marc* Fedeli[3], *Frederic* Boeuf[4], *Bertrand* Szelag[3] and *Laurent* Vivien[2]

[1]Dept. Multimedia and Information-Communication Technologies, University of Žilina, ,01026 Žilina, Slovakia
[2]UniversitéParis-Saclay, CNRS, Centre de Nanosciences et de Nanotechnologies, 91120 Palaiseau, France
[3]University Grenoble Alpes and CEA, LETI, 38054 Grenoble, France
[4]STMicroelectronics, Silicon Technology Development, 38923 Crolles, France

Abstract. Optical photodetectors are at the forefront of photonic research since the rise of integrated optics. Photodetectors are fundamental building blocks for chip-scale optoelectronics, enabling conversion of light into an electrical signal. Such devices play a key role in many surging applications from communication and computation to sensing, biomedicine and health monitoring, to name a few. However, chip integration of optical photodetectors with improved performances is an on-going challenge for mainstream optical communications at near-infrared wavelengths. Here, we present recent advances in heterostructured silicon-germanium-silicon p-i-n photodetectors, enabling high-speed detection on a foundry-compatible monolithic platform.

1 INTRODUCTION

Germanium photodetectors integrated with sub-micron silicon waveguides are essential building elements for many applications. This includes sensing, health monitoring, biomedicine or optical interconnects and communications. Silicon-germanium photodetectors are now mature components and their opto-electrical performances are rather compelling. A wide variety of chip-integrated photodetectors have been proposed and demonstrated in recent years [1]. Heterostructured silicon-germanium-silicon photodiodes with a butt-waveguide-coupling scheme and a lateral p-i-n electrical junction are promising to detect light on a silicon chip [2-5]. Cross-sectional schematics and an optical microscopy image of such heterostructured p-i-n photodetectors are shown in Fig. 1a and 1b, respectively. This unique integration strategy facilitates the development of compact devices with properly engineered waveguide features. Such photodiodes benefit from improved modal confinement and overall control over the intrinsic region. They are also easier to fabricate [2,3]. Such a photodiode configuration also yields enhanced performances (lower dark-current levels, better responsivities and faster operations) compared to full-germanium architectures [6] or metal-semiconductor-metal [7] alternatives. Moreover, heterostructured p-i-n photodiodes with silicon-germanium-silicon heterojunction leverage the maturity of foundry-based silicon processes and germanium epitaxial growth. A robust monolithic integration platform is then obtained, with complex passive and active chip functionalities simultaneously achieved.

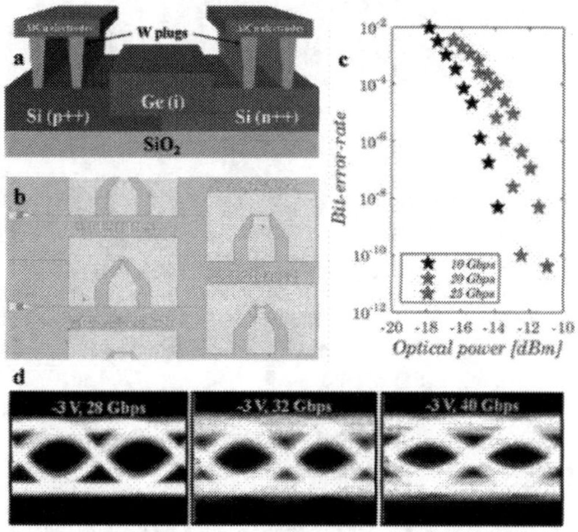

Fig. 1. (a) Schematic view of a p-i-n waveguide photodetector with a lateral silicon-germanium-silicon heterojunction. (b) Optical microscopy image of fabricated p-i-n photodetectors. (c) Bit-error-rate measurements as a function of optical input power for silicon-germanium-silicon photodetector at conventional 10 Gbps, 20 Gbps and 25 Gbps transmission bit rates and a 3 V reverse bias. (d) Eye diagram apertures at 28 Gbps, 32 Gbps and 40 Gbps, under the same low-voltage operation (with respective scale adapted).

*Corresponding author: daniel.benedikovic@c2n.upsaclay.fr & daniel.benedikovic@uniza.sk

Herein, we present recent progress in waveguide-integrated p-i-n photodetectors with lateral silicon-germanium-silicon hetero-junctions. Fabricated p-i-n photodiodes were operated under low-voltage (*"p-i-n regime"*) and high-voltage (*"avalanche regime"*) conditions, respectively, delivering high-speed and low-noise detection at a mainstream telecommunication waveband centered at 1.55 μm.

2 Opto-electrical performance of heterostructured photodetectors

Heterostructured silicon-germanium-silicon photodetectors of different widths and different lengths were fabricated on 200-mm silicon-on-insulator substrates using industrial-scale silicon processes, aiming at both low-voltage and avalanche operations.

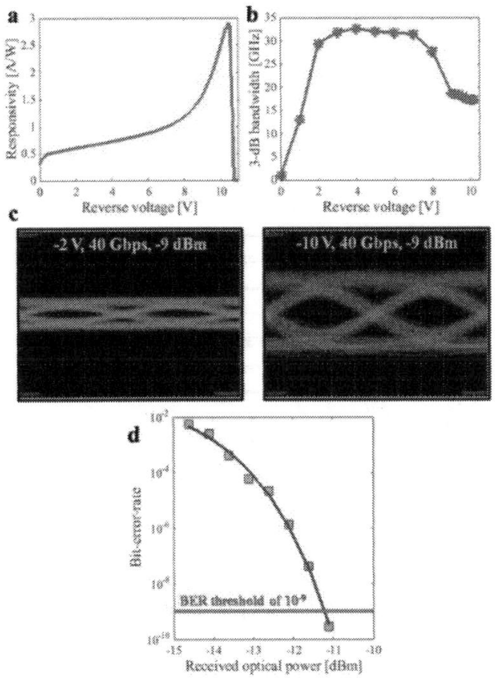

Fig.2. (a) Responsivity-voltage and (b) bandwidth-voltage characteristics of heterostructured silicon-germanium-silicon photodetector operated in an avalanche regime. (c) Output eye diagrams (with same vertical scale) for a 40 Gbps bit rate input signal at 1.55 μm carrier wavelength and different reverse bias conditions of 2 V and 10 V, with static avalanche multiplication gains of 1.4 and 5.5, respectively. (d) Bit-error-rate as a function of a received optical power for a silicon-germanium-silicon photodetector operated under an avalanche regime measured with a 40 Gbps signal, a reverse operating voltage of 10 V and an avalanche net-light gain of 7.5.

Under low-voltage operation, silicon-germanium-silicon photodetectors exhibited dark currents in the 5 nA to 150 nA range for the smallest up to the largest device geometries [5]. The corresponding dark-current densities were thus in the 0.404 A/cm^2 to 0.808 A/cm^2 range. Device responsivities of 0.20 A/W and 1.2 A/W were measured at 0.5 V reverse bias for the shortest and the longest photodetector designs [3,4]. Interestingly, the photodiode responsivities reached their maxima at reverse voltages as low as 0.5 V. These trends were consistently observed on all tested devices. Beyond a 0.5 V reverse bias, the photo-responsivity remained virtually flat, with small voltage dependence [4]. The strong built-in electrical field at low bias states showed up the great ability of such heterostructured p-i-n devices to sweep out the majority of the electron–hole pairs within their lifetime. Moreover, the -3 dB bandwidth was inversely proportional to the intrinsic region width, while the bandwidth remained unchanged for devices of different lengths and a fixed width [4,5]. This may indicate that the frequency response of the heterostructured p-i-n photodiode was not limited by the resistance-capacitance-delay. On the other hand, the device speed limit stemmed mainly from the transit time of photo-carriers. Measured -3 dB bandwidths were in the 7 GHz to 35 GHz range [5].

Figure 1c shows bit-error-rate results obtained for 1 μm wide by 40 μm long silicon-germanium-silicon photodiode. Measurements were carried out with a pseudo-random bit sequence (PBRS) at 10 Gbps, 20 Gbps, and 25 Gbps transmission bit rates, yielding optical power sensitivities of -14 dBm, -13 dBm, and -11 dBm for a 10^{-9} error-rate threshold [4]. In addition, Fig. 1d illustrates low-voltage operations, precisely here at 3 V reverse bias with electrical eye diagram apertures that remained open at 28 Gbps, 32 Gbps, and 40 Gbps. Such results pave a way for an on-chip detection of high-speed optical signals using silicon-germanium-silicon p-i-n photodetectors driven with low operating voltages.

However, the photo-responsivity is limited in conventional p-i-n photodiodes, because of a reduced external quantum efficiency. Optical power sensitivities are then modest for high-bit-rate optical signals and additional amplification circuits are mandatory for use in transmission networks and systems [1].

Avalanche photodetectors are very good alternatives to such device structures, as they can tackle many challenges. Avalanche photodetectors indeed leverage, under generated high electrical fields, an impact ionization effect, resulting in a significant internal multiplication gain. This internal gain is then harnessed to improve the diode performances in terms of device sensitivity, operation speed and generated noise [8-15].

Figures 2a and 2b show the photo-responsivity and the -3 dB bandwidth as a function of reverse bias for a 0.5 μm wide and 40 μm long silicon-germanium-silicon photodetector. Operating the silicon-germanium-silicon p-i-n photodiode in an avalanche regime substantially improves its photo-responsivity, from a nominal value of 0.5 A/W at low reverse voltages up to almost 3 A/W close to avalanche breakdown. This improvement corresponds to an avalanche multiplication net-light gain of about 6. The transition from an absorption regime towards an avalanche regime is otherwise rather gentle. On the other hand, the -3 dB bandwidth of the device is reduced from its maximum of about 33 GHz (for 3V up to 7V reverse bias) down to 16 GHz (for reverse bias of 10V). This device bandwidth drop is caused by the avalanche built-up time. Operating the diode in an avalanche regime provides the means to improve the signal detection. This is clearly shown in Fig. 2c. Here, electrical eye diagrams for 40 Gbps optical signals and two different bias conditions (2 V with a net-light gain of only 1.4 and 10 V with a net-light gain of 5.5) are depicted.

Moreover, silicon-germanium-silicon photodetectors are also capable of high-speed and low-noise operation, with a gain-bandwidth product up to 210 GHz and a low excess noise factor with an effective ionization ratio of about 0.25 [15]. The low noise characteristics of silicon-germanium-silicon photodetectors are attributed to the strong impact ionization processes occurring close to the silicon-germanium interfaces. In mainstream full-germanium p-i-n photodetectors, such processes take place in germanium, hindering a low-noise operation. Indeed, silicon-germanium-silicon photodiodes benefit from the low noise properties of silicon, with an extremely low ionization ratio. The intrinsic avalanche multiplication decreases the avalanche excess noise, and thus reducing the effective ionization coefficient through the dead space effect.

Finally, Fig. 2d shows bit-error-rate measurements versus received optical power for 40 Gbps optical signals. An optical power sensitivity of about -11 dBm was obtained with such a silicon-germanium-silicon photodetector for an error-rate level of 10^{-9}. Further device sensitivity improvements were prevented by the large dark current close to the avalanche breakdown. The dark current at the avalanche breakdown of ~11 V indeed reached 600 μA.

3 Conclusions

In this work, we presented recent achievements on chip-integrated silicon-germanium-silicon lateral p-i-n photodetectors. Devices were capable of 40 Gbps on-chip detection in low-voltage ("*p-i-n regime*") and high-voltage ("*avalanche regime*") conditions. Devices capitalized on the combination of butt-waveguide coupling with lateral silicon-germanium-silicon p-i-n junctions. Devices thus had a reduced footprint, a definitely reduced fabrication complexity and significantly improved opto-electrical performances. These results open up opportunities to deploy silicon-germanium-silicon photodetectors into high-speed and cost-effective on-chip networks, targeting many emerging photonic applications in health monitoring, biomedicine, data centres, or optical interconnects.

Acknowledgement

This work was supported by the Slovak Research and Development Agency under the project APVV-17-0631 CONSENS and PP-COVID-20-0100: DOLORES.AI: The pandemic guard system.

References

1. D. Benedikovic, *et al.*, Nanophotonics**10**, 1059 (2021)
2. H. Chen, *et al.*, Opt. Express **24**, 4622 (2016)
3. L. Virot, *et al.*,Opt. Express **25**, 19487 (2017)
4. D. Benedikovic, *et al.*, Photon. Res. **7**, 437 (2019)
5. D. Benedikovic, *et al.*, IEEE J. Quantum Electron. **56**, 8400409 (2020)
6. J. H. Nam, *et al.*, Opt. Express **23**, 15816 (2015)
7. H. Pan, *et al.*, Opt. Express **20**, 18145 (2012)
8. S. Assefa, *et al.*, Nature **464**, 80 (2010)
9. Y. Kang, *et al.*, Nat. Photonics **3**, 59 (2009)
10. L. Virot, *et al.*, Nat Commun. **5**, 4657 (2014)
11. H. T. Chen, *et al.*, Opt. Express **23**, 815 (2015)
12. Z. Huang, *et al.*, Optica**3**, 793 (2016)
13. X. Zeng, *et al.*, Optica**6**, 772 (2019)
14. S. A. Srinivasan, *et al.*, IEEE/OSA J. Light. Technol. **38**, 3044 (2020)
15. D. Benedikovic, *et al.*, Optica **7**, 774 (2020)

https://doi.org/10.1051/epjconf/202125501003

SOI multimode waveguide bend with radially bridged SWGs for broadband (de)multiplexing -INVITED

Kevan K. MacKayt[†,*] and *Winnie N.* Ye[†]

[†]Carleton University, Department of Electronics, 1125 Colonel By Dr. Ottawa, Canada

Abstract. A novel broadband multimode waveguide bend is proposed that supports the propagation of multiple TE modes on a silicon-on-insulator platform. The gradient curvature bend utilizes trapezoidal subwavelength grating (SWG) segments, connected by adiabatically tapered radial strips to achieve efficient mode (de)multiplexing. The inclusion of the radial strips offers an extra degree of design freedom, allowing the realization of a multimode bend with only one single full etch step. The access waveguide has a width of 2.075 µm with an effective radius of 10 µm. Propagation loss for all modes remains below 2.96 dB, and intermodal crosstalk has a maximum of -19 dB across a broad bandwidth of 100 nm, centred at 1550 nm. This work presents an excellent design choice for broadband mode-division multiplexing operations.

1 Introduction

As the demand for increased signal capacity in optical networks grows, many solutions have been explored to increase the number of possible waveguide channels. In particular, mode-division multiplexing (MDM) has become a popular option that transmits a signal through each spatial mode of a waveguide. The main drawback of MDM systems is their large footprint, which often requires large waveguide widths, with even larger bend radii on the order of millimeters. Small radius bends induce large mode propagation losses from the modal mismatch at the straight-bent interface, as well as the high radiation leakage of the higher order modes. Furthermore, higher order modes suffer from the generation of asymmetric outputs due to the asymmetry of their modal profiles within the multimode waveguide bend (MWB).

A common implementation of multimode bends makes use of a mode converter (MC) region that deforms the mode profiles prior to entering the bend [1,2]. Most MC assisted bends are optimized to two particular mode orders and cannot be easily scaled up. 45-degree corner bends that utilize total-internal reflection have been shown to effectively scale up to high order spatial modes simply by increasing the width of the bus waveguide [3]. The main drawback in this approach is that they require significantly larger widths than the minimum width required to propagate a given mode order, negating any benefit from footprint reduction.

By far the most promising implementation of MWBs is a combined solution of gradient curvature and gradient index bends. Replacing a standard circular bend profile with an alternate scheme such as Euler or Bezier curves reduces the index mismatch at the straight-bent transition interface [4], while smoothly varying the curvature to a maximum at the midpoint of the bend. The optimal bend on the silicon-on-insulator (SOI) platform followed an empirically derived curve [5] based on the minimization of the propagation loss.

To compensate for the radial asymmetry of the bend's index profile on-chip transformation optics can be applied. Early attempts at this approach employed a custom grayscale lithography process [6], varying the height of a 500 nm thick silicon layer such that the inner edge was taller than the outer edge. Although successful, this approach is severely limiting due to the custom fab process utilized, and has largely been superseded by SWG segment engineering.

Trapezoidal subwavelength grating (SWG) segments are commonly employed when designing SWG bends, where the inner radius has a larger duty cycle than that of the outer radius [7]. It has been used extensively for single-mode bends but has recently found use in multimode bends as well, successfully maintaining the symmetry of modes up to TE_2/TM_2 at a radius as small as 10 µm [8,9]. However, these multimode bends require an extra partial etch step for the SWG layer in order to keep propagation losses low.

In this work, we propose a novel design that adopts the trapezoidal SWG segments, connected by adiabatically tapered radial strips. The radially bridged SWG design provides a greater degree of freedom in tailoring the index gradient, making it possible to realize the bends with only one single etch step.

2 Device Structure and Simulation

The widely available 220 nm SOI platform was chosen. A width of 2.075 µm was chosen to ensure the first 5 propagating modes were of TE polarization. The target radius was selected to be 10 µm to achieve negligible bend

* Kevan MacKay: kevanmackay@cmail.carleton.ca

radiation loss. The pitch of the SWG segments was selected to be 220 nm to ensure that the device operates in the subwavelength regime when centred at 1550 nm, while leaving a reasonable buffer above the minimum feature size of a typical e-beam lithography process.

Fig. 1 illustrates the design geometry of the proposed multimode bend. The MWB was engineered by first calculating an optimal curve with an effective radius of 10 µm, and segmenting the curve into coordinates spaced exactly 220 nm apart.

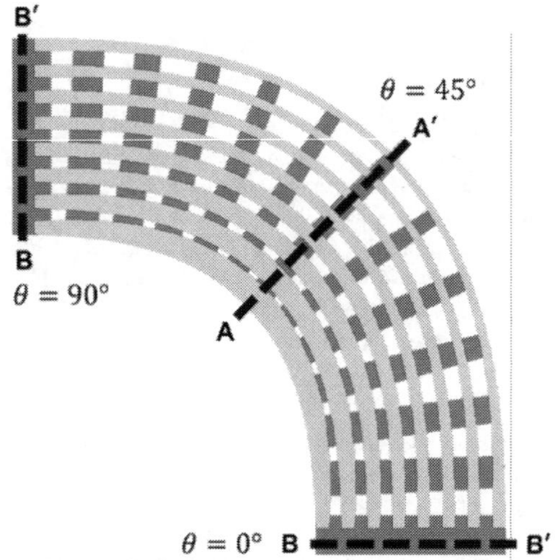

Fig. 1. A schematic of the multimode bend geometry illustrating the changing dimensions of the trapezoidal segments (red) and the adiabatically tapered radial strips (green), from θ=0° to θ=90°. The point of the maximum curvature along line AA' at θ=45°.

Each of these coordinates has a corresponding curvature which reaches a maximum at the midpoint, as picture in Fig. 2. As the curvature changes along the bend it is used as a variable to slowly vary both the trapezoidal SWG duty cycles as well as the widths of the radial strips.

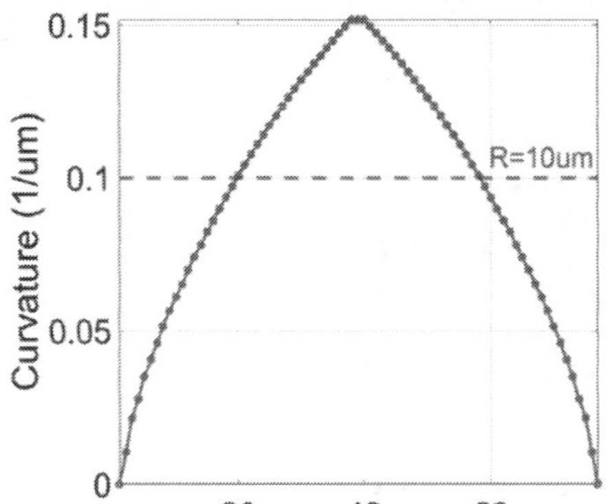

Fig. 2. Curvature at each SWG segment along the bend. A circular bend with an equivalent radius has a constant curvature denoted by the red dashed line.

By increasing the dimensions of the SWG segments on the inner radii and decreasing the dimensions of the segments on the outer radii, an index gradient could be tailored to compensate for the bend asymmetry. The starting SWG duty cycle is 60% for both the inner and outer radii. At the point of the maximum curvature (as shown by the dashed line AA' in Fig. 1), the duty cycle has been tapered down to 50% and 42% for the inner and outer radii, respectively, and is depicted in Fig. 3.

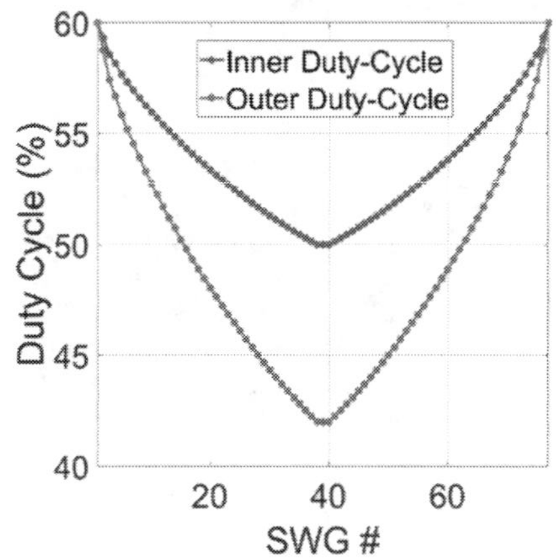

Fig. 3. Variation in SWG duty-cycle along the inner and outer edges along the curve.

The duty cycle ratio was set to ensure compliance with a minimum feature size of 60 nm for feasible fabrication. All the radial strips begin and end with a width of 170 nm at BB', with the inner four strips increasing in width until the point of maximum curvature AA', and the outer four strips decreasing in width. The maximum inner width and the minimum outer width of the strips were optimized through parameter sweeps. The innermost strip has a maximum width of 180 nm, while the outermost strip has a width of 130 nm, as depicted in Fig. 4.

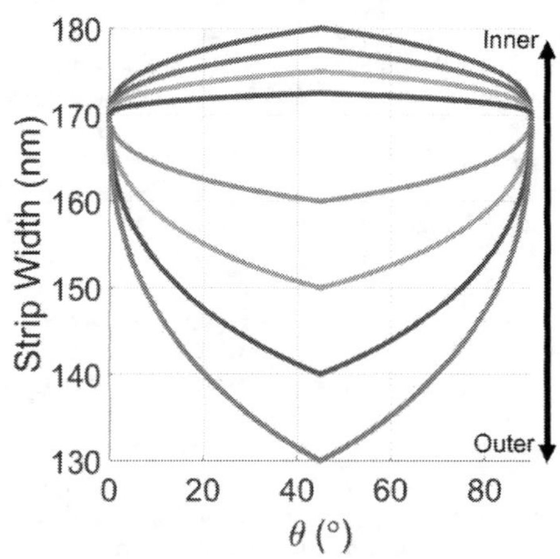

Fig. 4. Variation in radial strip width along the curve. The inner four strips increase in width and the outer four decrease in width.

Lumerical FDTD was used to simulate the field propagation of the first four TE modes through the bend geometry. The bend was optimized by selecting the geometry that maximized the mode overlap ratio for each mode at the output port of the bend. The field profile distributions for the first four TE modes are shown in Fig. 5.

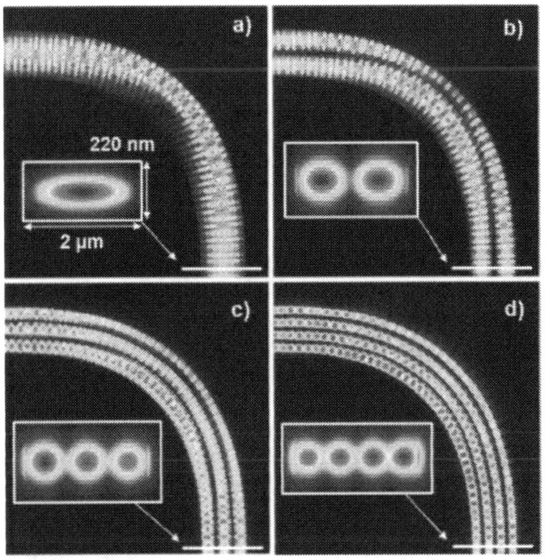

Fig. 5. Field distributions of the first four TE modes propagating through the proposed multimode waveguide bend: **a)** TE_0, **b)** TE_1, **c)** TE_2, **d)** TE_3.

Fig. 6 illustrates the transmission spectra centred at 1550 nm across a bandwidth of 100 nm when injecting each of the first four TE modes. It can be seen that the propagation loss remains below 2.14 dB, 1.64 dB, 2.96 dB, and 2.81 dB for TE_0, TE_1, TE_2, and TE_3, respectively. The maximum intermodal crosstalk is -19.65 dB, -19.01 dB, -19.05 dB, and -27.76 dB for TE_0, TE_1, TE_2, and TE_3, respectively.

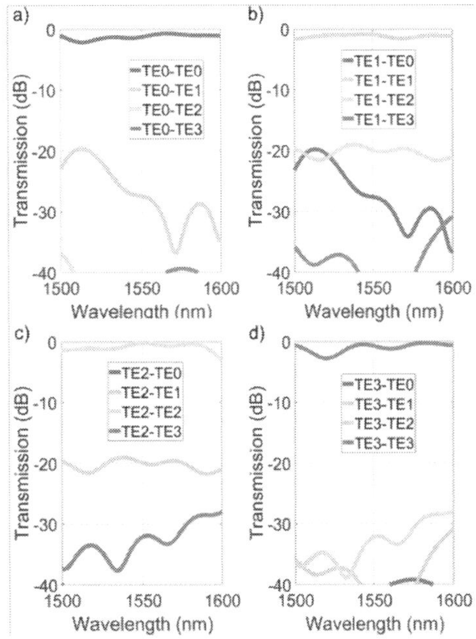

Fig. 6. Simulated transmittance spectra for the first four TE modes: **a)** TE_0, **b)** TE_1, **c)** TE_2, **d)** TE_3.

3 Conclusion

In conclusion we have successfully demonstrated an efficient and broadband multimode bend design for the first four TE modes, using radially bridged trapezoidal SWG segments. The bends have been shown to maintain the symmetry of the field profiles with low propagation losses and intermodal crosstalk over a bandwidth of 100 nm. This design could be easily scaled up to include additional TE and TM modes. Compact multimode bends are an essential building block for creating dense photonic circuits. High capacity mode (de)multiplexing networks will benefit greatly from the reduced footprint afforded by these bends.

References

1. C. Sun, Y. Yu, G. Chen, X. Zhang, Opt. Letters, **42**, 15 (2017)
2. W. Chang, L. Lu, D. Liu, M. Zhang, OFC, (2018)
3. Y. Wang, D. Dai, Opt. Exp., **28**, 7 (2020)
4. B. Dorin, W. Ye, Opt. Exp, **22**, 4 (2014)
5. M. Bahadori, M. Nikdast, Q. Cheng, K. Bergman, JLT, **37**, 13 (2019)
6. L. Gabrielli, D. Liu, S. Johnson, M. Lipson, Nat. Comm., **3**, 1217, (2012)
7. Z. Wang, X. Xu, D. Fan, Y. Wang, Scientific Reports, **6**, 24106 (2016)
8. H. Wu, C. Li, L. Song, H. Tsang, J. Bowers, D. Dai, Laser Photonics Rev., **13**, 1800119, (2019)
9. Y. Wang, D. Dai, JLT, **38**, 15 (2020)

https://doi.org/10.1051/epjconf/202125501004

Integrated circular optical phased array

Qiankun Liu[1,*], *Tom* Smy[1], *Ahmad* Atieh[2], *Pavel* Cheben[3], *Alejandro* Sánchez-Postigo[1,4] and *Winnie N.* Ye[1]

[1]Department of Electronics, Carleton University, 1125 Colonel By Drive, Ottawa, ON K1S 5B6, Canada
[2]Optiwave Systems Inc, Ottawa, ON, K2E 8A7, Canada
[3]Advanced Electronics and Photonics Research Center, National Research Council Canada, 1200 Montreal Road, Ottawa, ON K1A 0R6, Canada
[4]Currently at Telecommunication Research Institute (TELMA), Universidad de Málaga, CEI Andalucía TECH, Louis Pasteur 35, 29010 Málaga, Spain

Abstract. Existing OPAs are typically based on 2D rectangular arrays or 1D linear arrays. Both approaches present a limited field-of-view (FOV) due to the presence of the grating lobes when the element spacing is larger than λ/2. To address the need for an increased steering range, we propose a new design strategy of an OPA system utilizing a 2D circular phased array, with a substantially increased FOV. We present a circular OPA using a demonstrated antenna element design, with an 820-element array. A steering range Ω_{SR} calculated as a solid angle of 0.51π sr, and an angular beamwidth of 0.22°, was achieved. The array exhibits a sidelobe suppression larger than 10 dB, and a FOV of 2π sr. Although the performance is limited by the far field pattern of the individual antenna we chose, our circular OPA achieved, to the best of our knowledge, the largest steering range reported to date compared to the state-of-the-art integrated optical phased arrays reported in literature.

1 Introduction

In the last decade extensive research has been reported on static integrated optical phased arrays (OPAs) for solid-state beam steering with fast scanning and high resolution. These OPA systems typically use either a 2D rectangular array of antennas [1, 2], or a 1D linear array of antennas with wavelength sweeping to achieve a two-dimensional beam steering [3-7]. The far field radiation pattern of a rectangular or linear array shows unavoidable strong grating lobes, when element spacing is larger than λ/2. Therefore, these arrays demonstrate a restricted aliasing-free beam steering range, exhibiting a limited field-of-view (FOV). To suppress the grating lobes, we propose an OPA system using a 2D circular configuration. By implementing a circular array, the grating lobes are reduced hence increasing the system FOV and the steering range.

The circular OPA provides to date, to the best of our knowledge, the largest steering range compared to other reported integrated OPAs, while keeping the sidelobe suppression larger than 10 dB [8]. In this work, we present a circular array design: using 820 elements in total to show the superior performance compared to the reported OPAs with a compatible number of elements. As a unique feature of circular arrays, the design shows a theoretical maximum FOV of 2π sr, due to the effective suppression of grating lobes.

2 Beam formation and steering

A schematic view of our proposed circular array configuration is shown in Fig. 1, the blue dots represent the nano antennas. The circular array contains N concentric rings whose radius is given by:

$$R_n = R_0 + (n-1)dr \tag{1}$$

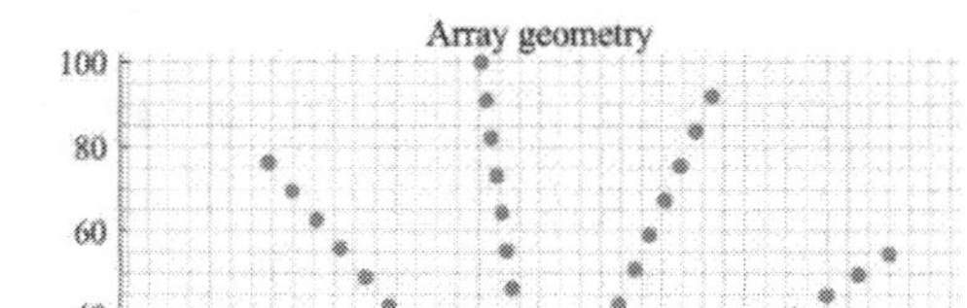

Fig. 1. Schematic view of circular array configuration.

where R_0 is minimum radius of the first center ring, n is the number of rings and dr is the delta radius between two adjacent rings. There are M antennas locating in each ring

* Corresponding author: vincentliu3@cunet.carleton.ca

with an equal angular spacing between two adjacent antennas in the same ring. In total, $M \times N$ elements are radially arranged in this circular array. Here we assume isotropic antennas, where each identical antenna element emits signal uniformly in all directions. The far-field radiation pattern obtained for the circular antenna array is defined as the array factor, which can be expressed as [9]:

$$A(\theta,\phi) = \sum_{m=1}^{M}\sum_{n=1}^{N} a_{mn} e^{-i\frac{2\pi}{\lambda}\sin(\theta)\cos\left(\phi-\frac{2\pi(m-1)}{M}\right)R_n} \tag{2}$$

where a_{mn} is the element weight coefficient which is equal to $1/(M \times N)$ for each element, N is the total number of rings, λ is the wavelength, θ and ϕ are the elevation and azimuth angles, respectively. As a comparison, in a rectangular array, the array factor is given by Eq. 3 where M is the number of elements in each row, N is the number of elements in each column, dx and dy are the elements spacing in x and y axis [9].

$$A(\theta,\phi) = \sum_{m=1}^{M}\sum_{n=1}^{N} a_{mn} e^{-i\frac{2\pi}{\lambda}[d_x \sin(\theta)\cos(\phi)+d_y \sin(\theta)\sin(\phi)]} \tag{3}$$

The array factors are calculated for a wavelength of 1550 nm. Fig. 2a shows the calculated amplitude of array factors, with $N = 20$, $M = 41$, $R_0 = 60$ μm, $dr = 9$ μm, in the uv space where $u = \sin(\theta)\cos(\phi)$, $v = \sin(\theta)\sin(\phi)$. The scanning region corresponds to an entire hemisphere with 1° resolution to reduce the simulation time (θ = 0°:1°:90°, ϕ = 0°:1°:360°). The elevation cut at ϕ = 0°, calculated with 0.01° resolution to measure the angular beamwidth, for the array is shown in Fig. 2b. The inset is the zoom-in view of the array factor when elevation angles are in the range of -5° < θ < 5°. The sidelobe suppression is measured to be 19.16 dB, while the angular beamwidth can be measured as 0.22°. In the applications for autonomous vehicles, it corresponds to a recognition of an object of 38 cm in size at a distance of 100 m away from the OPA. In the contrast, considering a same number of elements $N = 20$, $M = 41$ and choosing the minimum distance between adjacent elements in the circular array as the element spacing in the rectangular array, i.e., $d_x = d_y = dr = 9$ μm. Fig. 2c shows the calculated array factor of such rectangular array, a large amount grating lobes can be observed for the rectangular array.

The array far-field pattern (AFP) of the circular array can be calculated as the far-field of each individual antenna $E(\theta, \phi)$ multiplied by the circular array factor $A(\theta, \phi)$ [1, 8]:

$$AFP = E(\theta,\phi) \times A(\theta,\phi) \tag{3}$$

Using a commercial finite difference time domain (FDTD) tool, by adapting our recent published grating coupler [11, 12], we have designed a nano antenna on a silicon-on-insulator waveguide with a 2 μm thick upper and buried SiO2 cladding. It consists of 5 periods, each period contains 4 segments with lengths (L1 = 280 nm, L2 = 130 nm, L3 = 196, L4 = 134 nm). The first segment is subwavelength engineered to have a refractive index of 2.32 in order to reduce the back reflection at the interface of the waveguide and antenna. The grating lines have a circular curvature with a wedge opening angle of 26°. The footprint of this nano antenna is only 5.5 μm × 2.5 μm. It shows an upward radiation coefficient of -1.55 dB and a back-reflection as low as -19.94 dB at 1550 nm. A 3D schematic view of the nano antenna and its far-field radiation pattern are presented in Fig. 3.

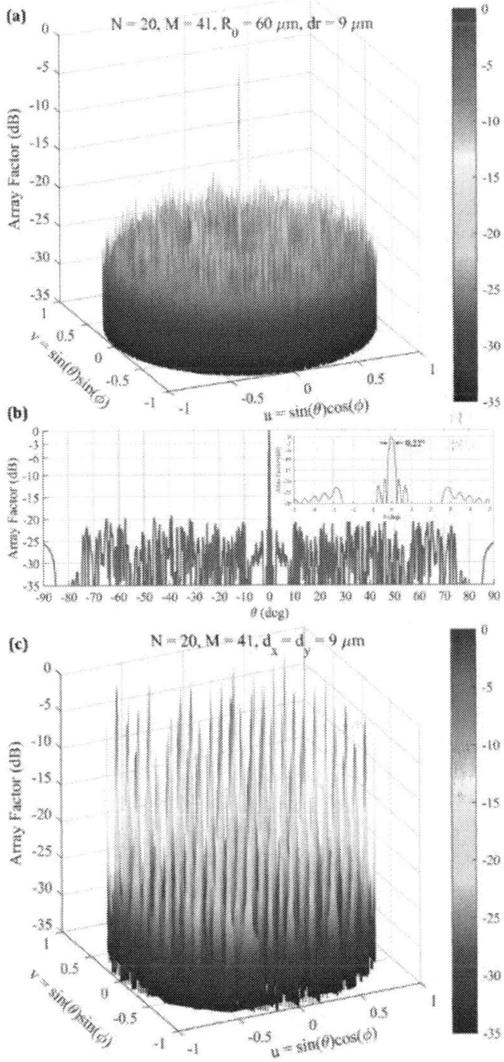

Fig. 2. (a) Simulated amplitude of the array factor in dB in uv space for Array with $N = 20$, $M = 41$, $R_0 = 60$ μm, $dr = 9$ μm at θ = 0°, ϕ = 0°; (b) Elevation cut (ϕ = 0°) of the array factor of Array 1; inset in (b) shows the zoom-in view of the array factor when elevation angles are in the range of -5° < θ < 5°; (c) Simulated array factor in dB of rectangular array with $N = 20$, $M = 41$, $d_x = d_y = 9$ μm.

Constructing the circular arrays using the above nano antenna, and collaborating with independent controlled phased shifters, we can achieve beam steering over the area within the antenna radiation far field pattern. Fig. 4

shows the steering range of the array with sidelobe suppression larger than 10 dB. The white circles correspond to the elevation angle θ from 0° (surface normal) to 90°, while the straight lines crossed center marked by outer black number represents the azimuthal angle ϕ from 0° to 360°. Because of the irregular shape of the steering range, the steering range is calculated as a solid angle:

$$\Omega_{SR}=\iint_{steering\ range} sin\theta d\theta d\phi \quad (4)$$

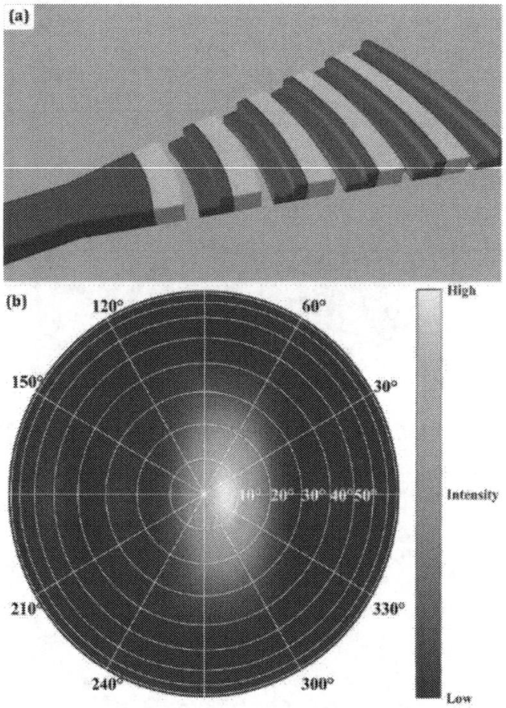

Fig. 3. (a) Schematic 3D view of nano antenna; (b) The corresponding far-field radiation pattern of the optical nanoantenna at λ = 1550 nm, calculated from the near-field emission using the near-to-far-field transformation.

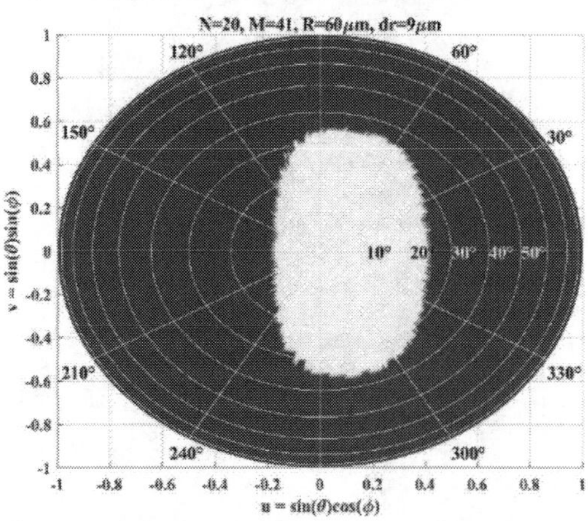

Fig. 4. Steering range of circular array operating at 1550 nm. The yellow area represents the set of all steerable angles where the sidelobe suppression is larger than 10 dB, and the blue area represents non-steerable angles.

According to Eq. 4, we can obtain the steering range of such circular array Ω_{SR} of 0.51π sr. It covers more than 20% over the full hemisphere in the forward direction by a single OPA. Moreover, the angular beamwidth of the OPA is as low as 0.22°.

3 Conclusion

In this work, we present a new circular optical phased array design which provides an array factor with effectively suppressed grating lobes resulting an overall sidelobe suppression larger than 10 dB. Although theoretically it is possible for the circular OPA to achieve the full FOV of 2π sr at a specific wavelength by using ideal isotropic individual nano antenna elements, we use practical pre-designed and demonstrated antenna elements for our numerical studies. With 820 elements in total, the circular OPA provides a Ω_{SR} as large as 0.51π sr covering more than 20% over a hemisphere in the forward direction. We believe that the proposed circular OPA design offers a promising solution to achieve large 2D beam steering range and high angular resolution demanded by applications such as integrated photonic lidars for autonomous vehicles.

References

1. J. Sun, E. Timurdogan, A. Yaacobi, E.S. Hosseini, and M. R. Watts, Nature **493**, (2013)
2. R. Fatemi, A. Khachaturian, and A. Hajimiri, IEEE J. Solid-State Circuits **54**, (2019)
3. C.V. Poulton, M. J. Byrd, P. Russo, E. Timurdogan, M. Khandaker, D. Vermeulen, M. R. Watts, IEEE J. Sel. Top. Quantum Electron **25**, (2019)
4. S. A. Miller, Y. Chang, C. T. Phare, M. C. Shin, M. Zadka, S. P. Roberts, B. Stern, X. Ji, A. Mohanty, O. A. Jimenez Gordillo, U. D. Dave and M. Lipson, Optica **7**, (2020)
5. J. C. Hulme, J. K. Doylend, M. J. R. Heck, J. D. Peters, M. L. Davenport, J. T. Bovington, L. A. Coldren, and J. E. Bowers, Opt. Express **23**, (2015)
6. D. N. Hutchison, J. Sun, J. K. Doylend, R. Kumar, J. Heck, W. Kim, C. T. Phare, A. Feshali and H. Rong, Optica **3**, (2016)
7. T. Komljenovic, R. Helkey, L. Coldren, and J. E. Bowers, Opt. Express **25**, (2015)
8. Q. Liu, T. Smy, A. Atieh, P Cheben and W. N. Ye, IEEE J. Sel. Top. Quantum Electron (submitted)
9. S. J. Orfanidis, *Electromagnetic Waves and Antennas*, 3rd ed. (2004)
10. M. Kamandar Dezfouli, Y. Grinberg, D. Melati, P. Cheben, J.H. Schmid, A. Sánchez-Postigo, A. Ortega-Moñux, G. Wangüemert-Pérez, R. Cheriton, S. Janz, and D.-X. Xu, Opt. Lett. **45**, (2020)
11. S. Khajavi, D. Melati, P. Cheben, J.H. Schmid, Q. Liu, D.X. Xu, and W.N. Ye, Opt. Express **29**, 7003 (2021)

Oscillator Finite-Difference Time-Domain (O-FDTD) electric field propagation model: integrated photonics and networks

Ricardo M. R. Adão[1], *Manuel* Caño-Garcia[1], *Christian* Maibohm[1], *Bruno* Romeira[1], and *Jana B.* Nieder[1,*]

[1]INL – International Iberian Nanotechnology Laboratory, Av. Mestre José Veiga s/n, 4715-330 Braga, Portugal

Abstract. The recently developed Lorentz Oscillator Model-inspired Oscillator Finite-Difference Time-Domain (O-FDTD) is one of the simplest FDTD models ever proposed, using a single field equation for electric field propagation. We demonstrate its versatility on various scales and benchmark its simulation performance against theory, conventional FDTD simulations, and experimental observations. The model's broad applicability is demonstrated for (but not limited to) three contrasting realms: integrated photonics components on the nano- and micrometer scale, city-wide propagating radiofrequency signals reaching into the hundreds of meters scale, and for the first time, in support of 3D optical waveguide design that may play a key role in neuromorphic photonic computational devices.

1 Introduction

Photonic simulation algorithms are crucial tools to understand, engineer, and optimize the optical properties of photonic structures in myriad applications. Discrete time-domain electrodynamic simulation models include Finite-Difference Time-Domain (FDTD), Finite-Element Time-Domain (FETD), Discontinuous Galerkin Time-Domain (DGTD), among others [1], most of them either solving or applying Maxwell's field equations over time. Still, novel system-specific or more efficient methods are developed continuously [2]. Here we use the new, simplified approach for dielectric and semiconductor materials, called Oscillator Finite-Difference Time-Domain (O-FDTD) [3]. The approach considers a mesh of coupled oscillators that allow electric field waves to propagate through space and matter. We aim to demonstrate its applicability versatility, ranging from planar photonic micro and nanostructures to RF signal coverage in cityscape environments [4].

Planar photonic devices embody countless applications ranging from biosensing [5] to photonic computing [6]. Most dielectric and semiconductor devices rely on coherent effects such as waveguiding, interference, and resonances. To showcase O-FDTD's applicability, we select a few of the most widely used photonic building blocks, such as Multimode Interferometers (MMI), photonic crystals, and microring resonators. We aim to simulate the device's performance and compare it against theoretical predictions and simulations using established FDTD software.

The interest in lightwave propagation simulations is not limited to the micro/nanoscale. With the raging demand of Smart Cities and the Internet of Things (IoT) comes the urgent need for efficient long-range wireless technologies such as Long Range (LoRa) [7], and the ability to accurately predict the signal strength and coverage towards effective network planning. We aim to push the boundaries of conventional photonic simulations and explore O-FDTD's applicability to complex, large-scale systems in the RF range by simulating a LoRa network coverage in a cityscape environment [4].

In the current fast-forward technological age, computation power and efficiency are vital, where photonic computation is the most promising fast and low-power alternative to conventional computing. In particular, photonic neuromorphic platforms promise hardware-based artificial intelligence implementations such as artificial neural networks [6]. Alongside the spiking neuron device, the on-chip interconnection of complex neuron architectures avoiding channel superposition and crosstalk remains challenging. At the same time, polymeric structures with unique properties and 3D fabrication processes arise as promising photonic elements [8], and recent works have reported on the use of 3D polymeric waveguides to interconnect photonic neural networks [9]. We aim to break O-FDTD's 2D limitation to study 3D polymeric waveguides towards micro-LED-based spiking neuron interconnections.

2 O-FDTD Theoretical model

O-FDTD addresses propagating electric field waves similarly to mechanical oscillations. Hendrik Lorentz first proposed using electron oscillations to quantify electric fields in the widely known Lorentz Oscillator Model (LOM) [10]. Instead of isolated, material-limiting electrons, one can consider a mesh of massless coupled oscillators. The oscillators move proportionally to the electric field amplitude, thus allowing waves to propagate as mechanical waves propagate over liquid surfaces (see water pond-like illustration in **Fig. 1**).

* Corresponding author: jana.nieder@inl.int

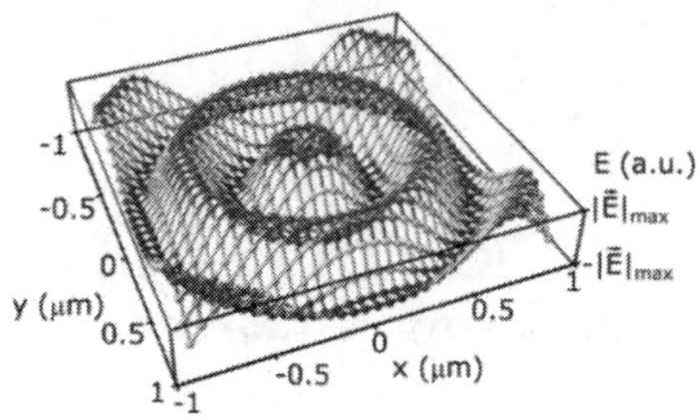

Fig. 1. O-FDTD electric field spatial distribution illustration. Each oscillator is coupled to its neighbors, allowing electric field wave propagation. (modified from [2]).

In O-FDTD, the electric field E temporal discretization is achieved via a simple leapfrog time backward differentiation

$$E_i = E_{i-1} + \frac{\mathrm{d}E_{i-1}}{\mathrm{d}t}\delta t + \frac{1}{2}\frac{\mathrm{d}E_{i-1}}{\mathrm{d}t}\delta t^2, \quad (1)$$

where d/dt and δt are the time derivative and time-step, respectively, and E_i is the electric field amplitude in iteration i. From an analogy with the LOM, we express the wave equation of the coupled oscillators as:

$$\frac{\mathrm{d}^2 E_{i-1}}{\mathrm{d}t^2} = \Gamma\left(E_{i-1}^{Neigh} - E_{i-1}\right) - \gamma\frac{\mathrm{d}E_{i-1}}{\mathrm{d}t}, \quad (2)$$

where $\gamma \propto 4\pi\kappa$ relates to the material absorption coefficient (imaginary part of the refractive index). The coupling strength Γ defines the electric field transfer rate between adjacent couplers and thus the speed of light in the medium,

$$\Gamma = \frac{8}{3n^2\delta t^2}, \quad (3)$$

where n is the real part of the material refractive index. E^{neigh} is the average neighborhood electric field, calculated by a unitary Ω kernel matrix (box blur) convolution of the electric field E_{i-1}. The final O-FDTD formulation can be expressed as

$$\frac{\mathrm{d}^2 E_{i-1}}{\mathrm{d}t^2} = \frac{8(\Omega * E_{i-1} - E_{i-1})}{3n^2\delta t^2} - \frac{\sqrt{2}c}{n}\frac{4\pi\kappa}{\lambda}\frac{\mathrm{d}E_{i-1}}{\mathrm{d}t}, \quad (5)$$

Where c and λ are the vacuum speed of light and wavelength. A simple O-FDTD simulation algorithm can be implemented by defining a rectangular-meshed refractive index map, light sources, and boundary conditions. In this proof-of-concept, light sources are implemented by localized sinusoidal electric field inputs $E_i^{source} = E_i + E_0 \sin(2\pi c t_i/\lambda)$, where E_0 and t_i are the input electric field amplitude and the time. Perfectly Matched Layer (PML) boundary conditions are implemented by surrounding the simulation space with artificial layers of linearly increasing absorption coefficient κ. Detailed implementation descriptions can be found in the Supplemental Document of ref [3].

3 Results

Here we demonstrate O-FDTD's broad applicability to photonic structures and wavelength ranges by showcasing examples from planar integrated circuit building blocks **(3.1)** to RF signals propagating in urban environments **(3.2)**. We furthermore present a symmetry-based approximation to extend O-FDTD's 2D model to out-of-plane light extraction using 3D polymeric waveguides.

3.1 Planar dielectric photonics

Fig. 2 shows the electric field amplitude maps for three of the most relevant PIC building blocks. **Fig. 2(a)** shows a simple Si/SiO_2 core/cladding 1x4 rectangular MMI that splits an input signal into four equal outputs, with a 90% efficiency. **Fig. 2(b)** shows a 2D Si/air pillar array-based photonic crystal with a pillar radius r = 100 nm, lattice constant a = 635.5 nm, wavelength λ = 1.55 μm, and angle θ = 60°. The photonic bandgap was calculated using MIT Photonic-Bands [11]. A "z-shaped" photonic crystal waveguide (gap) induces two 120° sharp bends, unsupported by "conventional" waveguides. **Fig. 2(c)** shows a 3 μm radius Si/SiO_2 core/cladding microring resonator driven at 1945 and 1970 nm. The two wavelengths excite off and on-resonance states, where the light is outputted via the *Through* (top) and *Drop* (bottom) waveguides, respectively.

Other O-FDTD applications to photonic microstructures can be found in ref [3], where the results are validated against theory, conventional FDTD simulations, and experimental observations. Having demonstrated O-FDTD's applicability to micro and nanostructures in the visible and infrared wavelength regimes, many other electromagnetic simulation applications are conceivable, as explored in the following section.

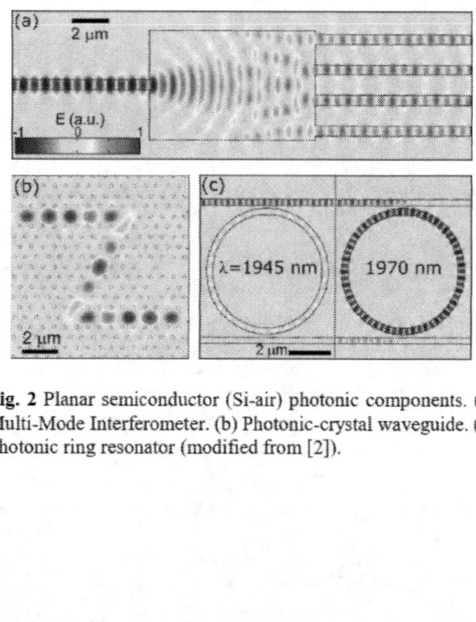

Fig. 2 Planar semiconductor (Si-air) photonic components. (a) Multi-Mode Interferometer. (b) Photonic-crystal waveguide. (c) Photonic ring resonator (modified from [2]).

3.2 Cityscape LoRa network coverage

RF antennas in urban environments (both indoors and outdoors) can be seen as dipole electric field sources surrounded by different dielectric constant materials in complex arrangements. **Fig. 3(a)** shows the O-FDTD-simulated electric field power (Receiver Signal Strength Indicator (RSSI)) propagating in a city-wide environment [4]. The simulation models concrete walls via their dielectric constant and real-world building data from Google Maps. The binary (on-off) coverage area map is obtained by considering a -124 dBm receiver sensitivity threshold and is compared against the experimental characterization in **Fig. 3(b)**. The red, green, and yellow channels represent the isolated simulated, experimental, and overlapping areas. The simulation-experiment agreement can be expressed by coverage area overlap, **Fig. 3(b-c)**, or edge variation, **Fig. 3(d)**. We obtain a considerable area overlap of 84%. However, this measure underestimates the rich O-FDTD-predicted reflection and interference pattern since all coverage models predict good coverage near the emitter. Most path loss-based coverage models only provide an average coverage distance and thus a circular coverage area. **Fig. 3(d)** compares the angle-dependent coverage (emitter-to-edge) distance experimental-simulation difference obtained with O-FDTD and the best-fitting circle. We obtain root mean square (RMS) errors of 24 and 41 m, respectively. Considering that the best-fitting circle (error minimizer) is an experiment-biased result, it is reasonable to assume that unbiased predictions will only yield larger errors. These results indicate that such simplistic path-loss models would yield at least twice the average O-FDTD coverage edge error.

Despite O-FDTD's 2D formulation, the following section describes how symmetry assumptions extend the model to some 3D architectures.

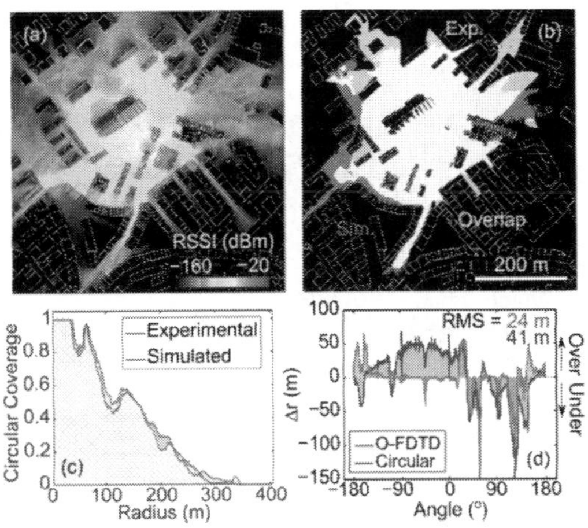

Fig. 3. Cityscape LoRa Propagation. (a) Simulated RSSI map. (b) Simulation versus experimental coverage areas. Quantitative area (c) and (d) edge comparison (modified from [3]).

3.3 3D interconnects for neural networks

The proposed interconnected photonic neural network concept is illustrated in **Fig. 4(a)**. Semiconductor (GaAs / AlGaAs) epilayer stacks aim to provide light sources with neuron-mimicking spiking properties interconnected into artificial neural networks by 3D polymeric waveguides. As a starting point, we consider 4 μm wide micropillar LEDs using SiO_2 as a passivation layer and Au as an electrical contact.

Assuming that a 3D waveguide extends over a single plane, we can estimate its waveguiding properties by simulating the electric field propagation along its longitudinal section. **Fig. 4(b)** shows the O-FDTD E^2 simulation of the micro-LED light emission without **(b1)** and with **(b2)** a 3D polymeric (OrmoCore, n = 1.56) outcoupling waveguide. **Fig. 4(b1)** reveals two main emission directions or components of similar intensity: a vertical component that escapes towards air and a horizontal component trapped within the SiO_2 layer. Simple polymeric structures can be designed for vertical or horizontal outcoupling. For ease of fabrication, we choose the horizontal. **Fig. 4(b2)** shows the efficient extraction of the horizontal component otherwise trapped in the SiO_2 layer. **Fig. 4(c)** shows the propagation across an entire waveguide structure, where some support structure-induced scattering can be observed.

Discussion

O-FDTD is an unprecedented simple, single-equation electrodynamic model that can readily implement photonic simulation algorithms. Empirical simulations demonstrate that the complex refractive index-dependent field modulation accurately reproduces bulk and interface light-matter interactions in some of the most widely used photonic device components. However, O-FDTD's simplicity comes at the cost of a few limitations. Firstly, the present formulation assumes orthogonality between the electric field oscillation and the propagation plane, thus limiting it to TM polarization.

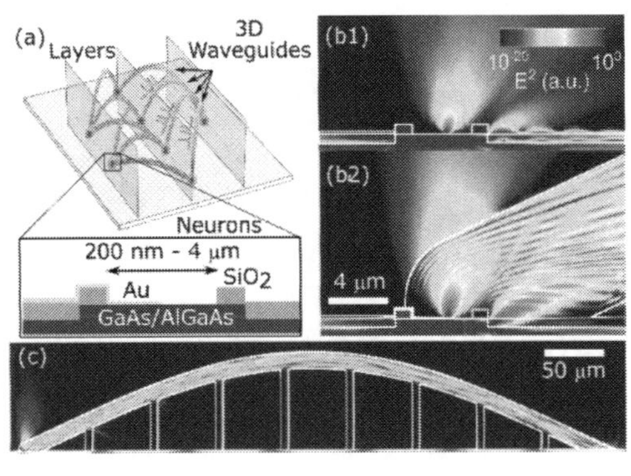

Fig. 4. 3D waveguides for photonic neuromorphic computing. (a) 3D-interconnected planar neural network concept. (b) Light outcoupling from GaAs/AlGaAs micro-LED at λ = 850 nm. (c) Light propagation in a supported waveguide.

Secondly, this orthogonality requirement limits O-FDTD to 2D simulations, as extrapolating it to 3D would lead to unrealistic (soundwave-like) propagation-direction electric field oscillations. Thirdly, the present O-FDTD formulation disregards the electric conductivity, making it less than optimal for metals. Furthermore, Eq. (5) shows that $n < 1$ materials (e.g., metals in the UV) can lead to enlarged coupling strengths and algorithm divergence. Higher mesh densities (resolution) help reduce the divergence at the cost of longer simulation times. Notwithstanding, the simulation results presented in this work demonstrate its vast wavelength range and system applicability.

The core/cladding refractive index contrast in Si/SiO_2 devices leads to single and multi-mode waveguiding and interference patterns matching the typical self-imaging symmetric MMI interference predicted theoretically. Cavity geometry optimization and input/output tapering can minimize the splitting losses but are beyond the scope and intent of this work. The photonic crystal waveguiding in **Fig. 2(b)** demonstrates the photonic crystal bandgap concept in periodic subwavelength structures. The add-drop microring simulations demonstrate resonant interference effects.

However, O-FDTD's potential is not restricted to the micro/nanoscale. Since the laws of electromagnetism hold for any wavelength range, we put O-FDTD's applicability to the test in a 360000 m^2 cityscape environment. **Fig. 3(d)** compares O-FDTD against the best-fitting path loss model. However, these are the simplest and most limited models. More sophisticated real-world data-based coverage prediction algorithms such as ray-tracing [12] and diffraction-based [13] models take map-specific effects into account, such as reflections, scattering, and diffraction. Other models, such as the Irregular Terrain Model (ITM), consider terrain elevation and climacteric conditions. Each approach has pros and cons and optimal application conditions. To name a few challenges: The quality of ray-based models is usually statistics or ray-density dependent; diffraction models are usually limited to simplified geometries; ITM is best suited for large-scale networks but provides low inter-building resolution; wave propagation-based models are forbiddingly computationally demanding for city-wide areas. On the other hand, the single-equation O-FDTD approach holds the promise for next-generation wave-propagation methods for indoor and outdoor high-precision network coverage predictions. This work opens broad perspectives for high-precision electromagnetic signal distribution predictions in wireless networks, particularly in modern Smart City and Internet IoT technologies.

Lastly, we report a sectioning approximation to extend O-FDTD's applicability to specific 3D problems. It is worth mentioning that this method does not allow the complete simulation of the 3D system. Notwithstanding, in this example, it provides vital information on the microLED emission angle and extraction achievable with standing polymeric waveguides. Significantly, it is observed that a fraction (~50%) of otherwise lost emission to the passivation layer can be collected. Strategies for off-vertical axis light extraction from the biomimicking neurons can be obtained by simulation-powered 3D waveguide optimization, including parameters such as core dimensions, outcoupling angle, bending radii, and support structures. The dimensions and number of support structures depend strongly on the mechanical stability of the TPP polymer and need to be optimized using mechanical models or experimental trials.

Conclusion

Our recently developed O-FDTD method [3] can model electric field wave propagation in different media by modeling matter as a mesh of oscillators coupled with a complex refractive index-dependent coupling strength. The O-FDTD model's accuracy is validated against theory, simulation, and experimental observations in various contexts, in and beyond photonics. Despite its polarization, dimensionality, and material limitations, we demonstrate its applicability to a vast set of commonly used photonic building blocks, both 2D and 3D. The model holds tremendous potential for dielectric and semiconductor photonics, PICs, and other wave propagation technologies, from UV to RF.

This work was funded by the European Commission (H2020-FET-OPEN No. 828841 "ChipAI") and ERDF (0181_NANOEATERS_1_EP). RA acknowledges the Laser Photonics & Vision Ph.D. program, U. Vigo.

References

1. U. S. Inan and R. A. Marshal, *Numerical Electromagnetics*, 1st ed. (Cambridge University Press, 2011).
2. Q. Sun, R. Zhang, Q. Zhan, and Q. H. Liu, IEEE Trans. Antennas Propag. **67**(8), 5469–5476 (2019).
3. R. M. R. Adão, M. Caño-García, C. Maibohm, and J. B. Nieder, Opt. Express **29**(8), 11903–11916 (2021).
4. R. M. R. Adão, E. Balvís, A. V Carpentier, H. Michinel, and J. B. Nieder, Sensors **21**(8), 2717 (2021).
5. W. Bogaerts, P. de Heyn, T. van Vaerenbergh, K. de Vos, S. Kumar Selvaraja, T. Claes, P. Dumon, P. Bienstman, D. van Thourhout, and R. Baets, Laser Photonics Rev. **6**(1), 47–73 (2012).
6. B. Romeira, J. M. L. Figueiredo, and J. Javaloyes, Nanophotonics **9**(13), 4149–4162 (2020).
7. U. Raza, P. Kulkarni, and M. Sooriyabandara, IEEE Communications Surveys & Tutorials **19**(2), 855–873 (2017).
8. M. C. Oh, W. S. Chu, J. S. Shin, J. W. Kim, K. J. Kim, J. K. Seo, H. K. Lee, Y. O. Noh, and H. J. Lee, Opt. Commun. **362**, 3–12 (2016).
9. J. Moughames, X. Porte, M. Thiel, G. Ulliac, L. Larger, M. Jacquot, M. Kadic, and D. Brunner, Optica **7**(6), 640-646 (2020).
10. A. F. J. Levi, "The Lorentz oscillator model," in *Essential Classical Mechanics for Device Physics*, 1st ed. (Morgan and Claypoll Publishers, 2016), p. 101.
11. S. G. Johnson and J. D. Joannopoulos, (2015).
12. F. Firdaus, N. A. Ahmad, and S. Sahibuddin, Sensors (Switzerland) **19**(24), 5546 (2019).
13. S. A. Vavilov and M. S. Lytaev, IEEE Trans. Antennas Propag. **68**(5), 3859–3877 (2020).

"First time right" - calculating imaging systems from scratch - INVITED

Fabian Duerr[1,*], *and Hugo Thienpont*[1]

[1]Brussels Photonics, Department of Applied Physics and Photonics, Vrije Universiteit Brussel, Pleinlaan 2, 1050, Brussels, Belgium

Abstract. Freeform optics can be used to greatly extend the functionalities, improve performance, and reduce the volume and weight of optical systems. Today, the design of imaging systems largely relies on efficient ray tracing and optimization algorithms. Such a "step-and-repeat" approach to optical design typically requires considerable experience, intuition, and eventually "trial-and-error" guesswork. This time-consuming process applies especially to freeform optical systems. In this work, we present a deterministic direct optical design method for freeform imaging systems based on differential equations derived from Fermat's principle and solved using power series. The method allows calculating all optical surface coefficients that ensure minimal image blurring for each individual order of aberrations. We demonstrate the systematic, deterministic, scalable, and holistic character of our method with several catoptric and catadioptric design examples.

1 Introduction

For more than 150 years, scientists have advanced aberration theory to describe, analyse and eliminate imperfections that disturb the imaging quality of optical components and systems. Simultaneously, they have developed design methods and manufacturing techniques for imaging systems with ever-increasing complexity and performance up to the point where they are now including optical elements that are unrestricted in their surface shape. These so-called freeform elements offer degrees of freedom that can greatly extend the functionalities and further boost the specifications of state-of-the-art imaging systems. However, the drastically increased number of surface coefficients of these freeform surfaces poses considerable challenges for the optical design process.

Today the design of optical systems largely relies on efficient raytracing and optimization algorithms for which a variety of (commercial) software and optimization algorithms are available. During an optical design cycle, different parameters and variables of the optical system are varied to optimize a defined merit function that indicates the image quality for a given field of view. These merit functions are typically "wild" with many local minima, and there is no guarantee that local or global optimization algorithms will lead to a satisfactory design solution. A successful and widely used optimization-based optical design strategy therefore consists of choosing a well-known optical system as a starting point (e.g. from literature) and steadily achieving incremental improvements. However, this "step-and-repeat" approach to optical design does require considerable experience, intuition, and "trial-and-error" guesswork. This applies especially to freeform optical systems.

In this work, we review the recently presented "first time right" optical design method for freeform imaging systems. It is based on solving differential equations derived from Fermat's principle of least time by using a power series approach. The method allows calculating all optical surface coefficients that ensure minimal image blurring for each individual order of aberrations. As such, it offers a disruptive methodology to design optical imaging systems from scratch to facilitate a fast-track uptake of freeform elements in the next generation of high-end imaging systems. Further mathematical details can be found in a recently published article [1].

2 Mathematical formulation

There are numerous possibilities to describe aberrations in freeform optical systems according to their symmetry: (1) two orthogonal planes of symmetry; (2) one plane of symmetry; (3) without any symmetry. Here, we will focus on systems with one plane of symmetry as it is the most common category. Following an ansatz similar to that used in most aberration theories, we assume an arbitrary ray that is emitted from an object at infinity with a field direction vector **H** for field angles H_x and H_y that passes through the pupil of the system (e.g., first surface in Fig. 1) at $(x_p, y_p, f_1(x_p,y_p))$, and that finally intersects the image plane at $\mathbf{h} = (h_x, h_y, D)$. This intersection can be written in vector form as a series expansion $\mathbf{h} = \mathbf{h_0} + \epsilon^{(1)} + \epsilon^{(2)} + \epsilon^{(3)} + \ldots$ where $\mathbf{h_0} = (g(H_x), g(H_y), D)$ denotes the ideal (aberration-free) image location as a function of H_x and H_y. Deviations from the ideal image location are then described by ray aberrations polynomials $\epsilon^{(i)}=(\epsilon_x,\epsilon_y,0)$ with the transverse ray aberration series $\epsilon_x(x_p,y_p,H_x,H_y)$ and $\epsilon_y(x_p,y_p,H_x,H_y)$, in x- and y-direction, respectively.

* Corresponding author: fduerr@b-phot.org

 https://doi.org/10.1051/epjconf/202125502001

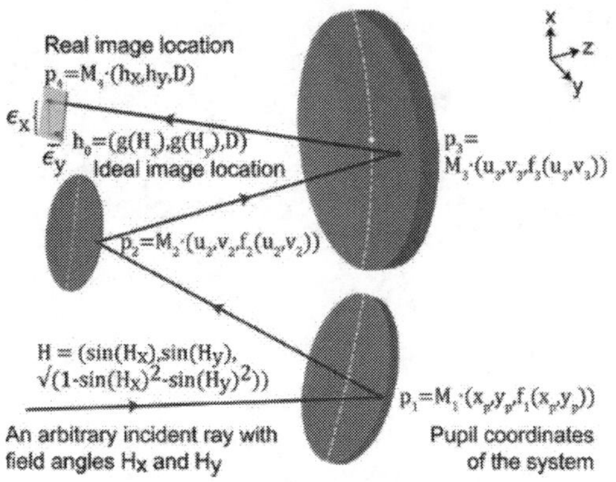

Fig. 1. Schematic layout of a typical plane-symmetric freeform optical system. The ray path of an arbitrary ray is shown including the introduced mathematical description of the initial ray direction and intersections with each mirror and the local transverse ray aberrations ϵ_x and ϵ_y in the image plane.

We now consider a sequence of N refractive and/or reflective optical surfaces f_i (i=1…N) (in Fig. 1 these are three mirrors) aligned along a principal ray path. To describe their surface functions, we use a XY-polynomial representation $f_i(x,y)$ with series coefficients $f_{i,st}$. Next, we introduce ray mapping functions (u_i,v_i) in x- and y-direction that describe where an arbitrary ray, described by variables (x_p,y_p,H_x,H_y), will intersect each optical surface. For individual surface orientations, we define the tilts of all elements through rotation matrices $\mathbf{M_i}$. The ray path from the object to the image plane then consists of (N+1) segments $d_1…d_{N+1}$ that describe the optical path length distances for an arbitrary ray. Vector geometry enables us to express these distances weighted by the refractive indices $n_{i,i+1}$ of the surrounding materials. By applying Fermat's principle, we can derive two sets of differential equations for all defined distances d_i pairwise from object to image space

$$D_{i,x} = \partial_{u_i}(d_i + d_{i+1}) = 0 \quad (i = 1 \ldots N) \qquad (1)$$
$$D_{i,y} = \partial_{v_i}(d_i + d_{i+1}) = 0 \quad (i = 1 \ldots N) \qquad (2)$$

An optical system consisting of N surfaces is thus fully described by N differential equations $D_{i,x}$ and N differential equations $D_{i,y}$ for i=1...N, for a given but arbitrary pupil plane.

Suppose that the functions h_x, h_y, u_i, v_i and f_i are analytic and smooth solutions of the differential equations $D_{i,x}$ and $D_{i,y}$, then Taylor's theorem implies that the functions must be infinitely differentiable and have power series representations. We can employ a power series approach to seek solutions to the derived differential equations. This method substitutes the power series into the differential equations to determine the values of the series coefficients. To calculate the coefficients, we differentiate Eqs. (1) and (2) with respect to (x_p,y_p,H_x,H_y), evaluated at $x_p = y_p = H_x = H_y = 0$, such that

$$\lim_{x_p,y_p,H_x,H_y\to 0} \frac{\partial^j}{\partial x_p^j}\frac{\partial^k}{\partial y_p^k}\frac{\partial^l}{\partial H_x^l}\frac{\partial^m}{\partial H_y^m} D_{i,x} = 0, (i = 1..N) \quad (3)$$
$$\lim_{x_p,y_p,H_x,H_y\to 0} \frac{\partial^j}{\partial x_p^j}\frac{\partial^k}{\partial y_p^k}\frac{\partial^l}{\partial H_x^l}\frac{\partial^m}{\partial H_y^m} D_{i,y} = 0, (i = 1..N) \quad (4)$$

Once all system specifications such as the entrance pupil diameter, focal length, field of view and design wavelength are specified, we can establish the differential equations and initiate the method as follows: (1) solve the non-linear first order case (j+k+l+m=1) using a standard non-linear solver; (2) solve the linear systems of equations in ascending order (j+k+l+m=2,3,…) by setting unwanted aberrations to zero or by minimizing a combination thereof as required by the targeted specifications of the imaging system. It is important to highlight that these two steps are virtually identical for all freeform designs. We calculate the surface coefficients up to the sixth order, and the mapping and aberration coefficients up to the eighth order throughout this work.

3 Advanced freeform design examples

The first design example is a catoptric imaging system proposed by Bauer et al. for the visible spectrum with x-z-plane symmetry and the object at infinity [2]. It consists of three freeform mirrors with the stop at the first mirror, and a targeted system volume of 60 liters. Thus, the layout of the optical components can be described by three distances and four rotation angles (one for each mirror, and one for the image), which provides seven geometrical degrees of freedom for the optical designer. These distances and angles can now be (manually) adjusted to create an unobscured starting geometry that meets certain geometrical constraints such as the target volume. The directly calculated design that we obtained by only optimizing the seven degrees of freedom of the geometry is shown in Fig. 2a,

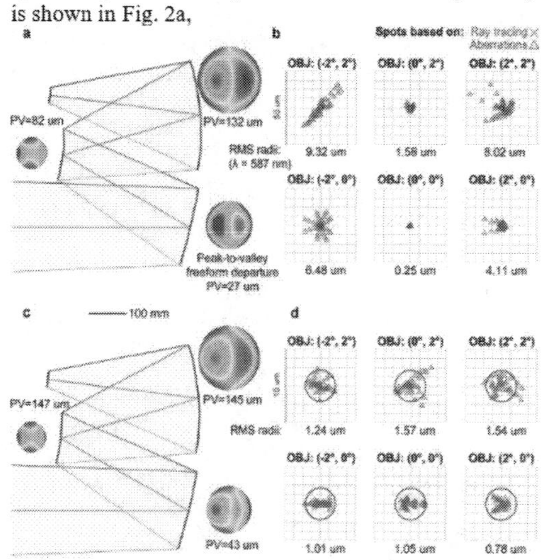

Fig. 2. "First time right" results for the three-mirror imager: **a** Cross-section of the directly calculated system. **b** Spot diagram for six fields based on aberration calculations and ray tracing in comparison. **c** Cross-section of the further optimized system. **d** Spot diagram for the identical six fields.

https://doi.org/10.1051/epjconf/202125502001

where the system layout cross-section is combined with the full 3D peak-to-valley freeform departures (PV) from the best-fit base sphere for each mirror, respectively. Fig. 2b shows the spot diagrams for six selected fields based on aberration calculations. With an average RMS spot radius of about 5 micrometers, our directly calculated system provides an already well-corrected "first time right" solution that can be readily further optimized. Thus, all forty previously calculated surface coefficients plus the initial seven degrees of freedom are used as variables for optimization, e.g. using MATLAB's lsqnonlin solver. After ten iterations and few minutes, the system already reaches diffraction-limited performance for almost the full field of view. The optimized design shows slightly increased and moderate freeform departures distributed among the three mirrors. The results are shown in Fig. 2c and 2d accordingly. The aberration-based performance estimation of our method was found to be in excellent agreement with spot diagram data from classical ray tracing (calculated using Zemax, overlaid red cross symbols in Fig. 2b and 2d).

A second example is a monolithic freeform objective for a very compact infrared camera with four optical surfaces [3]. We chose this example to illustrate the scalability of our design method, well beyond the capabilities of most present-day direct freeform design approaches. We furthermore highlight its holistic character, since in this case we are dealing with a catadioptric system with freeform and aspherical surfaces. Here, the objective consists of three aspheres and one freeform at the second surface, with the stop placed at the first surface. We follow this reference as closely as possible: an F/1.4 design covering a 37×25-degree field-of-view, an 8.4 mm entrance aperture, made of optical-grade germanium and operating in the long-wave infrared region (LWIR) from 8-12 um. The layout of the system can be described by four distances and five angles, defining the principal ray path from object to image and the respective positions and orientation of all surfaces and corresponding rotation matrices. Optimizing the 9 initial degrees of freedom of the geometry yields an already well-corrected system. Fig. 3a shows the configuration of the system of comparable size to the cited reference design. The peak-to-valley freeform departure (PV) from the best-fit base sphere for the second freeform surface is added next to it. Fig. 3b shows the corresponding spot diagrams for six selected fields (we used 15 fields for the optimization) based on our aberration calculations. With an average RMS spot radius of about 51 micrometers, we achieved an excellent starting point for further optimization. All surface coefficients are now set as variables for further optimization using MATLAB's lsqnonlin solver. The results are shown in Fig. 3c and 3d. Within few minutes, the system reaches a close to diffraction-limited performance in the LWIR band for the full field of view with an acceptable distortion of less than 5%. Chromatic aberrations are very well controlled due to the relatively low dispersion of Germanium in the LWIR band. This spot diagram-based performance is in excellent agreement with spot diagram data from classical ray tracing.

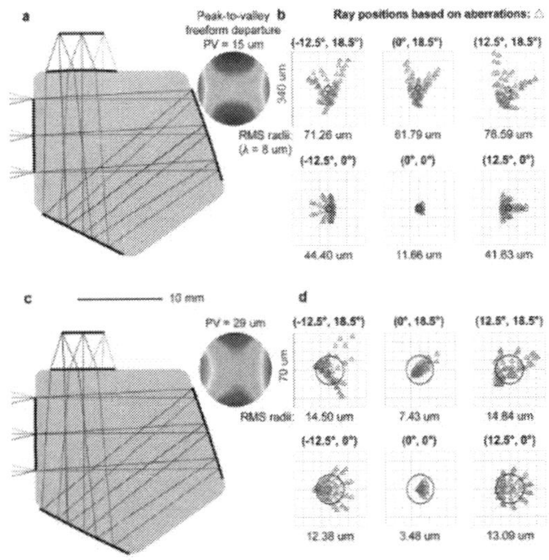

Fig. 3. "First time right" results for the monolithic LWIR design: **a** Cross-section of directly calculated system. **b** Spot diagram for six fields based on aberration calculations. **c** Cross-section of the further optimized system. **d** Spot diagram for identical six fields.

This catadioptric example clearly demonstrates yet again the highly effective nature of our proposed design and evaluation method. In addition, it highlights two further important features: (1) our method allows to combine all types of refractive and/or reflective surfaces of spherical, aspheric or freeform shapes; (2) the method enables to increase the number of calculated surfaces of an optical system without considerably increasing the computational complexity of the problem. This direct path to scaling the number of calculated surfaces is a result of the fact that the optical path length distance expressions do not depend on the number of considered surfaces.

Despite faster F-numbers and larger FOVs, freeform four-mirror designs bring in extra complexity to both mirror fabrication and system assembly compared to three-mirror designs. As a promising alternative, we have investigated a revived type of multi-reflection mirror systems (see Fig. 4), where at least one of the mirrors is passed twice in the ray path with desired overlap (not mutually centered to provide more flexible designs) [4].

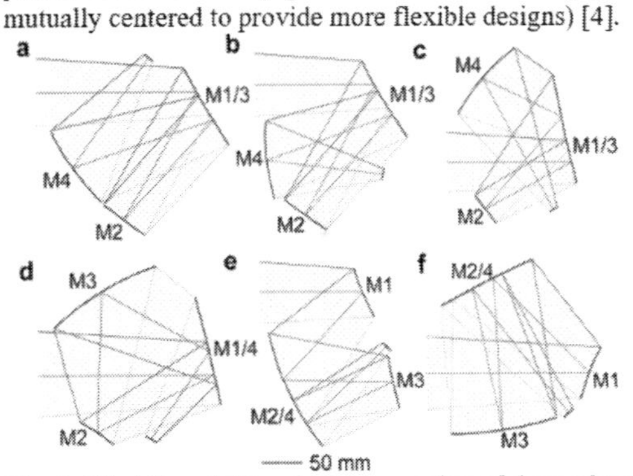

Fig. 4. "First time right" unlocks a new class of three-mirror, four-reflection systems with three different joint mirror options: **a-c** Joint M1&M3. **d** Joint M1&M4. **e-f** Joint M2/M4.

This contrasts the well-known Offner designs where two individual mirrors share separate parts (no overlap) of a joint substrate without a desired overlap in their footprints. Regarding the four-reflection cases, there are three possibilities to generate the "double-pass surface", respectively mirrors M1/3, M1/4 and M2/4. Several showcases of each type have been calculated with the proposed "first time right" method, summarized in Fig. 4. Please note that all layouts are based on real raytracing with a focal length of 250mm, F/2.5 and a full field of view of 7.2°×7.2°. To conclude, the three-mirror, four-reflection concept can yield high-performing yet compact designs, opening a largely extended design space to explore and seek new competitive freeform systems [5].

4 Discussion

Equipping lens- and mirror-based optical systems with freeform optical surfaces makes it possible to deliver highly original imaging functionalities with superior performance. Compared to their traditional (a)spherical counterparts, freeform systems can offer an enhanced field-of-view, increased light-collection efficiencies, larger spectral band and higher compactness. Until now mathematical models and design strategies for freeform optics remained limited and failed to provide streamlined solutions. In particular, the identification of a suitable initial design has often proven to be a painstaking and time-consuming "trial-and-error" process.

In this article, we reviewed the first deterministic direct design method for freeform optical systems that is not restricted by the aberration terms that can be controlled and corrected. The method allows a highly systematic generation and evaluation of directly calculated freeform design solutions that can be readily used as excellent starting points for further optimization. As such, this new method allows the straightforward generation of "first time right" initial designs that enable a rigorous, extensive, and real-time evaluation in solution space when combined with available local or global optimization algorithms.

The authors acknowledge funding by VUB-OZR, BELSPO-IAP, FWO [12C9817N], IOF, Hercules/FWO and the Methusalem foundations.

References

1. F. Duerr, H. Thienpont, Light Sci. Appl. **10**, 95 (2021)
2. A. Bauer, E.M. Schiesser, J.P. Rolland, Nat. Commun. **9**, 1756 (2018)
3. S.R. Kiontke, Proc. SPIE **9575**, 95750G (2015)
4. D.R. Shafer, Appl. Opt. **16**, 1178 (1977)
5. Y. Nie, D.R. Shafer, H. Ottevaere, H. Thienpont, F. Duerr, Opt. Express **29**, 17227 (2021)

Freeform design of a two-reflector system to collimate and shape a point source distribution

A.H. van Roosmalen[1,*], *M.J.H.* Anthonissen[1], *W.L.* IJzerman[1,2], and *J.H.M.* ten Thije Boonkkamp[1]

[1]CASA, Department of Mathematics and Computer Science, Eindhoven University of Technology, PO Box 513, 5600 MB Eindhoven, The Netherlands
[2]Signify Research, High Tech Campus 7, 5656 AE Eindhoven, The Netherlands

Abstract. We present a method to design a freeform two-reflector system to collimate and shape a beam from a point source. An important generalization compared to previous research is that the output beam can be in an arbitrary direction. The design problem is based on a generalized Monge-Ampère equation. This equation is solved using a least-squares algorithm for non-quadratic cost functions. We test our algorithm on two cases, first, uniform source and target distributions, and second, an elliptic Gaussian intensity of a laser diode to a ring-shaped illuminance. We are able to obtain good solutions in both cases.

1 Introduction

In this article we will give a summary of our paper in Optics Express [1]. We consider the design of an optical system with two freeform reflector surfaces for collimating and shaping the distribution from a point source. This is regularly linked to beam shaping for lasers. The source of a laser is often a laser diode, which can be modeled as a point source. In general, its output is collimated with a lens and subsequently shaped. Combining these two functions into one optical system of two reflector surfaces reduces the number of optical surfaces needed. To achieve this, we adapt a method developed by Prins [2], Yadav [3], Romijn [4] et al. for other optical systems.

2 Mathematical model

Our goal is to construct an optical map $\boldsymbol{m}$ from the source domain $\mathcal{S}$ to the target domain $\mathcal{T}$ such that the source density f is mapped to a target density g. The set $\mathcal{S}$ is a subset of the unit sphere and $\mathcal{T}$ contains two-dimensional position vectors on the target plane, which is at a distance l from the point source. This target plane is perpendicular to the direction of the outgoing beam, given by the vector $\hat{\boldsymbol{a}}_3$. This vector can be anywhere on the unit sphere, which gives freedom in designing optical systems. We regard $\hat{\boldsymbol{a}}_3$ as a rotation from the z-axis with a rotation matrix $\boldsymbol{A}$, defined by the polar angle φ and azimuthal angle θ, so $\hat{\boldsymbol{a}}_3 = \boldsymbol{A}\hat{\boldsymbol{e}}_3$. For $\hat{\boldsymbol{s}} \in \mathcal{S}$, the first reflector $\mathcal{R}_1$ is defined by $\boldsymbol{r}_1(\hat{\boldsymbol{s}}) = u(\hat{\boldsymbol{s}})\hat{\boldsymbol{s}}$, where u is the distance from the point source to the reflector. Similarly, for $\boldsymbol{y} \in \mathcal{T}$, the second reflector $\mathcal{R}_2$ is located at a distance $w = w(\boldsymbol{y})$ from the target plane. The reflector is then defined by $\boldsymbol{r}_2 = \boldsymbol{A}(y_1, y_2, l - w(\boldsymbol{y}))^\mathsf{T}$. The distance between the reflectors measured along a ray is denoted by d. See Fig. 1 for more details. The optical

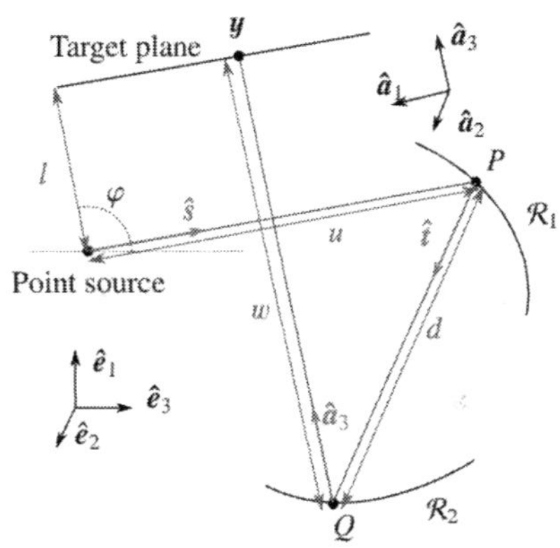

Figure 1: A cross-section of the optical system in the x,z-plane.

path length, L, is given by

$$L = u(\hat{\boldsymbol{s}}) + d + w(\boldsymbol{y}), \tag{1}$$

and is constant in $\hat{\boldsymbol{s}}$ and $\boldsymbol{y}$ as a consequence of the Malus-Dupin theorem [5, Sec. 4.4.3]. Using that d is the distance between the reflectors, i.e., $d = |\boldsymbol{r}_2 - \boldsymbol{r}_1|$, we can express it as a function of u and w. We parameterize the unit direction vectors $\hat{\boldsymbol{s}}$ by the stereographic projection $\boldsymbol{x}$, given by

$$\boldsymbol{x} = \frac{1}{1+s_3}\begin{pmatrix} s_1 \\ s_2 \end{pmatrix}, \tag{2}$$

and introduce $\mathcal{X} = \boldsymbol{x}(\mathcal{S})$. We use this to derive an equation of the form

$$u_1(\boldsymbol{x}) + u_2(\boldsymbol{y}) = c(\boldsymbol{x}, \boldsymbol{y}). \tag{3}$$

*e-mail: a.h.v.roosmalen@tue.nl

The functions u_1 and u_2 are related to the first and second reflector, respectively, given by

$$u_1 = \log\left(\frac{\beta}{u} + \hat{s}\cdot\hat{a}_3 - 1\right), \tag{4a}$$

$$u_2 = \log\left(-\frac{w}{\beta} - \frac{1}{2\beta^2}\left[|y|^2 - L^2 + l^2\right]\right), \tag{4b}$$

while c is called the cost function. It is given by

$$c(x,y) = \log\left[-\frac{1}{\beta}\hat{s}(x)\cdot A_2 y + \frac{1-\hat{s}(x)\cdot\hat{a}_3}{2\beta^2}|y|^2 + \frac{1}{2}\left(\hat{s}(x)\cdot\hat{a}_3 + 1\right)\right], \tag{5}$$

where $\beta = L - l$ is the reduced optical path length and A_2 is the matrix containing the first two columns of A.

Eq. (3) has many possible solutions. This gives the possibility to choose a c-concave (or c-convex) pair u_1, u_2 as a solution. This means, for c-concave, that

$$\begin{aligned} u_1(x) &= \min_{y\in\mathcal{T}}\left(c(x,y) - u_2(y)\right), \\ u_2(x) &= \min_{x\in\mathcal{X}}\left(c(x,y) - u_1(x)\right). \end{aligned} \tag{6}$$

Alternatively, for a c-convex pair, there is a maximum instead of the minimum. As a result, u_2 is at a stationary point in x and we require

$$\nabla_x c(x,y) - \nabla u_1(x) = 0. \tag{7}$$

From this, we can in principle compute a mapping $m(x) = y$, however, we do not. Instead, we substitute the relation $y = m(x)$ into Eq. (7) and subsequently differentiate w.r.t. x to obtain the condition

$$C\mathrm{D}m = \mathrm{D}^2 u_1 - \mathrm{D}_{xx}c =: P, \tag{8}$$

where $\mathrm{D}m$ is the Jacobian matrix of m and $C = \mathrm{D}_{xy}c = \left(\frac{\partial^2 c}{\partial x_i \partial y_j}\right)$, the mixed derivative matrix, which is a function of x and $m(x)$. The matrix $-P$ is equal to the Hessian of $c(.,y) - u_1$. Therefore, if we choose u_1, u_2 as a c-concave or c-convex pair, we need P to be symmetric negative definite (SND) or symmetric positive definite (SPD), respectively.

The map should also satisfy local energy conservation. For any subset $\tilde{\mathcal{A}} \subset \mathcal{S}$ there is a corresponding set $\mathcal{A}$ of stereographic coordinates. The energy conservation is given by

$$\iint_{\tilde{\mathcal{A}}} f(\hat{s})\,\mathrm{d}S(\hat{s}) = \iint_{m(\mathcal{A})} g(y)\,\mathrm{d}A(y). \tag{9}$$

This says that energy should be conserved through the optical system. The energy contained in a part of the source should be equal to that of the corresponding part on the target. When $\tilde{\mathcal{A}} = \mathcal{S}$, we have global energy conservation, meaning that the total flux of the source and target should be equal. Using integration by substitution, we write Eq. (9) in differential form as

$$\det(\mathrm{D}m) = J(x)\frac{f(x)}{g(m(x))} =: F(x, m(x)), \tag{10}$$

where J is the Jacobian resulting from coordinate transformation of the stereographic projection $\hat{s} \to x$, given by

$$J(x) = \frac{4}{(|x|^2+1)^2}. \tag{11}$$

The boundary condition for this problem is given by the transport boundary condition $m(\partial\mathcal{X}) = \partial\mathcal{T}$, stating that the boundary of the source should be mapped to the boundary of the target. Consequentially, all energy is mapped from the source to the target.

Summarizing, we can derive a mapping m from Eq. (8) subject to the transport boundary condition, where P satisfies $\det(P) = F\det(C)$, due to Eq. (10). Subsequently, we calculate u_1 and u_2 from Eq. (7) and (3), respectively, and from those we compute the optical surfaces from Eq. (4).

3 Least-squares solver

We solve the problem mentioned above iteratively with a least-squares algorithm. A brief explanation will be given here, for more details, see [1]. We enforce the equality in Eq. (8) by minimizing the functional J_I defined as

$$J_\mathrm{I}[m, P] = \frac{1}{2}\iint_{\mathcal{X}} \|C\mathrm{D}m - P\|_\mathrm{F}^2\,\mathrm{d}x. \tag{12}$$

The norm $\|.\|_\mathrm{F}$ is the Frobenius norm. To enforce the boundary condition we minimize the difference between the (given) boundary of $\mathcal{T}$ and the mapping of the boundary of the source:

$$J_\mathrm{B}[m, b] = \frac{1}{2}\int_{\partial\mathcal{X}} |m - b|^2\,\mathrm{d}s, \tag{13}$$

where $b : \partial\mathcal{X} \to \partial\mathcal{T}$. We combine these two functionals into a weighted average with parameter $\alpha \in [0,1]$ given by

$$J[m, P, b] = \alpha J_\mathrm{I}[m, P] + (1-\alpha)J_\mathrm{B}[m, b]. \tag{14}$$

All of these functionals are defined on the following spaces

$$\begin{aligned} \mathcal{P}(m) &= \left\{P \in [C^1(\mathcal{X})]^{2\times 2} \mid \det(P) = F\det(C),\ P \text{ SPD}\right\}, \\ \mathcal{B} &= \left\{b \in [C^1(\partial\mathcal{X})]^2 \mid b(x) \in \partial\mathcal{T}\right\}, \\ \mathcal{M} &= [C^2(\mathcal{X})]^2. \end{aligned}$$

In common parlance, $\mathcal{P}$ contains 2×2 matrices with differentiable entries such that the stated constraints are fulfilled and $\mathcal{B}$ contains vector-valued functions from the source boundary to the target boundary whose components are differentiable. Similarly, $\mathcal{M}$ contains vector-valued functions, that are twice differentiable.

We cover the domain $\mathcal{X}$ by a grid with gridpoints x_{ij}. The algorithm to find m is initialized by a guess m^0 for the mapping. With this mapping we compute the matrix $C^0 = C(x, m^0)$. Then, we iteratively perform the next steps, either for a fixed number of iterations or until a stop-

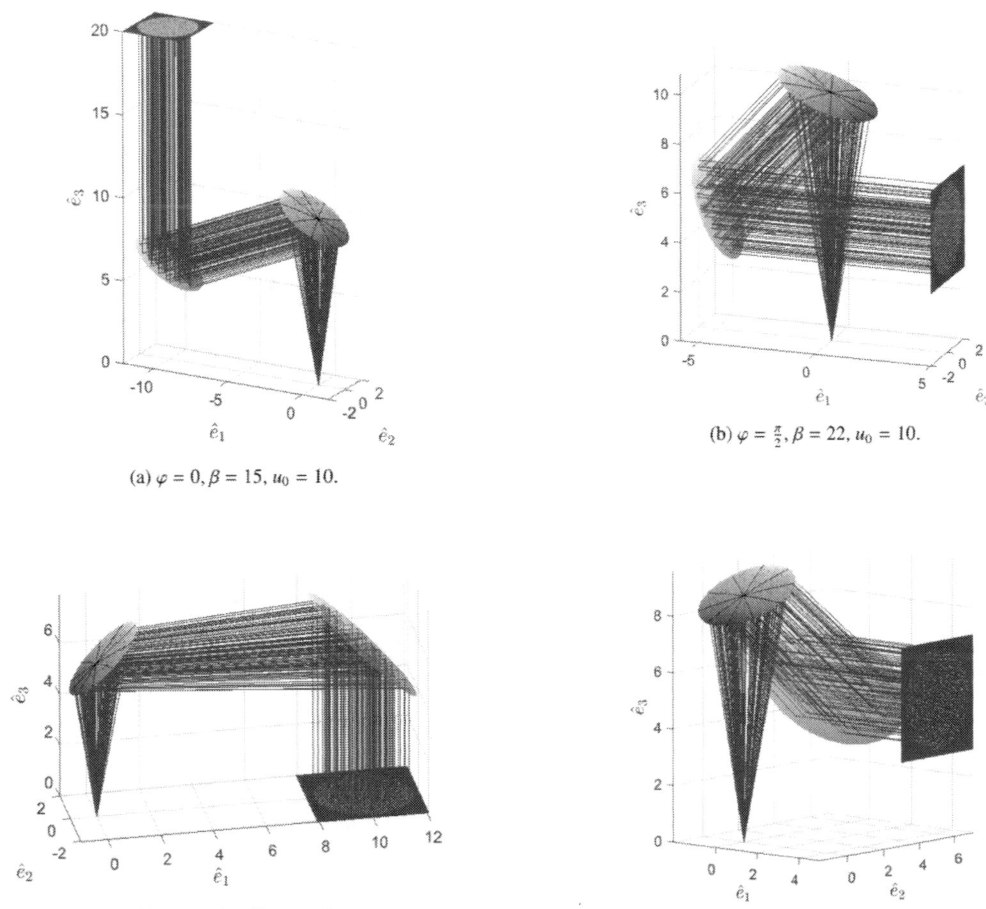

(a) $\varphi = 0, \beta = 15, u_0 = 10$.

(b) $\varphi = \frac{\pi}{2}, \beta = 22, u_0 = 10$.

(c) $\varphi = \pi, \beta = 22, w_0 = 6$.

(d) $\varphi = \frac{\pi}{2}, \beta = 15, w_0 = 5$.

Figure 2: Some examples of layouts constructed with our algorithm having a uniform source and target distribution.

ping criterion is met,

$$\boldsymbol{P}^{n+1} = \underset{\boldsymbol{P}\in\mathcal{P}(\boldsymbol{m}^n)}{\operatorname{argmin}}\, J_{\mathrm{I}}[\boldsymbol{m}^n, \boldsymbol{P}], \tag{15a}$$

$$\boldsymbol{b}^{n+1} = \underset{\boldsymbol{b}\in\mathcal{B}}{\operatorname{argmin}}\, J_{\mathrm{B}}[\boldsymbol{m}^n, \boldsymbol{b}], \tag{15b}$$

$$\boldsymbol{m}^{n+1} = \underset{\boldsymbol{m}\in\mathcal{M}}{\operatorname{argmin}}\, J[\boldsymbol{m}, \boldsymbol{P}^{n+1}, \boldsymbol{b}^{n+1}], \tag{15c}$$

$$\boldsymbol{C}^{n+1} = \boldsymbol{C}(\boldsymbol{x}, \boldsymbol{m}^{n+1}). \tag{15d}$$

The first two steps can be solved pointwise, since they do not contain derivatives of $\boldsymbol{P}$ and $\boldsymbol{b}$, respectively. The minimizer for (15a) can be found analytically and exactly, as shown in [3]. The step (15b) is explained in [4].

The computation of the mapping is a bit more intricate, since the functional J includes derivatives of $\boldsymbol{m}$. The minimization is done using calculus of variations [1]. By setting the first variation of J with respect to any $\boldsymbol{\eta} \in \mathcal{M}$ equal to 0, we can eventually derive the following boundary value problem:

$$\operatorname{div}(\boldsymbol{C}^T \mathrm{C} \mathrm{D}\boldsymbol{m}) = \operatorname{div}(\boldsymbol{C}^T \boldsymbol{P}), \qquad \text{for } \boldsymbol{x} \in \mathcal{X}, \tag{16a}$$

$$\alpha \boldsymbol{C}^T \boldsymbol{C}(\mathrm{D}\boldsymbol{m})\hat{\boldsymbol{n}} + (1-\alpha)\boldsymbol{m} = \alpha \boldsymbol{C}^T \boldsymbol{P}\hat{\boldsymbol{n}} + (1-\alpha)\boldsymbol{b} \quad \text{for } \boldsymbol{x} \in \partial\mathcal{X}, \tag{16b}$$

where div is the divergence operator applied to each row of a matrix and $\hat{\boldsymbol{n}}$ is the outward unit normal of $\mathcal{X}$. This system is discretized using a finite volume method and the resulting linear system is solved using a QR-decomposition. After convergence, we calculate the surface shapes from the mapping $\boldsymbol{m}$. This is done by using Eq. (7), finding a u_1 that minimizes

$$I[\phi] = \frac{1}{2} \iint_{\mathcal{X}} |\nabla_{\boldsymbol{x}} c(.,\boldsymbol{m}) - \nabla\phi|^2 \, \mathrm{d}\boldsymbol{x}. \tag{17}$$

The surface shapes are then calculated from u_1 using Eq. (4). This calculation introduces an extra degree of freedom, which we use to define the position of one of the two surfaces along the central ray, i.e. u_0 or w_0.

In this design problem, there are some degrees of freedom in the choice of input parameters. Obviously, f and g are given as input, as well as the angles φ and θ for the direction of the outgoing beam. However, for a rotationally symmetric f, the angle θ does not matter and can be chosen equal to 0. Furthermore, two out of L, l and β are input parameters, and either u_0 or w_0.

4 Numerical results

We tested our method with two cases. First, we consider the case of a uniform source and target. This is used to show the possibilities of designing optical systems with different layouts. A common layout has the output beam parallel to the symmetry axis of the source output, see Fig. 2a. After a Monte Carlo ray trace with ten million rays we obtain a standard deviation of 4.4% of the mean flux in a bin. This simulation was done on a 200×200 grid in 200 iterations, taking 167 seconds. As mentioned, we can change the output direction. Since our source is rotationally symmetric, only the polar angle φ of $\hat{\boldsymbol{a}}_3$ matters. By varying this angle, the reduced optical path length β and the central position of the first or second mirror (u_0 and w_0, respectively), we can construct optical systems such as in Fig. 2.

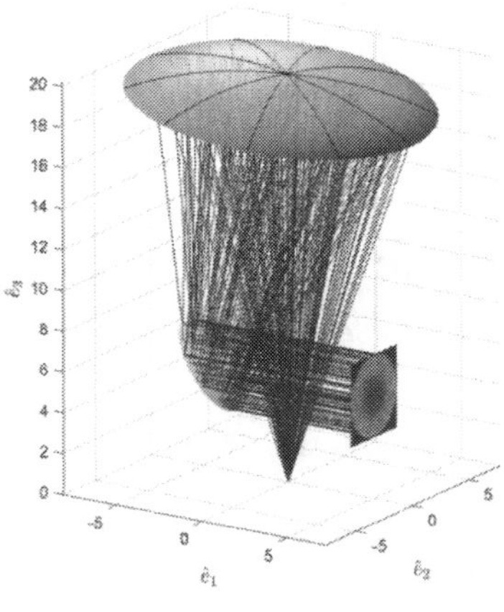

Figure 3: The layout of our test case using a laser diode and an annular output.

The second test case is a model of a laser diode source and a ring-shaped target. This specific laser beam output can be used in practical applications. The laser diode is modeled by an elliptical Gaussian intensity from the point source having $1/e^2$ intensity angles $\theta_x = 45°$ and $\theta_y = 13°$. The ring-target has an inner radius of 1 and an outer radius of 2. The layout of the optical system is given in Fig. 3. The polar angle of the output direction is given by $\varphi = \pi/2$. Furthermore, $\beta = 30$, $l = 5$, the distance of the first reflector along the central ray is 20 and the center of the ring is shifted by 5 along the $\hat{\boldsymbol{e}}_3$-axis.

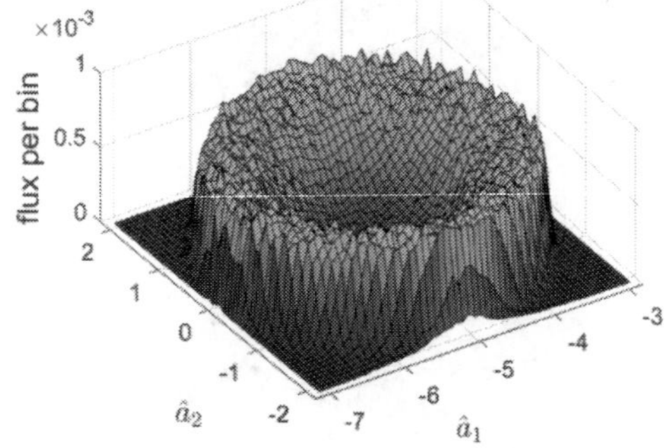

Figure 4: Raytraced resulting output illumination for the second test case

Using our algorithm to design an optical system and applying ray tracing gives an output illumination as in Fig. 4. This test case is more elaborate than the uniform case and therefore takes longer to converge. Using a 400×400 grid and 1000 iterations takes approximately 3000 seconds.

5 Conclusion

To conclude, we presented a method to design two-reflector systems shaping and collimating light for a wide range of input and output distributions. Because the output direction can be chosen arbitrarily, there is a high flexibility in choosing the layout of the system.

References

[1] T.A.H. van Roosmalen, J.H.M. ten Thije Boonkkamp, M.J.H. Anthonissen, W.L. IJzerman, Optics Express **29** (2021), preprint arXiv:2104.04468

[2] C.R. Prins, R. Beltman, J.H.M. ten Thije Boonkkamp, W.L. IJzerman, T.W. Tukker, SIAM Journal on Scientific Computing **37**, B937 (2015)

[3] N.K. Yadav, Ph.D. thesis, Eindhoven University of Technology (2018)

[4] L.B. Romijn, J.H.M. ten Thije Boonkkamp, W.L. IJzerman, Journal of Computational Physics **408** (2020)

[5] E. Hecht, *Optics* (Pearson, 2016), ISBN 9781292096933

An Inverse Model for Two-dimensional Reflectors With Scattering

Vì C.E. Kronberg[1,*], *Martijn J.H.* Anthonissen[1], *Jan H.M.* ten Thije Boonkkamp[1], and *Wilbert L.* IJzerman[1,2]

[1]Department of Mathematics and Computer Science, Eindhoven University of Technology (TU/e), PO Box 513, 5600 MB Eindhoven, The Netherlands
[2]Signify Research, High Tech Campus 7, 5656 AE Eindhoven, The Netherlands

Abstract. We present a novel approach of modelling surface light scattering in the context of freeform optical design. Using energy conservation, we derive an integral relation between the scattered and specular distributions. This integral relation reduces to a convolution integral in the case of isotropic scattering in the plane of incidence for cylindrically and rotationally symmetric problems.

1 Introduction

Freeform reflector design answers the question of transforming a given source light distribution into a given target light distribution. A contemporary technique involves solving a Monge-Ampère equation for the shape of the reflector [1]. If the system is cylindrically or rotationally symmetric, the procedure may be performed in two dimensions and the resulting differential equation is significantly easier to solve [2, Ch. 2-3]. Irrespective of the symmetry of the problem, these approaches rely on the law of reflection from geometrical optics, and are thus only valid under the assumption of perfect specular reflectors. When surface scattering is present, this no longer holds. In order to compute specular reflectors using known methods, whilst taking into account light scattering, we propose a scheme of converting the prescribed diffuse target distribution into an intermediate specular distribution — see Sec. 2, followed by a specular reflector computation — see Sec. 3 and Sec. 4 for an outline of the procedure. Some results are presented in Sec. 5.

2 Scattering Model

The derivations in this section are detailed in [3].

2.1 Cylindrically Symmetric Reflectors

Restricting our attention to scattering in the plane of incidence, consider Fig. 1 showing the cross-section of a cylindrically symmetric reflector $\mathbf{r}(\varphi) = u(\varphi)\hat{\mathbf{e}}_r$, where $u(\varphi)$ is the radius function and $\hat{\mathbf{e}}_r = (\cos(\varphi), \sin(\varphi))^{\mathsf{T}}$ is the radial unit vector in polar coordinates, and $\varphi \in (-\pi, \pi)$ is measured counter-clockwise with respect to the positive x-axis.

A source ray $\hat{\mathbf{s}}$ is emitted from the line-source in O at an angle φ and strikes the reflector at $\mathcal{P}$. The specular ray $\hat{\mathbf{t}}$ with angle $\psi \in (-\pi, \pi)$ (not shown, analogous to $\gamma \in (-\pi, \pi)$ in the figure) is given by the familiar law of reflection (LoR), $\hat{\mathbf{t}} = \hat{\mathbf{s}} - 2(\hat{\mathbf{s}} \cdot \hat{\mathbf{n}})\hat{\mathbf{n}}$, where $\hat{\mathbf{n}}$ is the unit normal at $\mathcal{P}$. By convention, we take $\hat{\mathbf{s}} \cdot \hat{\mathbf{n}} < 0$, i.e., the normal points back towards the source. The scattered ray $\hat{\mathbf{u}}$ is rotated by a stochastic parameter $\alpha \in (-\pi, \pi)$ with respect to $\hat{\mathbf{t}}$, and its direction is denoted $\gamma \equiv \psi + \alpha$. That is,

$$\hat{\mathbf{u}} = \mathbf{R}(\alpha)\hat{\mathbf{t}}, \quad \mathbf{R}(\alpha) = \begin{pmatrix} \cos(\alpha) & -\sin(\alpha) \\ \sin(\alpha) & \cos(\alpha) \end{pmatrix}. \tag{1}$$

In terms of mappings, let $m(\varphi) = \psi$ be the law of reflection yielding the direction of $\hat{\mathbf{t}}$, $s(\psi;\alpha) = \gamma$ be the so-called scattering map representing the direction of $\hat{\mathbf{u}}$, and $a(\psi,\gamma) = \alpha$ be a map yielding the stochastic parameter α, for fixed ψ and γ. A schematic summary is shown in Fig. 2. Additionally, we shall require the inverse maps $m^{-1}(\psi) = \varphi$ and $s^{-1}(\gamma;\alpha) = \psi$ to exist. Explicitly, the scattering maps are

$$\begin{aligned} s(\psi;\alpha) &= \psi + \alpha \equiv \gamma, \\ s^{-1}(\gamma;\alpha) &= \gamma - \alpha \equiv \psi, \\ a(\psi,\gamma) &= \gamma - \psi \equiv \alpha. \end{aligned} \tag{2}$$

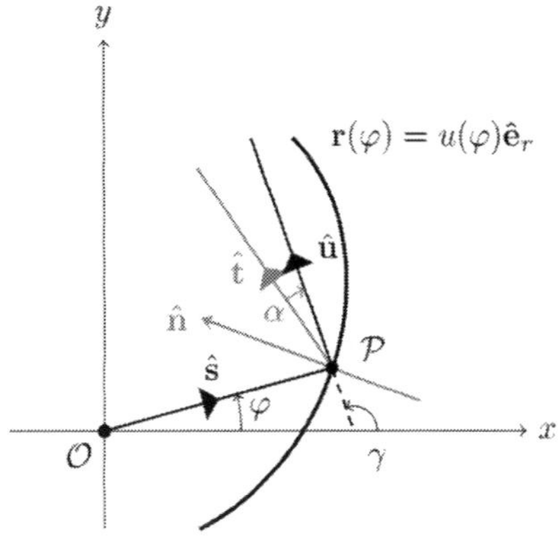

Figure 1. Cross-section of a diffuse cylindrically symmetric reflector in the plane of incidence.

*e-mail: s.kronberg@tue.nl

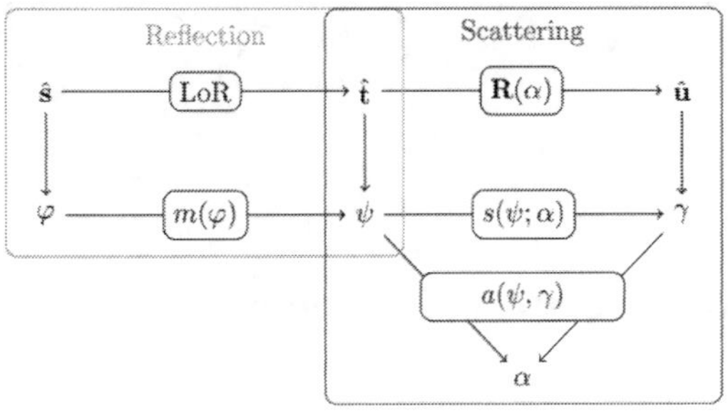

Figure 2. Mappings between unit vectors and angles.

Let f: $[\varphi_1, \varphi_2] \to \mathbb{R}^+$ be the source distribution, h: $[\gamma_1, \gamma_2] \to \mathbb{R}^+$ be the prescribed diffuse target distribution, and g: $[\psi_1, \psi_2] \to \mathbb{R}^+$ be an intermediate specular target distribution [lm/rad]. Additionally, let p: $[\alpha_1, \alpha_2] \to \mathbb{R}^+$, be a probability density function (PDF) representing the scattering characteristics of the surface. Since p is a PDF, we have $\int_{\alpha_1}^{\alpha_2} p(\alpha)\,\mathrm{d}\alpha = 1$. Finally, let ρ: $[\psi_1, \psi_2] \times [\gamma_1, \gamma_2] \to \mathbb{R}^+$, with the properties

$$\int_{\gamma_1}^{\gamma_2} \rho(\psi, \gamma)\,\mathrm{d}\gamma = g(\psi), \quad \int_{\psi_1}^{\psi_2} \rho(\psi, \gamma)\,\mathrm{d}\psi = h(\gamma), \tag{3}$$

be a density of the form $\rho(\psi, \gamma) = p(a(\psi, \gamma))g(\psi) = p(\gamma - \psi)g(\psi)$, which gives the light in direction γ when integrated with respect to ψ, and vice versa. Then the following convolution integral relations follow from the second relation in Eq. (3):

$$h(\gamma) = \int_{\psi_1}^{\psi_2} p(\gamma - \psi)g(\psi)\,\mathrm{d}\psi =: (p * g)(\gamma), \tag{4a}$$

$$h(\gamma) = \int_{\alpha_1}^{\alpha_2} p(\alpha)g(\gamma - \alpha)\,\mathrm{d}\alpha =: (g * p)(\gamma), \tag{4b}$$

where we transformed the integration variable ψ to α to reach the second relation. Physically, the density ρ represents a redistribution of light in direction ψ due to scattering via the scattering probability density function p. Note that mirrors mostly reflect in the specular direction, so that p should peak for $\alpha = 0$ and fall off rapidly to zero for most relevant surfaces.

2.2 Rotationally Symmetric Reflectors

Following an analogous procedure for rotationally symmetric problems, i.e., problems where we can consider not a translationally invariant cross-section of the reflector, but rather a rotationally invariant cross-section, the integral relations for the diffuse distribution in Eq. (4) become

$$\tilde{h}(\gamma) = \int_{\psi_1}^{\psi_2} p(\gamma - \psi)\tilde{g}(\psi)\,\mathrm{d}\psi, \quad \tilde{h}(\gamma) = \int_{\alpha_1}^{\alpha_2} p(\alpha)\tilde{g}(\gamma - \alpha)\,\mathrm{d}\alpha, \tag{5}$$

where $\tilde{h}(\gamma) := h(\gamma)\sin(\gamma)$ and $\tilde{g}(\psi) := g(\psi)\sin(\psi)$, and where ψ, $\gamma \in [0, \pi]$ are now polar angles, making the sine terms area elements in spherical coordinates. The situation is depicted in Fig. 3, where the source ray along $\hat{\mathbf{s}}$ emitted from the point source in $\mathcal{O}$, strikes the reflector at $\mathcal{P}$, where the normal is given by $\hat{\mathbf{n}}$. The specular direction $\hat{\mathbf{t}}$ is characterised by ψ (not shown), and the scattered direction $\hat{\mathbf{u}}$ is given by the angle γ. We note that the coordinate system is such that the radial unit vector $\hat{\mathbf{e}}_r = (\sin(\varphi), \cos(\varphi))$, $\varphi \in [0, \pi)$.

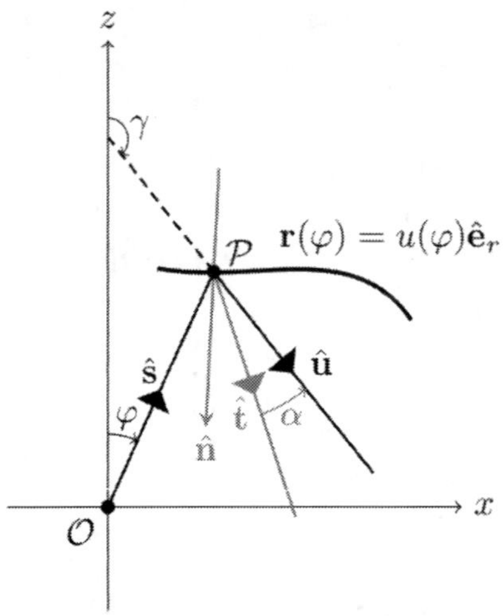

Figure 3. Cross-section of a diffuse rotationally symmetric reflector along the plane of incidence.

3 Specular Reflector Design

Our goal is to determine a specular reflector which transforms the given source distribution into the given specular target distribution. The approach we have chosen involves solving two ordinary differential equations – one for the radius function $u(\varphi)$ and one for the mapping $m(\varphi)$ – which together fully characterise the reflector. This is similar to the approach outlined in [2, Ch. 3.3] and [3].

3.1 Cylindrically Symmetric Reflectors

From the law of reflection, by computing $\hat{\mathbf{n}}$ in terms of u and u', and by fixing $u(\varphi_1) = 1$, we get the following initial value problem for the radius function u:

$$\frac{u'(\varphi)}{u(\varphi)} = \cot\left(\frac{m(\varphi) - \varphi}{2}\right), \quad u(\varphi_1) = 1. \tag{6}$$

There are many valid choices for the specular map m. Consider first a monotonically increasing $m := m_{\text{div}}$, resulting in a diverging ray-bundle. Then the following integral relation and equivalent initial value problem must hold for all $\varphi \in [\varphi_1, \varphi_2]$:

$$\int_{\varphi_1}^{\varphi} f(\tilde{\varphi})\,\mathrm{d}\tilde{\varphi} = \int_{\psi_1}^{m_{\text{div}}(\varphi)} g(\tilde{\psi})\,\mathrm{d}\tilde{\psi}, \tag{7a}$$

$$m'_{\text{div}}(\varphi) = \frac{f(\varphi)}{g(m_{\text{div}}(\varphi))}, \quad m_{\text{div}}(\varphi_1) = \psi_1. \tag{7b}$$

Similarly, for a monotonically decreasing $m := m_{\text{conv}}$, yielding a converging ray-bundle,

$$\int_{\varphi_1}^{\varphi} f(\tilde{\varphi})\,\mathrm{d}\tilde{\varphi} = \int_{m_{\text{conv}}(\varphi)}^{\psi_2} g(\tilde{\psi})\,\mathrm{d}\tilde{\psi}, \tag{8a}$$

$$m'_{\text{conv}}(\varphi) = -\frac{f(\varphi)}{g(m_{\text{conv}}(\varphi))}, \quad m_{\text{conv}}(\varphi_1) = \psi_2. \tag{8b}$$

3.2 Rotationally Symmetric Reflectors

In the case of rotationally symmetric reflectors, we measure φ clockwise from the positive z-axis, in the plane of incidence. The reflector is now parametrised by $\mathbf{r}(\varphi) = u(\varphi)\hat{\mathbf{e}}_r$, where $\hat{\mathbf{e}}_r = (\sin(\varphi), \cos(\varphi))^\top$ is the radial unit vector in this particular polar coordinate system. Following an analogous procedure to the one employed in the case of cylindrically symmetric reflectors, we get the following initial value problem for the monotonically increasing and decreasing maps, respectively:

$$m'_{\text{div}}(\varphi) = \frac{\tilde{f}(\varphi)}{\tilde{g}(m_{\text{div}}(\varphi))}, \quad m_{\text{div}}(\varphi_1) = \psi_1, \tag{9}$$

and

$$m'_{\text{conv}}(\varphi) = -\frac{\tilde{f}(\varphi)}{\tilde{g}(m_{\text{conv}}(\varphi))}, \quad m_{\text{conv}}(\varphi_1) = \psi_2. \tag{10}$$

The initial value problem for the radius function u remains the same.

4 Solution Algorithm

The solution algorithm we propose is as follows. Given the source distribution f, diffuse target distribution h and scattering probability density function p,

1. Deconvolve g from Eqs. (4) for cylindrically symmetric problems, or $\tilde{g}$ from Eqs. (5) for rotationally symmetric problems. Call this deconvolved distribution g_{dc} or $\tilde{g}_{\text{dc}}$. We used Matlab's `deconvlucy`.
2. Compute the reflector(s) using f and g_{dc} or $\tilde{f}$ and $\tilde{g}_{\text{dc}}$ by solving the relevant initial value problem in Eq. (7), (8), (9), or (10). We solved all of the above using Matlab's `ode15s`.
3. Compute the "reconvolved" diffuse distribution $h_{\text{rc}} := g_{\text{dc}} * p$ or $\tilde{h}_{\text{dc}} := \tilde{g}_{\text{dc}} * p$. We used Matlab's `conv`. This step is optional, but it will help us validate our prediction for the scattered distribution.
4. Validate the reflector(s) using raytracing.

The validation via raytracing uses our own raytracer in two dimensions, which implements scattering in accordance with our model, i.e., rotation of the reflected ray by a stochastic parameter α sampled from p.

5 Results

The results for a pair of cylindrically symmetric reflectors is shown in Fig. 4, where f is a block function, p is a Gaussian and the raytraced diffuse distribution h^*_{rc} is very close to our prescribed h. The raytraced reflector is m_{conv} and the root mean squre (RMS) error $\varepsilon(h, h^*) = \sqrt{\frac{1}{N}\sum_{n=1}^{N}\left|h_n - h^*_n\right|^2}$ goes to zero $\propto N_{\text{r}}^{-1/2}$, where N_{r} is the number of rays traced and the sum runs over the number of collection bins N, as expected from Monte Carlo raytracing [4, p. 9]. Turning to Fig. 5, we see an example of a rotationally symmetric reflector. For this example, we do not know $\tilde{g}$ exact, but we show $\tilde{g}_{\text{dc}}$, which is convolved with p — this time a Lorentzian (Cauchy) distribution — to form h_{rc}, representing the predicted diffuse light distribution. The raytraced reflector is once more m_{conv}, which is also shown in three dimensions for the sake of completion. The convergence of the raytraced distribution follows the expected $N_{\text{r}}^{-1/2}$ behaviour.

6 Conclusions

Our novel modelling approach to include surface scattering in the design of reflectors as part of optical systems reduces to a convolution problem for cylindrically and rotationally symmetric problems. Solving the convolution integral yields a specular target distribution which can be used in the mature framework of specular reflector design.

Funding: This work was partially supported by the **Nederlandse Organisatie voor Wetenschappelijk Onderzoek** through grant P15-36.

References

[1] L.B. Romijn, J.H.M. ten Thije Boonkkamp, W.L. IJzerman, Journal of Computational Physics **408** (2020)
[2] M.J.J.J.B. Maes, Ph.D. thesis, University of Amsterdam (1997)
[3] V.C.E. Kronberg, M.J.H. Anthonissen, J.H.M. ten Thije Boonkkamp, W.L. IJzerman, Journal of the Optical Society of America A (2021), submitted
[4] C. Filosa, Ph.D. thesis, Eindhoven University of Technology (2018)

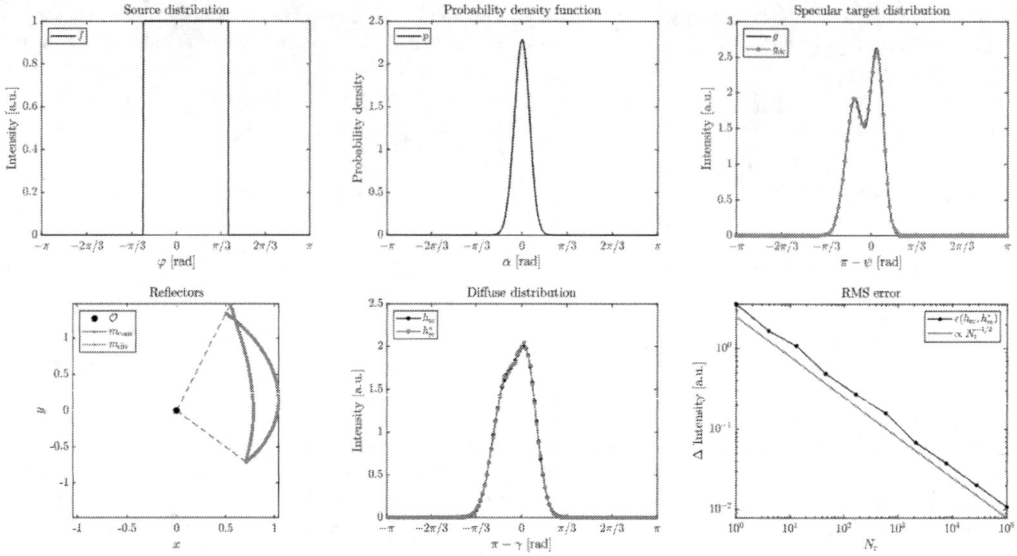

Figure 4. Example of cylindrically symmetric reflector design procedure. The prescribed diffuse distribution h is achieved, and the raytraced distribution converges with the expected square-root behaviour.

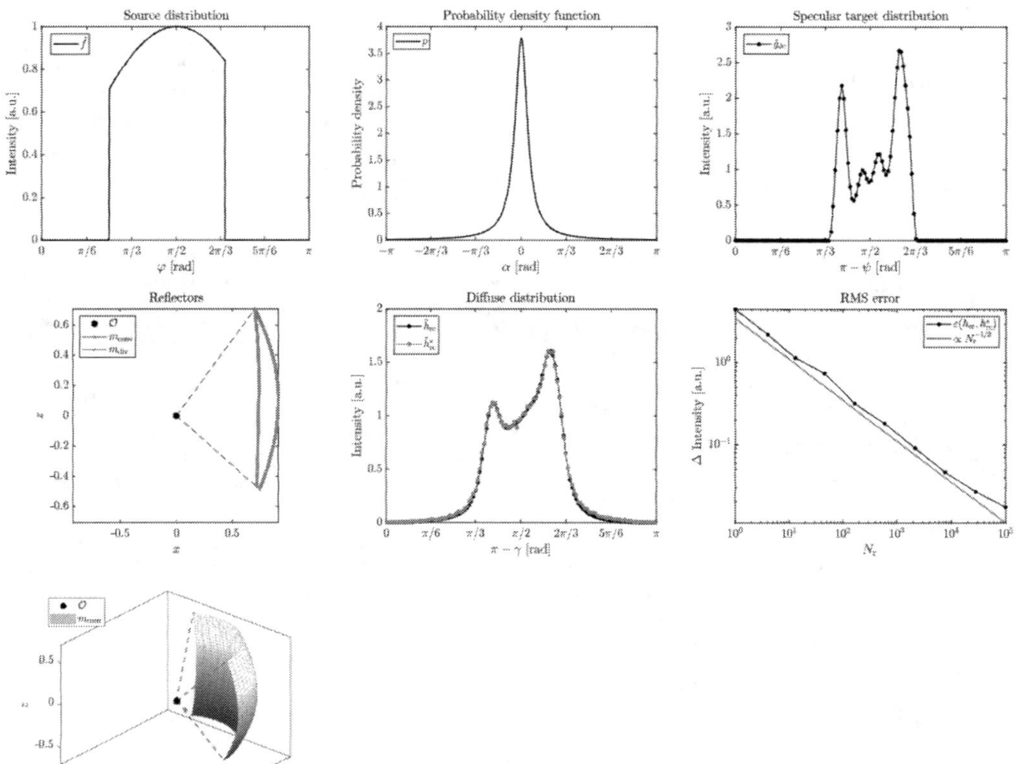

Figure 5. Example of rotationally symmetric reflector design procedure. The prescribed diffuse distribution $\bar{h}_{rc}$ is achieved, and the raytraced distribution (roughly) converges with the expected square-root behaviour. The 3D representation of m_{conv} is included for the sake of completion.

 https://doi.org/10.1051/epjconf/202125503001

Pre-heating by defocusation of the CO_2-Laser polishing beam: an experimental report from the lab-floor -INVITED

Jens Bliedtner[1,*], *Oliver* Faehnle[2], *Anne-Marie* Layher[1], *Robin* Hassel[1,*]& *Andrea* Barz[1],

[1]Ernst-Abbe University of Applied Sciences Jena, Carl-Zeiss-Promenade 2, 07745 Jena, Germany

[2] OST – Ostschweizer University of Applied Sciences, Werdenbergstrasse 4, 9471 Buchs, Switzerland

Abstract. The laser beam polishing for glass and plastics is a purely thermal process and melts the ground or lapped structures to a depth of limited extent. This results in a smoothing of the surface, whereby the 1st - 4th order shape deviations can be corrected very well and transparent surfaces are created. The process is excellently suited for quartz glasses and other optical glasses with a low coefficient of expansion α. Furthermore, thermoplastics or metallic molds for injection molding and precision molding applications can also be polished with the laser beam. On the other hand, special measures are required for glasses with a high α, e.g. preheating of the component. For the investigations, a defocused laser beam was used for the defined preheating of glasses with high linear expansion coefficients. After reaching the material-specific preheating temperature, the laser beam was focused and the polishing process started. A defined cooling process follows again with a defocused beam. In this way, a ground biconvex lens made of boron crown glass was successfully polished. The laser-polished surfaces have an RMS value of 1- 3 nm. The polishing process can be used very flexibly. Likewise, very differently shaped optical components can be polished.

The newly developed polishing regime is transferable to other optical glasses with high linear expansion coefficients.

1 Introduction

Thermal polishing processes represent a special form of interaction on optical surfaces. Two processes, laser polishing and plasma polishing, are suitable for this form of energy input to smooth optical functional surfaces. A microwave-excited Ar/He plasma jet can be used to smooth finely ground and lapped surfaces in relatively short process times. Average plasma powers of 100 W-150 W are required to achieve a temperature of approx. 1600 °C on the glass surfaces as a result of thermal heating. This temperature range is required for smoothing the high-melting quartz glasses. [1]

Laser polishing of glasses is also a purely thermal process and melts the surface structure to a limited depth. Due to the reduction of the viscosity in the near-surface zone to approx. $\eta = 2 - 3$ dPa-s, flow occurs in the depressed zone. This flowing and rearrangement of material fractions leads to a smoothing of the surface. [2] In [3] and [4] the polishing process could be developed very successfully for fused silica. Micro-roughnesses comparable to those of mechanical polishing were produced. An important process parameter is the temperature as well as the temperature distribution. In [5] simulation results for the temperature distribution during laser polishing were presented, which allow to derive important process models. The shape correction and polishing of optical functional surfaces with laser radiation could also be successfully demonstrated in the works [6] and [7]. However, a process disadvantage is the formation of mid spatial frequencies, which are somewhat more pronounced compared to conventional polishing processes. [8]

Mostly, laser polishing investigations are carried out for glasses with a low coefficient of thermal expansion. These glasses, such as fused silica and borosilicate glasses, are very suitable due to their high thermal shock resistance. Glasses with a high coefficient of thermal expansion, on the other hand, require special measures to be able to polish them with the laser beam. This polishing process and the results obtained are presented below.

2 Material properties and experimental setup

For the laser polishing process, the knowledge of the material parameters softening point, thermal shock resistance is a prerequisite to determine the required polishing temperature. Table 1 shows these parameters for three different types of glass. These are very different for the three glasses to be processed. The low thermal shock resistance of borosilica and boroncrowns glass makes it necessary to take special precautions for the polishing process.

Table 1. Relevant material parameters for laser polishing.

Parameter	**Fused silica**	**Borosilica (BF33)**	**Boron crowns glass (BK7)**
Softening point	1660°C	820°C	719°C
thermal shock resistance	2292 K	338 K	133 K
Process temperature	1950 – 2000 °C	1150 – 1350 °C	1150 – 1350 °C

Figure 1 shows the basic temperature-time curves for thermal forming processes, e.g. precision molding. The typical curves show three areas, preheating, forming and postheating. In particular, the postheating region determines the residual stresses remaining in the glass

* Corresponding author: jens.bliedtner@eah-jena.de

volume.

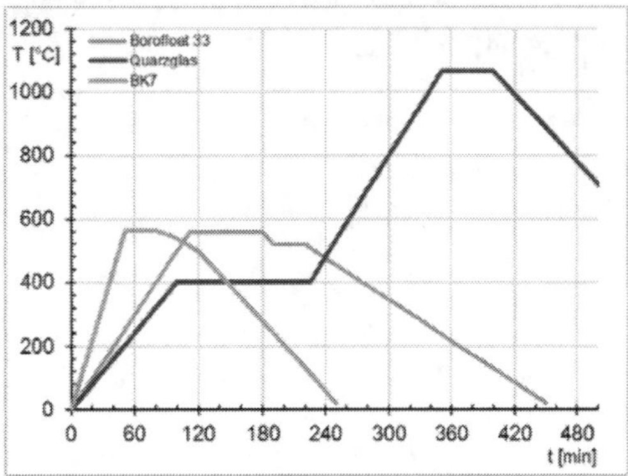

Fig. 1. Temperature-time curves for different glasses.

Due to the low thermal shock resistance of most optical glasses, the glasses must be heated to the softening point by a controlled temperature-time curve. This is necessary so that the glass does not exceed the critical range of internal stresses during the polishing process. In the investigations carried out, a defocused laser beam is used for the preheating phase in order to distribute the line energy over a larger volume of material. This prevents the glass sample from heating up too quickly and the risk of breakage. Figure 2 shows the range of beam defocusing required.

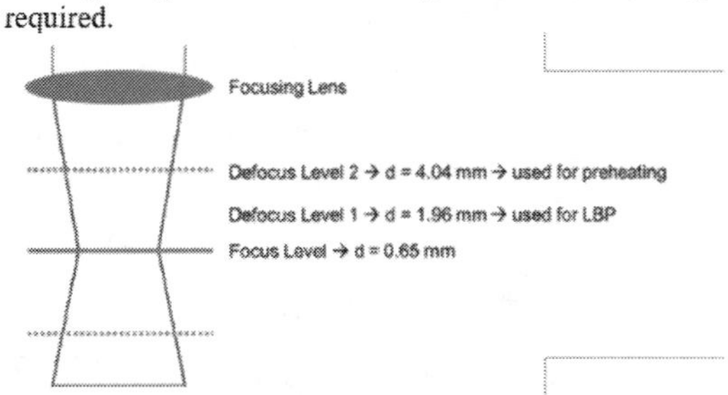

Fig. 2. Defocusing setup of the CO_2 laser beam for preheating

To achieve fast switching between the focused and defocused beam, a 3D scanner with dynamic z-position control can be used.

During the laser-based processes, the influence of the following parameters is investigated:

- preheating temperature [°C]
- values of laser power [W]
- values of scanning speed [m·s^{-1}]
- laser scanning strategies.

Finally, the influence on the following output parameters is examined:

- process temperature [°C]
- contour width and length
- surface and edge quality
- state of thermal stress.

3 Experiments and Results

In the experiments conducted, the focus was on processing the glasses that have a low thermal shock resistance value. These must be specifically processed through the 3 phases of (1) preheating, (2) laser beam polishing and (3) postheating. Figure3 shows a typical temperature-time curve used for laser beam polishing of boroncrown glass. This functional curve was recorded with a special pyrometer suitable for the material glass.

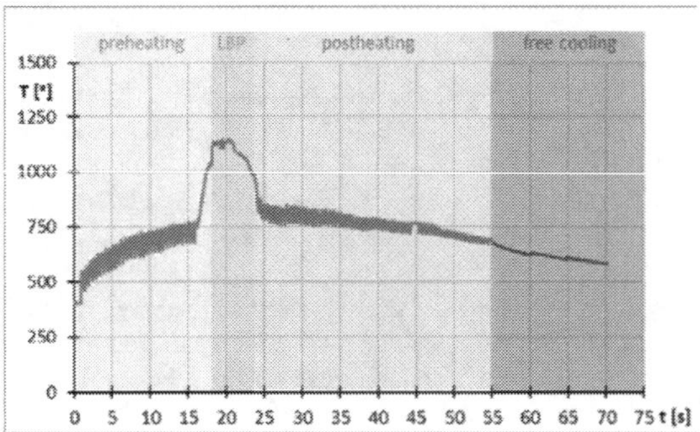

Fig. 3. Temperature-time curves for boroncrown glass (BK7)

The comparison of the T-t-function of boroncrown glass to the temperature-time curve of fused silica is illustrated in Figure 4. The comparison shows very clearly the considerably longer process times for pre- and postheating in the case of boroncrown glass.

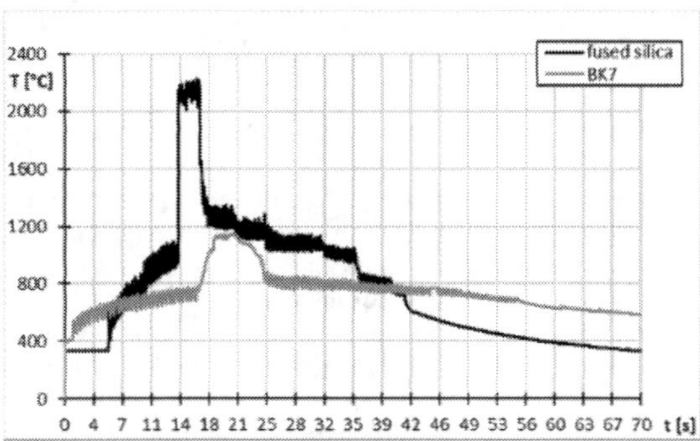

Fig. 4. Temperature-time curves for glasses with with high and low thermal shock resistance

Depending on the optical component to be processed, suitable scanning strategies and temperature-time function must be selected. Figure 5 compares thermographic images during the three phases using the example of a convex-concave lens with a diameter of 30 mm.

The scan direction and scan strategy for this example were chosen to allow the lens to be heated as homogeneously as possible by scanning in a meandering. The lens support was provided from a plate of fused silica.

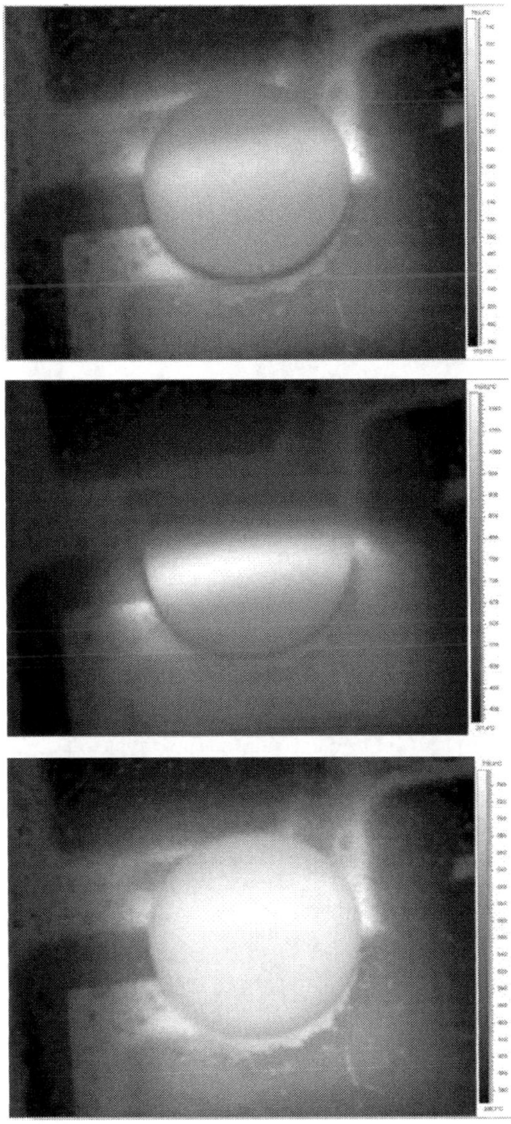

Fig. 5. Thermographic images during the three phases.
a) preheating, b) laser beam polishing and c) postheating.

Compared to mechanical-chemical polishing processes, laser polishing does not produce any disturbing edge effects.

Figure 6 shows the achievable micro-roughnesses for boroncrown glass BK7. The values were recorded according to DIN standard 4287 using two different measuring methods. Polishing with different line energies of the laser radiation was investigated. For the optimum parameter set (value 11), RMS values of less than 1 nm are obtained.

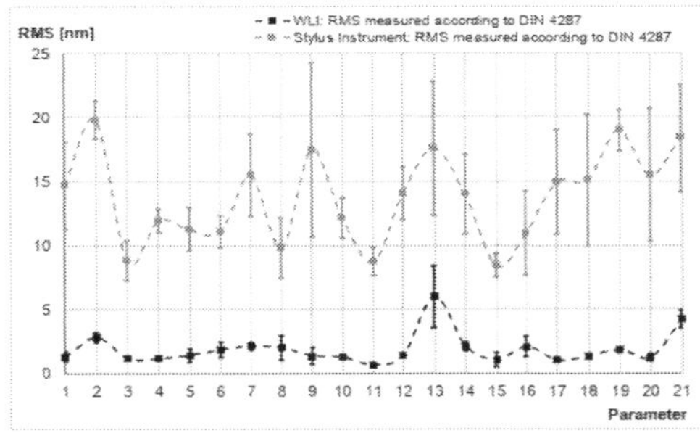

Fig. 6. Representation of the measured roughness values as a function of the parameter sets for the change in line energy.

Further information on the shape deviation, waviness and roughness parameters is provided by the PSD functions in Fig. 7. The measurement results were recorded for two phases. Firstly, after preheating and secondly, after postheating. As expected, significantly higher values are illustrated by the curve after the preheating phase. The increased mid spatial frequencies values after postheating, which are typical for the laser polishing process, can also be seen.

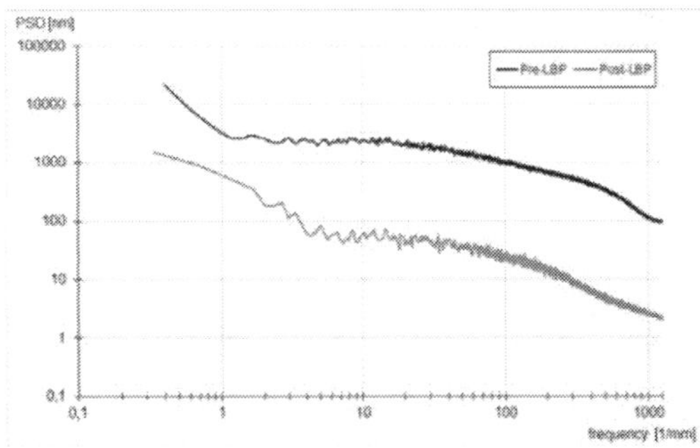

Fig. 7. Power spectral density functions for BK7 after preheating and postheating.

Fig. 8. Functional proof of the laser-polished convex-concave lens.

During the experimental investigations, the laser polishing results were applied to different optical geometries. Figures 8 and 9 illustrate two selected optical components made of broncrown glass.

Fig. 9. Functional verification on a laser-polished prism.

The laser polishing process presented can also be used in principle for structured or almost arbitrarily shaped surfaces. However, the angle of inclination between the laser beam and the glass surface should not exceed 45°. Figure 10 illustrates an array structure with rectangular domes that could be polished with shape retention.

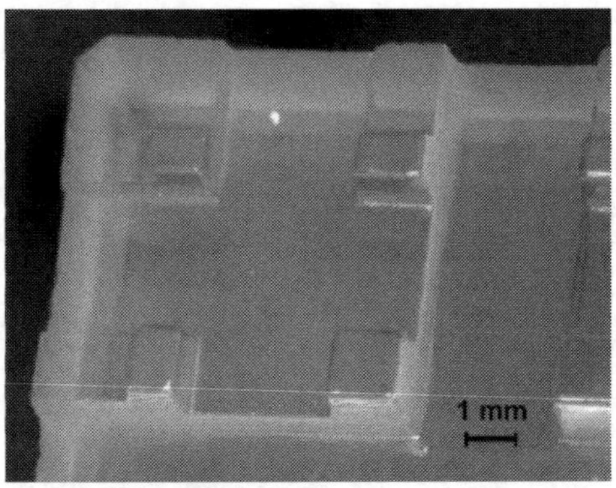

Fig. 10. Functional verification on a laser-polished array structure.

4 Conclusion and Outlook

The presented investigations allow the use of the laser polishing process also for glasses which have only a low thermal shock resistance. A prerequisite is the determination and optimization of the phases preheating, laser polishing and postheating. This can be done advantageously with the rapid variation of the laser beam diameter.

Fig. 11. Laser polishing of a biconvex lens made of Boroncrown glass (BK7).

In further investigations, the aim is to extend the process to other optical glasses, such as flint glasses.

References

1. Arnold, T.; Boehm, G.; Paetzelt, H.: Precision asphere and freeform optics manufacturing using plasma jet machining technology. In: Proc. SPIE 10448. Optifab 2017
2. Bliedtner, J.: *Optiktechnologie: Grundlagen - Verfahren - Anwendungen - Beispiele*. 3., überarbeitete und erweiterte Edition. Carl Hanser Verlag GmbH & Co. KG. (to be published)
3. Hecht, K.: *Entwicklung eines Laserstrahlpolierverfahrens für Quarzglasoberflächen*. Dissertation, TU Ilmenau. 2012
4. Richmann, A.: *Polieren von Gläsern und Kunststoffen mit CO_2-Laserstrahlung*. Dissertation. RWTH Aachen. 2013
5. Hildebrand J., Hecht, K., Bliedtner, J. & Müller, H.: *Laser Beam Polishing of Quartz Glass Surfaces*. Physics Procedia Volume 12, Part A, 2011, Pages 452-461
6. Weingarten, C., Schmickler, A., Willenborg, E., Wissenbach, K.: *Laser polishing and laser shape correction of optical glass*. Journal of Laser Applications 29, 011702 (2017); https://doi.org/10.2351/1.4974905
7. Kessler, H., Botha, R. & Carsten Ziolek: *A new approach to laser polishing and form correction of optical components*. EPJ Web of Conferences 215, 09001 (2019). EOS Optical Technologies. https://doi.org/10.1051/epjconf/201921509001
8. Forbes, G.W.: *Never-ending struggles with mid-spatial frequencies*. Proc. SPIE 9525, Optical Measurement Systems for Industrial Inspection IX, 95251B. 22. June 2015

 https://doi.org/10.1051/epjconf/202125503002

Experimental results from face and lateral fine grinding of fused silica and BK7 using metal and resin bonded tools

Christian Schulze[1,*], *Sebastian* Henkel[1], and *Jens* Bliedtner[1]

[1]Ernst-Abbe University of Applied Sciences Jena, Carl-Zeiss-Promenade 2, 07745 Jena, Germany

Abstract The experimental setup shall compare face grinding with lateral grinding in a single processing step using fused silica and BK7 as materials. Resin and metal bonded tools are used in face and lateral grinding strategies. The paper presents the results in order to deepen the knowledge about those grinding technologies and to further improve the properties of fabricated components. Furthermore, the results confirm, that ultra-fine grinding is a technology, which can be used to process inorganic non-metallic materials with high quality surfaces with a low roughness and high flatness in a low processing time.

1 Experimental setup

A 5-axis CNC machine from company SAUER (DMG) with integrated ultrasonic assistance was used to conduct the experiments on both materials, fused silica as well as BK7. The experimental setup with an optimized sample geometry was already described in a previous paper of last year's EOSAM.[1] Two diamond tools with different bonds are used to compare face grinding with lateral grinding in a single processing step. The tool diameter is 14mm with an average grain size of 16 µm (D16, C50). The compared sample areas are either face grinded/lateral grinded only or processed with both technologies combined.

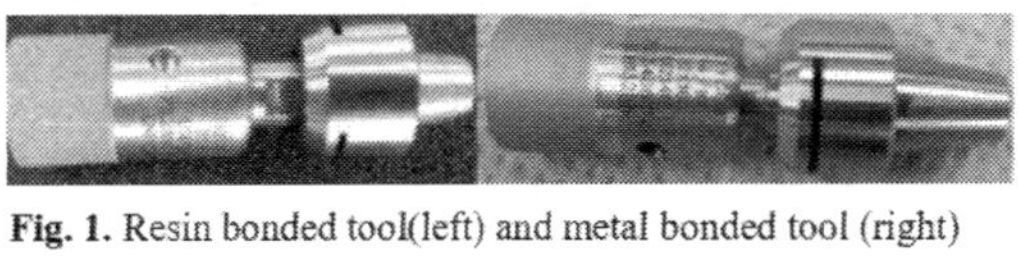

Fig. 1. Resin bonded tool(left) and metal bonded tool (right)

The process influence of the following parameters is investigated:

- 2 materials: fused silica, BK7
- 2 diamond tools: metal bond, resin bond
- 3 values of cutting depth ap
- 2 values of cutting width ae
- state of ultrasonic assistance (on, off)

Table 1. Grinding parameters

ultra fine grinding				
v_f [mm/min]	v_c [m/s]	n [min^{-1}]	a_e [mm]	a_p [mm+µm]
60	22	30012	0.003	2+3
			0.011	4+7
				8+11

The depth of cut is varied in three steps to gain knowledge about the influence of removal volume in regards to the cutting width, while regarding the special step like sample geometry. The stated number in mm represents the infeed of one lane, the number stated in µm represents the cutting depth for the last lane, where the tool is in contact with the material again for face grinding. The examined parameters are the roughness and waviness. Furthermore, these parameters are distinguished by their direction of measurement according to the direction of the feed rate (parallel, vertical) and the position of the measurement (single/combined processing and horizontal/vertical surface).

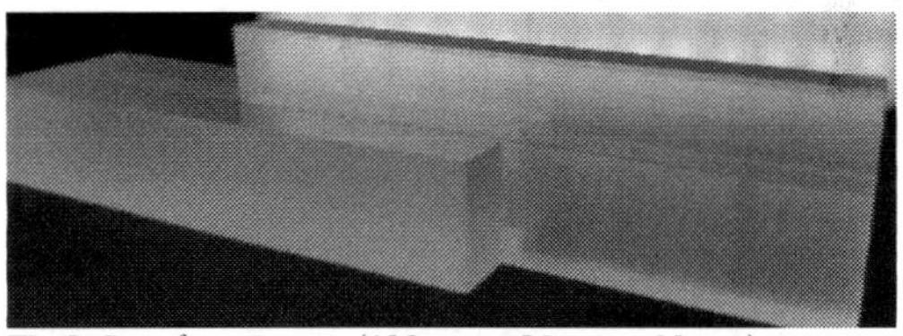

Fig.2. Sample geometry (100 mm x 20 mm x 20 mm)

2 Parameter significance

After conducting the experiments and analysing the results, it shows that the parameter with the highest significance is the bond type of the tool, followed by the usage of ultrasonic assistance. Only a shallow level of significance could be recognized for the cut of depth. The cut of width had almost no influence on the results.

Except for the vertical measurement direction on the vertical surface, there are no significant deviations for the measurement directions. There are no remarkable differences in processing the materials using a face/lateral grinding only strategy in comparison to a combined grinding strategy.

* Corresponding author: christian.schulze@eah-jena.de

3 Results

As shown in fig. 3 and 4, a higher surface quality can be achieved processing fused silica compared to BK7 using identical grinding conditions. Furthermore, using the resin bonded tool for both face and lateral grinding achieves a superior surface quality compared to the metal bonded tool. For lateral grinding the gap of attainable surface quality between both tools, decreases. Especially when processing BK7.

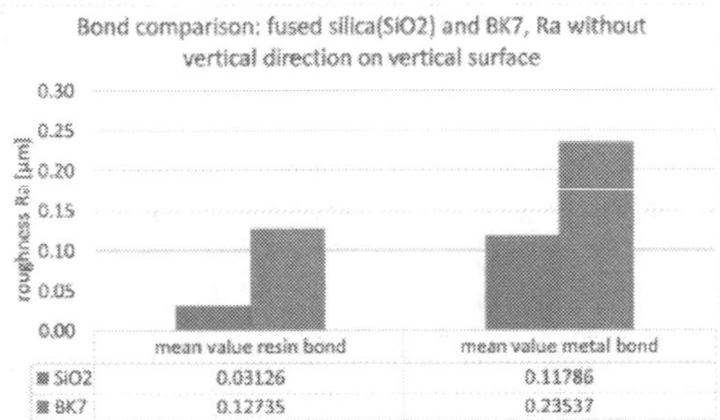

Fig. 3. Comparison: bond, mean value

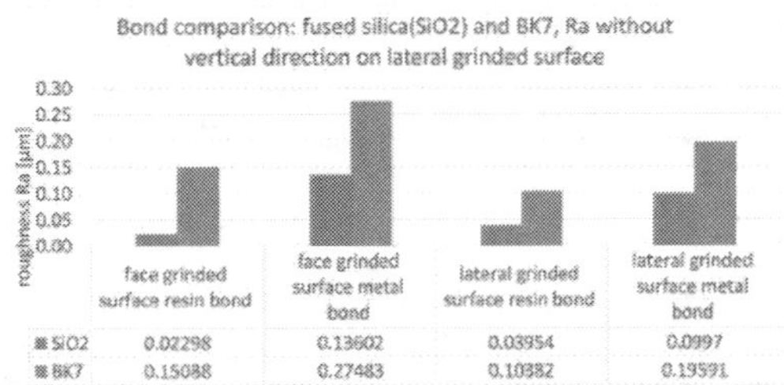

Fig. 4. Comparison: bond, face/lateral grinded surface

Fig. 5. represents the degradation of the surface quality when using ultrasonic assistance. While the mean value of roughness almost doubles for fused silica, it is increased by about 20% for BK7. Even though the overall mean value of the experiments increases, the ultrasonic assistance shows slight improvements of about 5% to 10% for the quality of the vertical surface for lateral grinding.

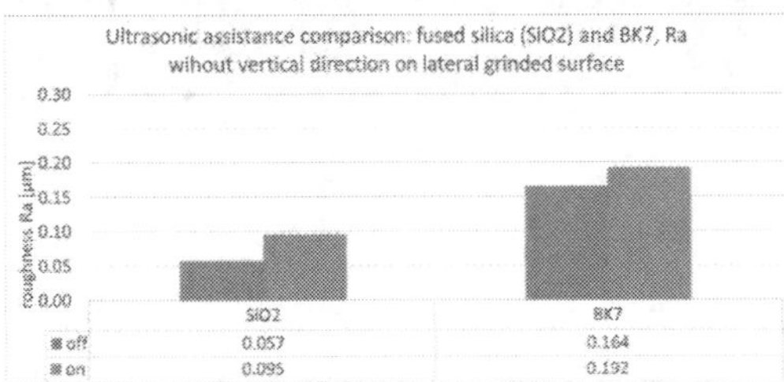

Fig. 5. Comparison: ultrasonic assistance

The achievable flatness lies within a range of about 0.4 µm to 0.8 µm without remarkable peculiarities.

The vertical measurement direction of the lateral grinded surface is not taken into consideration for the mean values, since its values are 10 to 20 times higher compared to the other measurement directions. A possible explanation can be found when looking at the footprint the tools produce on the surface with a white light interferometer (WLI). While face grinding creates a smooth surface with circular structures with depths around 1µm (fig. 6), lateral grinding generates a groove profile due to its kinematics, as seen in fig. 7. Those groove profiles can have a total height of more than 10µm. In general, the vertical direction of the vertical surfaces shows a deviant behaviour and needs to be further examined.

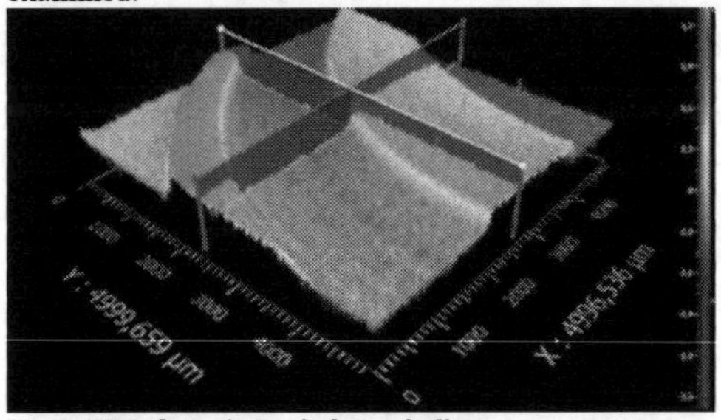

Fig.6. WLI: footprint tool: face grinding

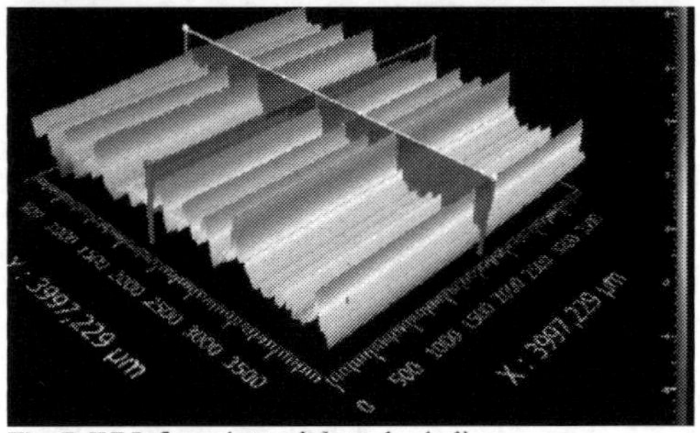

Fig. 7. WLI: footprint tool: lateral grinding

4 Conclusion

Overall the conducted experiments have shown, that the used technology is able to create high quality surfaces with a low roughness and high flatness.

The resin bond achieved a better surface quality in comparison to the metal bonded tool using identical tool geometries and grain sizes. Most of the time the usage of ultrasonic assistance resulted in a degradation of the surface quality except for some cases of the vertical surface. Generally speaking a better surface quality could be achieved for processing fused silica in comparison to BK7. Further investigations, especially for lateral grinding are needed to deepen the knowledge and improve the properties of fabricated components.

The authors gratefully acknowledge financial support by the German Federal Ministry of Education and Research, Funding reference: 13FH003IA6 "OpTec 4.0".

References

1. C. Schulze, S. Henkel, J. Bliedtner: Ultra-fine grinding of inorganic non-metallic materials using various types of bonds and processing strategies. *EOSAM 2020*, EPJ Web of Conferences Vol.**238**, article number 03011. (2020) *https://doi.org/10.1051/epjconf/202023803011*

 https://doi.org/10.1051/epjconf/202125503003

High-resolution measurement technology for the detection of complex process influences in machining operations

Till-Hendrik Hage[1,*], *Marco* Fritzsche[2], *Sebastian* Henkel[1], *Jens* Bliedtner[1],

[1]Ernst-Abbe University of Applied Sciences Jena, Carl-Zeiss-Promenade 2, 07745 Jena, Germany

[2]Polytec GmbH, Polytec-Platz 1-7, 76377 Waldbronn, Germany

Abstract. Inorganic non-metallic materials are widely used in many application areas such as precision optics, semiconductor technology, high-temperature technology or medical technology. This increasingly requires the application of precise multi-axis CNC machine technology and poses great challenges for the machining processes, especially with regard to the economic production of high-quality surfaces. In the field of optics technology, high surface qualities in the nanometer range are often required. The production machines designed for this purpose must therefore guarantee a high degree of accuracy and be constant with respect to their environmental factors. An influence on the quality of the machining results that should not be underestimated is the vibrations occurring on the machine, which are caused on the one hand by the operating vibrations generated by the running spindle, and on the other hand by resonance frequencies. A side effect of these vibrations can also be damage to the workpiece or the tool. In order to detect these influences, a modal analysis and a measurement of the operating vibration mode are carried out on an ultrasonic-assisted CNC grinding machine using a three-dimensional laser Doppler vibrometer system provided by Polytec GmbH. In addition, corresponding measurements are carried out by means of a one-dimensional laser Doppler vibrometer on a 5-axis CNC universal milling machine in order to establish comparability between an ultrasound-assisted and a conventional processing machine.

1 Introduction & Methods

In general, modal analyses serve the purpose of investigating the dynamic behavior of oscillatory systems with regard to their natural frequency, natural mode and modal mass. Since vibrations occur on the workpiece and tool side during machining in addition to the set variables feed rate, cutting speed and depth of cut, they directly influence the process and thus the process result. Basically, vibrations occur in a machine tool either from the cutting process through self-excitation or through external excitation mechanisms such as unbalance, drive and gear vibrations or through neighboring machines. [2] Especially in the field of optical technology, which deals with the production of complex components made of brittle-hard materials such as fused silica or the manufacture of monolithic components, high surface qualities and accuracies in the nanometer range are required. In order to find out the influences of these resulting vibrations, investigations on an ultrasound-supported CNC grinding machine "Ultrasonic 20 linear" from company DMG Mori are presented below in the form of a three-dimensional modal analysis using a scanning laser Doppler vibrometer system. To enable a 3-dimensional measurement, three laser doppler vibrometers are positioned in an isosceles triangle in front of the machine. The position of the measuring devices is determined by triangulation during calibration, so that a spatial image is generated. (see Figure 1). The Doppler vibrometers measure vibrations without contact, no additional mass on the measuring object is necessary, so that a non-invasive measurement is made by means of a laser beam.

Fig. 1. Measurement setup of a Laser Doppler interferometer from Polytec GmbH

* Corresponding author: till-hendrik.hage@eah-jena.de

When the laser beam comes into contact with a moving object, the back-reflecting laser beam undergoes a frequency or phase shift based on the principle of the Doppler effect. Using a heterodyne Mach-Zehnder interferometer and digital decoder technology, this frequency or phase shift is first recorded by a photodetector before being digitized and then demodulated by a digital decoder. As a result, the displacement (phase demodulation) and the velocity (frequency demodulation) of the moving object is measured over time by the Laser Doppler Vibrometer (see Figure 2).

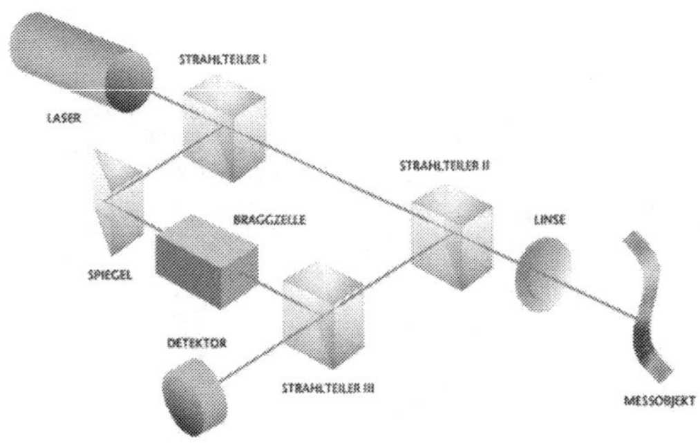

Fig. 2. Basic principle and schematic structure of a Laser Doppler interferometer from Polytec GmbH [3]

By an integrated camera and two scanning mirrors it is possible to define any number of measuring points on the object (see Figure 3). Then these measuring points are approached one after the other during the measurement. At each measuring point, the vibrations are recorded and then displayed in the evaluation in an animated vibration form with the video image superimposed. The measuring system ensures a high bandwidth and dynamics, as even high-frequency vibrations are reliably recorded and adapted to the existing amplitude of the vibration by means of different measuring ranges. The same measuring procedure is carried out on a CNC universal milling machine "DMU 50" from the company DMG Mori AG using a one-dimensional scanning laser Doppler vibrometer, with the measuring direction in the Y-axis.

Fig. 3. Overview of measuring points, which were set individually in the object

2 Examinations & Results

The results of the investigations are presented below. The ultrasonic CNC grinding machine "Ultrasonic 20 linear" has a speed spectrum of 0 - 40,000 revolutions per minute. For the detection of dominant vibrations, the speed is adjusted in steps of thousands. The vibration detection of the operational deflection shape is carried out by measuring the conspicuous amplitudes and vibration frequencies on the machines. By varying the speed, a comprehensive overview is obtained. To facilitate the evaluation, this is graphically supported by visualizing the occurring vibration through the previously set measuring points in an exaggerated way (see figure 4).

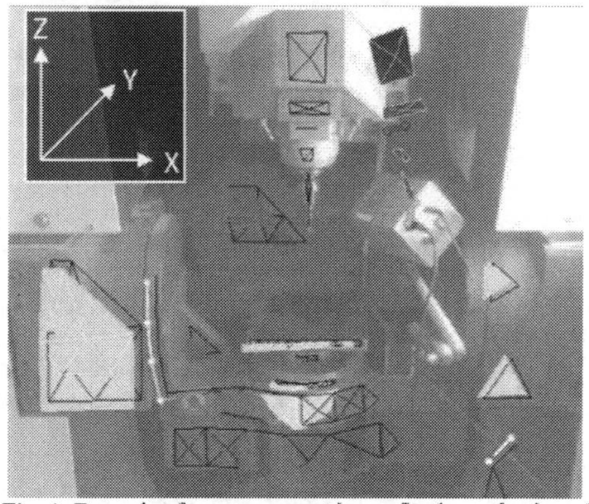

Fig. 4. Example of a measurement image for the evaluation of an occurring vibration

The strongest amplitudes in the frequency spectrum are defined as dominant vibrations. These are always to be noted at the respective natural frequencies of the speed values. For example, at a set speed of 24,000 rpm, the strongest amplitude is at 400 Hz. When considering all measurements, a proportional influence of the set speed on the resulting amplitude can be observed, apart from the superimposition with distinct amplitude peaks (see Figure 5). However, the maximum elongation smaller than 0.1 μm is to be considered small.

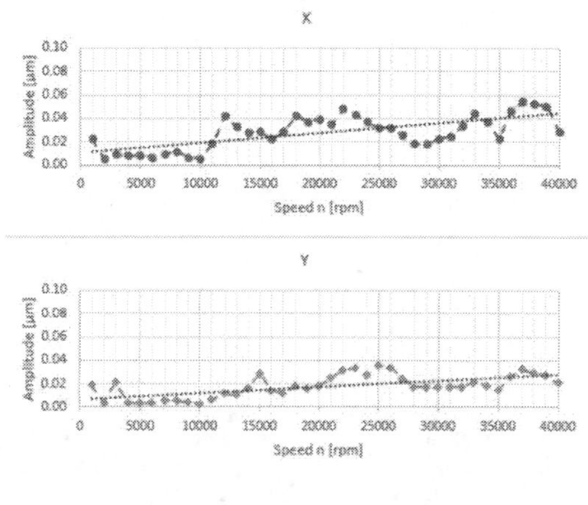

https://doi.org/10.1051/epjconf/202125503003

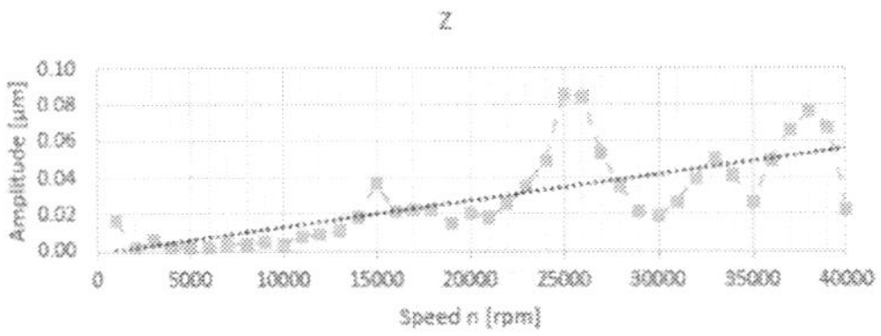

Fig. 5. Operational deflection shape: dominant vibrations in relation to speed of "Ultrasonic 20 linear"

In addition to the dominant oscillations, continuous oscillations can be noted, for example, an amplitude at 42.2 Hz, 112.5 Hz as well as 300 Hz occurs in all measurements. The latter frequency is due to vibrations of the machine table, the origin of the other two frequencies is unknown so far. Since the "Ultrasonic 20 linear" enables machining with an ultrasonic oscillation on the tool, it is also checked whether the induced vibrations have an effect on the spindle. For this purpose, the ultrasonic working frequency of the tool used is examined at varying rotational speeds in the spectrum mentioned. Continuous amplitudes can be observed over the entire speed range, which, however, do not represent significant vibration transmissions to the spindle due to their small size of less than 1 nm.

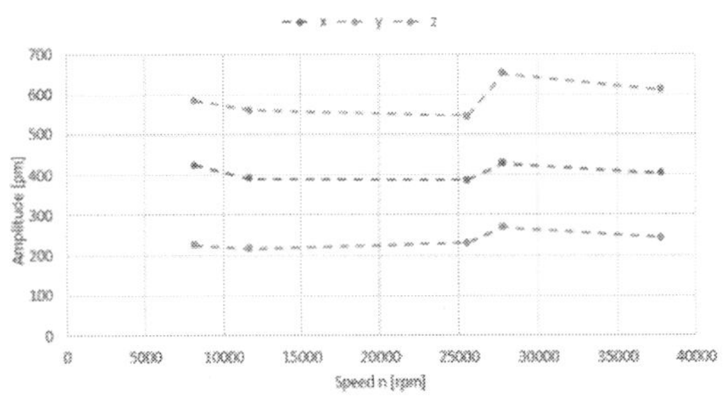

Fig. 6. Comparison of US frequency (25.4 kHz) at different rotation speeds of "Ultrasonic 20 linear"

To perform the modal analysis, resonant frequencies are determined using a modal hammer excitation and a Dirac pulse triggered by it; a scalable automated hammer is used to avoid variable excitations. Since the previous operating vibrations occurred at the spindle and at the table, excitation is performed at both locations. The resonant frequency determined here occurs at the spindle at 136 Hz, which corresponds to a speed of approx. 8000 rpm and should therefore be avoided for minimal spindle vibrations. The modal analysis with excitation at the table results in a resonant frequency at 490 Hz and thus shows a difference to the frequency determined in the operating mode (300 Hz). This is presumably due to a difference in the way the measurement is carried out, since the operating mode is determined with mains voltage switched on, while the modal analysis is determined without mains voltage and therefore without compressed air.

The measurements are repeated analogously with a scanning vibrometer on the CNC universal milling machine "DMU 50". This has a speed spectrum of 0 - 14,000 rpm, the speed is varied here in steps of 2,000. Here too, an increase in amplitude is evident with an increase in the set speed. The maximum amplitude in space exhibits a maximum elongation of less than 1.0 µm. In comparison with the "Ultrasonic 20 linear", it is clearly visible that the vibration amplitudes of the "DMU 50" are approx. a factor of 10 higher. Thus, despite its higher adjustable speed, the "Ultrasonic 20 linear" achieves lower vibration amplitudes than the "DMU 50".

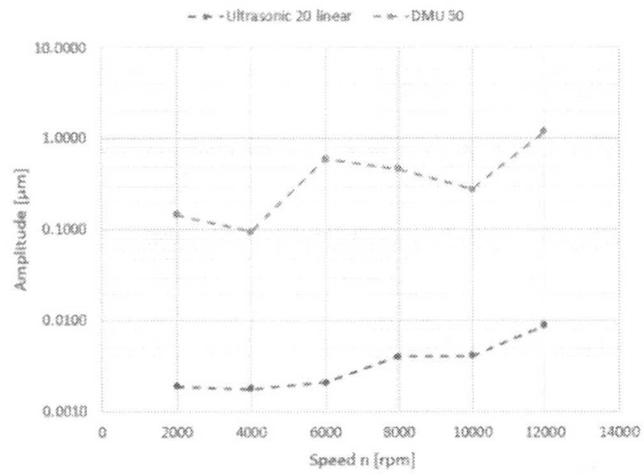

Fig. 7. Comparison of operational deflection shape of "Ultrasonic 20 linear" and "DMU 50"

3 Conclusion and Outlook

It was shown that higher vibration amplitudes are generated with increasing speed. These are very low for the "Ultrasonic 20 linear" with a value of less than 0.1 µm. The ultrasonic vibration generated on the tool is not significantly transmitted to the spindle or the entire system. The comparison between the two processing machines shows that the "Ultrasonic 20 linear" generates 10 times lower vibration amplitudes than the "DMU 50" despite its higher adjustable speeds. In the future, the modal analysis is to be tested in process operation. For this purpose, investigations will be carried out on a new type of simultaneous measuring system consisting of a vibrometer, a force measuring platform and a thermographic camera enabling a time-synchronized measurement of three main process parameters: vibrations, forces and temperatures. In addition, the influences of the vibrations on the workpiece and the quality of the machining will be further examined.

References

1. M. Möser: *Modalanalyse.* Berlin: Springer Vieweg. Fachwissen Technische Akustik (2020)
2. H.K. Tönshoff, U. Wasmann: Auswirkung erzwungener Schwingungen auf die Standzeit beim Stirnfräsen. In: Forschung im Ingenieurwesen **58**, pp. 283-289 (1992)
3. Polytec GmbH (2020): In Natur und Technik : Alles schwingt! [online] https://www.polytec.com/de/vibrometrie/technologie/laser-doppler-vibrometrie

Impact of Pupil Aberrations on Wavefront Manipulation

Thomas Nobis[1,*]

[1]Carl Zeiss AG, Corporate Research and Technology, Jena location, Carl-Zeiss-Promenade 10, 07745 Jena, Germany

Abstract. A systematic and quantitative analysis is given of the impact of pupil aberrations on the imaging performance in wavefront manipulation applications using adaptive optical elements. For the practical case of rotationally-symmetric types of wavefront corrections, such as defocus or spherical aberration, analytical expressions of the induced aberrations are derived including their pupil and field dependence. Each aberration is thereby related to the specific pupil aberration present at the adaptive element. The results can be used to specify the acceptable amount of pupil correction required for a specific magnitude and type of wavefront manipulation.

1 Introduction

In the recent decade, the technology of adaptive optical elements, such as spatial light modulators or deformable mirrors, has rapidly improved. Wavefront manipulation is an obvious application of these devices, where they, e.g., either generate or compensate for a specific phase profile in the imaging path. Especially in wide-field imaging, the adaptive element is typically placed at a conjugated pupil plane. The footprints of all ray bundles are identical there, and all field points are therefore affected in the same manner by the element.

For such application, the considered optical system must provide an accessible pupil plane. Depending on the specific application, either the image of the entrance pupil or the image of the exit pupil onto that element must reach a certain level of correction. In practical cases, however, the intermediate image of the pupil is not ideal, i.e., pupil aberrations are present. They can be classified in the same systematic way as aberrations in the image plane, as well as there are analytical Seidel surface contributions based on paraxial ray data similar as for the common imaging aberrations [1]. Usually, investigations of pupil aberrations aim for their direct relation to image correction. Doubtless, the most-famous case is the Abbe sine condition, a specific type of pupil distortion, and its relation to coma correction for object imaging [2]. Other known influences of pupil aberrations are, e.g., beam deformations [1], changes in relative illumination [3] or additional aberrations when conjugates are changed [4-6]. The occurrence of field-dependent aberrations caused by pupil aberrations in adaptive optics is also known [7,8]. However, a systematic and quantitative approach is still missing, describing both the induced types of image aberrations and the acceptable amount of pupil aberrations depending on the specific wavefront manipulation being done. In this paper, such systematic overview and a quantitative analysis is given of the impact of pupil aberrations on wavefront manipulation with adaptive phase elements.

2 Theory

Aberrations of rotational-symmetric optical systems can be expressed by the well-known Seidel series both as wavefront deviations W from the reference sphere in the exit pupil as well and transverse ray aberrations $\vec{T}$ from the ideal image point in the image plane. Therefore, a ray is considered starting in the object plane at object height h and hitting the entrance pupil at a radius r at an angle ϑ with respect to the meridional plane, as given in Fig. 1. The wavefront aberration, W, then reads [2]

$$W(h,r,\vartheta) = \underbrace{W_{020}r^2}_{\text{Defocus}} + \underbrace{W_{111}hr\cos\vartheta}_{\text{Magnification}} + \underbrace{W_{040}r^4}_{\text{Spherical Ab.}} + \underbrace{W_{131}hr^3\cos\vartheta}_{\text{Coma}} + \underbrace{W_{220}h^2r^2}_{\text{Sagittal field curv.}} + \underbrace{W_{222}h^2r^2\cos^2\vartheta}_{\text{Astigmatism}} + \underbrace{W_{311}h^3r\cos\vartheta}_{\text{Distortion}} + [\ldots], \quad (1)$$

where the coefficients W_{klm} are a measure of the overall amount of the specific aberration present. The curly brackets indicate the common names for the two first-order terms and the five third-order Seidel aberrations. Higher-order terms are not regarded, here.

For wavefront manipulation a thin phase element is considered, placed at an intermediate conjugated pupil plane, see Fig. 1. The phase element is supposed to add a specific optical phase $M(\vec{R})$ to the total optical path of each ray, depending on the position $\vec{R}$ where the ray hits the adaptive element assuming that the entrance pupil is imaged perfectly onto the adaptive element. $|M|$ is considered to be small, such that the overall ray path after the element is unchanged and no further induced aberrations are generated.

* Corresponding author: thomas.nobis@zeiss.com

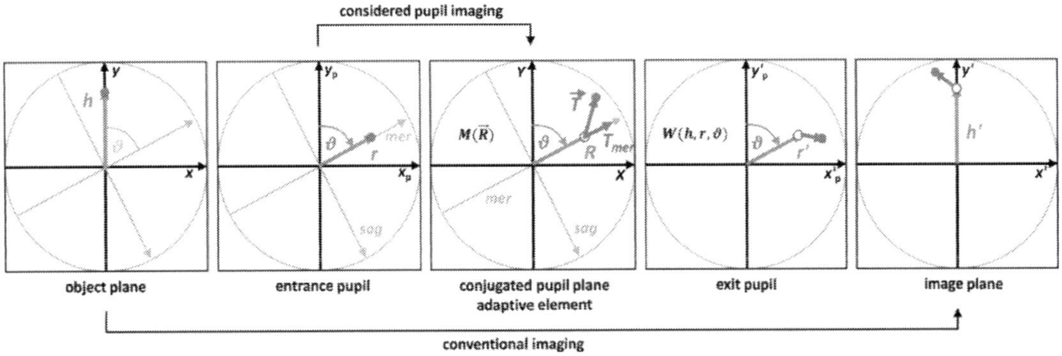

Fig. 1. Coordinate systems for object, pupil and image planes of a rotationally-symmetric optical system. Positions, where a selected ray hits the planes, are marked with a red dot, respectively. A solid dot gives the actual position, while a blank dot indicates the ideal position with no aberrations present. The vector between these dots yields the transverse ray aberration in that plane. The green vector indicates the coordinates of the ideal ray location in polar coordinates in each plane. For object imaging, the line connecting the optical axis with the object point determines the meridional direction by definition, which is the y-direction here. For pupil imaging, however, object and pupil planes exchange their role [1]. Therefore, the meridional direction for the Seidel series of pupil aberrations is determined in the entrance pupil plane by the line between the optical axis and the intersection of the ray in that plane. This direction together with the corresponding sagittal direction is indicated by the light-gray coordinate system.

However, pupil aberrations may lead to a transverse aberration $\vec{T}$ of the considered ray. Hence, an additional contribution ΔW is added to the wavefront, which is considered as an induced aberration for object imaging. If $|\vec{T}|$ is small, ΔW can be approximated by the first-order term of a Taylor series of M,

$$\Delta W = M(\vec{R}+\vec{T}) - M(\vec{R}) \approx \mathrm{grad}(M)\cdot\vec{T} \qquad (2)$$

i.e. as the scalar product of the gradient of M and $\vec{T}$. This already gives a first quantitative estimation of the maximum allowed $|\vec{T}|$, since $\mathrm{grad}(M)$ is known and the maximum acceptable ΔW is typically specified for object imaging. For further systematic investigation we restrict ourselves to rotationally-symmetric wavefront corrections, i.e.

$$M(\vec{R}) = M(R) = M_2\cdot R^2 + M_4\cdot R^4 + [\dots]. \qquad (3)$$

Such restriction is necessary to not break the rotational symmetry of the overall system. In this case, $\mathrm{grad}(M)$ only has a radial component

$$\mathrm{grad}(M) = \frac{\partial M}{\partial R}\cdot\vec{e_R}, \qquad (4)$$

and the scalar product yields the radial component of $\vec{T}$. From the viewpoint of pupil imaging, this is actually the *meridional* component of $\vec{T}$, i.e. T_{mer},

$$\Delta W \cong \mathrm{grad}(M)\cdot\vec{T} = \frac{\partial M}{\partial R}\cdot T_{\mathrm{mer}}, \qquad (5)$$

as indicated in Fig. 1.

T_{mer} can be expressed as a Seidel series similar to W, but for *transverse* aberrations, as e.g. in [9]. For the series expansion the identical ray is used as for W. Since we are now considering pupil imaging, however, the object plane and entrance pupil plane exchange their roles [1], and r now stands for the object height, while h refers the pupil radial coordinate. The series up to third-order aberrations [9] then translates into

$$T_{\mathrm{mer}} = \underbrace{\sigma_0\cdot h\cos\vartheta}_{\text{Defokus}} + \underbrace{\sigma'_0\cdot r}_{\text{Magnification}} + \underbrace{\sigma_1\cdot h^3\cos\vartheta}_{\text{Spherical Ab.}} + \underbrace{\sigma_2\cdot(1+2\cos^2\vartheta)\cdot h^2 r}_{\text{Coma}} + \underbrace{(3\sigma_3+\sigma_4)\cdot h\cos\vartheta\, r^2}_{\text{Astigmatism \& Field Curv.}} + \underbrace{\sigma_5\cdot r^3}_{\text{Distortion}} + [\dots]\,. \qquad (6)$$

Again, the common names of the terms are indicated by the curly brackets. The coefficients $\sigma_{1..5}$ are the well-known Seidel coefficients. The coefficients σ_0 and σ_0' refer to the respective coefficient for first-order aberrations.

Eqn. (5) now turns out as a product of two polynomial series. The systematic expansion into 6 × 2 terms is given in Table 1. The radius R is substituted by r using the paraxial pupil magnification β between the entrance pupil and the adaptive element, i.e. $R = \beta r$. By comparison with Eqn. (1), each term identifies a specific type of wavefront aberration. The quantitative amount W_{klm} of the induced aberration is given by the respective prefactor in each cell. Table 1 therefore gives the systematic impact of pupil aberrations in wavefront manipulation applications. If, e.g., the adaptive element is used for defocus compensation (1st column), and pupil imaging suffers from coma (4th line), object imaging will be affected by sagittal field curvature and astigmatism. Higher order corrections or pupil aberrations can be obtained in a similar way.

3 Example and Discussion

An example optical system demonstrating the discussed effects of pupil aberrations is given in Fig. 2. The system in Fig. 2(a) is a symmetrical relay lens imaging an object field diameter of 10 mm onto an image diameter of the same size.

* Corresponding author: thomas.nobis@zeiss.com

Table 1. Systematic influence of pupil aberrations on the effect of wavefront manipulation.

	Resulting induced wavefront aberration for object imaging while doing (a) or (b)	
Pupil transverse aberration	**(a) focus correction** $\partial M/\partial R = 2M_2 \cdot R$	**(b) correction of low-order spherical aberration** $\partial M/\partial R = 4M_4 \cdot R^3$
$\underbrace{\sigma_0 \cdot h\cos\vartheta}_{\text{Defokus}}$	$2M_2\beta \cdot \sigma_0 \cdot \underbrace{hr\cos\vartheta}_{\text{Magnification}}$	$4M_4\beta^3 \cdot \sigma_0 \cdot \underbrace{hr^3\cos\vartheta}_{\text{Coma}}$
$\underbrace{\sigma'_0 \cdot r}_{\text{Magnification}}$	$2M_2\beta \cdot \sigma'_0 \cdot \underbrace{r^2}_{\text{Defocus}}$	$4M_4\beta^3 \cdot \sigma'_0 \cdot \underbrace{r^4}_{\text{Spherical Aberration}}$
$\underbrace{\sigma_1 \cdot h^3\cos\vartheta}_{\text{Spherical Ab.}}$	$2M_2\beta \cdot \sigma_1 \cdot \underbrace{h^3r\cos\vartheta}_{\text{Distortion}}$	$4M_4\beta^3 \cdot \sigma_1 \cdot \underbrace{h^3r^3\cos\vartheta}_{\text{Field-cubic coma}}$
$\underbrace{\sigma_2 \cdot (1+2\cos^2\vartheta) \cdot h^2r}_{\text{Coma}}$	$2M_2\beta \cdot \sigma_2 \cdot \underbrace{(1+2\cos^2\vartheta) \cdot h^2r^2}_{\text{Sagital field curv. + Astigmatism}}$	$\underbrace{4M_4\beta^3 \cdot \sigma_2 \cdot (1+2\cos^2\vartheta) \cdot h^2r^4}_{\text{Spherical Ab. \& higher-order astigmatism}}$
$\underbrace{(3\sigma_3+\sigma_4) \cdot h\cos\vartheta\, r^2}_{\text{Astigmatism \& Field Curv.}}$	$2M_2\beta \cdot (3\sigma_3+\sigma_4) \cdot \underbrace{hr^3\cos\vartheta}_{\text{Coma}}$	$4M_4\beta^3 \cdot (3\sigma_3+\sigma_4) \cdot \underbrace{h\,r^5\cos\vartheta}_{\text{5th order coma}}$
$\underbrace{\sigma_5 \cdot r^3}_{\text{Distortion}}$	$2M_2\beta \cdot \sigma_5 \cdot \underbrace{r^4}_{\text{Spherical Aberration}}$	$4M_4\beta^3 \cdot \sigma_5 \cdot \underbrace{r^6}_{\text{5th order spherical aberration}}$

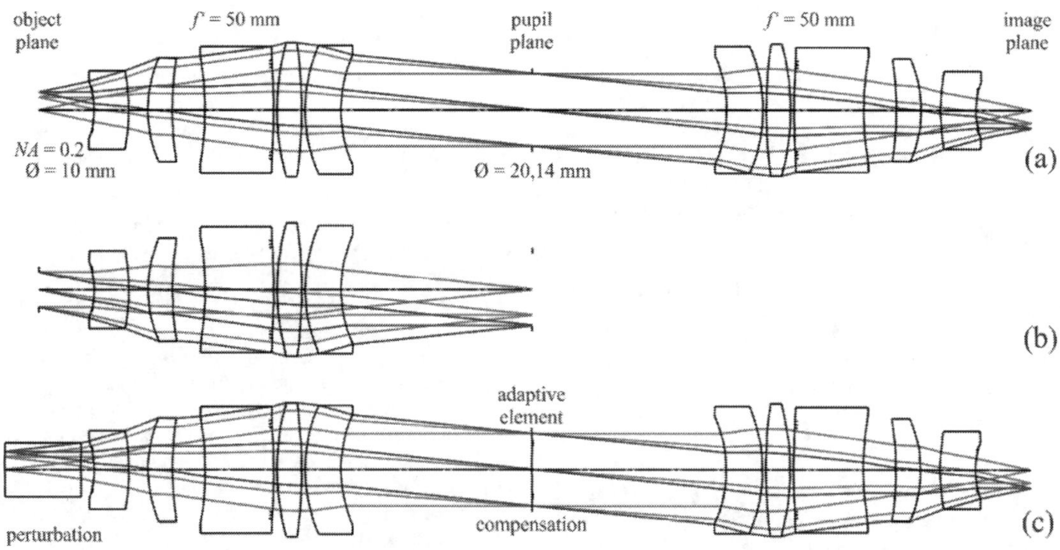

Fig. 2. Example system. (a) Relay system in its initial symmetric state without adaptive element. (b) Imaging of the entrance pupil at infinity to the conjugated pupil plane in the center of the relay system, where the adaptive element can be placed. (c) Relay system in its perturbed state with a glass plate inserted in object space. A thin adaptive element is placed at the conjugated pupil plane and compensates for the 3rd-order spherical aberration introduced by the glass plate. Different colors indicate different object points. On-axis object or pupil points are in red, zonal object or pupil points are in green, and edge points are in blue.

The focal length of the front and back group is 50 mm, and the object and image space numerical aperture is 0.2, respectively. While the entrance and exit pupil are each located at infinity, a conjugated pupil plane is right in the center of the relay, where a thin adaptive phase element is placed as given in Fig. 2(c). Pupil imaging from the entrance pupil to the conjugated pupil plane is given in Fig. 2(b). The system is corrected monochromatically for a wavelength of 550 nm. It consists of 2 × 5 lenses made of the Schott glass N-LASF44 with refractive index n = 1.8079 for the considered wavelength. The first and the

* Corresponding author: thomas.nobis@zeiss.com

sixth surface of the front group as well as their symmetrical counterparts in the rear group are aspherical, while all other surfaces are spherical. The system is considered as a pure example for the given theory and is specifically prepared to pronounce a selected type of pupil aberration. It has, however, no further application nor has it been manufactured. The performance of the system for object imaging in the initial state is given in Fig. 3 in terms of wavefront aberrations for three field heights. Strehl ratio for these field points are 0.999, 0.989 and 0.970, respectively, indicating that the system is highly diffract-tion-limited over the whole field of view.

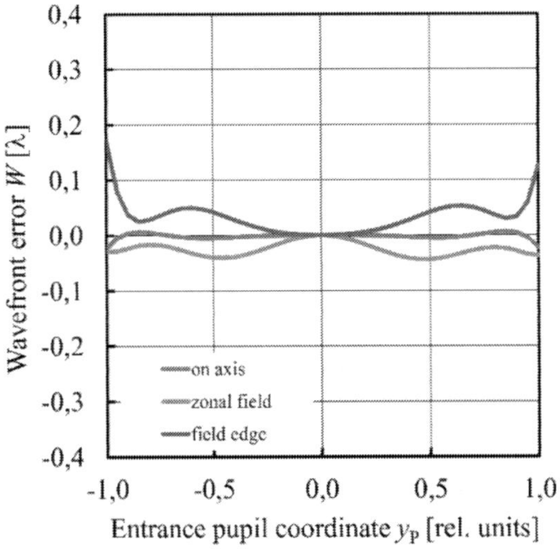

Fig. 3. Wavefront aberration W in meridional direction ($\vartheta = 0$) of the initial system given in Fig. 2(a) for three object heights $h = 0.0$ mm (on axis, red), $h = 3.54$ mm (zonal, green) and $h = 5.0$ mm (field edge, blue).

Further efforts, however, have been made to pronounce a specific type of pupil aberration while suppressing all others. The transverse aberration diagram for pupil imaging as shown in Fig. 2(b) is given in Fig. 4 for three pupil heights. As can be seen, all pupil points suffer from the same amount of 3rd-order spherical aberration

$$T_{mer} = \sigma_1 \cdot h^3 \cos\vartheta = T_{max} \cdot \frac{h^3}{h_{max}^3} \cos\vartheta \quad (7)$$

with $T_{max} = -116{,}95$ μm and $h_{max} = 5$ mm. The differences between the three curves are below 1 μm, and therefore two orders of magnitude lower. Note that, however, the actual diameter of the conjugated pupil plane is 20.14 mm. This is neither 20.41 mm, as it would be the case for zero tangent distortion in pupil imaging, nor is it 20.00 mm = 50 mm × 0.2 × 2, as it would be the case if the front group was corrected for coma in object imaging, and the sine condition was fulfilled in the conjugated pupil plane. The effects of the residual tangent distortion of −1.3% in that plane are discussed below.

The system of Fig. 2(a) is perturbed by inserting a glass plate in object space made of NLASF44 with a thickness of 22 mm, as shown in Fig. 2(c). The system has then been refocused to the paraxial image plane by adjusting the working distance in object space. Hence, from a paraxial point of view, the ray heights and ray angles of the perturbed system are identical to the one in the initial state, and so are the intrinsic Seidel contributions to object imaging. However, as a result of the inserted glass block, a specific amount of mainly 3rd-order spherical aberration is introduced to all ray bundles for object imaging. This specific aberration is compensated by a thin phase element adding a positive amount of spherical aberration

$$M = M_4 \cdot R^4 = M_{max} \cdot \frac{R^4}{R_{max}^4} \quad (8)$$

with $M_{max} = +3.16\ \lambda$ and $R_{max} = 10{,}07$ mm to the wavefront. The specific value of M_{max} was determined numerically by restoring the best-focus on axis in image space using optical design software, as it would be the case for an adaptive optics setup. Without this compensation, the diffraction-limited performance of the system would be completely gone.

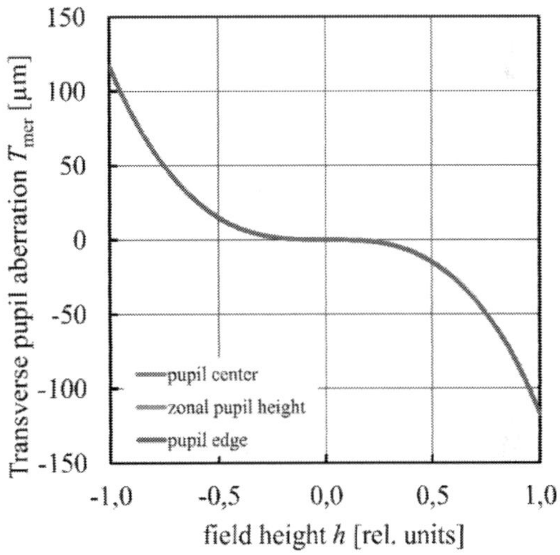

Fig. 4. Transverse pupil aberration T_{mer} in meridional direction for three resulting pupil heights, that refer to ray angles in object space, namely 0° (pupil center, red), 8,16° (zonal, green) and 11,54° (pupil edge, blue).

Using the results of Table 1, specifically the third line of the last column, one can now estimate the amount of residual induced aberration, i.e. field-cubic coma:

$$\Delta W = 4 \cdot M_4 \cdot \beta^3 \cdot \sigma_1 \cdot h^3 r^3 \cos\vartheta$$
$$= \underbrace{4 \cdot \frac{M_{max}}{R_{max}} \cdot T_{max}}_{W_{331,max}} \cdot \frac{h^3}{h_{max}^3} \frac{R^3}{R_{max}^3} \cos\vartheta \quad (9)$$

where $\beta \cdot r = R$ as well as Eqns. 7 and 8 have been applied. The resulting coma coefficient yields $W_{331,max} = -0{,}147\ \lambda$. The induced wavefront aberration ΔW is added to the wavefront aberration of the initial state and

https://doi.org/10.1051/epjconf/202125503004

compared to the actual wavefront aberration of the perturbed system. The results are given in Fig. 5.

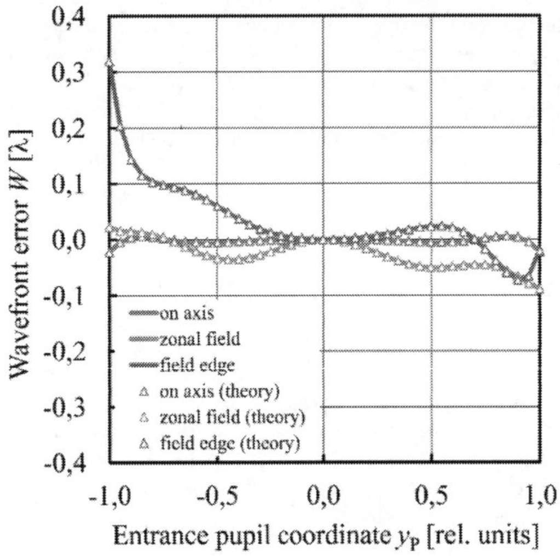

Fig. 5. Wavefront aberration W in meridional direction of the perturbed system given in Fig. 2(c) for three object heights $h = 0.0$ mm (on axis, red), $h = 3.54$ mm (zonal, green) and $h = 5.0$ mm (field edge, blue). Red, green and blue triangles indicate the theoretical prediction of the aberrations according to Eqn. 9.

The thin adaptive phase element has corrected the spherical aberration completely and therefore restored the on-axis performance of the initial system. Indeed, the glass plate itself also causes a small amount of about 0,08λ of 5th-order spherical aberration. In the given example, this is however compensated by induced 5th-order spherical aberration according to Table 1 because of the residual pupil distortion mentioned above. Therefore, a purely 4th-order phase correction is sufficient here.

According to the cubic field dependency of the induced coma aberration, no further influence is predicted for the on-axis object point. The zonal object point is only minor affected by the pupil aberration effect. However, the performance at the field edge is significantly changed. As can be seen, the theoretical estimation is in perfect agreement with the actual aberrations determined by optical design software.

4 Comments and Remarks

Pupil magnification deviations or pupil distortion both do not cause field-dependent aberrations according to Table 1. If the adaptive element provides enough degrees of freedom, the effects of these aberrations can therefore be compensated by the element itself. Hence, the specific kind of reference or ideal distortion plays a minor role.

First-order deviations, i.e. pupil defocus or pupil magnification deviations, are typically corrected for the reference wavelength in monochromatic applications. For broadband adaptive optical systems, however, they may play a significant role if axial and lateral color are not corrected for the pupil image. In such case, the respective aberrations of Table 1 are then induced for wavelengths other than the reference wavelength, and color aberrations arise. The series expansions of Eqn. 2, 3 and 6 can, of course, be extended also to higher orders. If transverse pupil aberrations are large, e.g, also the 2nd-order term of the Tayler series of M is required. For reasons of simplicity, this work only gives the lowest-order results, which have the highest practical relevance.

In the theoretical deviations, the adaptive element is assumed to be infinitely thin. In practical cases, this is unlikely to be true. As a consequence, the element itself may cause field-dependent aberrations once a specific phase is applied. If, e.g, a liquid lens is used for a parabolic or spherical phase, such lens will also intrinsically contribute to the Petzval field curvature. Other field-dependent aberrations may arise if an adaptive element, typically a reflective one, is used in an off-axis setup. These aberrations will then also be induced for perfectly-corrected conjugated pupil images and are therefore not the scope of this work. They may, however, occur in the same order of magnitude as the aberrations given in Table 1.

5 Conclusion

Adaptive elements used for wavefront manipulations are typically placed in a conjugated pupil plane, which may suffer from pupil aberrations. Depending on the actual shape of the phase profile, specific types of field-depending aberrations are induced. This work gives a systematic derivation of the resulting types of induced aberrations for the case of rotationally-symmetric wavefront corrections together with a quantitative estimation of the respective amount of aberration. An example optical system showed perfect agreement with the theoretical prediction. The theory allows for specifying the required pupil correction for a given wavefront manipulation task.

6 Acknowledgments

The optical systems have been designed and analyzed with the ZEISS in-house optical design software.

References

1. J. Sasian, Proc. SPIE **6342**, 634208 (2006)
2. M. J. Kidger, *Fundamental Optical Design* (SPIE Press, Bellingham, 2002)
3. D. Reshidko, J. Sasian, Opt. Eng. **55**, 115105 (2016)
4. T. Smith, Trans. Opt. Soc. 23, 139-153 (1921)
5. C. G. Wynne, Proc. Phys. Soc. B **65**, 429 (1952)
6. C.-H. Lin *et al.*, J. Opt. **14**, 065401 (2012)
7. B. J. Bauman, Proc. SPIE **5903**, 59030R (2005)
8. D.L. Fried, J. Opt. Soc. Am. A, **69**, 399-406 (1982)
9. R. Kingslake, R. B. Johnson, *Lens Design Fundamentals* (SPIE Press, Bellingham, 2010)

Investigation of the influence of plasma source power on the properties of magnetron sputtered Ta_2O_5 thin films

Manuel Bärtschi[1,3], *Daniel Schachtler*[1], *Silvia Schwyn-Thöny*[2], *Thomas Südmeyer*[3] *and Roelene Botha*[1]

[1] RhySearch, the Research and Innovation Center in the Alpine Rhine Valley, Werdenbergstrasse 4, 9471 Buchs SG, Switzerland
[2] Evatec Ltd., Hauptstrasse 1a, 9477 Trübbach, Switzerland
[3] Laboratoire Temps-Fréquence, Université de Neuchâtel, 2000 Neuchâtel, Switzerland

Abstract. To enable the production of sophisticated optical interference coating designs, coatings with very low absorption and stray light losses and excellent layer thickness deposition accuracy are required. The selection and optimization of suitable coating materials and deposition processes are consequently essential. This study investigated the influence of the plasma source power on the optical properties, layer uniformity and stress, scattered light behavior and optical losses of magnetron sputtered Ta_2O_5 thin films.

1 Introduction

In the field of optical thin-film technology, it is essential to characterize the properties of the dielectric material and the coating equipment used, to meet the requirements of the ever-increasing range of applications. Tantalum pentoxide (Ta_2O_5) is a widely used material for optical interference coatings due to its high refractive index [1,2]. Likewise, magnetron sputtering is a technology of choice for the mass production of optical coatings, offering high deposition rates and high throughputs [3]. This work consequently investigates the properties of Ta_2O_5 thin films and their production using the latest generation magnetron sputtering equipment.

2 Experimental setups

During the experiments, the Clusterline200 BPME, a magnetron sputter tool from Evatec AG, was used (Figure 1). It allows coating 15 substrates with a diameter of 20cm simultaneously. Besides load locks for loading and unloading the substrates, an aligner, a handler and a variety of other secondary equipment, the main part of the system is the coating vacuum chamber. It consists of a rotating table with 15 substrate carriers which rotate as well. Three plasma sources are located on the top of the coating vacuum chamber. For this work, one magnetron sputtering source with a tantalum target, one with a silicon target and one as an additional argon-oxide plasma source (PSC) was used. The magnetron sputter source is powered by a DC-generator and uses a rotating magnetic system to obtain an optimal distribution and enable the mixture of oxygen and argon to be adjusted.

Figure 1: The Clusterline200 BPME magnetron sputter tool.

The PSC works as a capacitively coupled RF-plasma source and can be operated with argon for increased energy during the coating process with oxygen to oxidize the coated layer (Figure 2). Various improvements in material properties have been achieved using the PSC for depositing silicon dioxide (SiO_2) and hafnium dioxide (HfO_2) [4]. The coating process is monitored by an optical broadband monitoring system (GSM) to calculate the actual layer thickness at a specific time during the deposition process by measuring the changing reflection from a monitoring sample coated together with the production substrates. By comparing the reflection spectrum with the expected spectrum for the specific growing layer, the optimal deposition time can be used to

* Corresponding author: manuel.baertschi@rhysearch.ch

ensure the best possible accordance of the measured layer spectrum with the desired design.

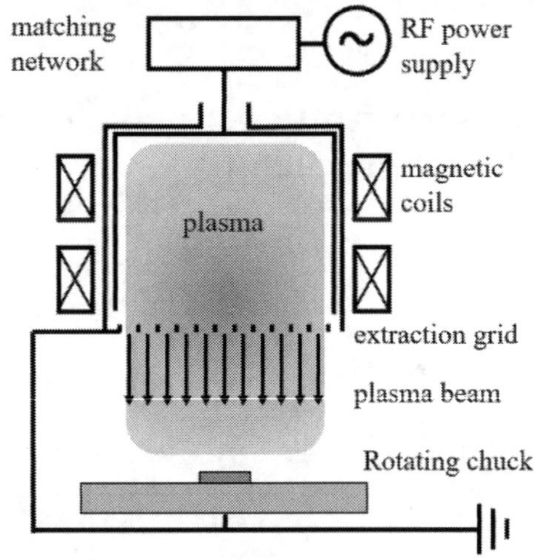

Figure 2: Schematic of the additional capacitive coupled RF-Plasma source (PSC) operating with argon and oxygen [4].

Transmission measurements of the deposited Ta_2O_5 thin films were performed using an Essent Photon RT spectrophotometer scanning the wavelength range from 250nm to 980nm in steps of 1nm. The layer uniformity was measured using a Sentech SENresearch ellipsometer, scanning the 200mm diameter in steps of 5mm with a 10mm edge exclusion. The results were fitted using a Cauchy model for Ta_2O_5 on a silicon wafer. Roughness measurement involved using a Park NX20 atomic force microscope, scanning a 0.5μm square area with 516 pixels edge length. The density of the Ta_2O_5 layers has been measured with a Bruker D8 Discover XRD system.

An initial parameter study was performed in order to identify a process window for manufacturing Ta_2O_5 thin films showing low optical losses without making use of the PSC. This was then used as a starting point to find a working point where the PSC RF-plasma and the plasma from the magnetron system don't affect each other, and stable operation is possible. Using these settings, 455nm thick Ta_2O_5 layers were produced at varying PSC power levels from 0kW to 3kW in steps of 500W. Three sets of the Ta_2O_5 layers were manufactured for this purpose. Thickness control was achieved using the above-described GSM system to provide a precise turn-off.

3 Measurements and results

The spectral measurements of the three sets of Ta_2O_5 layers were used to evaluate the optical losses in the range from 380nm to 450nm, calculated as follows:

$$Loss = 100\% - Transmission - Reflection \quad (1)$$

The resulting losses of the Ta_2O_5 layers can be seen in Figure 3. With the exception of two coatings, the losses were found to be below the resolution of the spectrometer with no significant identifiable trend. The two Ta_2O_5 samples' higher losses were also slightly different behavior in the broader absorption spectrum compared to the other samples. The significantly higher loss values compared to the other samples can however be explained by the chosen evaluation wavelength range which is close to the Ta_2O_5 absorption edge [5]. All the coated layers can consequently be used for optical coating applications. For the purpose of further evaluations in this study, only set 2 was used.

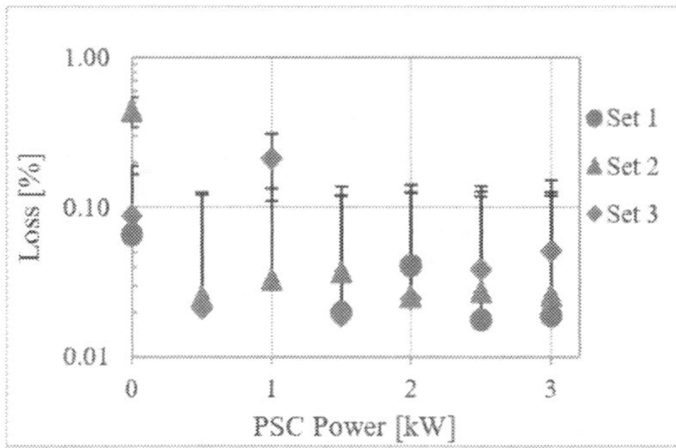

Figure 3: Calculated losses for the three sets of Ta_2O_5 layers.

The layer thickness measurements were normalized to 100% to eliminate any variations in thickness and then used to calculate the uniformity. For this purpose, the variation in thickness, d_{var}, was calculated from the maximum (d_{max}) and minimum (d_{min}) measured thicknesses as follows:

$$d_{var} = \frac{d_{max} - d_{min}}{d_{max} + d_{min}} \quad (2)$$

As shown in Figure 4, increasing the PSC power showed a positive effect on the uniformity distribution, reducing the thickness variation from 1.15% when not using the PSC to 0.31% when using 3 kW PSC power. To ensure the clarity of the results, the error bars for the uniformity plots have been omitted, but all lie in a close range between 0.19% to 0.21%. The interpretation of the resulting uniformity improvement must, however, take into account that it usually just lowers the edge and raises the center; hence with a different starting distribution, the outcome can vary. In general, the table to substrate distance and used magnetic system will be the main influences on the uniformity [6].

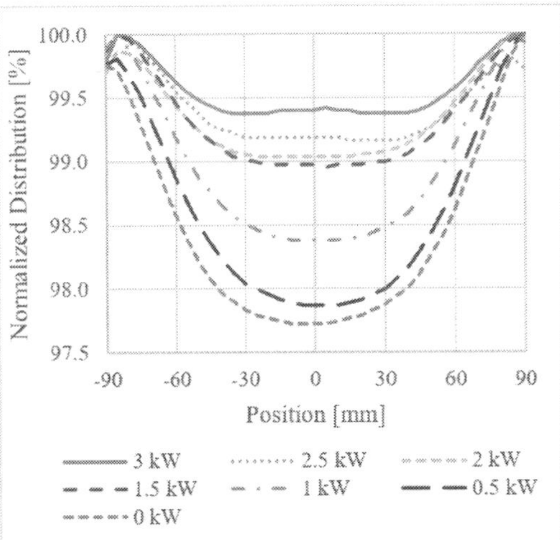

Figure 2: Normalized Ta_2O_5 layer uniformity for depositions performed at different PSC power levels.

The stress of the Ta_2O_5 layers was measured on single side polished silicon wafers before and after the coating process in order to take the substrate material variability into account. The compressive stress in the Ta_2O_5 layers was reduced with increased PSC power levels (Figure 5).

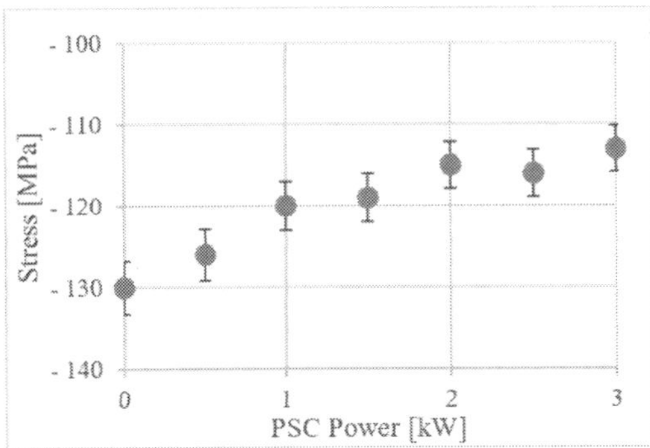

Figure 3: Stress in the Ta_2O_5 layers as a function of PSC power used during deposition.

An increase in PSC power during deposition was found to slightly reduce the roughness of the Ta_2O_5 layers, as can be seen in Figure 6. As all samples showed roughness levels in the lower Angstrom range, the measurement uncertainty is very large as the values are close to the detection limit of the AFM system.

Figure 7 shows the influence of the PSC power level on the refractive index of the Ta_2O_5 layers. In order to understand this, element analysis was performed to evaluate the oxygen and argon concentration in the films, which exhibited no trend (Figure 8).

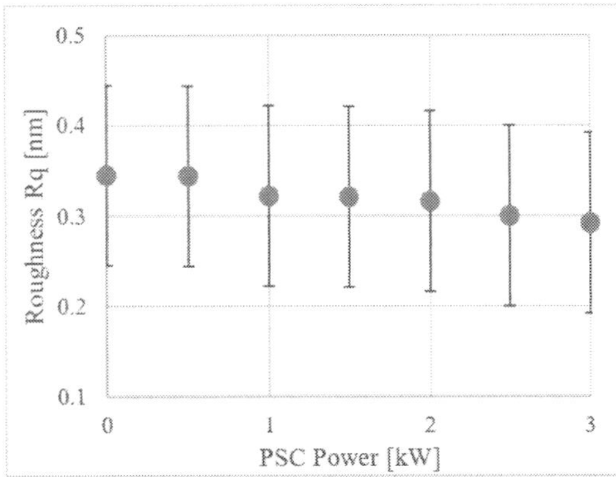

Figure 6: Root mean square roughness for Ta_2O_5 layers deposited at different PSC power levels.

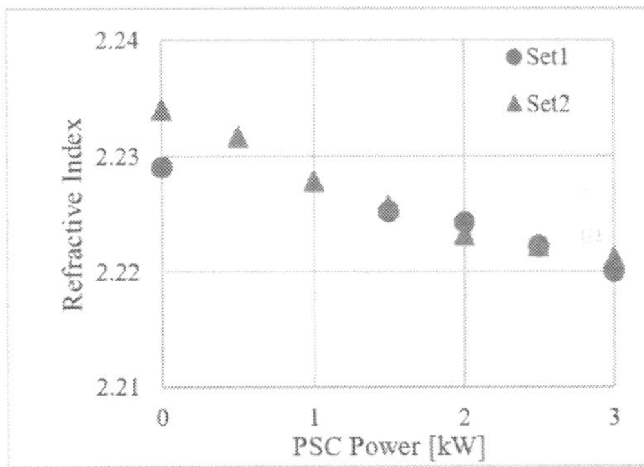

Figure 7: Observed change in refractive index with increasing PSC power during deposition.

Subsequent XRD measurements (Figure 9) showed a marked decrease in the density of the Ta_2O_5 layers when increasing the PSC power. A linear correlation could be observed when plotted against the refractive index, as shown in Figure 10.

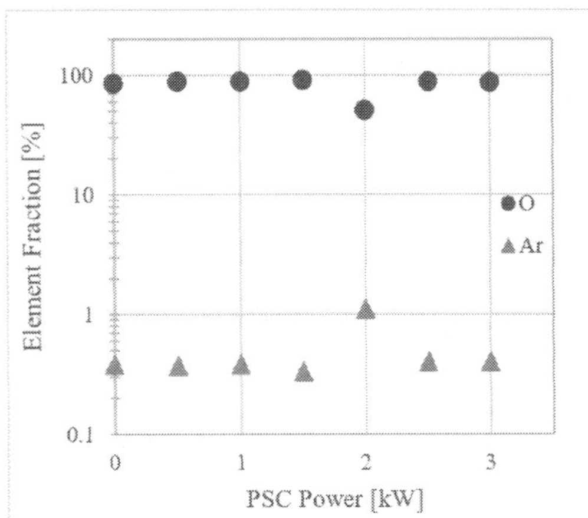

Figure 8: EDX element analysis of the Ta_2O_5 layers deposited at various PSC power levels.

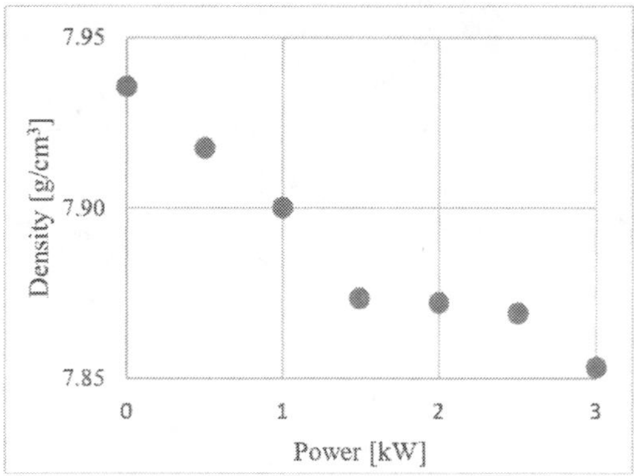

Figure 9: Density of the Ta_2O_5 layers as a function of increasing PSC power levels.

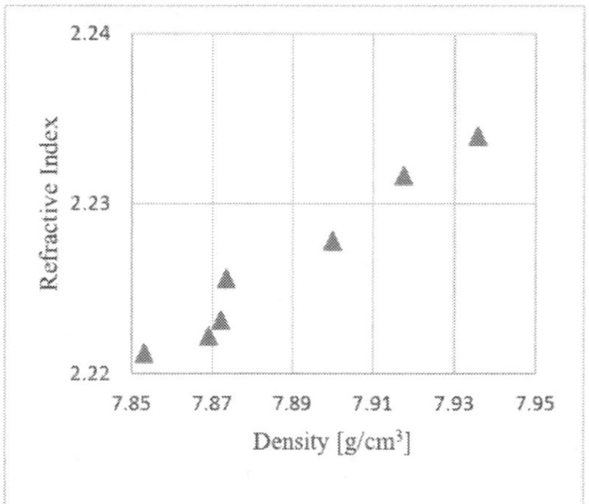

Figure 10: Refractive index as a function of the density of the Ta_2O_5 layers.

4 Conclusion

This study investigated the influence the PSC power level on the material properties of thin Ta_2O_5 layers. The strongest influence of the PSC was found to be on the stress in the thin Ta_2O_5 layers and the coating uniformity. The use of 3 kW power led to a decrease in the compressive stress in the Ta_2O_5 layers of approximately 10%.The thickness variation over a 200 mm wafer substrate was found to reduce from 1.15% to 0.31% using the PSC. Although this improvement is not seen as a flattening of the deposition profile uniformity, it is seen as a lowering of the deposition thickness in the outer areas of the substrate and increased thickness in the center and the uniformity itself mainly depends on the table-substrate distance. This distance, however, also influences the plasma stability and therefore the possibility to use more stable distances and improve the resulting uniformity distribution can be helpful. A surprising result was the decrease in refractive index with increasing PSC power. Initial results of density measurements showed a linear correlation with the refractive index. Intuitively a layer from a similar material with lower density would result in a lower refractive index. It is, however, not clear why Ta_2O_5 layers coated with higher PSC power should decrease in density. This will be the subject of further investigations.

5 Acknowledgment

Special thanks to André Mocker and Jan Allaart from the OST University of Applied Science in Buchs SG, Switzerland, for assisting with the element analysis.

6 Bibliography

1. C. Chaneliere, J. L. Autran, R. A. B. Devine, B. Balland, Materials Science and Engeneering Reports, *Tantalum pentoxide (Ta_2O_5) thin films for advanced dielectric applications*, Volume **22** Issue 6, (1998)
2. K. Chen, M, Nielsen, G. R. Yang, E. J. Rymaszewsky, T. -M. Lu, Journal of Electronic Materials, *Study of Amorphous Ta2O5 thin films by DC magnetron reactive sputtering*, Vol. **26**, Nr.4 (1997)
3. J. E. Greene, Journal of Vacuum Science & Technology, *Tracing the recorded history of thin-film sputter deposition: From the 1800s to 2017*, A **35**, (2017)
4. S. Schwyn-Thöny, S. Gees, E. Schüngel, OSA Publishing, *improving film stress and surface roughness by using a plasma source in magnetron sputtering*, WC.**4**, (2019)
5. L. V. Rodríguez-de Marcos, J. I. Larruquert, J. A. Méndez, J. A. Aznárez, OSA Publishing, *Self-consistent optical constants of SiO_2 and Ta_2O_5 films*, Vol. **6** Issue 11, (2016)
6. M. Chesaux, S. Schwyn Thöny, S. Gees, A. Frigg, OSA Publishing, *Rotating target Source: Novel shaperless concept for magnetron sputtering with excellent uniformity*, WA.**2**, (2016)

Investigations on the surface quality of grinded inner contours using different manufacturing strategies on fused silica

Marcel Binder*, *Sebastian* Henkel*, *Jens* Bliedtner,

Ernst-Abbe University of Applied Sciences Jena, Carl-Zeiss-Promenade 2, 07745 Jena, Germany

Abstract. In the field of optical manufacturing, fused silica has a high and constantly growing application potential. Its material advantages, such as low thermal expansion and high thermal shock resistance, as well as its high transparency from the ultraviolet to the infrared spectral range, result in a large number of application fields. For example, manufacturing processes in semiconductor technology require high-quality quartz materials throughout the wafer handling process to avoid non-permissible contamination and to withstand the high process temperatures. Another example are monolithic components for fiber preform manufacturing, where internal contours with high aspect ratios (e.g. component length to component diameter) and high surface qualities are required to draw fiber types with special properties. The demands on the complexity and accuracy of these components are constantly increasing, which is accompanied by the need to analyse and optimize modern CNC manufacturing techniques more and more. In the following, investigations on the grinding of internal contours with a high aspect ratio are presented, in which the influence of an ultrasonic assistance as well as different machining strategies are considered.

1 Introduction & Methods

For the production of complex elements with internal contours with high aspect ratios in brittle-hard materials, the use of modern CNC technology and special tools are necessary. Numerous investigations have already demonstrated the positive properties of ultrasonic-assisted grinding. With the help of the special kinematic principle, which is based on a high-frequency longitudinal oscillation, the process forces can be reduced and thus higher material removal rates or a more gentle machining for tool and material can be achieved. [1,2] The use of ultrasonics (US) in conjunction with special hollow drills thus enables the rapid and gentle insertion of long holes in brittle-hard materials. However, the surface quality inside of the bore also plays a major role in many areas of application. Conventional polishing processes using loose grain reach their limits due to the strongly limited accessibility of the inner contours and cannot be used in many cases. One possibility for surface finishing after grinding is the so-called ultra-fine grinding using resin-bonded tools. This enables average surface roughness that is already comparable with conventional pre-polishing. [3] Therefore, the diamond grains (D16) are embedded in a special resin bond, which has a certain elasticity. This can be compared with polishing pads and, due to these special tool properties, a kind of ductile grinding is presumably produced, which results in partially transparent glass surfaces with a low resulting roughness. It is therefore generally possible to carry out pre-, fine and ultra-fine grinding of internal contours in one machine.

In previous investigations on the insertion of holes of different diameters in the materials fused silica (Corning HPFS 79805F), ZERODUR® and CLEARCERAM®, which have already been presented in last year's abstract, it could already be shown that the surface roughness within the holes differ strongly in some cases, depending on the machining strategy used (deep drilling or helical drilling kinematics) and the measuring position (tool entry/exit). [4] In the investigations presented here, these are now examined in greater depth and extended to include the influence of ultrasonic support during pre-grinding and subsequent ultra-fine grinding with different helical infeed depths and feed speeds.

2 Examinations & Results

Following the mentioned investigations, the kinematic principles and grinding parameters (regarding cutting speed and feed rate) were kept constant and the influence of ultrasound on the surface roughness was analyzed on two different bore diameters (5 mm and 7.5 mm). In this case, all holes with a diameter of 5 mm are realized using a deep drilling process and the 7.5 mm holes are realized using helical drilling kinematics. The processed component is a fused silica cube (Corning HPFS 79805F, height 50 mm), shown in Fig. 1, in which 12 through holes are drilled using a metal bonded hollow drill (tool diameter 5 mm / grain size D64). These include six holes with a diameter of 5 mm and six holes with a diameter of 7.5 mm. For each diameter three of the bores are grinded

* Corresponding authors: Marcel.Binder@eah-jena.de, Sebastian.Henkel@eah-jena.de

with the US switched on (Frequency = 27.1 kHz, Amplitude = 5 µm) and three without US. Subsequently, the roughness inside the bores is measured tactile at the tool entry and exit points and at the bore center, using the stylus method according to ISO 4288.

Using the same pre-grinding parameters, additional holes with a diameter of 7.5 mm and a depth of 8 mm are created to investigate ultra-fine grinding. For this purpose, the existing bores are reworked with a helical infeed movement of a resin-bonded grinding tool (tool diameter 5 mm / grain size D16). Different helical infeed depths (5 µm, 50 µm and 150 µm) and feed speeds (60 mm/min and 300 mm/min) are selected in order to test their effect on the surface quality.

Fig. 1. component (cube - 50 x 50 x 50 mm³) with drilled holes of 50 mm depth

2.1 Ultrasonic assisted grinding (drilling)

The investigations on the holes with a depth of 50 mm show a tendency for ultrasonic to have a negative influence on the resulting surface roughness during deep drilling (Fig. 2) and a very low influence on the helically drilled holes (Fig. 3).

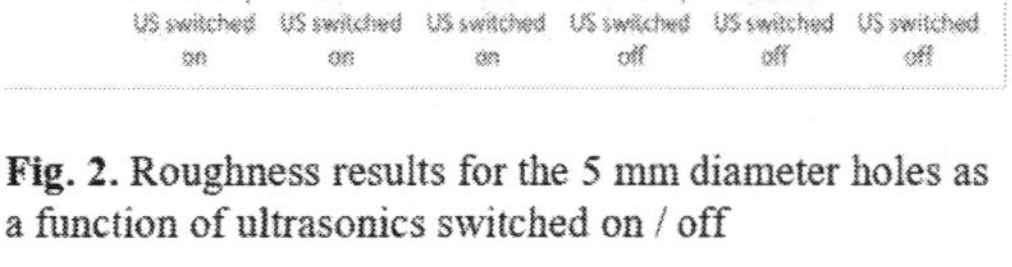

Fig. 2. Roughness results for the 5 mm diameter holes as a function of ultrasonics switched on / off

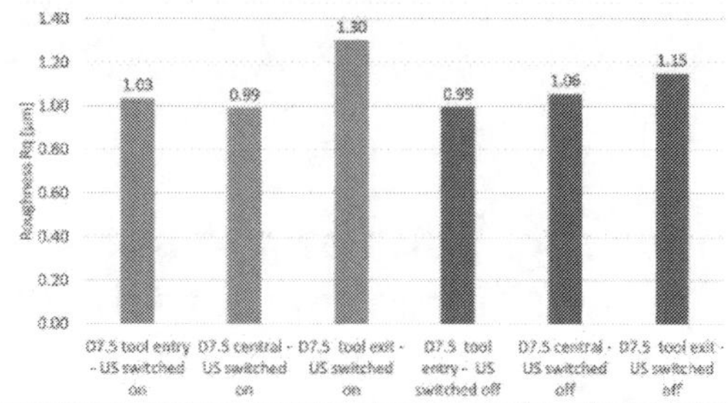

Fig. 3. Roughness results for the 7.5 mm diameter holes as a function of ultrasonics switched on / off

For the 5 mm diameter holes, average values of R_q = 0.60 µm (without US) versus R_q = 0.72 µm (with US) were determined. In contrast, the influence of ultrasound on the helically drilled holes with a diameter of 7.5 mm with R_q = 1.07 µm (without US) to R_q = 1.11 µm (with US) is to be classified as very low, as already written. Furthermore, differences within the individual bore sections are discernible in both cases. The roughness values, both with and without US, at the tool entry and exit points are in some cases significantly higher than those at the center of the bore. Using the example of the 5 mm diameter holes, R_q = 0.82 µm was determined with ultrasonic support at the tool entry, R_q = 0.70 µm at the tool exit and R_q = 0.61 µm at the bore center. It can be assumed that a different rotational stability of the long drill in the different sections is a reason for the partially strong deviations. Irrespective of the measurement location and ultrasonics, the kinematics itself show a significant influence on the surface quality, with the roughness increasing by almost a factor of 2 when the drilling kinematics are changed (deep drilling without US - R_q = 0.60 µm versus helical drilling without US - R_q = 1.07 µm).

2.2 Ultra-fine grinding

Regarding the process of ultra-fine grinding of the internal contours with resin-bonded tools, a strong improvement of surface quality in a short time can be shown. Furthermore, it was possible to demonstrate a strong dependency of the resulting roughness on the helical infeed and feed speed. Figure 4 and 5 show a comparison of the surfaces produced as a function of the tool bond used, the helical infeed (depth of infeed) and feed speed during ultra-fine grinding.

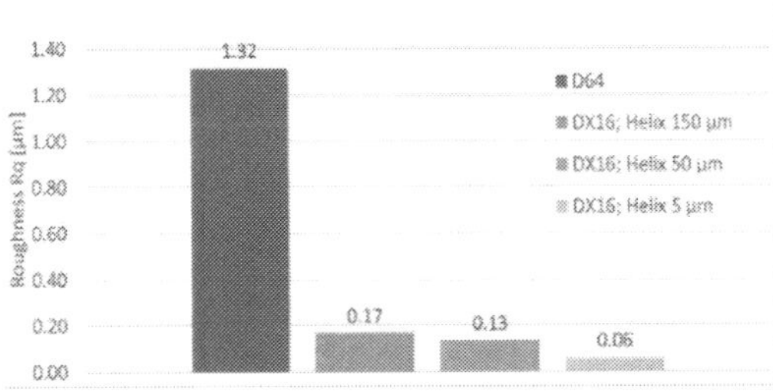

Fig. 4. Roughness results with different helical infeed depths at a feed speed of 60 mm/min

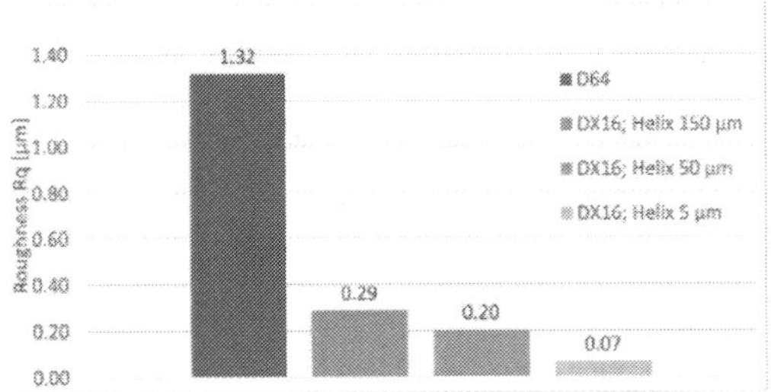

Fig. 5. Roughness results with different helical infeed depths at a feed speed of 300 mm/min

It can be seen that in both cases roughness improvements of up to about 95% can be achieved by using resin-bonded drilling tools (D64 - R_q = 1.32 µm to D16 - R_q = 0.06 µm / R_q = 0,07 µm). Furthermore, a strong relationship between the helical infeed used and the resulting roughness can be observed. By reducing the helical infeed depth from 150 µm to 5 µm, the roughness could be improved by about 65% (R_q = 0.17 µm to R_q = 0.06 µm) in case of the slow feed speed of 60 mm/min. With the faster feed speed of 300 mm/min, an improvement of 76% could even be achieved (R_q = 0.29 µm to R_q = 0.07 µm). An interesting aspect is the very low resulting roughness of 0.06 µm and 0.07 µm for the smallest helical infeed depth, which is almost identical in both cases, despite a speed increase by a factor of 5. The feed speed therefore has only a very small influence in the lower roughness range for the parameter ranges considered. This can be an essential factor for process duration and the associated process efficiency. Figure 6 shows (ultra-) fine grinded holes that were produced with the above parameters.

Fig. 6. Fused silica cuboid with ultra-fine grinded holes

3 Conclusion and Outlook

In the investigations on the insertion of long holes into a fused silica cube, it was shown that different machining strategies have a strong influence on the resulting surface qualities. When comparing the roughness achieved during pre-drilling, it was detected that a change in the machining kinematics from punctual deep drilling to helical drilling caused a deterioration of the roughness by a factor of about 2. Furthermore, negative influences of ultrasound on the process of deep drilling regarding surface roughness could be detected, whereas US had only a minor effect on the surface during helical drilling. In addition, quality differences could be found in different sections of the individual holes, which can be attributed to variable drill stability during machining. There was also evidence, that the ultrasonic assistance can improve the dimensional accuracy of the boreholes. All of this will be further investigated in future grinding and simulative analyses.

The investigations into the subsequent machining of generated holes by using ultra-fine grinding represent a promising approach for the rapid generation of internal contours with high surface quality, whereby the roughness could be improved by up to 95% through the use of resin-bonded tools. The detected and only very slight influence of the feed speed in the lower roughness range represents an interesting approach for further investigations into the duration and thus economic efficiency of the grinding processes considered.

The authors gratefully acknowledge financial support by the German Federal Ministry of Economy and Energy in the funding program ZIM (funding reference: ZF4039613TV8).

References

1. M. Rost, J. Bliedtner, S. Henkel: Einflüsse des ultraschallunterstützten Schleifens auf silikatische Werkstoffe. In: Eltmann: *Diamond Business* **1/2016**, pp. 68-77. (2016)
2. J. Zhang et al.: Study on Effect of Ultrasonic Vibration on Grinding force and Surface Quality in Ultrasonic Assisted Micro End Grinding of Silica Glass. In: *Shock and Vibration*, Vol. **2014**, Article ID 418059 (2014)
3. S. Henkel, A.-M. Schwager, J. Bliedtner, et al.: New surface smoothing technologies for manufacturing of complex shaped glass components. In: *Proceedings of SPIE* Vol. **10448**. (2017)
4. M. Binder, S. Henkel, A.-M. Schwager, et al.: Efficient and innovative polishing processes for smoothing complex surface geometries and internal contours on brittle-hard components. In: *EPJ Web of Conferences*, Vol. **238**. (2020)

Manufacturing risk management in optical design

O.*Faehnle1 and I.Livshits2*
[1] OST-University of Applied Sciences, Buchs, Switzerland,
[2]ITMO University, St. Petersburg, Russia

Abstract. This paper reports on strategies to determine optimal sets of optical fabrication technologies (OFT) for given optical elements, to be applied in optical fabrication chains. For this purpose, optical systems are categorized and linked to critical characteristics of optical fabrication technologies, which are determined by a methodical analysis of their processing parameters.

1. Introduction

Predictable and stable fabrication processes are essential for reliable cost and quality management in optical fabrication technology. This paper describes strategies to determine for given optical elements and systems optimum sets of optical fabrication technologies (OFT) to be applied in optical fabrication chains.

2. Optical systems generation and classification

Mankind has been applying light as a tool to solve daily life challenges such as detection, illumination and information transmittance. And to apply light as a tool, it takes optical systems. Consequently, optical systems, e.g., objectives, interferometers, optical computer mice, endoscopes, telescopes, or lasers, are good examples of today's light tools we apply to meet essential needs with which we are confronted.
During the generation of optical systems (composed of various optical elements properly mounted and functioning together), three authorities are sequentially engaged: "optical systems design," followed by "optical fabrication design" and finally its actual "manufacturing" (see Table 1).
It is the job of "optical designers" to translate customers' requirements into a well-toleranced optical system design comprising various optical elements (such as lenses, mirrors, and beam splitters) modulating the traversing light as desired.
Subsequently, it is the "optics fabrication designer" who translates the specifications of the finished optical system design and tolerances of the employed optical elements into optimized fabrication chains consisting of a series of subsequent fabrication steps taking available machinery and technologies into account ensuring optimum performance at minimum manufacturing costs. Furthermore, systems stability and durability as well as safety of manufacturers, the eventual users and the environment have to be taken into consideration. In the optimized fabrication chain, each fabrication step (e.g., polishing[1]) employs a specific optical fabrication process (e.g., bonnet polishing,[2] bowl feed polishing,[3] magnetorheological finishing,[4] ion beam figuring,[5] ultraform finishing,[6] single-point diamond turning,[7] and fluid jet polishing[8]) the choice of which depends on the right balance between quality, fabrication cost, and number of systems to be produced.

Finally, the optical system is being manufactured, applying the developed fabrication chain and processes not exceeding the designed tolerances generating a fully functioning light tool that meets customers' demands. Optical systems can be classified applying different strategies, a.o. by the markets they are applied to, e.g. astronomy, medical, lighting or lithography, or by the type of instruments they are applied in, e.g. microscopes, telescopes, endoscopes or interferometers.
But to be able to identify for a given optical element the optimum optical fabrication technologies, it is important to characterize optical elements by their four major characteristics relevant for fabrication, which are summarized in table2.

3. Optical fabrication technologies classification

While a limited number of optics fabrication technologies have been around for millenniums, their diversity level has been increasing tremendously since the first industrial production was started 1670 for Leeuwenhoek's microscope in Delft and due to the increasing demands in application and quality ever since (see Fig.1). Consequently, today there are more than 300 optical fabrication technologies existing, making the optimum choice for the generation of a given optical element a challenge.
Analyzing the optical fabrication technologies methodologically, we identify that they are based on only about 11 polishing techniques: Fresh feed polishing (FFP), ductile machining (DG), chemical polishing CP), bowl feed polishing (BFP), elastic emission machining (EEM), magnetorheological finishing (MRF), laser fire polishing (LP), ion beam figuring (IBF), abrasive slurry jet machining (ASJ), plasma assisted chemical etching (PACE) as well as laser-induced backside wet etching (LIBWE).
If we analyze the interaction between these different polishing techniques and surfaces being processed, we identify 5 different *wear approaches* as shown in Fig,2 which are based on (i) 2-body abrasion, (ii) 3-body abrasion, (iii) kinetic abrasion, (iv) energy transmission and (v) chemical reaction. Each of them featuring specific strong and weak characteristics in terms of performance and applicability (indicated by "+" and "-" in Fig.2.). Within these 5 wear approaches various combinations of five different *wear processes* are possible based on (see Fig.3) (a) brittle cracking, (b)

ductile flow, (c) chemical reactions, (d) heat and (e) sputtering.

Taking the above into account, it is possible to classify the 11 existing polishing techniques in terms of wear approaches possible and combinations of wear processes they exist of, as shown in table3.

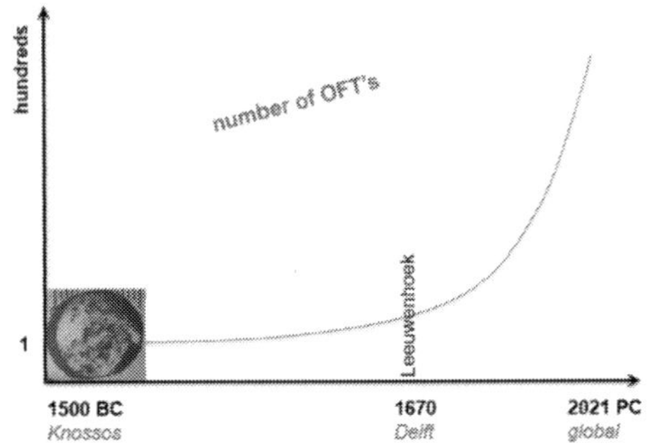

Fig.1. Sketch of development of number of optical fabrication technologies starting from ancient lenses found on Crete through 1670 the first industrial optical fabrication until today.

4. Matching optical elements with optical fabrication technologies

Taking the four key characteristics of optical elements into accout (see table2) and extending the methodological classification of polishing techniques (summarized in table3) towards e.g. characteristics of machining kinematics and types of tool-to-workpiece contacts (e.g. point, line and area contacting), enables the classification and modeling of optical fabrication technologies.

Consequently, a data based optimum selection of optical fabrication technologies for given optical elements can be carried out already during the design phase of optical systems generation.

Fig.3 and 4 show, as an example given, an application of the methodological analysis of optical fabrication technologies to identify optimum fabrication chains for a spherical lens, minimizing production risk.

References

[1] Cook, Lee M. "Chemical processes in glass polishing." Journal of non-crystalline solids 120.1, pp. 152-171, 1990;

[2] David Walker ; A. T. Beaucamp ; David Brooks ; Richard Freeman ; Andrew King ; Gerry McCavana ; Roger Morton ; David Riley ; John Simms; Novel CNC polishing process for control of form and texture on aspheric surfaces. Proc. SPIE 4767, Current Developments in Lens Design and Optical Engineering III, 2002;

[3] J.van Wingerden, H.J. Frankena and B.A. van der Zwan, "Production and measurement of superpolished surfaces", Optical Engineering 31, 1086-1092, 1992;

[4] William I. Kordonski and Stephen D. Jacobs, "Magnetorheological finishing", Int. J. Mod. Phys. B 10, 2837, 1996;

[5] Franz Thomas, and Thomas Hänsel, "Ion beam figuring (IBF) solutions for the correction of surface errors of small high performance optics." Optical Fabrication and Testing. Optical Society of America, 2008;

[6] Edd Fess ; J. Schoen ; M. Bechtold ; D. Mohring, "UltraForm finishing", Proc. SPIE 5786, Window and Dome Technologies and Materials IX, 305, doi:10.1117/12.606468, 2005;

[7] Jiwang Yan, J. Tamaki, K. Syoji, T. Kuriyagawa, "Single-point diamond turning of CaF2 for nanometric surface", The International Journal of Advanced Manufacturing Technology, Volume 24, Issue 9, pp 640-646, 2004

[8] O. Fähnle, H. van Brug and H. Frankena, "Fluid Jet Polishing of optical surfaces", Applied Optics 37(28), 6771-6773, 1998;

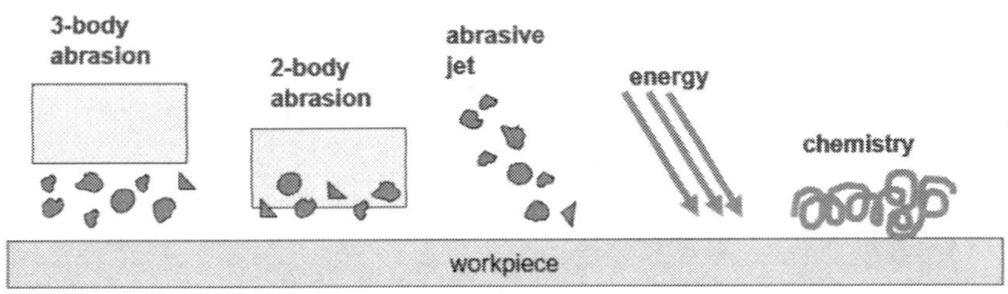

-	mech.contact, slow, tool wear	mech.contact, MSF	material density dependency	temperature, n, vapour, material stress	type ofreactions, residuals, material stress
+	highly accurate	fast, little tool wear	non-contact, highly accurate	non-contact, highly accurate	non-contact, uniform

Fig.2. Optical polishing techniques are applied in five different *wear approaches*, each with its strong ("+") and weak ("-") characteristics in terms of generating optical elements.

	Task	Result
customer	definition of requirements for light to be used as a tool ("light tool")	light tool specifications (e.g. MTF, image resolution, dB, LIDT)
optical system designer	***translation*** of product specifications into a well toleranced optical system design	well toleranced and specified optical system design
optical fabrication designer	***translation*** of optical system design into fabrication chain and fabrication processes parameters	fabrication chain and fabrication process parameters defined and documented (including measuring technology)
manufacturing department	manufacturing of optical system as designed by employing skilled employees and applying optimized machines at high levels of automation.	optical system (light tool); ready for subsequent mounting of further mechanical and electronical devices

Table1. Generation process of optical systems

optical elements' characteristics	description
commercial requirements	maximum fabrication cost allowed and throughput to be produced
geometry	any combinations of flat, spherical, aspherical, off-axis aspherical or freeform surfaces
dimension	ranging from submillimeter to meter class optics
quality	described by parameters such as surface roughness, shape accuracy, centering accuracy, power spectral density curve and material stress level

Table2. Classification of optical elements

	a) brittle cracking	b) ductileFlow	c) chemical reaction	d) heat	e) sputtering
-	*SSD cracks, stress*	*mechanical stress, n*	*residuals, minerals doped, environment, n, «switch-off time»*	*SSD, tension, MSW, n*	*dependency on starting roughness, very slow*
+	*fast*	*slow*	*uniform*	*non-contact, environmental ly clean*	*molekular accurate shaping*

Fig.3. Optical polishing techniques consist of combinations of five *wear processes*, each with its strong ("+" and weak ("-") characteristics in terms of generating optical elements.

polishing techniques	FFP	DG	CP	BFP	EEM	MRF	LP	IBF	ASJ	PACE	LIBWE
wear approaches applicable (see Fig.2)	i, ii, iii	i, ii	ii, v	ii, iii, v	ii, iii, v	iii	iv	iii, v	iii	iv, v	iv, v
based on these combinations of wear processes (see Fig.3)	b, c	b	c	b, c	b, c	b, c	b, d	e	b, c	c, d	c, d

Table3. classification of optical polishing techniques

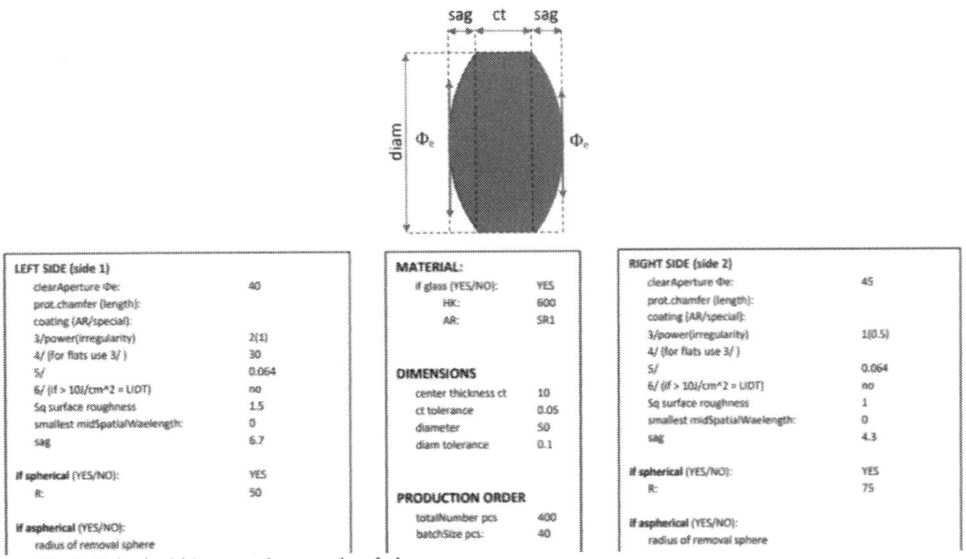

Fig.3. Spherical biconvex lens made of glass.

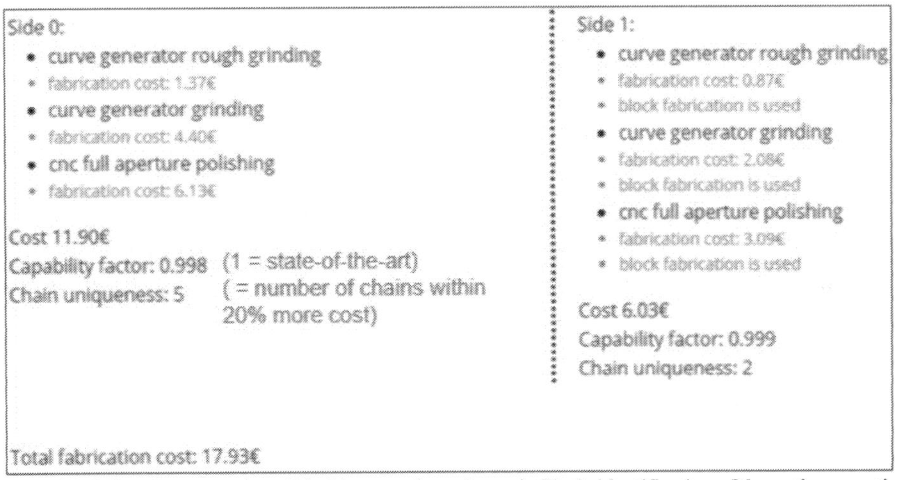

Fig.4. Modeling the generation of the biconvex lens shown in Fig.3: identification of the optimum optical fabrication technologies needed for the minimum-cost fabrication chain.

Preliminary study: Polishing force measurement by viscosity - the return of ketchup polishing

Max Schneckenburger[1,2], *Sven Höfler*[1,3,*], *Oliver Fähnle*[4], *and Rainer Börret*[1]

[1]Aalen University, Centre for Optical Technologies, Aalen, BW 73430, Germany
[2]Technische Universität Ilmenau, Ilmenau, TH 98693, Germany
[3]Glasgow Caledonian University, Scotland, United Kingdom
[4]Eastern Switzerland University of Applied Sciences, Optical Fabrication Technology, 9471, Switzerland

Abstract. Due to the advantages over conventional polishing strategies, polishing with non-Newtonian fluids are state of the art in precision shape correction of precision optical glass surfaces. The viscosity of such fluids is not constant since it changes as a function of shear rate and time. An example is during the shape correction by polishing with pitch or ice, where pitch flows slowly under its own weight and acts like a solid body during short periods of stress as its viscosity increases. The effect can be measured in the polishing gap with a viscometer. If there is a change in force or a process variation of the polishing pressure in the polishing gap, the viscosity also changes. Conversely, the viscosity value could be used to determine the process variation of the polishing force, at least quantitatively. It is to be expected that the distance of the sensor to the polishing gap (effective zone of the polishing force) and the associated change in the viscosity value has a decisive influence on the accuracy of the measurement resolution. First polishing results will be presented and a bowl feed polishing like approach will be presented

Due to the Covid-19 virus and the associated container congestion in Yantian, the construction could not be fully realised at the time of submission. The authors apologise for this. Nevertheless, the theoretical idea is so extensive that the theory and expectations are presented below.

1 Introduction

The material removal (MRR) in the polish depends primarily on the normal force (F_N) and the relative speed (v_{rel}), in addition to a number of other process parameters. Preston showed this in his deterministic polishing process and summarised the other process parameters in the Preston-coefficient (c_P). Preston defines material removal as follows:

$$MRR = c_P * \frac{F_N}{A} * v_{rel} \quad (1)$$

This model is accepted in polishing and has been extended by various authors [1].
Currently, the force of the polishing tool can only be measured with a lot of effort:

- If the position of the polishing head is not additionally measured, it is not possible to calculate back to the origin of the force.
- If the polishing tool is not congruent on the sensor (Δ), lateral tensile and compressive forces (F_x) are mapped onto the F_z(= $F_{x\ res\ Sensor}$) data. Which makes it impossible to determine the exact F_z values. This is shown schematically in Figure 1.

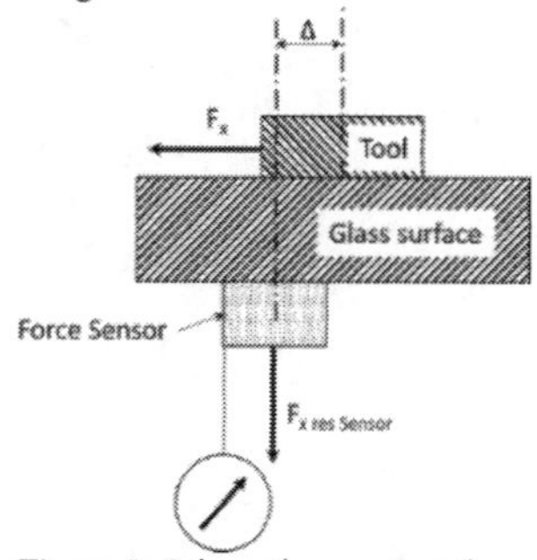

Figure 1: Schematic representation of the effects of the lateral forces on the sensor data

- Sensors directly at the polishing gap have the challenge that the cable is more likely to break due to the movement of the polishing tool and the associated alternating stress on the cable. This particularly affects force sensors directly above the polishing tool.
- The sensor values have drift, which has a visible influence on the measurement data when measured over several hours. According to the manufacturer, the drift can always be calculated, but in practice there are considerable measurement uncertainties.

These difficulties concern in particular polishing tools with an axis of rotation parallel to the normal force axis

* Corresponding author: sven.hoefler@hs-aalen.de

(eg: eccentric tool). The installation of a force sensor is not a challenge with a polishing wheel. On the other hand, there is the measurement uncertainty due to vibrations and the generation of direction-dependent structures by the polishing wheel. In addition, the force sensor is far away (~0.1 m, 4 inch) from the polishing gap.

One possibility of indirect force measurement is measurement above viscosity. In this case, the viscosity of the polishing fluid is measured during the test, which also changes when the force changes. The interesting thing about this is that only the normal force can be measured at a constant relative speed.
Viscosity is the coefficient of internal friction [2]. The internal friction (F_R) of a fluid that is to be moved parallel to a flat surface requires force and is equal to the amount of the frictional force. The internal friction force is defined as a function of the contact area (A), the relative velocity between the boundary surfaces (υ), the distance between the boundary surfaces (d) and the dynamic viscosity of the fluid (η). The dynamic viscosity is the toughness respectively the coefficient of internal friction. This can be applied analogously to the polishing gap, the area between the polishing tool and the glass workpiece (see Figure 1).

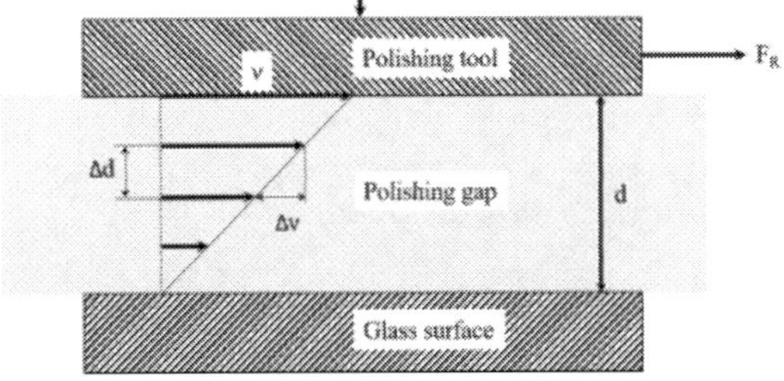

Figure 2: Schematic Polishing gap

$$F_R = \eta * A \frac{\Delta \upsilon}{\Delta \mathrm{d}} \quad (2)$$

The coefficient of friction (μ) can be used to relate the viscosity to the normal force:

$$F_N = \mu * F_R = \mu * \eta * A \frac{\Delta \upsilon}{\Delta \mathrm{d}} \quad (3)$$

The following relationship results via the shear rate ($\dot{\gamma}$) and the shear stress (τ):

$$\tau = \eta * \dot{\gamma} \quad (4)$$

Formula 2 shows that the viscosity could also be measured indirectly via the polishing gap height instead of measuring by a specific sensor. HABERLAND et al have already presented a suitable setup for measuring the polishing gap height [3]. This setup has the advantage over the setup presented later that the viscosity can be measured in-situ in the polishing gap.
The dynamic viscosity is strongly temperature-dependent (T), which is why the sensor used has a temperature sensor. B and C are empirical constants. However, a lot of frictional heat is generated in the polishing gap, which cannot be fully considered in the following set-up.

$$\eta = B e^{\frac{C}{T}} \quad (5)$$

MULLANY ET AL investigated the effects of viscosity in the planarization of wafers [4]. The main focus was the force distribution in the polishing fluid. However, the sensor was not applied in reverse to indirectly measure the force.

The planarization of wafers is very similar to the optical polishing of glasses. Opaline (Ce_2O) is usually used as a polishing agent, which is mixed with distilled water. The water serves as a carrier medium for the polishing agent, transports the heat out of the polishing gap and participates in the chemical reaction by dissociating itself. The cerium does not dissolve in water and is present as solid particles. In terms of viscosity, the polishing suspension behaves like a Newtonian fluid and similar to water (see **Figure 3**).

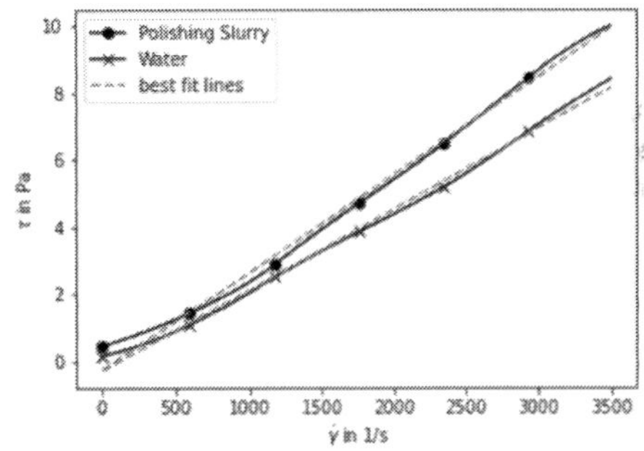

Figure 3: Shows the comparison of the viscosity of water and water with opalines (polishing slurries)

To see the effect of the normal force F=f(η) better in the polishing gap and to increase its height, the viscosity of the polishing liquid will be increased. Figure 4 shows the thixotropic behaviour of a pectin-containing liquid (in this case: ketchup). Under increasing load (in this case the shear rate), the polishing slurry liquefies. A liquid with a higher viscosity or a rheopexy liquid could also have been used. All these viscosity measurements were made with the Haake RotoVisco laboratory instrument.
Instead of using ketchup as in the previous paper [5], this time 100% pure apple pectin is used. Currently, an industrial viscosity sensor needs a working range of >10 mm because the measuring range has to move in the liquid. Currently, there is no polishing process that has such a polishing gap height, so measurements must be taken close to the polishing gap to avoid dead space. Usually, the polishing gap is less than 1 mm.

https://doi.org/10.1051/epjconf/202125503008

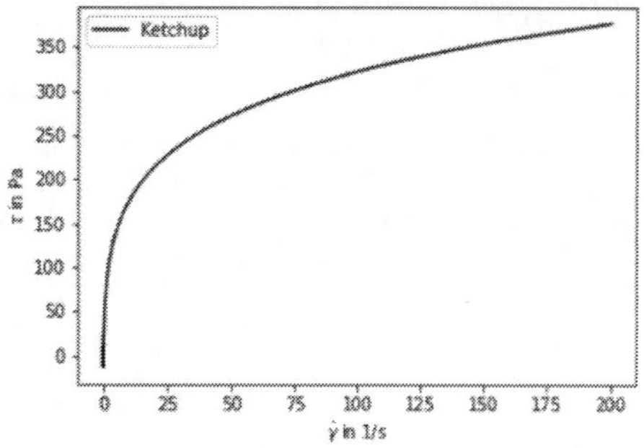

Figure 4: Viscosity measurement of a pectin-containing liquid

2 Proceeding

In the following, the experimental set-up and the planned realisation of the experiment are described.

2.1. Set up

Figure 5 shows a schematic of the experimental set-up. The robot polishing head has a rotary motor for the rotation speed of the polishing wheel and a direct drive for applying the force to the glass surface respectively to the polishing gap. It is polished in a polishing bath, i.e. the tool is below the polishing slurry surface. The SOFRASER MIVI sensor is used as the viscosity sensor [6]. Table 1 shows the key data of the sensor. A flat Zerodur lens with a shape deviation of <1 µm is used as the workpiece. To avoid dead space, the sensor is placed close to the polishing gap so that the polishing wheel rotates in this direction. This means that the polishing gap is not measured directly, but at least close to it. Preliminary tests with ketchup showed that the viscosity around the polishing wheel is affected by a force and not only in the immediate area. However, the rotation of the polishing wheel also causes a shearing effect on the process, so that the viscosity is also influenced.

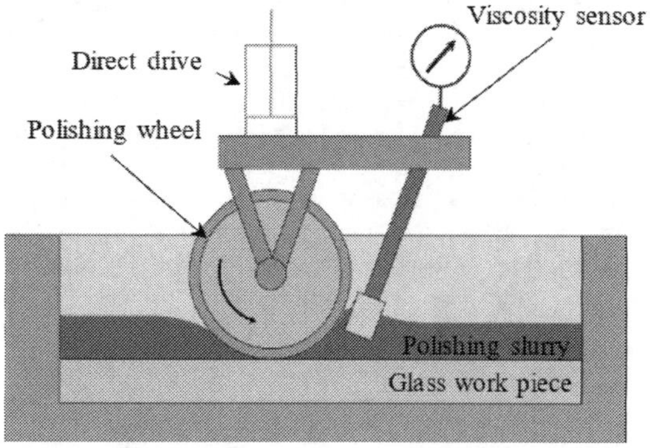

Figure 5: Schematic application of the set up

Table 1. Viscosity sensor key data

	Viscosity sensor
Measuring range	100.000 mPas
Viscosity precision	± 0,2 %
Viscosity accuracy	± 0,5 %
Operating temperature	0 … 200°C

A load cell above the polishing wheel and a force measuring stage below the glass workpiece are used to check the force applied by the direct drive. Only the force is used as the setting parameter. Varying the output viscosity or the speed will not take place. The speed and some other parameters (including the power of the motor) are monitored and stored during the tests.

3 Expected results

The change in force changes the viscosity and normal force can be measured. Despite many influences (including vibrations, process divergences, ...) this is a promising approach. In theory, the decisive factor will be the resolution system of the setup.

4 Summary and outlook

In the preliminary study, a setup for measuring the normal force during the polishing process was presented. At present, no comprehensively satisfactory solutions for force measurement in the polishing process have been found using conventional methods. Therefore, this preliminary study offers a fresh, new, but also unconventional approach.

Unfortunately, no tests could be carried out now, but they will be carried out as soon as possible. Currently, only the viscosity sensor is missing.

Acknowledgement

The work is part of a funded German research project named OpTec 4.0 with the goal of funding young scientist, contract number 13FH003IB6. The authors would like to thank the Federal Ministry of Education and Research for funding this project.

References

[1] LUO, Jianfeng ; DORNFELD, David: *Integrated modeling of chemical mechanical planarization for sub-micron IC fabrication : From particle scale to feature, die and wafer scales ; with 14 tables*. Berlin : Springer, 2004

[2] KUCHLING, Horst: *Taschenbuch der Physik : Mit zahlreichen Tabellen*. 21., aktualisierte Aufl., Jub.-Ausg. 50 Jahre. München : Fachbuchverl. Leipzig im Carl-Hanser-Verl., 2014

[3] HABERLAND, R. (Hrsg.); BLUME-WERRY. U. (Hrsg.); SEYFRIED, P. (Mitarb.); KUNZMANN, H. (Mitarb.); MCKEOWN, P. (Mitarb.); WECK, M. (Mitarb.) : *Measurement of Gap and Parameters in the Polishing Process of Optical Glass : Proceedings of the 6th International Precision Engineering Seminar (IPES 6)/2nd International Conference on Ultraprecision in Manufacturing Engineering (UME 2), May, 1991 Braunschweig, Germany*. Progress in Precision Engineering. Berlin, Heidelberg : Springer Berlin Heidelberg, 1991

[4] MULLANY, B. ; MAINUDDIN, M.: *The influence of process vibrations on precision polishing metrics*. In: *CIRP Annals* 61 (2012), Nr. 1, S. 555–558

[5] SCHNECKENBURGER, Max ; SCHIFFNER, Melanie ; BÖRRET, Rainer: High precision glass polishing with ketchup. In: HABERL, Alexander; FÜTTERER, Gerald; FÄHNLE, Oliver W.; RASCHER, Rolf (Hrsg.): *Seventh European Seminar on Precision Optics Manufacturing* : SPIE, 2020 - 2020, S. 16

[6] SOFRASER: *MIVI : Process Viscometer*. 2021. URL https://www.sofraser.com/app/uploads/2021/02/FT3-E-MIVI-134_12-1.pdf

SSD detection on LIDT tested coated Fused Silica samples

Heidi Cattaneo[1*], *Daniel* Schachtler[2], *Roelene* Botha[2], and *Oliver* Fähnle[1]

[1]OST Eastern Switzerland University of Applied Sciences, Institute for Microtechnology and Photonics, 9471 Buchs, Switzerland
[2]RhySearch, Optical Coating Division, 9471 Buchs, Switzerland

Abstract. Material changes and Sub-Surface Damage (SSD) under Laser Induced Damage Threshold (LIDT) testing sites were investigated on 3 diverse single layer coated transparent fused silica samples to obtain additional information on damage precursors on these samples. As a detection method, photothermal deflection technique utilizing a resonant UV laser beam was used. The local variations in UV absorption and probe beam transmittance due to previous exposure to high fluence laser pulses were strongly dependent on the coating itself and on the laser energy used during the LIDT testing. Also the obtained LIDT values differ from coating to coating. Detected effects on the coatings and substrate surface beneath ranged from small surface dislocations to complete coating damage with material transformations. Additional absorbing damage precursors were found close to the damaged sites.

1 Introduction

Sub-Surface Damage (SSD) such as cracks, contamination and local inhomogeneities in material's optical properties can cause enhanced surface absorption and lead to electrical and light field intensification in the material [1, 2]. These defects prone to act as initiators for laser-induced damage resulting in material's premature degradation [3, 4], which is relevant especially when using coated optical components in high-power laser applications. SSDs are residual defects that originate from manufacturing and handling steps and typically reside few µm below the polished substrate surface, but in brittle materials some cracks can extend even beyond 100 µm deep in the bulk material [1].

SSDs are typically detected using microscopic techniques. Many of these methods are based on scattering related effects of SSDs and are sensitive to local structural and refractive index inhomogeneities. Small laser-induced damage precursors, however, are mostly of absorptive nature and need to be detected accordingly. Photothermal microscopy is a measurement method that relies on absorption related changes in the material's optothermal properties. The radiation from a resonant pump beam causes local temperature and refractive index variations which can then be measured by their effects, such as beam deflection and defocusing, on a non-resonant probe beam.

Since the invention of the Q-switch, the obtainable peak pulse power of lasers has increased dramatically. These so-called giant pulses easily damage the optical components. The laser induced damaged threshold (LIDT) has become an important quantity to qualify optical components and coatings for high-power laser applications. [5]

In this study, photothermal deflection (PTD) method is used to detect SSD and other material variations in coated fused silica samples that have been previously subjected to strong laser irradiation during LIDT testing. The goal of the study is to investigate different laser induced damage sites and gain knowledge on the correlation between SSD and the resulting material changes. As some material and wavelength combinations show higher density of damage precursors or reduced resistance to laser induced damage, it's of interest to better understand the nature of these precursors, in order to comprehend the root cause of the damage and to improve the coatings.

2 Experimental methods

2.1 Samples

P4 polished (rms roughness of about 3 Å) fused silica substrates were used for the LIDT measurements. The substrates were cleaned by a multistep procedure at cleanroom facilities prior to the coating. The investigated coatings were a single-layer coating of 3QW 1) at 1064 nm of Ta_2O_5, 2) at 1064 nm and 3) at 355 nm of SiO_2. The coatings were produced with reactive magnetron sputtering at Evatec AG.

2.2 Laser-induced threshold testing

The LIDT measurements were performed using a Litron LPYG-450-100 diode-pumped laser. The LIDT was measured at 1064 nm and 355 nm for the SiO_2 layers and 1064 nm for Ta_2O_5. The laser operated at a repetition rate of 100 Hz with a pulse duration of 11 ns. The laser beam

* Corresponding author: heidi.cattaneo@ost.ch

was focused onto the sample surface and the different test sites where irradiated with 5000 pulses per site. The sample was positioned on a x-y-z motorized stage which allowed the automated irradiation of each test site. The effective beam diameter at the sample surface was around 230 µm. The online damage detection during the measurement was done with a photodiode which detected scattered light on the damaged site. The final offline damage detection was performed with differential interference contrast (DIC) microscope images of each test site. At least 150 sites were irradiated with appropriately chosen laser fluences over the transition regime between 0 % and 100 % damage probability.

2.3 SSD detection

A sensitive PTD microscope operating in transmission scheme with collinear pump and probe beams has been utilized to investigate SSD and material changes in optical samples. A detailed description of the setup can be found in Ref. 6. As a pump source, a frequency tripled Q-switched Nd-laser (TRUMPF TruMark 6350) having maximum average output power of 5 W and operating at 20 kHz repetition frequency was used. The average power was reduced approximately to 130 mW before entering the sample. Alternatively, an externally modulated (125 Hz) UV Diode laser (Toptica IBEAM-SMART) operating at 375 nm with maximum cw output power of 70 mW was used as pump source for samples coated with Ta_2O_5. The probe source was an intensity-stabilized HeNe laser having 1 mW optical power and power stability of ± 0.2 % over 8 h. The pump and probe beams were focused with the same 40 mm EFL UV-VIS achromatic objective to approximately 10-20 µm spots on the sample surface.

The UV-induced probe beam deflection was measured as AC variations occurring at the repetition rate of the pump laser in two orthogonal position signals of a position sensitive detector (New Focus 2901). The relative transmitted optical power of the probe beam was obtained with the same detector as well. To attain 2D transmission and PTD mappings, the sample was mounted on a XYZ translation platform and its surface was scanned across the beam foci using linear translation stages having sub-µm resolutions. Several 2D mappings were also measured inside the material at different depths from the front surface of the sample.

The PTD method is sensitive to the relative positions of the pump and probe beam foci on the sample surface. Maximum response was typically achieved when the beams were crossing each other with a small angle (max few degrees) and partially overlapping at the sample surface. The relative positions of the beam foci were first roughly adjusted using a laser beam profiler or knife edge -technique and finally, by maximizing the PTD-signals obtained from a reference sample (such as NBK7) showing larger intrinsic UV absorption.

3 Results

Three fused silica samples with different single-layer coatings (Ta_2O_5 for 1064 nm, SiO_2 for 1064 and 355 nm) were tested using both the LIDT and PTD methods described above. The PTD testing was conducted under the LIDT test sites that have at least been partially damaged during the testing.

3.1. LIDT

The investigation of the damages with DIC microscopy revealed that the damage sites look very different for the three different samples studied. For the Ta_2O_5 coated sample (LIDT at 1064 nm), the damages look like the coating was disintegrated around the centre of irradiation. There is also an additional less dramatic change of the material around the perimeter of the main damage zone.

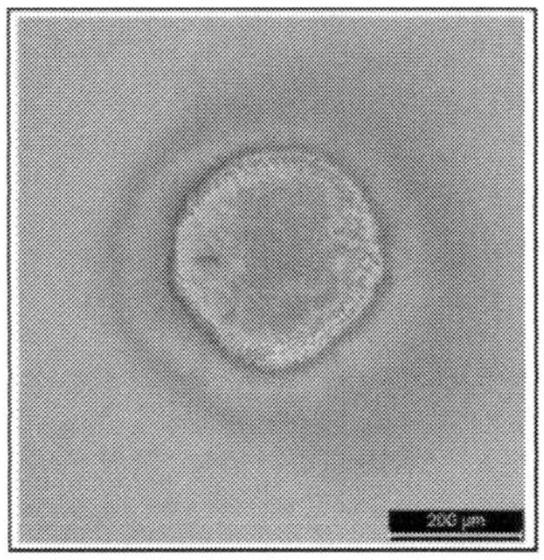

Fig. 1 DIC microscopy image of the investigated damage site (see also Fig. 4) of the Ta_2O_5 coated sample. The site was irradiated with an energy density of 8.8 Jcm^{-2}.

The damages of the SiO_2 coated sample (LIDT at 1064 nm) always appear around a damage precursor in form of a small defect at the centre. Around this precursor there is always quasi circular area which is visible with DIC microscopy. This circular area is called a plasma scald and it is caused by a laser-supported detonation wave, with its origin at the laser induced damage site [7].

Fig. 2 DIC microscopy image of the investigated damage site (see also Fig. 5) of the SiO_2 coated sample (1064 nm LIDT). The site was irradiated with an energy density of 130 Jcm^{-2}.

The damages of the SiO_2 coated sample (LIDT at 355 nm) appear as an agglomeration of small micron-sized defects. These defects in the laser irradiated area increase in density exponentially with increasing laser energy density. The damages have probably very small

https://doi.org/10.1051/epjconf/202125503009

precursors which are not visible by light microscopy and show laser induced growth. This laser induced growth was also observed via scattered light detection.

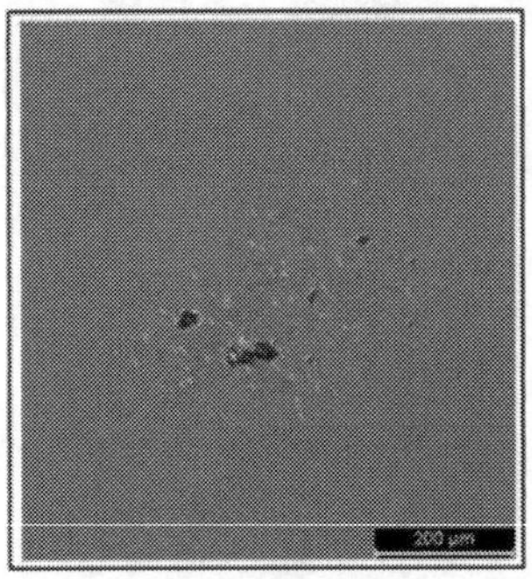

Fig. 3 DIC microscopy image of the investigated damage site of the SiO_2 coated sample (355 nm LIDT). The site was irradiated with an energy density of 9.8 Jcm^{-2}.

3.2 PTD

The PTD response of the coated samples depended strongly on the coating itself and on the type of the laser-induced damage. The first sample (Ta_2O_5 for 1064 nm) exhibited relatively strong UV absorption and produced the highest PTD signals. In order to prevent further damage on the coating during the measurements, UV diode laser was used as a pump source. The damaged areas showed increased transmission (+ 15 – 30 %) of the probe beam and slightly higher PTD signals (up to x10) than the surrounding undamaged coating, indicating coating removal with material alterations on the substrate surface as the pump energy was not sufficient to produce PTD signal on pure fused silica. The edges of the damaged areas produced the most intense PTD peaks (up to x50). Around the damage areas, the increase in transmission of the probe beam (up to 20 %) was visible in form of a circular ring or a halo. There was no increase in PTD signal. The increase in transmission can be explained as surface dislocation. During PTD scanning, other small absorption centres were found as well on regions without prior damage. An example of transmission and PTD mappings performed simultaneously at the sample surface over one of the LIDT sites is shown in Figure 4.

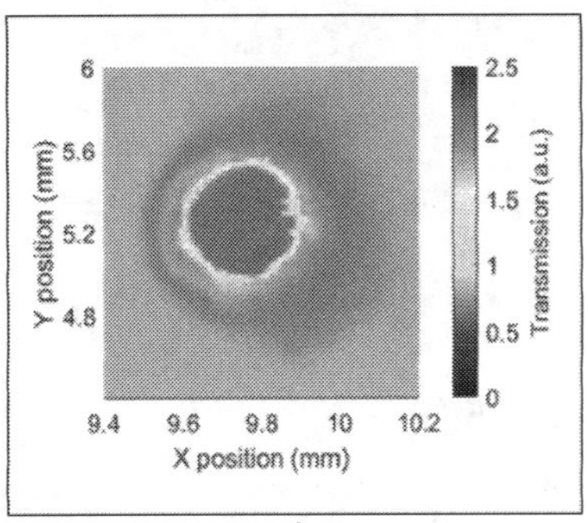

a)

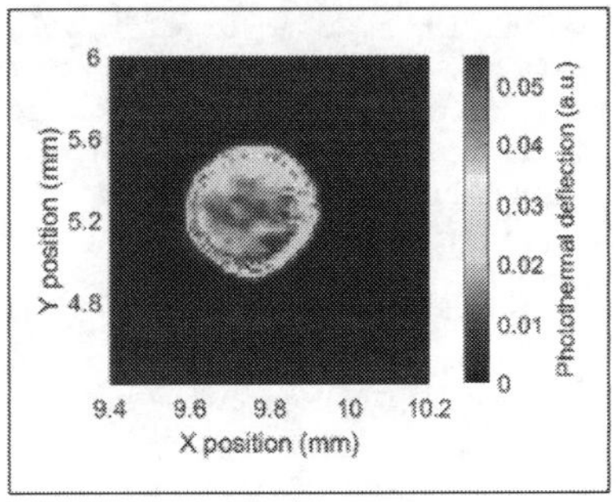

b)

Fig. 4. Example of a a) transmission and b) PTD mapping over a LIDT site in Ta_2O_5 coated fused silica sample. Scanning area (800 µm)^2, scanning step 10 µm, PSD gain 1, and average pump power 18 mW.

Few z-scans were performed as well and they revealed that LIDT testing has caused material deposits on the edges of the sites. Accordingly, maximum PTD signal was obtained few 100 µm above the sample surface.

The second sample (SiO_2 for 1064 nm) also produced faint interference fringes on probe beam transmission mappings over the LIDT sites indicating small surface dislocation. The magnitude of the dislocation increased with the radiation dosage used during the LIDT testing. For the UV absorption mapping, ns pulsed UV laser was used as a pump source to produce sufficient PTD signal as the SiO_2 coating did not show higher UV absorption than the substrate itself. The measurements were performed 100 – 200 µm below the surface to prevent further damage on the coatings due to the absorption-induced local heating from the pulsed pump laser. The PTD mappings showed slightly increased signal at the edges of the plasma scalds and occasionally also at the precursor centre. For larger energy dosages, increased PTD signal was obtained on a larger area around the damage centre. An example of a PTD mapping performed 200 µm below the sample surface over one of the LIDT sites is shown in Figure 5.

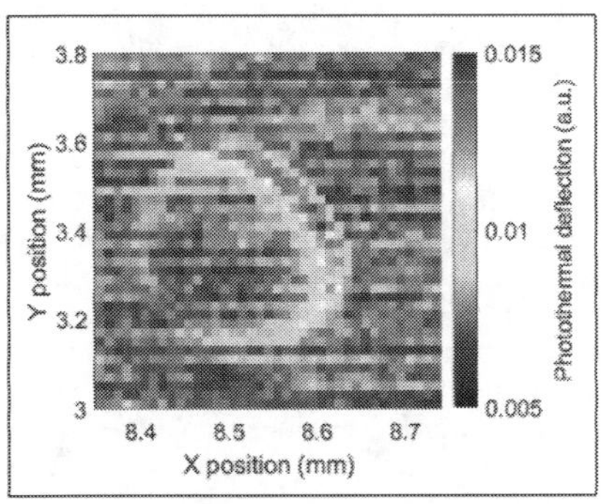

Fig. 5. Example of a PTD mapping over a LIDT site in SiO_2 (for 1064 nm) coated fused silica sample. Scanning area (400 µm)^2, scanning step 10 µm, PSD gain 10, and average pump power 130 mW.

The coating itself was not damaged, but the increased PTD signal indicates possible delamination of the coating on the edges of the plasma scalds. According to the z-scans there was no material deposit on the sample surface and the PTD signal was at maximum at the immediate close to the surface.

The other SiO_2-coated samples showed increased transmission only around the completely damaged LIDT sites. No increase in UV-absorption was observed neither under the LIDT sites. The small damage markings were purely of scattering nature and caused locally reduced transmission and hence reduced photothermal deflection as well compared to the signals obtained from intact areas. Few small absorbing centres were incidentally found close to the irradiated sites.

4 Conclusions

Three different single-layer coatings were tested using LIDT S-on-1 with S=5000 method. The obtained LIDT values varied between 3.5 and 80 Jcm^{-2} and depended on the coating material and design wavelength. Similarly, the optical appearance of the damage sites differed greatly from coating to coating. This is also an indication that the underlying physical cause of the damages varies for different materials and photon energies. The appearance of the damages also partially depended on the energy dosage used during the testing, e.g. the visible defect density of the SiO_2 coated samples tested at 355 nm increased exponentially with increasing energy density. The damage sites were additionally analysed using a combination of transmission and PTD measurements. Several lateral and in-depth scans were performed over the LIDT sites using UV pump and VIS probe laser beams. The laser induced effects on the coatings varied from small surface dislocations to complete coating disintegration with material alterations. Few additional small UV absorbing damage precursors were found close to the damage sites as well. No additional cracks nor other scattering type SSD were found below the coatings in the vicinity of the damage sites indicating that most of the small damage precursors were of absorbing nature. The damage precursors might be metallic or other absorbing particles coming from the coating process.

Acknowledgements

The authors thank Innosuisse – Swiss Innovation Agency for financial support (Project 39986.1 IP-ENG Coating technology for the high throughput, high yield production of dielectric optical interference filters for multispectral imaging). We wish to thank Evatec AG for providing the coatings.

References

1. J. Wang, Y., Li, J. Han, Q. Xu, Y. Guo, J. Eur. Opt. Soc.-Rapid **6**, 1-16 (2011)
2. M. Chen, M. Li, C. An, L. Zhou, J. Cheng, Y. Xiao, W. Jiang, Jpn. J. Appl. Phys. **52**, 1-8 (2013)
3. L. Hongjie, H. Li, W. Fengrui, Z. Xinda, Y. Xin, Z. Xiaoyan, S. Laixi, J. Xiaodong, S. Zhan, Z. Wanguo, Opt. Express **21**, 12204-12217 (2013)
4. J. Neauport, L. Lamaignere, H. Bercegol, F. Pilon, J.-C. Birolleau, Opt. Express **13**, 10163-10171 (2005)
5. Ristau, Detlev, ed. *Laser-induced damage in optical materials* (CRC Press, Hannover, 2014)
6. H. Cattaneo, R. Botha, C. Ziolek, *Proc. SPIE* **11352**, 1135205-1 (2020)
7. X. Liu, Y. Zhao, D. Li, G. Hu, Y. Gao, Z. Fan, J. Shao, Appl. Opt. **50**, 4226-4231 (2011)

EPJ Web of Conferences **255**, 03010 (2021) https://doi.org/10.1051/epjconf/202125503010
EOSAM 2021

What can optical engineers learn from Leonardo?

I.L. Livshits1 and O.Faehnle2
1 ITMO University, Saint Petersburg, 197101, Russian Federation
2 OST-University of Applied Sciences, Buchs, 9471, Switzerland)

Abstract: Optics in general is an old science. Many brilliant researchers have created both theory and practice in this field. Some of their achievements are still in use, some are unjustly forgotten. In this paper we tried to rethink some points of ancient scientists to show that they had «seeds» which are still important ourdays. In particular, the applicability of the "golden section" from nature to the modern, computational process of optical system design was tested experimentally. Design principles, current challenges, possible solutions and some applications are discussed. They could be used, for example, in optical design, at the stage of starting scheme or help making decision of optical system producibility.

1. Indroduction

State-of-the-art optical technologies are characterized by the highest levels of "precision" possible today. At the same time, the step towards optical applications of tomorrow (e.g. for telescopes, endoscopes, lithography machines or lidar systems) requires the synthesis of know-how of a wide variety of technology fields, such as materials science, optical surface generation, metrology and chemistry as well as mechanical engineering and software development. Consequently, it is essential to develope and apply innovative research methods that are capable of enabling such syntheses [1-4]. Following that trail, the analysis of the morphology of the word re-search itself might be of help. We find it to be a combination of two morphemes: the latin words "re" (meaning "back") and "circare" (meaning "go about"). If we now apply both parts literally and travel back in time to mankind's origins and start searching there for multi-disciplinary solutions, we find ourselves observing nature herself. Doing so, mankind converted already several solutions from nature into high end technology applications in industry: e.g, gecko's feet's van-der-Waals' contact while walking up walls has been applied to bonding components; chamaeleon's gentle grip applied to industrial robots handling delicate parts or, last not least, butterfly's beautifully colored wings inspiring us to developing diffractive optical elements.

One omnipresent effect in nature is the golden ratio (1.618), firstly described by Euclid [5]. It describes the shape of nautilus shells, hurricans and galaxies as well as the hunting pattern of hawks. In general, it seems to be often the golden solution to solve multi-trandisciplinary and complex situations in nature. Therefore we assume the golden ratio a matter, worthwhile analyzing even from the perspective of leveling-up "precision" towards tomorrows optical technology solutions. This paper reports on authors first curious steps analyzing the impact of natures golden ratio, phi, onto high end optical systems design. When we look back at historical times of mankind and if we are interested in optics, our attention is attracted by the time of the renaissance epoch where brilliant and gifted people worked. At that time called «golden age» attention of many brains were directed to natural and precise sciences. Natural philosophy was developing and science was considered as a single entity. First we would like to mention Leonardo da Vinci, who was a genius involved at the same time into many activities: a.o. painting, sculpture, architecture, engineering, medicine and fortification. A lot of time Leonardo spent in optical technology, e.g. mirror production. He created special equipment to produce spherical and parabolic concave mirrors. The heritage of Leonardo's engineering activity is very big, but not all of his inventions were produced. Time-to time somebody tries to reconstruct them, and it works! Leonardo applied the golden ratio in many of his works and paintings, e.g. in the painting "The Annunciation", located in the Uffizi, Florence, Italy. In this paper we apply the idea of a "golden ratio" to modern lens design.

2. Golden ration based optical design

In the following and without any claim to absoluteness, the possible approach and impact of the use of the golden ratio in optical systems design is tested experimentally. To that aim, a single thin meniscus lens A has been designed where radii are connected by golden ratio: R2/R1 = 33.87/20.93 = 1.618. Lens A looks well balanced and is free of astigmatism, distortion and all high order aberrations (see Fig.1). ***Optical performance stability testing:*** Without any claim to absoluteness, Table 1 presents results of tests carried out to check if the optical performance of the *golden radii ratio* lens A reported in Fig.1. was a lucky shot or a position of local minimal aberrations. To that aim, two different lenses have been designed by changing R1 in such a way, that the *radii ratio* yields 1.718 for lens B and 1.818 for lens C, respectively. For better comparison, both lenses are scaled to the same F'=100 mm and same aperture as the *golden radii ratio* lens A. The order of aberration is determined by the degree of the coordinate in the expansion of the transverse aberration in a series. We consider 3-d and 5-th order of aberrations, and we see that 5th order aberrations are almost zero because meniscus lens like this is in optimum shape and does not get high order aberrations. Third order aberrations of the meniscus with R2/R1=1.618 (the golden ratio) are better than for lens B and lens C. While, in comparison to the *golden radii ratio* lens A, the absolute differences in levels of aberrations of lenses B an C are not so big, the *golden radii ratio* lens features a clearly better performance confirming that golden ratio works...in this case.

3. Conclusions

In nature, beauty is often to be equivalent with being strong in survival skills, reproduction and protection. The perception of beauty herself can be found in the golden ratio which mathematically describes humans evolutionary perception of the "best solution" while interacting as a being within nature. Therefore, we might define our inclination towards beauty as an evolutionary "optimization method generated by nature". This paper demonstrated that it is worthwhile going back in time, challenging modern methodology of research methods by analyzing solutions generated by nature herself. Without any claim to absoluteness, first steps along this trail have been made experimentally and promising insights have been generated applying the "golden ratio approach" to State-of-the-Art optical systems design. The golden ratio works for the reported lens A design: if radii of a meniscus lens are connected with golden ratio number (1,618); this lens has no astigmatism, distortion or all monochromatic high order aberrations and a clearly better performance than lenses featuring radius ratios located next to the golden ratio. Although the golden ratio approach to optical design reported in this paper generated impressive preliminary results, we must of course take into account that all "optimization techniques" installed by nature do not contain the claim to absoluteness but to evolutionary trial and error survival of the best. The golden ratio approach is in authors conviction a perfect complementary method to the application of modern high technology optical design software tools which are based on Snellius law, only. So it is true, that by looking back into the past we gain insights that might help us to design an optimum future; even in optical technologies

References

[1] U.Östlund, L.Kidd, Y.Wengström, N.Rowa-Dewar, "Combining qualitative and quantitative research within mixed method research designs: A methodological review", International Journal of Nursing Studies, Volume 48, Issue 3, 2011, ISSN 0020-7489,

[2] Williams, C. (2007). Research Methods. Journal of Business & Economics Research (JBER), 5(3),

[3] O.Faehnle, "Process optimization in optical fabrication", SPIE Journal on Optical Engineering, doi:10.1117/1.OE.55.3.035106., 2016

[4] Z.Mehrjerdi, (2011), "Six-Sigma: methodology, tools and its future", Assembly Automation, Vol. 31 No. 1, pp. 79-88..

[5] Euclid; sketch and explanation of the golden ratio, book2 proposition5, Oxyrhynchus Papyri, Egypt (28°32'N 30°40'E); ca. 100 AD.

[6] Francesco Melzi, "Leonardo da Vinci's "Treatise on Painting", 1632, France.

Table 1. testing optical performance stability of the golden radii ratio lens A in comparison to lens B and lens C

N	R2/R1	SA3	CO3	TI3	SI3	PETZ	DI3	S'Fm m	F'mm
1	20.93/33.87= 1.618	-0.079	0.037	-0.0385	-0.0385	-0.0385	0.0062	79.12	100.0
2	22.27/37.77= 1.718	-0.08	0.039	-0.047	-0.044	-0.042	0.0069	77.93	100.0
3	24.17/43.95= 1.818	-0.08	0.037	-0.060	-0.050	-0.045	0.0077	76.76	100.0

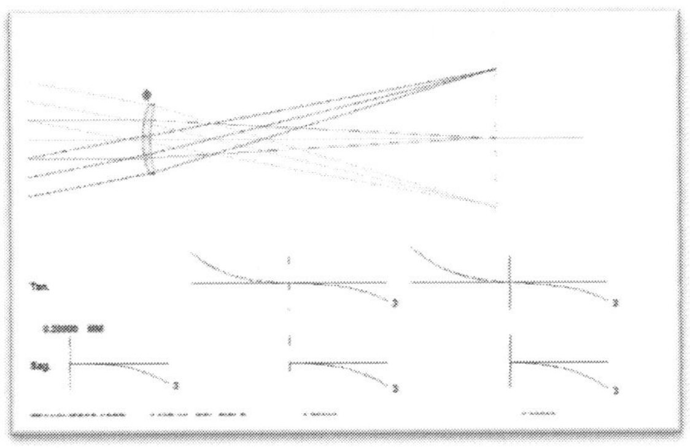

Fig. 1. Optical scheme of meniscus under goldenratio and plots of its transverse aberrations

EPJ Web of Conferences **255**, 04001 (2021) https://doi.org/10.1051/epjconf/202125504001
EOSAM 2021

Design of erbium doped silicon nanocavities for single photon applications

Matteo Di Giancamillo[1,2], *Paolo* Biagioni[2], *Vito* Sorianello[3] and *Enrico* Prati[1*]

[1]Istituto di Fotonica e Nanotecnologie, Consiglio Nazionale delle Ricerche, 20133 Milano, Italy
[2]Dipartimento di Fisica, Politecnico di Milano, 20133 Milano, Italy
[3]Photonic Networks and Technologies Lab, Consorzio Nazionale Interuniversitario per le Telecomunicazioni (CNIT), 56124 Pisa, Italy

Abstract Silicon-based quantum communication technologies are becoming a factual reality. However, the challenges related to an earth-space unifying technology are several, and nowadays an integrated source compatible with the CMOS technology is still missing. Here we present the design of a weak photon source consisting of a LED able to emit directly into the optical circuit and obtained through the doping of a portion of a silicon waveguide with ErO_x complexes. To enhance the radiative emission, the source is placed inside a resonant cavity delimited by two waveguide Bragg mirrors. A study on the performance of the device is carried out as a function of different parameters, such as the geometry of the cavity and of the contacts used to electrically excite the defects, the doping level, and the characteristics of the mirrors. We design a prototype that guarantees a Purcell factor in the order of tens, emitting ideally 10^7-10^8 photons per second. The simulations provide a promising ground to further develop fully integrated single photon sources in silicon photonic circuits.

1 Introduction

The goal of the research is to design a LED weak photon source for silicon-based quantum communications in space at a wavelength of 1540 nm, through the radiative de-excitation of ErO_x complexes generated by erbium implant in a portion of a silicon waveguide, and to evaluate the total number of photons emitted per second as a function of different parameters. The aim of the source is to generate photons directly in the optical guide and to use materials compatible with the CMOS technology, making it competitive in terms of costs and performance. Nowadays, a commercial device that satisfies all the demands for a ground-space technology and where the native generation and detection of single photons and their polarization states is integrated on the same chip does not exist [1, 2], even if some of the Authors already explored nano-LEDs based on Er [3]. The use of silicon has the potential to combine the advantages and knowledge of electronic control based on CMOS microelectronic technology with those deriving from the integration of silicon photonics, increasing the tolerance to temperature variations that could undermine the performance of non-monolithic architectures [1, 2].
Fig. 1 represents a top-view sketch of the proposed device. A potential difference is applied to two contacts made of doped Si, which are placed at the sides of the waveguide and excite the erbium atoms through carrier-mediated excitation [4]. To enhance the radiative emission, the doped region, which acts as the source, is placed inside a cavity tuned to be resonant with the emitted wavelength and obtained through two waveguide Bragg mirrors [5]. The waveguide core, with a width of 480 nm and a height of 220 nm, is realized on top of the silicon oxide layer of a SOI wafer and surrounded by a thin layer of 100 nm thickness of the same oxide material.

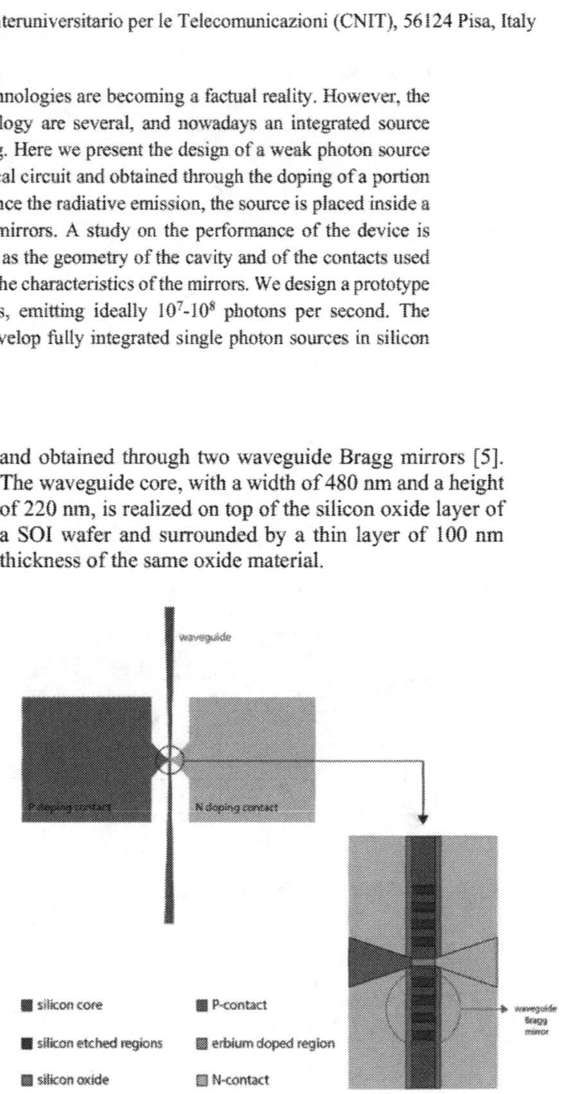

Fig. 1. Sketch of the device.

The present work focuses on the electromagnetic design of the waveguide and of the optical cavity, in order to

[*] Corresponding author: enrico.prati@cnr.it

https://doi.org/10.1051/epjconf/202125504001

optimize their geometry. It is carried out using two simulation software packages, Ansys Lumerical [6] and Comsol Multiphysics [7], taking all the technical requirements and the manufacturing constraints into account. To model the erbium-doped silicon material, we related the concentration of dopants to the variation of the real and imaginary part of the refractive index of the host material following Ref. 8.

2 Results and discussion

The results of the simulations are divided into two sections: the first one is focused on the propagation within the cavity, while the second one concerns the characteristics and properties of the cavity resonance.

Propagation The following results are obtained with the Comsol Multi-physics wave optics module. An automatic mesh controlled by the software and based on the physics of the problem has been selected. A mode analysis of the cavity is performed to evaluate the effective mode index of the cavity, which results to be n_1 = 2.402 for the fundamental mode and n_2 = 1.681 for the second mode, both represented in Fig. 2. By exploiting a vectorial representation of the electric field, one can appreciate that the fundamental mode of the waveguide possesses a dominant TE character and is polarized perpendicular to the second mode, which is mainly TM. There are no other propagating modes. As we deal with resonant effects in the waveguide, one should also notice that the resonant mode in the cavity only involves the TE mode, while the TM is suppressed along propagation in the optical circuit.

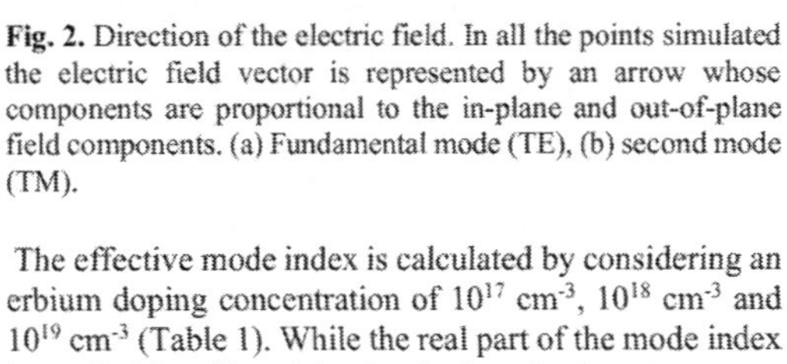

Fig. 2. Direction of the electric field. In all the points simulated the electric field vector is represented by an arrow whose components are proportional to the in-plane and out-of-plane field components. (a) Fundamental mode (TE), (b) second mode (TM).

The effective mode index is calculated by considering an erbium doping concentration of 10^{17} cm^{-3}, 10^{18} cm^{-3} and 10^{19} cm^{-3} (Table 1). While the real part of the mode index is negligibly affected by the doping, the imaginary part increases up to 10^{-2} for the highest concentration, still it does not represent the main source of losses in the cavity, whose performance is dominated by mirror losses and scattering losses, as discussed later on. It is also important to notice that the real part of the refractive index of the erbium doped area is very close to the undoped one. This is a fundamental prerequisite, since it avoids significant mismatch between mode indices inside and outside of the doper area. A larger mismatch would cause scattering and reflection of the light. We can conclude, from this analysis, that the doping level is irrelevant for this major aspect, and the optical properties of the material are not sensibly modified, especially considering that the maximum investigated doping value of 10^{19} cm^{-3} is difficult to obtain using standard techniques.

Doping concentration	10^{17} cm^{-3}	10^{18} cm^{-3}	10^{19} cm^{-3}
Effective mode index	2.4133-8.2E-5i	2.4123-0.0012i	2.4022-0.0180i

Table 1. Effective mode index for three different erbium doping levels of the silicon waveguide.

The mode effective index of the etched waveguide (70 nm etching depth) for the realization of the Bragg mirror is n_{etch} = 2.060. Such a value is exploited to design the periodicity of the fibre Bragg mirror. Considering now the full 3D structure, an analysis on the losses of the cavity, due to the presence of the dopants and the contacts, was performed. In Fig. 3 the transmission along the doped region of the waveguide is plotted as a function of the length of the doped area (and therefore of the lateral contacts), for a cavity with and without contacts. The graph shows that the doping level, which is fixed at 10^{18} cm^{-3}, does not significantly influence the overall losses of the cavity, which are instead determined mainly by the presence of the contacts.

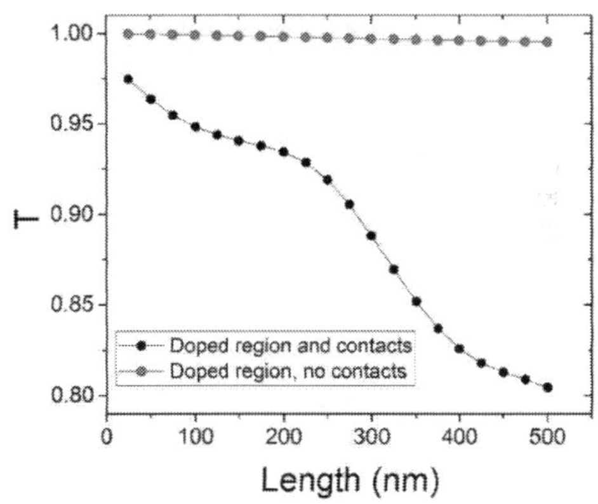

Fig. 3. Transmission coefficients from the input to the output port of the cavity as a function of the length of the doped area and therefore of the lateral contacts.

Cavity analysis The cavity is designed considering two waveguide Bragg mirrors as reported in Fig. 4, which are characterized by a periodic variation in the mode index of the waveguide core obtained by etching the height of the core by 70 nm. According to the mentioned effective mode indices and the etching depth, the grating periodicity for a stopband mirror centred at 1540 nm is around 344 nm. The reflection and transmission coefficients of the mirror, which is reported in Fig. 5

together with the norm of the electric field along the longitudinal cut plane, starts to saturate with a number of periods roughly equal to 40, with a value approximately equal to 96%.

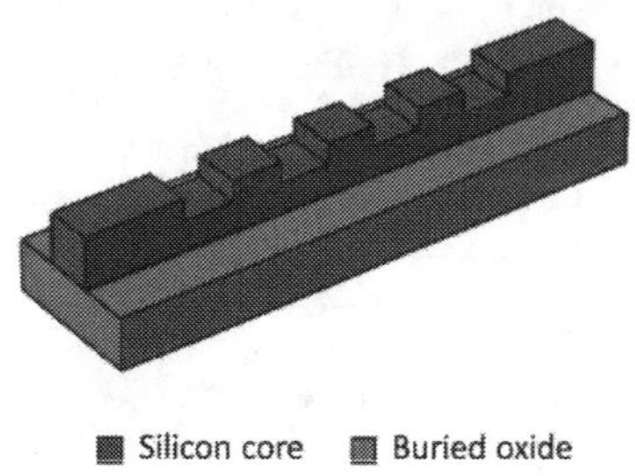

Fig. 4. Waveguide Bragg mirror after the etching process. The silicon core is shaped to obtain a periodic discontinuity of the refractive index. In this conceptual view the upper cladding of 100 nm thickness is not drawn, but it is present in all the simulations.

The mirror effectively dampens the electric field and the reflectivity saturates after a specific length of the mirror because no electric field is present within the structure after a sufficiently large number of periods.

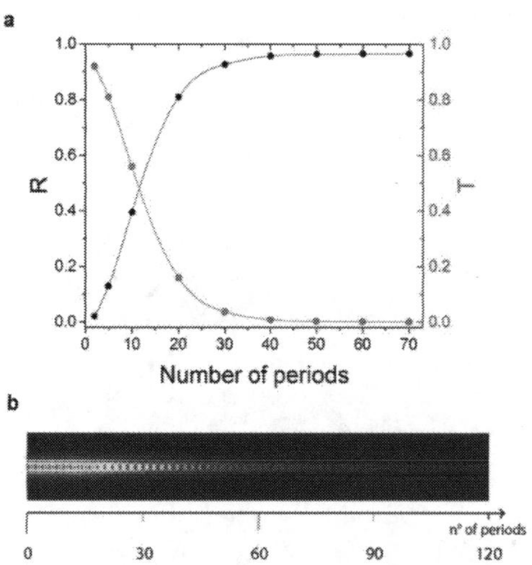

Fig. 5. (**a**) The trend of the reflection and transmission coefficients as a function of the number of periods, (**b**) the norm of the electric field along the longitudinal cut plane.

The sum between R and T is in the order of 97%. This is because the core discontinuities, due to the etching, scatter part of the light towards the lower or upper cladding. Finally, the spectral response for a mirror with 35 periods is reported in Fig. 6. The spectrum is well centred and it has a reflectivity in the order of 90 % between 1520 and 1560 nm. This stopband, which is determined by the mismatch between the two effective mode indices and therefore by the etching depth, is matched to the width of the erbium emission peak at room temperature. It cannot be enlarged further since a deeper etching corresponds also to larger scattering losses.

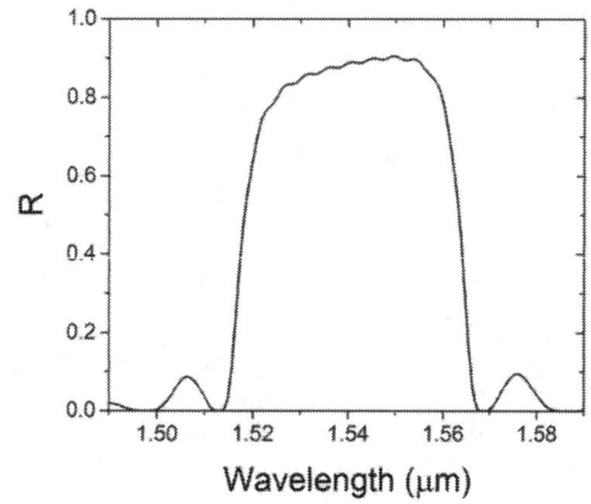

Fig. 6. Spectral response of the waveguide Bragg mirror with 35 periods, a periodicity of 344 nm, and an etching depth of 70 nm.

In order to evaluate the Purcell factor, which returns the enhancement of the spontaneous emission rate of a quantum system by its environment, an electric dipole source emitting at 1540 nm is placed in the cavity, coupled to the TE mode, and surrounded by a transmission box to evaluate the difference of the emission with respect to the free space, as shown in Fig. 7. A parametric study of the length of the cavity was carried out with Ansys Lumerical to obtain the lengths that maximize this effect, i.e. those resonant with the emission wavelength, which are 344 nm (half-wavelength cavity) and 644 nm (one-wavelength cavity).

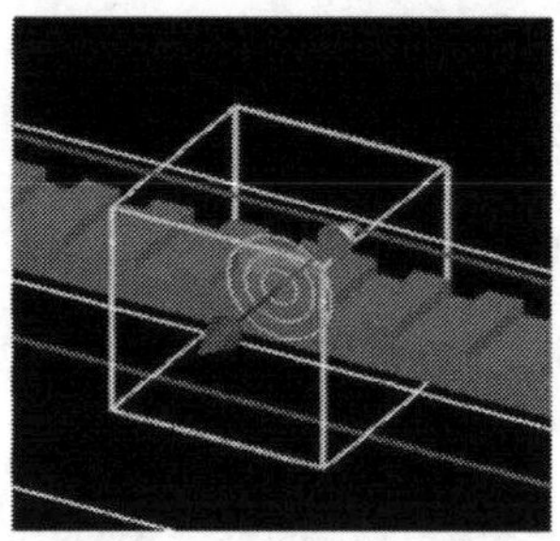

Fig. 7. Electric dipole source surrounded by the transmission box.

The results obtained with these simulations must be divided by two to consider the standing-wave modulation of the density of states along the cavity and then divided further by 3 to take into account all the possible spatial orientations of the dipole, ending up with a Purcell factor approximately equal to 10.

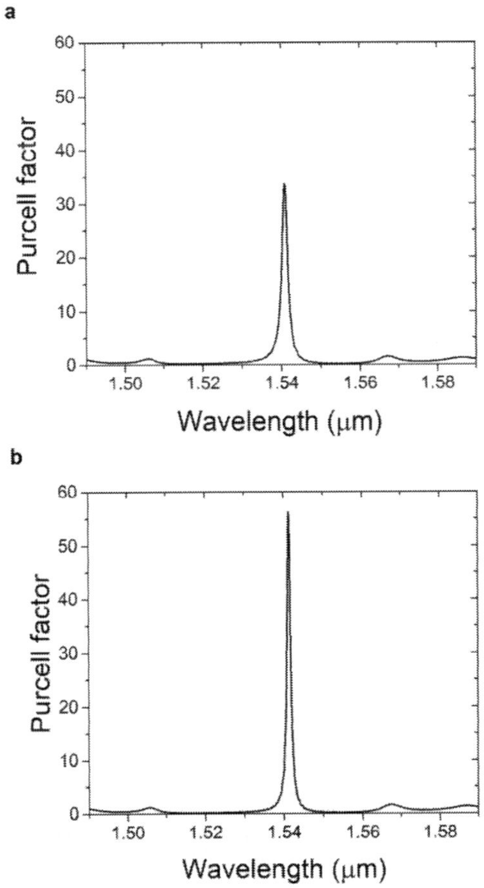

Fig. 8. Purcell factor of a broadband source mimicking Er emission for a cavity length of 344 nm (upper plot) and 644 nm (lower plot).

The difference between the two graphs could be due to different scattering losses when the mode meets the mirror. In fact, some of the scattered light can be 'recycled' into the cavity [9] and this phenomenon could be different for different cavity lengths. Finally, it is necessary to highlight that the simulations in Fig. 8 did not take doping into account, since the numerical code becomes unstable when the emitting dipole is immersed in an absorbing region.

3 Conclusions

This work provides a first systematic assessment of the perspective performance of a nanocavity LED emitter. Considering a volume of the doped region equal to 480 x 220 x 100 nm^3 and a doping concentration of 10^{18} cm^{-3} we obtain a total number of emitting ions in the order of 10^4. Since about one atom out of ten typically emits, considering the previously calculated Purcell factor and a radiative lifetime of 0.1-1 ms, we expect a rate of emitted photons at least of the order of 10^7-10^8 s^{-1} in continuous-wave operation and even higher by pulsing the LED. This supports the feasibility of a device that meets the requirements of operation and compatibility with the manufacturing processes and growth of silicon in modern electronics.

Acknowledgements

This research was part of the QUASIX project funded by the Italian Space Agency (http://www.quasix.space/).

References

1. I. Aharonovich, D. Englund, M. Toth, Nat. Phot., 10, 631–641 (2016).
2. F. Cavaliere et al., Quantum Reports 2.1, 80-106 (2020)
3. T. Fujimoto et al., IEEE Silicon Nanoelectronics Workshop (SNW), 123-124 (2020)
4. A. Kenyon, Semicond. Sci. & Tech., 20 (2005).
5. M. Pruessner, T. Stievater, W. Rabinovich, Opt. Lett., 32, 533-5 (2007)
6. Inc. C. COMSOL [Internet]. 2020. Available from: http://www.comsol.com/products/multiphysics/
7. Lumerical Inc. FDTD: 3D Electromagnetic Simulator
8. M. Nedeljkovic, R. Soref, G. Z. Mashanovich, IEEE Phot. J., 3, 1171-1180 (2011)
9. Ph. Lalanne, S. Mias, and J.P. Hugonin, Opt. Exp. 12, 458-467 (2004)

https://doi.org/10.1051/epjconf/202125504002

Efficient coupling of free propagating light into Whispering Gallery Modes

Xavier Zambrana-Puyalto[1,2,*], *Davide* D'Ambrosio[2], and *Gianluca* Gagliardi[2]

[1]Istituto Italiano di Tecnologia, Via Morego 30, 16136 Genova, Italy
[2]Consiglio Nazionale delle Ricerche, Istituto Nazionale di Ottica (INO), via Campi Flegrei 34—Comprensorio A. Olivetti, 80078 Pozzuoli (Na), Italy

Abstract. Whispering Gallery Mode resonators are dielectric structures with cylindrical symmetry. They are typically excited with an evanescent field leaking out of a tapered fiber or a waveguide. It is also known that they can be excited with free propagating beams. In this work, we use a recently developed analytical model which quantitatively describes the coupling of free propagating beams into Whispering Gallery Modes for spherical particles. Using this model, we have been able to theoretically quantify the mode purity and the coupling efficiency of a resonant Whispering Gallery Mode of an order $j^* = 1456$. We have observed that the transverse position of the beam plays a crucial role in determining the mode purity and coupling efficiency. Last but not least, we have verified that the coupling efficiency as well as the Q-factor predicted by our model are in an outstanding agreement with the experimental values measured on a microresonator of the same dimensions as the simulated one.

1 Introduction

Whispering Gallery Mode (WGM) resonators are dielectric structures with cylindrical symmetry. Their sizes tend to range from ten to a few hundreds of micrometers. These optical resonators have incredibly high Q factors and their use is widely spread across the photonic sciences [1, 2]. Most of the WGM resonators are excited via evanescent coupling, yet it has also been demonstrated that they can be excited via direct scattering [3, 4]. Recently, we have theoretically demonstrated that the WGM excitation in direct scattering set-ups is due to angular momentum (AM) matching [5]. We showed that the AM content of the WGMs induced in a spherical microresonator must match the AM content of the excitation beam. To do that, we used the analytical Generalized Mie Theory [6, 7] in the transverse angular momentum (TAM) basis [8]. The choice of the TAM basis is critical in order to be able to simulate particles with diameters greater than 100 μm [5] at optical frequencies. Here, we focus our attention on the mode purity (MP) and the coupling efficiency (CE) of the excited WGMs. Moreover, we compare our theoretical predictions with the experimental data collected for a microdroplet resonator with a diameter of 300 μm. As it will be shown, the match of our theoretical and experimental data is outstanding, not only in terms of CE, but also in terms of resonance spectrum, Q-factor of resonances, and EM field distribution.

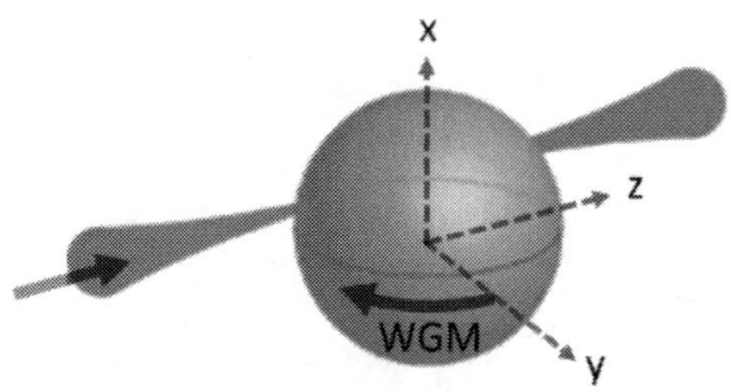

Figure 1. Schematics of a direct WGM excitation. An incident beam propagating along the z axis is displaced a distance d in the y axis, and it tangentially hits the microresonator. As a result, a WGM that oscillates around the x axis is excited.

2 Theoretical model

Our theoretical model is developed within the Mie Theory framework. That is, we consider an isotropic, homogeneous sphere embedded in a homogeneous and non-absorbing medium. The sphere has an index of refraction n_r and a radius R. The sphere is illuminated by a focused Gaussian beam with a well-defined helicity [7] at a transverse distance d from the center of the sphere (see Figure 1). The EM field is decomposed in three sub-fields: the incident $\mathbf{E}^{\mathrm{i}}$, the scattered $\mathbf{E}^{\mathrm{sca}}$ and the interior $\mathbf{E}^{\mathrm{int}}$. All the fields are decomposed into multipolar fields whose origin is in the center of the sphere [7]. We express the multipolar basis as $\left\{\mathbf{A}_{jm_z}^{(y)}\right\}$, with $(y) = (e)$ being the electric multipoles and $(y) = (m)$ the magnetic ones. Due to the symmetries of the problem, the analytical expressions of $\mathbf{E}^{\mathrm{sca}}$ and $\mathbf{E}^{\mathrm{int}}$ are readily obtained when the expression of $\mathbf{E}^{\mathrm{i}}$ is known. The expression of focused Gaussian beam with a well-defined helicity propagating along the z axis at a distance $\mathbf{d}$ of the

*e-mail: xavislow@protonmail.com

https://doi.org/10.1051/epjconf/202125504002

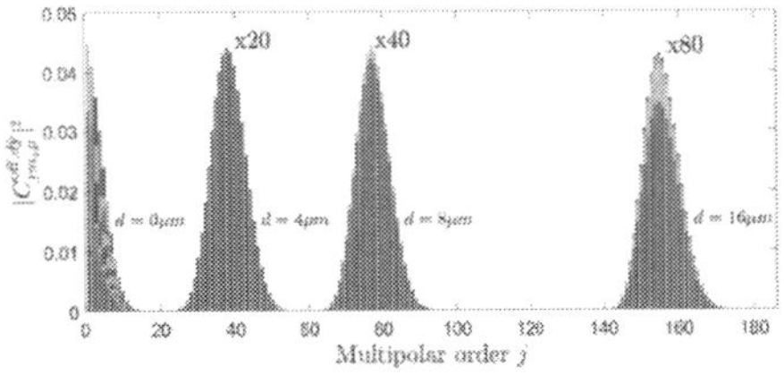

Figure 2. Displaced beam shape function $|C^{\text{off},d\hat{\mathbf{y}}}_{j,m_x,p}|^2$ for the rotated multipoles $\mathbf{A}^{(y)}_{jm_x}$. The plots are done for NA= 0.4 and $d = 0, 4, 8, 16\mu$m. The blue color displays the weight of the multipolar modes with $m_x = j$, the yellow color shows the weights of $m_x = j - 1$, etc. The plot with $d = 0\ \mu$m corresponds to the on-axis beam shape function with $m_z^* = 1$. The wavelength is $\lambda = 640$ nm.

center of the particle is given by [5]:

$$\mathbf{E}^{\text{i}}(\mathbf{d}) = \sum_{j=1}^{\infty}\sum_{m_z=-j}^{j} C^{\text{off},\mathbf{d}}_{j,m_z,p}\left(\mathbf{A}^{(e)}_{jm_z} + p\mathbf{A}^{(m)}_{jm_z}\right) \quad (1)$$

where the expression of $C^{\text{off},\mathbf{d}}_{j,m_z,p}$ is given in [9, 10]. The most common configuration in order to excite WGMs with direct scattering can be seen in Figure 1, where $\mathbf{d} = d\hat{\mathbf{y}}$. In order to make the calculations simpler and the simulations faster, we particularize $\mathbf{d} = d\hat{\mathbf{y}}$ and express Eq.(1) in the TAM basis. For that, we do a rotation of $-\pi/2$ around the y axis. This rotation is done with the Wigner matrix, yielding the following final expression:

$$\mathbf{E}^{\text{i}}(d\hat{\mathbf{y}}) = \sum_{j=1}^{\infty}\sum_{m_x=-j}^{j} C^{\text{off},d\hat{\mathbf{y}}}_{j,m_x,p}\left(\mathbf{A}^{(e)}_{jm_x} + p\mathbf{A}^{(m)}_{jm_x}\right) \quad (2)$$

$$C^{\text{off},d\hat{\mathbf{y}}}_{j,m_x,p} = \sum_{n=-\min(j,j')}^{\min(j,j')}\sum_{L=0}^{\infty}(2L+1)(-i)^L j_L(kd) \quad (3)$$
$$\langle j,n;L,0|j',n\rangle\langle j,p;L,0|j',p\rangle$$
$$\sum_{j'=m_z^*}^{\infty}\sum_{m_z=-j}^{j} \mathrm{d}^j(-\pi/2)^{m_x}_{m_z} D^j(\hat{\mathbf{d}})^{m_z}_n D^{j'}(\hat{\mathbf{d}})^{m_z^*}_n C^{\text{on}}_{j'm_z^*p}$$

with $j_L(kd)$ being the spherical Bessel function; $\langle j_1, m_1; j_2, m_2|J, M\rangle$ a Clebsch-Gordan coefficient; and $D^j(\hat{\mathbf{r}})^m_n$ being the Wigner matrix. The analytical expression of $C^{\text{off},d\hat{\mathbf{y}}}_{j,m_x,p}$ given by Eq.(3) is the main theoretical result in [5]. It shows us how the multipolar decomposition of a focused Gaussian beam aligned with respect to the center of a sphere $C^{\text{on}}_{j'm_z^*p}$ is transformed by means of a translation along the y axis. The net result of such transformation is that of i) shifting the multipolar decomposition towards larger j's by an amount which is proportional to $2\pi d/\lambda$, and ii) turning the initial AM along the z axis into transverse AM along the x axis [8]. This can be observed in Figure 1, where different multipolar decompositions are plotted for different $d = 0, 4, 8, 16\ \mu$m. Note that the expression of $C^{\text{on}}_{j'm_z^*p}$ for a focused Gaussian beam with a well-defined helicity can be found in some previous works [7, 11–13].

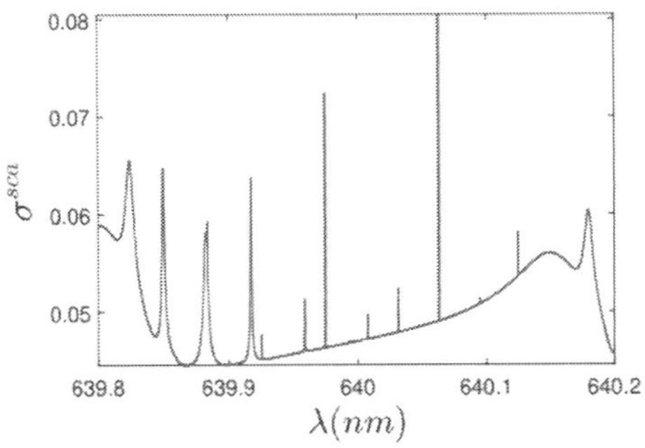

Figure 3. Scattering cross section σ^{sca} as a function of λ for a sphere with $R = 145\mu$m, and $n_r = 1.3989$. The sphere is illuminated with a displaced focused Gaussian beam with a well-defined helicity. The beam is displaced $d = 150\ \mu$m and the lens has an NA = 0.4.

3 Results

Thanks to the semi-analytical formulation given by Eqs.(2-3), we have been able to analytically compute the total EM field around a sphere with a radius $R = 145\mu$m, and a relative index of refraction $n_r = 1.3989$. The sphere is tangentially illuminated with a focused Gaussian beam with a well-defined helicity (or handedness). The focusing aplanatic lens has a numerical aperture NA = 0.4, and it focuses the beam at a distance $d = 150\ \mu$m of the center of sphere. In order to mimic a scattering experiment, first we compute the scattering cross section as a function of the wavelength, for a wavelength scan of $\Delta\lambda = 0.4$ nm centred at $\lambda = 640$ nm. In Figure 3, we observe that a very narrow λ scan of $\Delta\lambda = 0.4$ nm is enough to find many resonances. Here, we are interested in studying the coupling efficiency (CE) as well as the mode purity (MP) of one of these resonances. We will study the resonance that maximizes σ^{sca} in this range, which happens at $\lambda = 640.063295$ nm. Our choice is motivated by the experimental measurements, where a greater scattering signal will make the measurement of CE easier. Thanks to our semi-analytical code, we can find the fundamental mode which is causing the sphere to resonate at this wavelength: in this case, it is the magnetic Mie coefficient b_{1456}. However, not all the scattered power is given by this mode. Some other modes also give some partial contributions to the scattering. We can quantify the purity of a resonant mode by computing its contribution to the total scattering. Specifically, we compute the MP as the scattered power given by a targeted single mode (the WGM given by b_{1456} in this case), divided by the total scattered power. Analytically, it

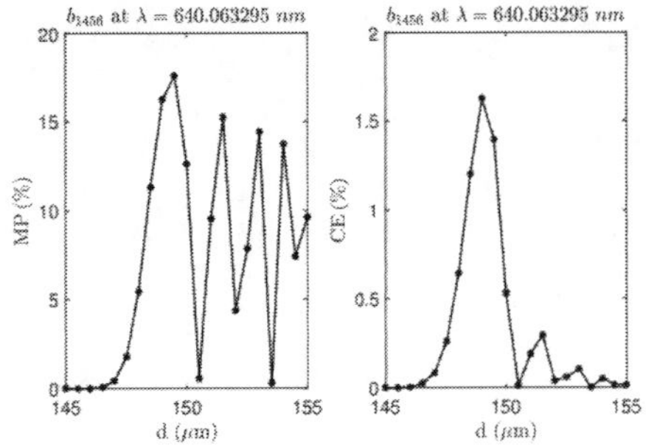

Figure 4. Mode Purity (MP) and Coupling Efficiency (CE) as a function of the transverse position d for a WGM defined by the magnetic Mie coefficient b_{1456}. The spherical resonator has $R = 145\mu$m and $n_r = 1.3989$. The mode is excited at $\lambda = 640.063295$ nm.

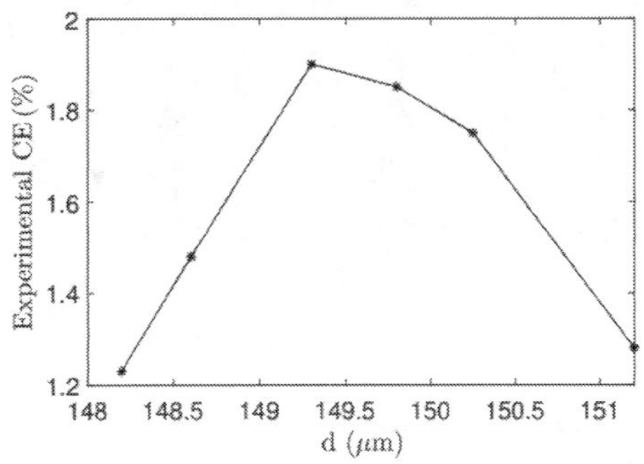

Figure 5. Experimental CE for an equivalent b_{1456} mode as a function of the transverse distance d. The pendant droplet has a radius of $R = 156.5$ μm, and it is made of silicone oil with $n_r = 1.3989$. The laser is works at 640 nm.

can be expressed as:

$$\mathrm{MP}^{(e)}_{j^*}(d) = \frac{|a_{j^*} C^{\mathrm{off},d\hat{y}}_{j^*,m_x=j^*,p}|^2}{\sum_{j=1}^{\infty}\sum_{m_x=-j}^{j}|C^{\mathrm{off},d\hat{y}}_{j,m_x,p}|^2\left(|a_j|^2+|b_j|^2\right)} \quad (4)$$

$$\mathrm{MP}^{(m)}_{j^*}(d) = \frac{|b_{j^*} C^{\mathrm{off},d\hat{y}}_{j^*,m_x=j^*,p}|^2}{\sum_{j=1}^{\infty}\sum_{m_x=-j}^{j}|C^{\mathrm{off},d\hat{y}}_{j,m_x,p}|^2\left(|a_j|^2+|b_j|^2\right)} \quad (5)$$

where Eq.(4) quantifies the MP of electric modes, and Eq.(5) does the same for magnetic modes. We have used Eq.(5) to compute the MP of the magnetic mode with $j^* = 1456$ resonating at $\lambda = 640.063295$ nm, as a function of the displacement d. The theoretical results are shown in Fig.4. We observe that despite measuring the MP at the resonant wavelength of the b_{1456} mode, the values can oscillate in between $0 - 20$ %. It is clear that the transverse position of the beam d places a crucial role regarding the excitation of a pure mode. In contrast, the CE is computed as the total scattered power given by a targeted single mode, divided over the total incident power. Analytically, we can express it as:

$$\mathrm{CE}^{(e)}_{j^*}(d) = |a_{j^*} C^{\mathrm{off},d\hat{y}}_{j^*,m_x=j^*,p}|^2 \quad (6)$$

$$\mathrm{CE}^{(m)}_{j^*}(d) = |b_{j^*} C^{\mathrm{off},d\hat{y}}_{j^*,m_x=j^*,p}|^2 \quad (7)$$

where the function $C^{\mathrm{off},d\hat{y}}_{j,m_x,p}$ needs to be such that the incident power is normalized to 1:

$$\sum_{j=1}^{\infty}\sum_{m_x=-j}^{j}|C^{\mathrm{off},d\hat{y}}_{j,m_x,p}|^2 = 1 \quad (8)$$

The theoretical results of the CE computed using Eq.(7) can also be seen in Figure 4. We observe that the CE is much more sensitive to the transverse distance d than the MP. It can also be observed that both MP and CE peak at different d values. Also, note that the peak of both MP and CE happen for a value of d which is larger than the radius of the particle. That is, it is clear that the excitation of WGM needs to be done a position d such that most of the beam does not directly hit the surface of the sphere. Taking all these theoretical aspects into account, we have experimentally verified that the CE retrieved from our theoretical model matches the experimental conditions. In order to do that, we have measured the backscattering of a spherical pendant droplet made of silicone oil with $n_r = 1.3989$. The radius of the pendant droplet has been measured to be $R = 156.5$ μm, and the wavelength of the laser is centred at $\lambda = 640$ nm. The laser beam is focused with a lens with an NA = 0.4. The wavelength scans are done by varying the temperature of the diode laser. We have studied a resonance with a Q factor of the same order of magnitude as the one given by the Mie mode b_{1456}. We experimentally determine the CE as the ratio between the resonantly-coupled optical power and incident optical power. In Figure 5, we plot the measured CE for the pendant droplet. It can be seen that the matching between the experimental value and the simulated one is outstanding. Not only the absolute values are very similar, but also the functional shape of CE as a function of d is almost equal. Besides, we observe that both the experimental and theoretical CE(%) is of the order of 2 %. Note that this value is clearly within the boundaries of a theoretical threshold calculated in a previous work [14].

4 Conclusion

We have developed an analytical theoretical method that quantitatively predicts the wavelength, Q-factor, coupling efficiency and mode purity of WGMs excited with tangentially free propagating beam. Thanks to this model, we have demonstrated that the efficient excitation of WGMs with free propagating beams needs to be performed at transverse distances d which are greater than R. We have computed the MP of a resonant b_{1456} WGM, and have seen that its value is below 20 %. We have also observed that the transverse distance d has a huge impact in the MP, varying it from 0 to nearly 20 % for the same resonant wavelength. Finally, we have also compared the experimental and theoretical CE for a 300 μm resonator, yielding very similar results, of the order of 2 %. We foresee that our method can be applied both to understand current experimental undertakings and to predict new effects.

References

[1] A. Chiasera, Y. Dumeige, P. Feron, M. Ferrari, Y. Jestin, G. Nunzi Conti, S. Pelli, S. Soria, G.C. Righini, Laser & Photonics Reviews **4**, 457 (2010)

[2] L. He, Ş.K. Özdemir, L. Yang, Laser & Photonics Reviews **7**, 60 (2013)

[3] A. Giorgini, S. Avino, P. Malara, P. De Natale, G. Gagliardi, Sensors **19**, 473 (2019)

[4] D. D'Ambrosio, M. Capezzuto, S. Avino, P. Malara, A. Giorgini, P. De Natale, G. Gagliardi, Optics Letters **46**, 3111 (2021)

[5] X. Zambrana-Puyalto, D. D'Ambrosio, G. Gagliardi, Laser & Photonics Reviews **15**, 2000528 (2021)

[6] G. Gouesbet, G. Gréhan, *Generalized Lorenz-Mie Theories* (Springer, 2011)

[7] X. Zambrana-Puyalto, X. Vidal, G. Molina-Terriza, Opt. Express **20**, 24536 (2012)

[8] I. Fernandez-Corbaton, X. Zambrana-Puyalto, N. Bonod, C. Rockstuhl, Physical Review A **94**, 053822 (2016)

[9] W.K. Tung, *Group Theory in Physics* (World Scientific, Singapore, 1985)

[10] D.W. Mackowski, Journal of Quantitative Spectroscopy and Radiative Transfer **109**, 770 (2008)

[11] X. Zambrana-Puyalto, G. Molina-Terriza, J. Quant. Spectrosc. Radiat. Transf. **126**, 50 (2013)

[12] X. Zambrana-Puyalto, X. Vidal, M.L. Juan, G. Molina-Terriza, Opt. Express **21**, 17520 (2013)

[13] X. Zambrana-Puyalto, X. Vidal, P. Woźniak, P. Banzer, G. Molina-Terriza, ACS Photonics **5**, 2936 (2018)

[14] C.L. Zou, F.J. Shu, F.W. Sun, Z.J. Gong, Z.F. Han, G.C. Guo, Optics express **21**, 9982 (2013)

 https://doi.org/10.1051/epjconf/202125504003

Hybrid Confinement of Visible Light in a Nanophotonic Resonator

Tommaso Perani[1,*] and *Marco* Liscidini[1,2]

[1]Dipartimento di Fisica, Università di Pavia, Via Agostino Bassi 6, 27100 Pavia (PV), Italy
[2]Istituto di Fotonica e Nanotecnologie, Consiglio Nazionale delle Ricerche (IFN-CNR), Piazza Leonardo da Vinci 32, 20133 Milano (MI), Italy

Abstract. We report on the design of a novel nanoresonator operating at visible wavelengths, in which light confinement is achieved by a hybrid mechanism based on total internal reflection and photonic band gap. We show that this structure can support resonant nanophotonic modes with mode volumes on the order of one cubic wavelength, and Q factors exceeding several tens of thousands. Its properties make it ideal for controlling and enhancing the light-matter interaction at sub-wavelength scales.

1 Introduction

Nanoresonators are extremely powerful in increasing the light-matter interaction in dielectric systems, as they can provide high quality factors (Q) and small mode volumes (V). In this respect, structures based on photonic crystals (PhCs) are known to support resonant modes characterized by small V, even smaller than a cubic wavelength of light [1]. However, such structures are usually quite challenging to fabricate, because they require a sufficient refractive index contrast of the constituent materials and are typically realized in suspended membranes, which limits their mechanical stability. Finally, the large field enhancement achievable in these systems is often obtained inside the structure, reducing their utility in those applications that require the enhancement of the light-matter interaction near the surface.

Bloch surface waves (BSWs) propagating at the truncation interface of a periodic multilayer are a valuable approach for surface optics, with surface field enhancements larger than those achievable in dielectric slabs [2]. BSWs provide large freedom in terms of constituent materials, from organic compounds to semiconductors, and operate in a wide spectral range, remarkably at visible wavelengths, for which there is paucity of transparent materials having a strong refractive index contrast. The possibility of guiding BSWs by simply realizing a polymeric ridge waveguide makes them appealing for the development of an etchless, all-dielectric integrated photonic platform [3]. A full 3D confinement of BSWs can be readily achieved, e.g., by bending the ridge on itself to obtain high-Q ring resonators [4].

However, so far, the development of BSW-based ridges has been hampered by relatively high propagation loss (of the order of dB/mm or higher) that cannot be explained in terms of surface scattering or material absorption. Plus, 3D confinement via ring resonators or other whispering-gallery-mode solutions [5] usually comes at the price of relatively large V, due to the large bending radii allowed by the gentle in-plane confinement of light.

Here, we demonstrate a photonic crystal nanobeam cavity (PhCNC) in a PhC ridge supporting BSWs [6], thus combining the small V and footprint size of a PhC cavity with the surface field enhancement typical of BSWs. We also track down the origin of the disappointing performance of PhC ridge waveguides in terms of propagation loss, and suggest a general design principle to reduce it.

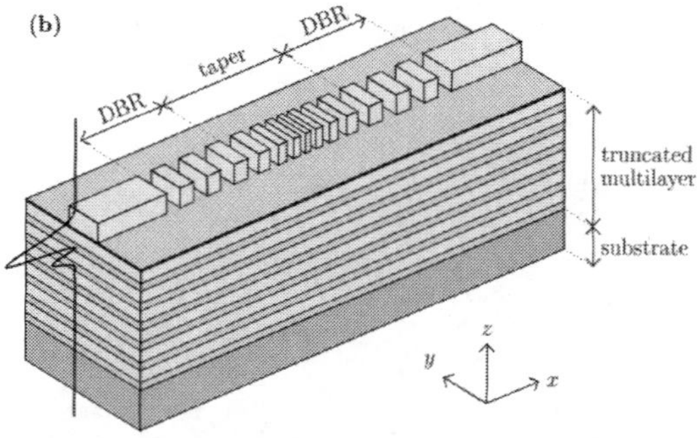

Figure 1. (a) Cross section of the PhC ridge. (b) Sketch of the BSW-based PhCNC.

*e-mail: tommaso.perani01@ateneopv.it

2 Results

The structure under investigation is a truncated $TiO_2(a)/SiO_2(b)$ periodic multilayer having a unit cell composed of two layers of thickness d_a = 100 nm and d_b = 182 nm, and refractive index n_a = 2.67 and n_b = 1.46, respectively. The multilayer has N = 15 periods and is truncated with a TiO_2 layer of thickness d_σ = 11 nm. The planar structure is loaded with a PMMA (n_{ridge} = 1.49) ridge of width w = 1.0 µm and thickness t = 0.4 µm in air (n_{clad} = 1), as illustrated in Fig. 1(a). We operate at a target wavelength λ_0 = 532 nm. Here, light confinement relies on the hybrid combination of total internal reflection (TIR) from the homogeneous cladding and reflection within a (polarization-dependent) photonic band gap (PBG) from the multilayer. The geometry above guarantees the existence of a TE-like guided BSW that propagates along the ridge with estimated propagation loss of about 5.4 dB/m.

The PhCNC is obtained by patterning two Bragg mirrors (DBRs) along the PMMA ridge, each including N_{Bragg} = 30 repetitions of a PMMA rectangular stack/air slit unit cell having lattice constant Λ_{Bragg} = 189 nm and PMMA filling fraction f = 0.476, as sketched in Fig. 1(b). This geometry greatly simplifies the fabrication procedure, which can be done by standard lithography. In addition, the system can be modeled within an effective index (EI) framework to limit the computational cost of its optimization, which usually requires to investigate a large parameter space.

Here, the EI method is applied twice. First, we determine the effective refractive index of the guided mode supported by the ridge (n_{eff}) by modeling its cross section as an effective slab waveguide, where the cladding and core refractive indices are those of the modes supported by the 1D bare (n_{bare}) and PMMA-loaded (n_{load}) multilayers, respectively. Second, the in-plane cavity is treated as an effective 1D multilayer of alternating layers of indices n_{eff} and n_{bare} for each PMMA stack and air slit, respectively. All these tasks are easily solved, e.g., via the transfer matrix method, which can take only minutes, if not seconds, on a standard personal computer.

In our design, the Bragg mirrors are adiabatically tapered by quadratically resizing a number N_{taper} = 30 of unit cells in both directions away from the center of the cavity [7], with the lattice constant Λ_i of the i-th cell of the taper given by $\Lambda_i = \Lambda_0 + (\Lambda_{\mathrm{Bragg}} - \Lambda_0)(i/N_{\mathrm{taper}})^2$, where Λ_0 = 181 nm is the minimum lattice constant.

Tapering the Bragg mirrors is crucial to mitigate the effect of diffraction loss [8], which is expected to be the leading loss mechanism when one neglects scattering loss from fabrication imperfections and finite size effects of the multilayer. This can be understood by looking at the mode field distribution in Fourier space. The field Fourier components that fall inside the air light cone or outside the multilayer PBG are not guided and can couple to the continuum of radiation modes. A smooth modification of the nanobeam geometry gives rise to a gentle variation of the electric field, and therefore, to a narrower distribution of its Fourier components, which can be piloted away from the leaky region. We notice that, as compared to free-standing or on-insulator nanobeams, here the leaky region extends not only to the air light cone, but also to the photonic bands associated with the underlying periodic multilayer.

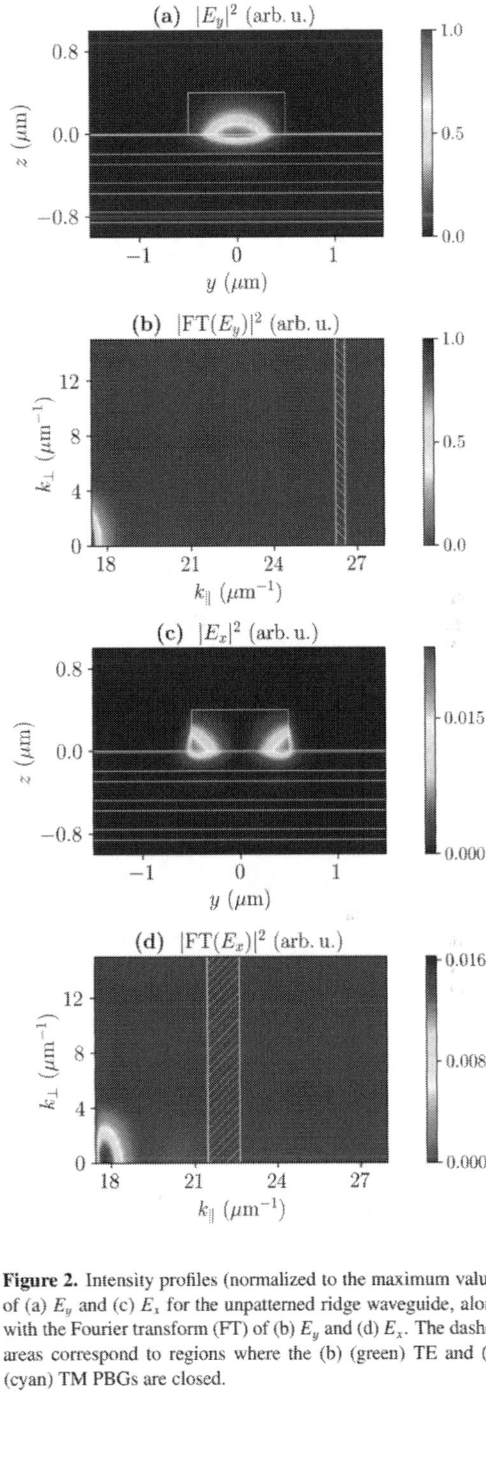

Figure 2. Intensity profiles (normalized to the maximum value) of (a) E_y and (c) E_x for the unpatterned ridge waveguide, along with the Fourier transform (FT) of (b) E_y and (d) E_x. The dashed areas correspond to regions where the (b) (green) TE and (d) (cyan) TM PBGs are closed.

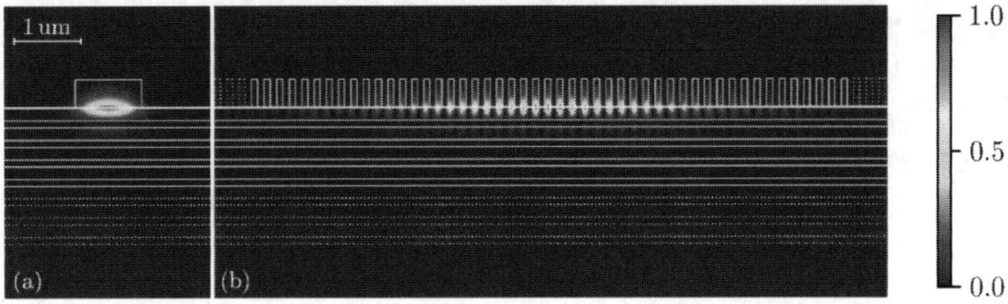

Figure 3. Intensity profile (normalized to the maximum value) of the electric field of the fundamental mode as obtained from FDTD simulations. (a) PhC-ridge cross section. (b) Nanobeam cross section.

In this respect, light polarization plays an essential role. We remark that, due to the confinement in the lateral direction, the mode supported by the PhC ridge is not purely TE polarized, but TE-like polarized, in that it exhibits all six nonvanishing field components, with E_y being the dominant one and E_x, E_z being weak but nonzero. Hence, the simultaneous presence of both TE and TM PBGs is always required to have each field component's Fourier distribution well-located within the appropriate PBG. A careful choice of the PhC ridge geometry can ensure wide and well-overlapping TE and TM PBGs in the spectral region of interest.

An analogous polarization-wise strategy also allows one to obtain low-loss guided modes in the unpatterned ridge without resorting to ultra-wide ridges, thus ensuring small mode areas in a single-mode regime [9]. Low-loss waveguide modes are particularly desirable for the realization of more complex devices, e.g., PhCNC arrays. In Figs. 2(a) and 2(c) we show the intensity profiles for the electric field components E_y (TE component) and E_x (TM component), respectively, of the TE-like mode guided by the unpatterned ridge (E_z is not shown because negligible). Figs. 2(b) and 2(d) show the corresponding Fourier transforms (FTs) as a function of the in-plane wavevector $k_\parallel = \sqrt{k_x^2 + k_y^2}$ and the out-of-plane wavevector $k_\perp = k_z$. The dashed areas denote the leaky regions in which either TE or TM PBG is closed. We observe that both E_y and E_x FTs are distributed almost entirely within the respective guiding regions, which results in reasonably low propagation loss $\alpha = 5.4\,\mathrm{dB/m}$. One can show that even a slight variation of the multilayer composition (e.g., changing its filling fraction or its period or both) can determine a considerable modification in the spectral position and extension of the TE/TM PBGs, which can result in a dramatic increase in propagation loss.

In Fig. 3 we show the results of a full 3D FDTD simulation for the PhCNC with the geometry parameters reported above, which are obtained from an EI-based optimization. 3D confinement of the BSW mode is associated with a Q factor of about 5×10^4 and $V \sim \lambda_0^3$. This demonstrates the existence of the confined mode and also assesses the efficacy of our EI framework, which proves to be robust to some variation of the model parameter that accounts for the variability of the actual fabrication process.

3 Conclusion

We demonstrated a new PhC resonator based on BSWs. This structure, characterized by a high Q/V and a field enhancement near the structure surface, has a small footprint size, is mechanically stable, is flexible in terms of constituent materials, and can be fabricated with etching-free technologies, e.g., from low-index polymeric ridges on commercially available multilayers. It can be implemented in a wide spectral range, from infrared to visible wavelengths.

All these properties, which are not easily obtained at once in other conventional photonic cavities, make this platform appealing for a number of applications that require large enhancement of surface light-matter interaction, from optical sensing to quantum nanophotonics.

References

[1] K. J. Vahala, *Nature* 424, 839 (2003).
[2] D. Aurelio and M. Liscidini, *Phys. Rev. B* 96, 045308 (2017).
[3] E. Descrovi et al., *Nano Lett.* 10, 2087 (2010).
[4] G. A. Rodriguez, D. Aurelio, M. Liscidini, and S. M. Weiss, *Appl. Phys. Lett.* 115, 011101 (2019).
[5] R. Dubey et al., *J. Eur. Opt. Soc.-Rapid Publ.* 13, 5 (2017).
[6] T. Perani, D. Aurelio, and M. Liscidini, *Opt. Lett.* 44, 5133 (2019).
[7] Q. Quan and M. Loncar, *Opt. Express* 19, 18529 (2011).
[8] Y. Akahane, T. Asano, B.-S. Song, and S. Noda, *Nature* 425, 944 (2003).
[9] T. Perani and M. Liscidini, *Opt. Lett.* 45, 6534 (2020).

Numerical analysis of plasmonic metasurfaces for fluorescence enhancement

Roxana Tomescu, *Florin* Nita, *Stefan* Caramizoiu, *Veronica* Anastasoaie, *Dana* Cristea

National Institute for Research and Development in Microtechnology – IMT Bucharest, 126A, Erou Iancu Nicolae Street, 077190, Voluntari, Ilfov, ROMANIA

Abstract. The paper presents an extensive numerical analysis performed by three-dimensional (3D) simulations using the finite difference in time and space (FDTD) method to identify the optimal geometry, size and configuration of the nano-antennas that constitute a plasmonic metasurface. The aim was to achieve the highest resonance at various wavelengths (NIR-VIS), for local enhancement of the excitation field and collection efficiency of emitted photons. We investigated ten different types of metals, two shapes (disks and U-shape resonators) and various geometrical parameters for the nano-resonators composing the metasurface. The best results for Rhodamine 6G excitation and emission were obtained using silver resonators with 105 nm diameter of the cylinder elements in a rectangular array with a 110 nm period, and with 110 nm long U-shape placed at a period of 40 nm.

1. Introduction

Metasurfaces are arrays of sub-wavelength plasmonic structures that can concentrate the electromagnetic fields in small volumes, well below the diffraction limit, through the excitation of localized surface plasmons [1]. In the last years it was demonstrated that the metasurfaces have the potential to control both the light propagation and emission at the nanoscale, showing thus a great potential for tailoring the excitation and emission processes of fluorescent molecules for quantum emitters and biosensing application [2-4].

The paper presents an extensive numerical analysis performed by three-dimensional (3D) simulations using the finite difference in time and space (FDTD) method to identify the optimal geometry, size and configuration of the nano-structures that constitute a plasmonic metasurface, focusing on achieving the highest resonance at various wavelengths (NIR-VIS), for local enhancement of the excitation field and collection efficiency of emitted photons. The final goal is to attain a considerable enhancement of fluorescent emission of a variety of fluorophores from the visible (VIS) to the near infrared (NIR) spectral domains, based on metasurfaces and explore its applicability in biosensing using metasurface platforms.

2. Model

To achieve the purpose of this investigation, we analysed two different geometries for the nanostructures that compose a plasmonic metasurface.
The first proposed structure is composed of metallic nano-disks in a rectangular array patterned on a fused-silica substrate. The second analysed geometry is an array of U-shaped structures. We investigated the influence of the nano-resonator dimensions and material on the local enhancement of the excitation field and collection efficiency of emitted photons.

The geometrical parameters analysis and material studies of the nano-structures were performed applying 3D - FDTD method using the OptiFDTD commercial software developed by Opti Wave Canada. We employed periodic boundary conditions (PBC) on the planes parallel to the propagation direction of radiation. The incident radiation is a Gaussian pulse centred at 532 nm wavelength (excitation wavelength for Rhodamine 6G) and 552 nm (emission wavelength) respectively. The refractive index of the metals is modelled by a standard Lorentz-Drude formula. More details about the conditions we use to model and simulate a plasmonic metasurface structure are given in [5].

The elementary cell of the first structure under investigation, together with its geometric characteristics is presented in Figure 1 where L_x^0 and L_y^0 are the transverse dimensions, H_s is the thickness of the substrate, h and ϕ are the height and the diameter of the nano-cylinder, respectively.

We studied the influence of ten different metals and varied the diameters of the resonators from 50 nm up to 105 nm with a 5 nm step. The metal disks are patterned in a rectangular lattice with a constant of 110 nm. The height of the nanostructures is set at 100 nm.

* Corresponding author: stefan.caramizoiu@imt.ro

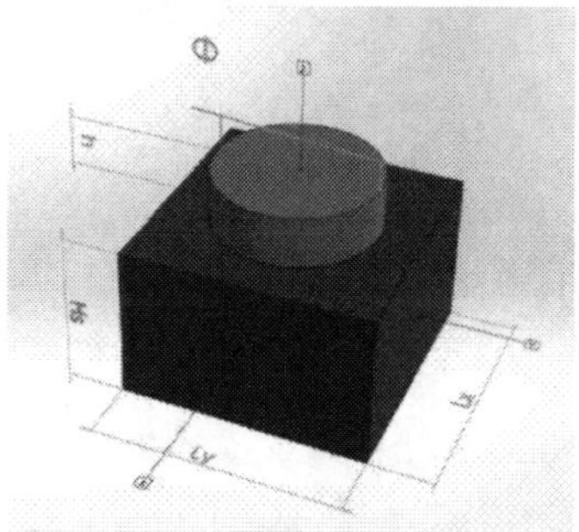

Fig 1. The elementary cell through which can be obtained a metasurfaces of variable dimensions by translation in the two directions *x* and *y*. The propagation direction of the electromagnetic radiation is the *z* direction.

In our second numerical analysis we propose a metasurface composed of an array of U-shaped nano-structures. The materials and simulation conditions employed here are the same as in the first case. The elementary cell of this structure is presented in Figure 2.

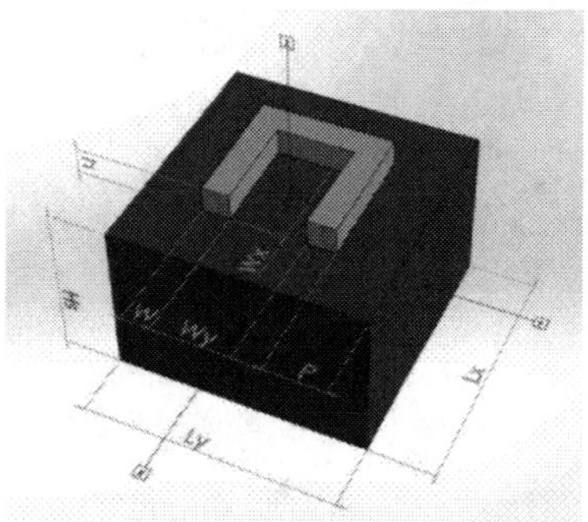

Fig 2. The elementary cell through which can be obtained a metasurfaces with U-shape nanostructures of variable dimensions by translation in the two directions *x* and *y*. The propagation direction of the electromagnetic radiation is the *z* direction.

For this type of structures, the local enhancement of the excitation field was studied by varying the dimensions of the structures, or the array period. The geometries employed in three promising sets of 3D simulations are presented in table 1 – where Ly, Lx and Hs are the geometrical parameters of the substrate, H is the height of the metallic structure, P is the distance between the resonator and the boundary of the simulation domain, W (thickness), Wx (arm length) and Wy (base length) are the parameter of the U-shape structure illustrated in Figure 2.

Table 1. Geometrical parameter used in 3 sets of 3D numerical investigations

No. Set	Ly [nm]	Lx [nm]	Hs [nm]	H [nm]	P [nm]	W [nm]	Wx [nm]	Wy [nm]
1	200	200	600	20	50	20	80	60
2	200	200	600	20	20	20	140	120
3	200	200	600	20	20	10	110	90

3. Results and discussions

To identify the best possible result, we analysed the electromagnetic radiation intensity on three different planes: a) interface air/resonators; b) middle of resonator; c) interface resonators/substrate. Examples of results obtained for the excitation wavelength of 532 nm are given in Figures 3-7.

In Figure 3 one can observe the electromagnetic field intensity function of ϕ, for the nano-disks metasurface, for all investigated metals. The increase of the field intensity for silver is over 6 times and about 5 times for aluminium at the interface air/resonator for a metasurface with 105 nm diameter disks.

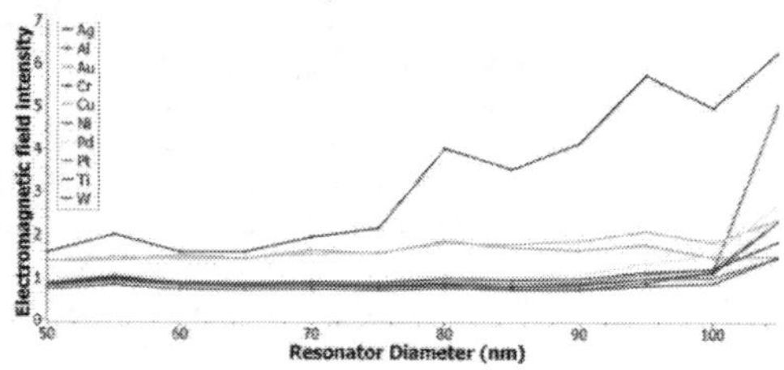

a)

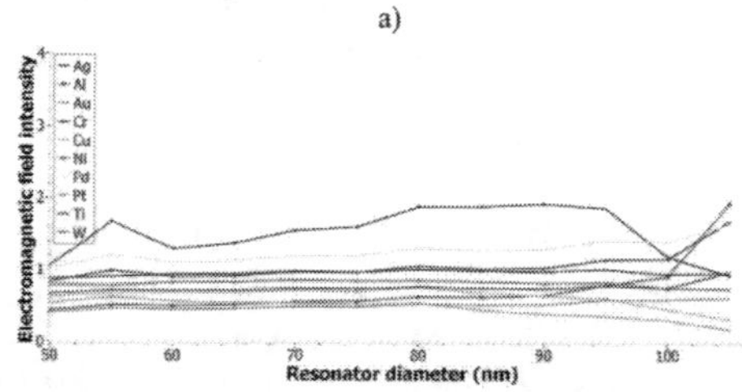

b)

Fig. 3. Electromagnetic field intensity: a) interface medium/resonators; b) interface resonators/substrate.

From the results presented in Figure 3 one can conclude that the appropriate metal for localized fluorescent enhancement is silver which offers optimum increase values for the field intensity for the desired application. In Figure 4 is presented the intensity obtain by modelling a silver metasurface on the three analysed planes.

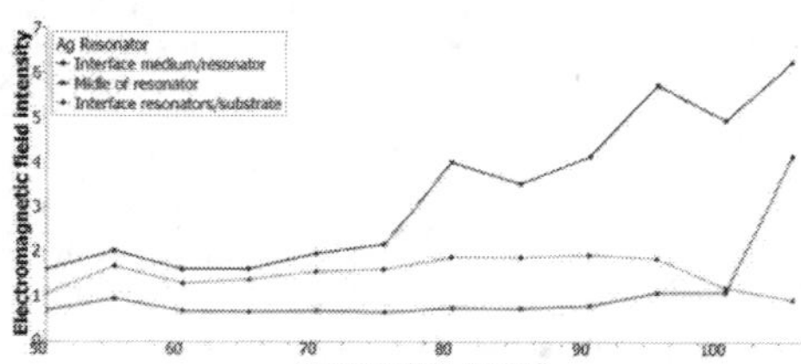

Fig. 4. Electromagnetic field intensity for silver nano-disks metasurface.

Figure 5 shows the electromagnetic field at the two planes for metasurfaces with a 2D squared array of silver nano-disk resonators 105 nm-diameter and 100 nm- high.

https://doi.org/10.1051/epjconf/202125504004

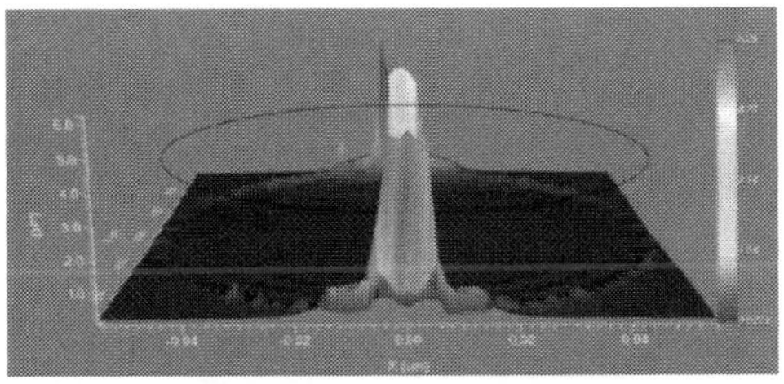

a)

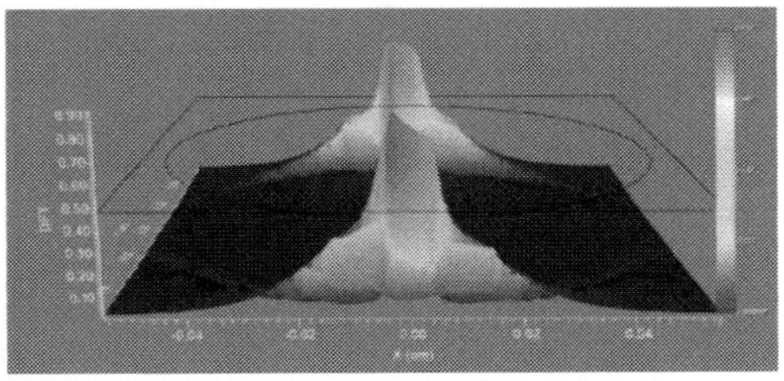

b)

Fig. 5. Electromagnetic field intensity for silver nano-disks metasurface: a) interface medium/resonators; b) on the substrate around the resonator – 105 nm-diameter and 100 nm-high.

One can see from Figures 4 and 5 an important field enhancement both at the interface medium/resonator and on the substrate around the resonator. These results offer the possibility of employing the designed structures in the development of fluorescent based bio-sensing devices.

Following the steps utilised in the first part of our investigation, we performed 3D simulations of a metasurface structure with U-shape nano-elements as meta-atoms.

The results obtain during this study are presented in figures 6-7. In Figure 6 are given the best electromagnetic field intensities obtain during the investigation performed on the ten selected metals. The graphs represent the results attained performing the third set of simulation for silver and aluminium. Once again, the best choices concerning the materials for the meta-atoms are silver (Figure 6a) and aluminium (Figure 6b). The electromagnetic field localized enchantment goes to values as high as 4.47 on the substrate around a silver U-shape resonator, and around 4 at the medium/resonators for aluminium meta-atoms.

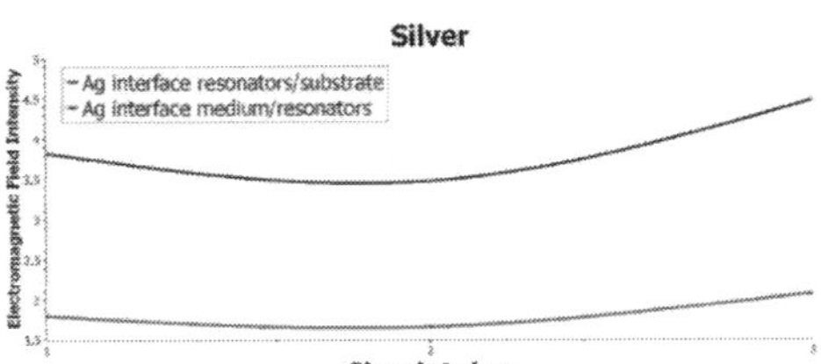

a)

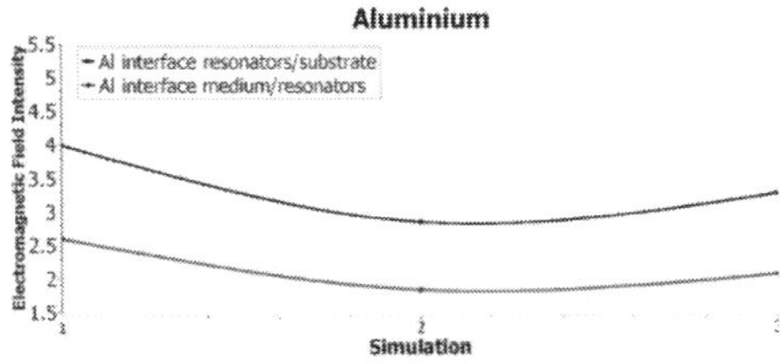

b)

Fig. 6. Electromagnetic field intensity on the two observation planes for U-shape resonators: a) silver U; b) aluminium.

Figure 7 presents the electromagnetic field at the two planes (at the interface medium/resonators and on the substrate around the resonator respectively) for metasurfaces with a 2D squared array of silver U-shape elements corresponding to simulation set 3 from table 1.

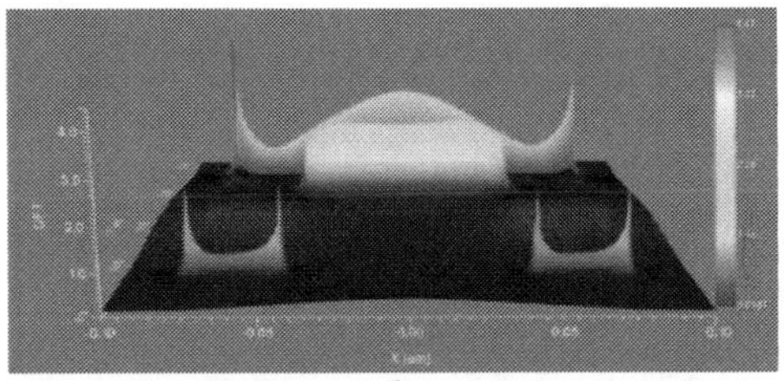

a)

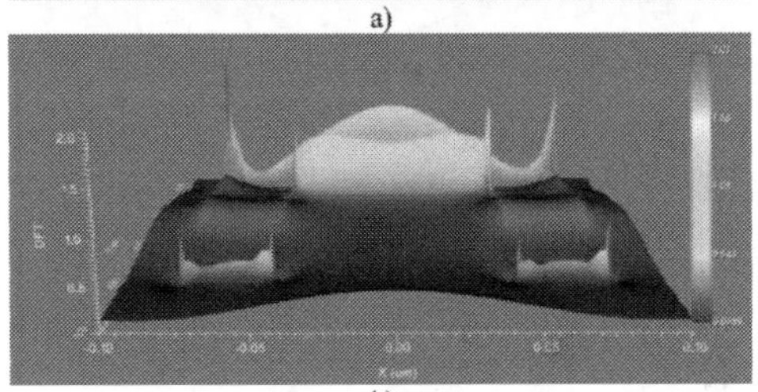

b)

Fig. 7. Electromagnetic field intensity for silver U-shape elements metasurface: a) interface medium/resonators; b) on the substrate around the resonator – W = 10 nm, Wx = 110 nm; Wy = 90 nm.

Similar results were obtained for the emission wavelength of Rhodamine 6G - 552 nm. Here, the increase is around 4.48 times, meanwhile the increment at the medium/resonators interface is 2.25 times. Figure 8 present the simulation results at the medium/resonators interface.

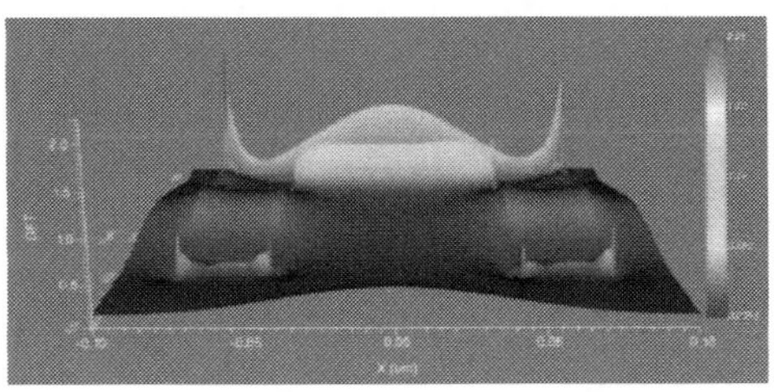

Fig. 8. Electromagnetic field intensity on the substrate around the resonator for silver U-shape elements metasurface with – W = 10 nm, Wx = 110 nm; Wy = 90 nm parameters; at the Rhodamine 6G emission wavelength.

4. Conclusions

In this paper we present an ample numerical analysis method to identify the optimum geometry, size and the best metal of the nano-antennas that compose a plasmonic metasurface, in order to obtain localized fluorescence enhancement. We investigated ten different types of metals, two shapes and various geometrical parameters for the nano-resonators composing the metasurface.

The best result for Rhodamine 6G excitation and emission, when the meta-atoms of the metasurface are nano-disks with 105 nm diameter in a rectangular array with a 110 nm period, are obtained using silver nano-resonators. The disadvantage of this geometry is the high aspect-ratio of the structures, thus making it almost impossible to fabricate.

In the case of U-shape nano-resonators, the best results are obtained with 90 nm base length, 110 nm arm length configured in a rectangular array with 20 nm period. This type of structure can be easily fabricated using electron beam lithography (EBL) and lift-off technique.

The promising local enhancement results and the available technology process make this type of structure suitable for biosensing application with possibility of binding the molecule labelled with fluorophores both on top of the nano-resonators and between them.

1. J. Hu, S. Bandyopadhyay, Y. Liu, L. Shao, Front. Phys. **8**, 586087 (2021)
2. A. Vaskina, R. Kolkowski, A. Femius Koenderink, I. Staude, Nanophotonics **8** (7), 1151–1198 (2019)
3. D. Bouchet et.al, Phys. Rev. Applied **6**, 064016 (2016)
4. R. H Siddique, S. Kumar, V. Narasimhan, H. Kwon, H. Choo, ACS Nano **13**, 13775−13783 (2019)
5. R. Tomescu, C. Kusko, D. Cristea, R. Calinoiu, C. Parvulescu, SSELJ **2**, 146-150, (2020)

Acknowledgment: This work was supported by a grant of the Romanian Ministry of Research, Innovation and Digitalization, CCCDI - UEFISCDI, project number PN-III-P2-2.1-PED-2019-1300, within PNCDI III

Enhanced photorefractivity and rare-earth photoluminescence in SnO_2 nanocrystals-based photonic glass-ceramics

Thi Ngoc Lam Tran[1,2,3,*], *Anna Szczurek*[1], *Stefano Varas*[1], *Cristina Armellini*[1], *Alessandro Carpentiero*[1], *Andrea Chiappini*[1], *Erica Iacob*[4], *Gloria Ischia*[5], *Simone Berneschi*[6], *Gualtiero Nunzi Conti*[6], *Monica Bollani*[7], *Francesco Scotognella*[2,8], *Giancarlo C. Righini*[6], *Paweł Głuchowski*[9], *Alessandro Chiasera*[1], *Anna Lukowiak*[9], *Maurizio Ferrari*[1]

[1]IFN-CNR CSMFO Lab. and FBK Photonics Unit, Via alla Cascata 56/C, 38123 Povo, Italy;
[2]Department of Physics, Politecnico di Milano, Piazza Leonardo da Vinci 32, 20133 Milano, Italy;
[3]Department of Materials Technology, Faculty of Applied Science, Ho Chi Minh City University of Technology and Education, Vo Van Ngan Street 1, Thu Duc District, 720214 Ho Chi Minh City, Vietnam;
[4]Fondazione Bruno Kessler, Sensors and Devices, Micro Nano Facility, Via Sommarive 18, Povo, Trento, 38123, Italy;
[5]Department of Industrial Engineering, University of Trento, Via Sommarive 9, Povo, Trento, 38123, Italy;
[6]MiPLab, IFAC-CNR, Via Madonna del Piano 10, 50019 Sesto Fiorentino, Italy;
[7]CNR-IFN, P.zza Leonardo da Vinci 32, Milano, 20133 Italy;
[8]Center for Nano Science and Technology@PoliMi, Istituto Italiano di Tecnologia (IIT), Via Giovanni Pascoli, 70/3, 20133, Milan, Italy;
[9]Institute of Low Temperature and Structure Research, PAS, ul. Okólna 2, 50422 Wroclaw, Poland;

Abstract. This work presents state of the art rare-earth activated SnO_2 nanocrystals - based transparent glass-ceramics. With combined enhancements in both photorefractivity and rare-earth photoluminescence, the glass-ceramic has unique benefits as a lasing material. It exhibits high photorefractivity with UV induced refractive index modifications in the order of 10^{-3}. Exploiting its high photorefractivity, optical gratings are fabricated on the glass-ceramic under an energy-efficient direct UV writing process. Furthermore, SnO_2 semiconductor nanocrystals are also employed as efficient rare-earth sensitizers enhancing drastically the rare-earth photoluminescence.

1 Introduction

Optical materials and photonic structures are strategic for many applications, e.g., sensing, integrated photonics, light sources, quantum technologies, and green energy systems. Such research activities pave the way for tackling current challenges, opening new scientific knowledge and technological advancements. To design and tailor various interesting functionalities, nanocomposites are among the more appealing materials. Transparent glass-ceramic are two-phases systems constituted by nanocrystals distributed in a glass matrix [1,2]. When the nanocrystals are activated with rare earth ions, photonic glass-ceramics exhibit unique properties such as [1,3](i) the high absorption and emission cross-sections, (ii) the reduction of the non-radiative relaxation processes thanks to the lower phonon cut-off energy, and (iii) the tailoring of the ion-ion interaction by the control of the rare-earth ion partition[3]. Furthermore, the desired properties can be tuned on the base of composition, material, and respective volume fractions of crystalline and amorphous phases.

In this work, Er^{3+}-activated SiO_2-SnO_2 glass-ceramics are discussed. This system uniquely merges two interesting properties: (i) high photorefractivity, and (ii) SnO_2 nanocrystals act as efficient rare-earth luminescent sensitizers. Therefore, this photorefractive glass-ceramic can settle the issue of rare-earths' low absorption cross-section. Furthermore, it can provide strategic fabrication approach for defining channel waveguides and optical gratings with minimal losses using direct UV writing technique[4-6].

2 Sol-gel derived SnO_2 nanocrystals-based photonic glass-ceramics

SnO_2-doped glasses and glass-ceramics can be fabricated by several techniques [7-11]. However, issues of SnO_2 volatilization during the high temperature preparation [9], or α-Sn nanoclusters and nonstoichiometric SnO_x nanoparticles [12,13] raised a challenge of finding a reliable synthesis approach. Sol-gel route showed profitability in synthesizing such glasses with higher SnO_2 content [3,10,12]. Thus, in this work, we present a versatile and reliable sol-gel synthesis route to obtain transparent glass-ceramics with high content of SnO_2. We fabricated SiO_2-SnO_2 transparent glass ceramics as monoliths with a SnO_2 content up to 12.5 mol% and as planar waveguides with a SnO_2 content up to 30 mol%.

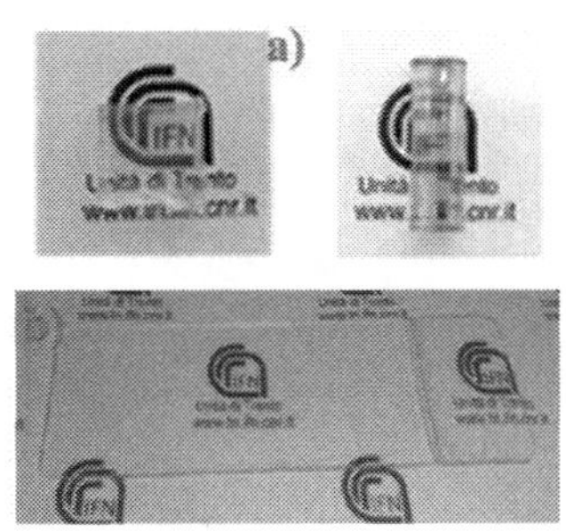

Fig. 1. Pictures of the densified SiO_2-SnO_2:Er^{3+} transparent glass-ceramics in forms of a) monoliths [15] and b) planar waveguides[7].

* Corresponding author: lam.tran@ifn.cnr.it

3 Enhanced photorefractivity

Attracted by potentials of SnO_2-doped glasses and glass-ceramics for applications in optical fibers and integrated photonics, numerous researches have investigated their photorefractivity. However, due to the low amount of SnO_2 content in the glass matrices (maximum up to 5 mol%), the reported UV-induced refractive index changes are limited to the order of 10^{-4}[16]. In the last years, we have demonstrated: a) that we can fabricate SiO_2-SnO_2 transparent glass ceramics with high content of SnO_2 nanocrystals (from 20 to 30 mol%), and b) that those achieve highly efficient photorefractivity[16-18] (see Fig.2), sufficient for applying direct UV writing technique to fabricate channel waveguides and optical gratings. In such glass-ceramics, refractive index modification, induced by UV irradiation, is $4x10^{-3}$ [5,16-18] (see Fig.2). The fabrication of optical gratings on SiO_2:SnO_2:Er^{3+} glass-ceramics using UV direct writing technique is also recently realized [5]. It is worthy to note that the energy employed to write a grating on SiO_2:SnO_2:Er^{3+} glass-ceramics is much lower compared to those necessary to write the gratings on hydrogen loaded germanoborosilicate glasses [5].

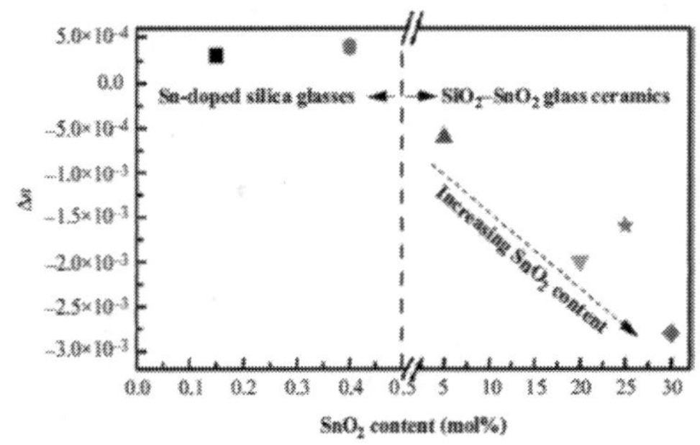

Fig. 2. The refractive index change under UV irradiation of the sol–gel derived tin-dioxide-based glass-ceramics and the tin-doped silica glasses as a function of SnO_2 content[18].

4 Enhanced rare-earth luminescence

The low absorption cross section of the electronic states of the rare earth ions is an important drawback for their application in photonics. In this contest, employing luminescence sensitizers is a robust approach. As reported in our works on different rare-earth-activated SiO_2-SnO_2 glass-ceramic systems, e.g., monoliths, thin films, and waveguides [3-5,7,15-16], we have proved the role of SnO_2 nanocrystals as efficient rare-earth sensitizers. As an example, Figure 3a shows the excitation spectrum recorded at 1530 nm emission of Er^{3+} of a SiO_2-SnO_2:Er^{3+} glass-ceramic monoliths containing 10 mol% of SnO_2 and activated by 0.5 mol% Er^{3+}. This spectrum demonstrates that when the glass-ceramic monolith isexcited at SnO_2 bandgap centered at 360 nm, Er^{3+} luminescence is drastically enhanced in comparison with directly exciting at Er^{3+} electronic states, e.g., $^2H_{11/2}$, $^4F_{7/2}$, or $^2F_{9/2}$. Furthermore, in Figure 3b, the 1.5 μm emission spectrum of Er^{3+} under SnO_2 bandgap excitation (λ_{ex} = 360 nm) reveals Stark splitting and narrow peaks. This again puts in evidence the location of Er^{3+} ions in SnO_2 crystalline local environment as well as the energy transfer from SnO_2 to the rare-earth ions[3,4,15].

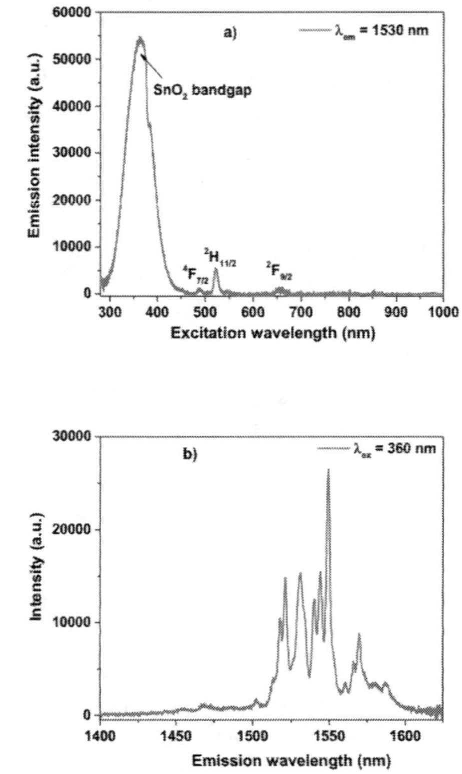

Fig. 3. a) Excitation spectrum monitored at emission λ_{em} =1530 nm of Er^{3+} and b) emission spectrum under excitation at UV SnO_2 bandgap (λ_{ex} = 360 nm) using Edinburgh FLS980 spectrofluorometer of the SiO_2-SnO_2:Er^{3+} glass-ceramic monolith containing 10 mol% SnO_2 and activated by 0.5 mol% Er^{3+}

Acknowledgements

The research activity is performed in the framework of the projects CNR-PAS "Flexible Photonics" (2020-2021), PRIN 2019 "NOMEN" (2020-2022), and ERC-H2020 PAIDEIA GA 816313.

References

1. M. Ferrari, G.C. Righini, Int J Appl Glass Sci **6**, 3, 240-248 (2015).

2. A. de Pablos-Martin, M. Ferrari, M. J. Pascual, G. C. Righini La Rivista del Nuovo Cimento **7-8**, 311-369 (2015).
3. L. Zur, L.T.N. Tran, M. Meneghetti, A. Lukowiak, A. Chiasera, D. Zonta, M. Ferrari, G.C. Righini, Opt. Mat. **63**, 95 (2017).
4. T.N.L. Tran, C. Armellini, S. Varas, A. Carpentiero, A. Chiappini, P. Głuchowski, E. Iacob, G. Ischia, F. Scotognella, M. Bollani, A. Lukowiak, G. C. Righini, M. Ferrari, A. Chiasera, Ceram. Int., **47**, 4, 5534-5541 (2021).
5. T.N.L. Tran, S. Berneschi, C. Trono, G.N. Conti, L. Zur, C. Armellini, S. Varas, A. Carpentiero, A. Chiappini, A. Chiasera, J. Gates, P.J. Sazio, M. Bollani, A. Lukowiak, G.C. Righini, M. Ferrari, Opt. Mat. X, **7**, 100056 (2020).
6. Q.S. Ahmed, P.C. Gow, C. Holmes, P.L. Mennea, J.W. Field, R. H. Bannerman, D.H. Smith, C. BE Gawith, P. GR Smith, J.C. Gates, Electron. Lett., **57**, 8, 331-333 (2021).
7. L. Zur, L.T.N. Tran, D. Massella, A. Vaccari, A. Chiappini, A. Chiasera, S. Varas, C. Armellini, A. Carpentiero, B. Boulard, D. Dorosz, Proc. SPIE, **10914**, 1091411 (2019).
8. L. Dong, J.L. Cruz, L. Reekie, M.G. Xu, D.N. Payne, IEEE Photon. Technol. Lett., **7**, 9 (1995).
9. G. Brambilla, V. Pruneri, L. Reekie, Applied Physics Letters **76**, 7 (2000).
10. B.S. Bhaktha, C. Kinowski, M. Bouazaoui, B. Capoen, O. Robbe-Cristini, F. Beclin, P. Roussel, M. Ferrari, S. Turrell, J. Phys. Chem. C, **113**, 52 (2009).
11. T.T. Van, S. Turrell, B. Capoen, M. Ferrari, D. Ristic, L. Boussekey, C. Kinowski, J. Mater. Sci., **49**, 24 (2014).
12. Y. Yu, D. Chen, P. Huang, H. Lin, A. Yang, Y. Wang, J. Solid State Chem., **184**, 2 (2011).
13. A.F. Zatsepin, E.A. Buntov, V.S. Kortov, V.A. Pustovarov, H.J. Fitting, B. Schmidt, N.V. Gavrilov, J. Surf Invest-X-RAY+, **6**, 4 (2012).
14. T. Van Tran, S. Turrell, M. Eddafi, B. Capoen, B., M. Bouazaoui, P. Roussel, S. Berneschi, G. Righini, M. Ferrari, S.N.B. Bhaktha, and O. Cristini, J. Mol. Struct., **976**, 1-3 (2010).
15. L.T.N. Tran, D. Massella, L. Zur, A. Chiasera, S. Varas, C. Armellini, G.C. Righini, A. Lukowiak, D. Zonta, M. Ferrari, M., Applied Sciences, **8**, 8 (2018).
16. L.T.N. Tran, C. Armellini, R. Balda, M. Benabdesselam, S. Berneschi, W. Blanc, B. Boulard, B., A. Carpentiero, A. Chiappini, A. Chiasera, P. Dentella, D. Dorosz, S. Eaton, M.C. Falconi, J. Fernandez, M. Ferrari, J. Gates, P. Gluchowski, G. Ischia, A. Lukowiak, F. Mady, D. Massella, G.N. Conti, F. Prudenzano, B. Rossi, R. Ramponi, G.C. Righini, P.J. Sazio, G. Speranza, S. Varas, D. Zonta, L. Zur, Proc. SPIE, **11276**, 1127614 (2020).
17. A. Lukowiak, L. Zur, T.N.L. Tran, M. Meneghetti, S. Berneschi, G. Nunzi Conti, S. Pelli, C. Trono, S.B.N. Bhaktha, D. Zonta, S. Taccheo, G.C. Righini, M. Ferrari, Cryst., **7**, 2 (2017).
18. T.N.L. Tran, S. Berneschi, G. Nunzi Conti, M. Ferrari, Maurizio, Integrated Optics Volume 1: Modeling, material platforms and fabrication techniques, Chap. **9** (2020).

https://doi.org/10.1051/epjconf/202125505002

Ionic liquids as tunable refractive index fluids

Carlos Damián Rodríguez-Fernández[1,*], *Elena* López Lago[1], *Christian* Schröder[2], and *Luis M.* Varela[1]

[1]Nafomat group, Departamento de Física Aplicada e Departamento de Física de Partículas, Universidade de Santiago de Compostela, E-15782 Santiago de Compostela, Spain.
[2]University of Vienna, Department of Computational Biological Chemistry, Währingerstr. 17, A-1090 Vienna, Austria.

Abstract. In this contribution we simulate the refractive index of several ion combinations yielding ionic liquids. The results show that their structural tunability can be exploited to successfully design liquids with task-specific refractive indices over a wide range of values, even higher than 2.0. Some designing clues are provided, being charge delocalization and the presence of fluorine atoms key factors to reach ion combinations with the highest refractive index possible.

1 Introduction

Ionic Liquids (ILs) are a wide family of liquid materials exclusively composed by ions which present a set of unique properties such as negligible vapour pressure, high electrochemical and thermal stability or great solvation power. However, the most important feature of ILs is that their composing ions can be independently tailored to suit specific applications. In this regard, ILs are starting to attract attention as photonic materials since their ions can be tailored to make them show fluorescence [1], liquid crystal phases [2] or thermochromism [3]. As the number of possible ion combinations is huge, computational simulations are an excellent tool to study the properties of these materials and identify the key mechanisms to tailor them. In this contribution we aim to simulate the refractive index of several combinations of IL-forming ions and to provide some designing keys to tailor the refractive index of ILs.

2 Computational Details and Procedure

2.1 DFT Calculations Details

Polarizability and volume of the ions were calculated by means of density functional theory (DFT) at the theory level B3LYP/6-311++G(d,p) using the Gaussian 16 rev. C.01 program [4]. The calculations were carried out independently for anions and cations, both in the gas phase. The optimized structures were verified by means of a vibrational analysis and the polarizability was calculated at a finite wavelength of $\lambda = 589$ nm by means of the Couple-Perturbed Hartree-Fock Method (CPHF) [5].

2.2 From ions to ionic liquids

The polarizability of ILs, α_{IL} was calculated as the sum of the contributions α_i of their constituent ions i: $\alpha_{IL} = \sum \alpha_i$. However, their molecular volume was calculated using an empirically obtained quadratic summation rule: $V_{IL} = a_1 \cdot \sum V_i + a_2 \cdot (\sum V_i)^2$ with $a_1 = 1.0189$ and $a_2 = 3.635 \times 10^{-4}$ Å^{-3}. The IL refractive index was calculated by means of the Lorentz-Lorenz equation:

$$\frac{n^2 - 1}{n^2 + 2} = \frac{4\pi}{3} \frac{\alpha_{IL}}{V_{IL}} \tag{1}$$

The adimensional ratio (in cgs units), α_{IL}/V_{IL}, of the IL is related to that of its constituent ions, $\alpha_{IL}/V_{IL} \leq \alpha_i/V_i$, which is an indicator of the individual influence of each ion in the refractive index of the IL.

3 Studied Ions

The considered anions are some of the most commonly used in ILs: tetrafluoroborate $[BF4]^-$, trisfluorotris(perfluoroethyl)phosphate $[FAP]^-$, trifluoromethanesulfonate $[OTf]^-$ and thiocyanate $[SCN]^-$. The first two anions contain fluorine and lack of charge delocalization while the last two present resonances. In the case of $[SCN]^-$ all the bonds participate in the resonance while it is not the case for $[OTf]^-$, which also contains fluorine atoms. Regarding the cations, three families of cationic heterocycles of different alkyl chain lengths were considered: 1-alkyl-1-methylpyrrolidinium $[C_k\text{mpyrr}]^+$, 1-alkyl-methylimidazolium, $[C_k\text{mim}]^+$, and 1-alkylpyridinium $[C_k\text{py}]^+$. These families of cations differ in the number of carbons in their rings, but also in their aromatic character. While in pyridinium the heterocycle is clearly aromatic, in the imidazolium the heterocycle presents a more limited conjugation and it is clearly aliphatic in pyrrolidinium. A fourth family of cations with a conjugated chain of variable length was taken into account to explicitly evaluate the influence of charge delocalization, 1-polyenyl-3- methylimidazolium, $[uC_k\text{mim}]^+$. The abbreviation and structure of all the considered ions are shown in Table 1.

*e-mail: damian.rodriguez@usc.es

https://doi.org/10.1051/epjconf/202125505002

Table 1. IL-forming ions studied in this work

Anions		Cations	
$[FAP]^-$	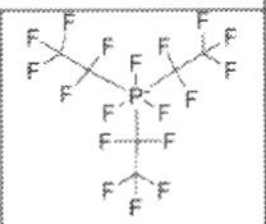	$[C_k mpyrr]^+$	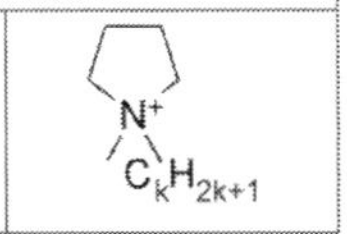
$[BF_4]^-$	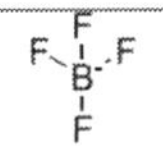	$[C_k mim]^+$	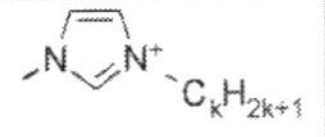
$[OTf]^-$	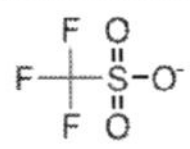	$[C_k py]^+$	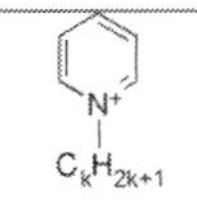
$[SCN]^-$	$^{-}S{\equiv}N$	$[uC_k mim]^+$	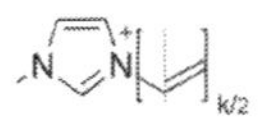

4 Results

4.1 Anionic Influence

The α_i/V_i ratio is very different for the different anions studied due to their large structural and compositional differences. Table 2 shows the ratios of selected anions.

Table 2. α_i/V_i ratio of selected anions.

Anion	α_i/V_i
$[BF_4]^-$	0.041
$[FAP]^-$	0.050
$[OTf]^-$	0.061
$[SCN]^-$	0.086

Inorganic anions without electronic charge delocalization and including fluorine atoms such as $[BF_4]^-$ or $[FAP]^-$ present the lowest ratios. The presence of fluorine atoms in a molecule is expected to decrease its overall polarizability since highly electronegative atoms produce tightly bonded electronic densities which are hardly polarizable. As a consequence, these anions are expected to provide low refractive index ILs. In contrast, those anions showing resonance due to charge delocalization throughout their π bonds such as $[OTf]^-$ or $[SCN]^-$ show higher α_i/V_i ratios. Among them, the $[SCN]^-$ anion, with all its bonds participating in the resonance, presents the highest one. In fact, the value of this ratio for $[OTf]^-$ is much more moderate than for $[SCN]^-$, since not all the bonds in the molecule participate in the resonance and there are fluorine atoms in its molecular structure. Hence, among the considered anions, the most promising candidate to yield high refractive index ILs is the $[SCN]^-$.

4.2 Cationic Influence

The α_i/V_i ratio of the considered cations for two different chain lengths, ($k = 2$) and ($k = 5$), is shown in Table 3. For a fixed chain length, $k = 5$, the α_i/V_i ratio increases with conjugation: $[C_k mpyrr]^+ < [C_k mim]^+ <$

Table 3. α_i/V_i ratio of considered cations.

Cation	α_i/V_i ($k = 2$)	α_i/V_i ($k = 5$)
$[C_k mpyrr]^+$	0.066	0.067
$[C_k mim]^+$	0.070	0.070
$[C_k py]^+$	0.074	0.073
$[uC_k mim]^+$	0.076	0.100

$[C_k py]^+ < [uC_k mim]^+$. Hence, higher refractive indices are expected for ILs with more extensive regions of conjugation. This is an unequivocal evidence of the large influence of charge delocalization in the polarizability of IL-forming ions. With respect to the side chain variation, its elongation produces different effects in heterocycles bearing alkyl chains or conjugated chains. When an alkyl chain is varied, it produces small changes in the α_i/V_i ratio, which could be used to finely tune the refractive index of derived ILs. However, when this variation is done in a polyenyl conjugated chain, the effect on polarizability is much more drastic than that observed in saturated alkyl chains. In order to shed more light on this, we show in Fig. 1 the polarizability as a function of the chain length for both $[C_k mim]^+$ and $[uC_k mim]^+$ families, which only differ in the conjugated character of their chains.

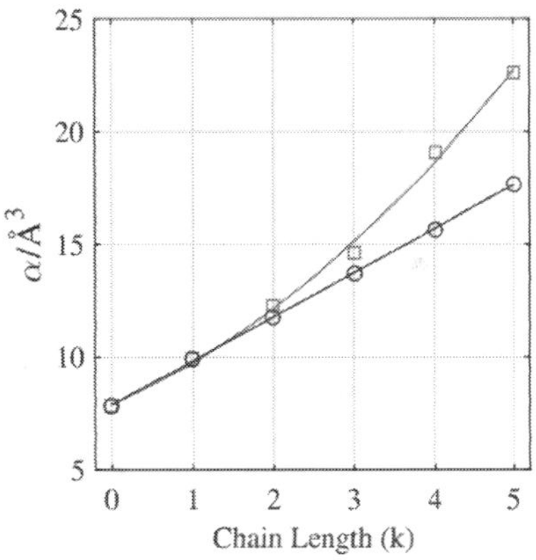

Figure 1. Electronic polarizability of $[C_k mim]^+$ (○) and $[uC_k mim]^+$ (□) cation families as function of the chain length.

The figure reveals that polarizability is a linear function of the saturated chain length but a supra-linear function of the conjugated one. According to theory [6], polarizability is proportional to L^4, the fourth power of the physical extension where electrons are allowed to move. In the saturated case, electrons are highly localized in the σ bonds of each $-CH_2-$ unit. Hence, adding new units to the alkyl chain produces a linear increase of polarizability since the contribution of the $-CH_2-$ unit is k times independently added to that of the imidazolium heterocycle. On the other hand, in the case of the conjugated unsaturated chain, the electrons of each $=CH_2-$ unit participate in the charge delocalization that takes place all over the conjugated π bonds of the side chain. Thus, in this case, with each new bond added to the side chain, the effective length where electrons are allowed to move in the molecule is enlarged (L), producing the observed non-linear increase of polarizability. According to this phenomenology, the most suitable cations to produce ILs with high refractive index are those that present large aromaticity in their heterocycles, such as $[C_k py]^+$, or in their side chains, such as those belonging to the $[uC_k mim]^+$ family.

4.3 Ionic Liquids

The refractive index of the combination of all the anions with those cations of chain length $k = 2$ and $k = 5$ are shown in Table 4. With only the considered ionic species, it is possible to synthesize ILs with refractive indices ranging from 1.36 to 1.69. The lowest values correspond to combinations of ions lacking of charge delocalization and containing fluorine atoms. On the other hand, the highest refractive indices belong to combinations where both ions present charge delocalization and lack of fluorine.

Table 4. Refractive index of ILs with cations bearing side chains of length $k = 2$ (first value) and $k = 5$ (second value).

	Anions			
Cations	$[FAP]^-$	$[BF_4]^-$	$[OTf]^-$	$[SCN]^-$
$[C_k mpyrr]^+$	1.36–1.37	1.39–1.41	1.43–1.43	1.50–1.49
$[C_k mim]^+$	1.37–1.38	1.40–1.42	1.44–1.45	1.52–1.51
$[C_k py]^+$	1.38–1.39	1.42–1.45	1.46–1.47	1.54–1.53
$[uC_k mim]^+$	1.39–1.46	1.44–1.59	1.47–1.60	1.55–1.69

As shown in the Table, the presence of delocalized charge regions and avoiding fluorine atoms are the key features to produce ILs with very high refractive index. In this regard, it is possible to perform extrapolations based on our calculations to predict the evolution of the refractive index as a function of the side chain length of the most promising cations, $[C_k py]^+$ and $[uC_k mim]^+$ together with the most promising anion $[SCN]^-$. Fig. 2 shows the evolution of $[C_k py][SCN]$ and $[uC_k mim][SCN]$ families of ILs as a function of the chain length from $k = 2$ to $k = 16$. At the shortest chain lengths, $k = 0, 1$, the $[C_k py][SCN]$ family yields higher refractive indices than $[uC_k mim][SCN]$ one, which is a consequence of the higher aromaticity of the pyridinium heterocycle in comparison with imidazolium. However, when increasing the chain length, the refractive index of the $[C_k py][SCN]$ family decreases while the refractive index of $[uC_k mim][SCN]$ monotically increases. In the case of $[C_k py][SCN]$ the low polarizability of the alkyl chain in comparison with that of the pyridinium heterocycle and the thiocyanate anion is the cause of the overall decreasing of the refractive index. On the other hand, as conjugated chains non-linearly increase their polarizability with their length, increasing the side chain in $[uC_k mim][SCN]$ leads to an overall increasing of refractive index, which reaches the threshold of 2.0 with $k = 11$.

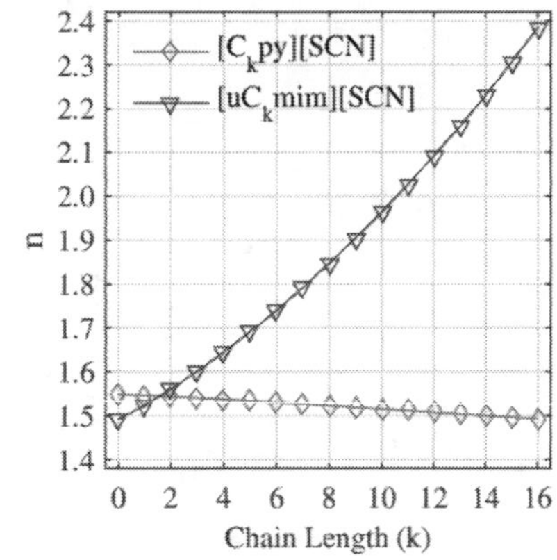

Figure 2. Refractive index extrapolation of $[C_k py][SCN]$ (◇) and $[uC_k mim][SCN]$ (▽) families of ILs as function of the cation side chain length.

5 Conclusions

ILs are materials with an inherent large degree of tunability and potential applications in photonics. Wise selection of IL's constituents provides refractive index fluids covering a wide range of refractive indices, from 1.36 to 1.69. Charge delocalization and presence of fluorine atoms are key aspects to produce combinations in both upper and lower limits of the previous range. According to our calculations, increasing the degree of charge delocalization in cations could provide ILs with refractive indices higher than 2.0, such as in $[SCN][uC_k mim]$.

References

[1] K. Lunstroot, K. Driesen, P. Nockemann, C. Görller-Walrand, K. Binnemans, S. Bellayer, J. Le Bideau, A. Vioux, Chem. Mater. **18**, 5711 (2006)

[2] K. Tanabe, Y. Suzui, M. Hasegawa, T. Kato, J. Am. Chem. Soc. **134**, 5652 (2012)

[3] S.J. Osborne, S. Wellens, C. Ward, S. Felton, R.M. Bowman, K. Binnemans, M. Swadźba-Kwäny, H.Q. Nimal Gunaratne, P. Nockemann, Dalt. Trans. **44**, 11286 (2015)

[4] M.J. Frisch, G.W. Trucks, H.B. Schlegel, G.E. Scuseria, M.a. Robb, J.R. Cheeseman, G. Scalmani, V. Barone, G.a. Petersson, H. Nakatsuji et al., *Gaussian 16, Revision C.01* (2016)

[5] J. Dodds, R. McWeeny, W. Raynes, J. Riley, Mol. Phys. **33**, 611 (1977)

[6] K. Rustagi, J. Ducuing, Opt. Commun. **10**, 258 (1974)

Near-IR transparent conductive amorphous tungsten oxide thin layers by non-reactive radio-frequency magnetron sputtering

Hao Chen[1,2*], *Alessandro* Chiasera[3], *Cristina* Armellini[3], *Giorgio* Speranza[3,4], *Stefano* Varas[3], *Osman* Sayginer[3], *Antonio* Alfano[2], *Marco* Cassinelli[2], *Mario* Caironi[2], *Raffaella* Suriano[5], *Mohamed* Zaghloul[1,2], *Alberto* Tagliaferri[1,2], *Maurizio* Ferrari[3] and *Silvia M.* Pietralunga[2,6]

[1]Politecnico di Milano, Department of Physics, Piazza Leonardo da Vinci, 32, 20133 Milano, Italy
[2]CNST@PoliMi, Istituto Italiano di Tecnologia (IIT), Via Giovanni Pascoli 70/3, Milano, Italy
[3]IFN-CNR CSMFO Laboratory and FBK Photonics Unit, Via alla Cascata 56/C Povo 38100 Trento, Italy
[4]Fondazione Bruno Kessler (FBK)—Centro Materiali e Microsistemi (CMM), Via alla Cascata 56/C, 38123 Trento, Italy
[5]Dipartimento di Chimica, Materiali e Ingegneria Chimica "Giulio Natta", Politecnico di Milano, Piazza L. da Vinci, 32, 20133 Milano, Italy
[6]Institute for Photonics and Nanotechnologies (IFN)—National Research Council (CNR), Piazza L. da Vinci, 32, 20133 Milano, Italy

Abstract. Key assets for transparent electric contacts in optoelectronic applications are high conductivity and large transparency over extended spectral range. Indium-Tin-Oxide and Aluminium-doped-Zinc-oxide are commercial examples, with their electrical conductivity resembling those of metals, despite, their transparency being limited up to 1.5μm. This work introduces smooth and compact amorphous thin films of *n*-type semiconducting WO_{3-x} prepared by RF-sputtering followed by annealing in dry air, as optical layers of tailorable dielectric properties. We evaluate Figure of Merit, combining electrical conductivity and optical transparency, and rate the performances as a transparent conductive layer.

1 Introduction

The demand for simultaneous light transmission and electrical conduction in thin films, namely transparent conductive films (TCFs), is dramatically increasing among advanced applications such as information displays (e.g. liquid crystal displays [1] and organic light-emitting diodes [2]), and energy technologies (e.g. photovoltaics [3] and architectural/window glass [4]). Transparent and conductive metal oxides (TCOs) are the representative category that is well-studied and widespread in industry [5]. Indium tin oxide (ITO) [6], an alloy of indium oxide (In_2O_3) and tin oxide (SnO_2), is probably the most outstanding and well-known TCO, due to the excellent performance at high optical transmittance (>85%) combined with low electrical resistance (sheet resistance R_s ranging from 5 to 100 $\Omega\cdot sq^{-1}$). However, some inherent shortcomings, such as high cost of raw material and production, fragile nature in lifetime caused either by mechanical cracking at low strain, or by degradation upon cyclic operation, somehow hinder the sustained use of ITO [7]. Besides, optical transparency of ITO and of other popular TCOs, such as aluminium-doped and gallium-doped zinc-oxide (AZO and GZO) [8] and fluorine-doped tin oxide (FTO) [9] is frustrated in the Near-InfraRed (NIR) range, where free-carrier absorption arises.

Research into alternative TCOs is therefore becoming necessary. Studies have been dived into organic-based materials and metallic nanostructures, etc. [10], but they seldom reach the ideality for low-cost commercial fabrication and long-lifetime duration. In this study, we analyse amorphous tungsten oxide (WO_{3-x}) thin films, prepared by Radio-Frequency (RF) magnetron sputtering followed by thermal annealing in dry air, as an additional choice to the existing TCFs, especially towards the NIR spectral range. Among the various thin-film deposition methods, RF-sputtering has the advantage of cost-effectiveness and upgradability at industrial production scale, while being a flexible, controllable and stable process [11].

Tungsten oxide is a versatile wide bandgap semiconductor, featuring wide polymorphism and compositional ratios. Different stoichiometries and stable crystalline phases have been found, depending on growth conditions [12-15]. In particular, oxygen deficient WO_{3-x} phases can show remarkable electrical conductance in the presence of considerable optical transparency [16, 17]. In this experiment, we introduce sub-stoichiometric amorphous WO_{3-x} thin films deposited by RF-sputtering, that show good optical transparency in the NIR range. We rate their electrical conductivity and discuss their performances as transparent and conductive layers in the NIR.

* Corresponding author: hao.chen@polimi.it

2 Film fabrication and testing

Amorphous tungsten oxide WO_{3-x} thin films were deposited by non-reactive RF magnetron sputtering in Ar atmosphere [18,19], onto different substrates, namely p-doped Si (100), fused silica (SiO_2) and ITO-coated borosilicate glass (R_s=8~12 $\Omega\cdot sq^{-1}$). Before deposition, substrates were thoroughly cleaned, by ethanol rinsing followed by heating at 120 °C for 30 minutes, while the target of stoichiometric WO_3 was pre-sputtered for 10 minutes. During deposition, substrates and target were constantly kept at room temperature (RT), and a working pressure at 5.4×10^{-3} mbar was stabilized in the deposition chamber.

By changing the deposition time, two sets of film were obtained, respectively about 160nm and 50nm thick. Post-growth annealing was carried out in a conventional tubular oven [19] for 8 hours in dry air at 300°C, below the threshold for crystallization [20]. The process provides four film systems to be tested, each one deposited on all the three substrates. Thicker and thinner sets are respectively {S1, S2} and {S3 S4}, being S1 and S3 *as-deposited*, and S2 and S4 annealed.

Thickness d (nm) and morphology of the films were characterized by High-Resolution Scanning Electron Microscopy (HR-SEM) and profilometry. Surface topology and roughness R_a (nm) were analysed by Atomic Force Microscopy (AFM). The amorphous structure of the films deposited on Si substrate was assessed by grazing-index X-Ray Diffractometer (XRD) θ-2θ measurements in Bragg-Brentano geometry. The stoichiometry at surface was analysed by X-ray Photoelectron Spectroscopy (XPS). The semiconducting self-doping n-type character of the films was revealed by the hot probe method [21]. DC sheet resistance R_s ($\Omega\cdot sq^{-1}$) was tested at RT by 4-probe method in Van der Pauw configuration [22] on samples deposited on SiO_2 substrate. Electrical conductivity was correlated to the position of the Fermi level E_F and to the value of work function (Φ) by Kelvin Probe (KP) measurements (KPSP020, KP Technologies Inc).

Optical absorbance spectrum in the range 300 nm-2500 nm was measured on samples deposited onto SiO_2 substrate and the amplitude of the optical bandgap E_g (eV) was estimated by detailed Tauc plot analysis [23] to an accuracy of 10 meV. The complex dielectric function of the films deposited on Si substrate was spectrally investigated by performing Variable Angle Spectroscopic Ellipsometry (VASE) over the range 350 nm $<\lambda<$ 1700 nm.

In principle, the two key parameters for TCFs, which are light transmission across the spectral range of interest, and electrical conductivity, grow at the expense of each other. Therefore, a tradeoff between them exists, and a proper Figure of Merit (FOM) for TCFs can be introduced to provide quantitative comparison of performances for TCFs. Among the various evaluation methods [24, 25], the most popular is the FOM proposed by Haacke [26],

$$\Psi(\lambda) = T(\lambda)^{10}/R_s \qquad (1)$$

where $T(\lambda)$ is the transmittance and R_s the sheet resistance of the film. Higher values of $\Psi(\lambda)$ represent better transparent-conductive performances.

3 Results and discussion

The cross-sectional HR-SEM images of samples S1 and S2 deposited on Si substrate are shown in Figure 1, which highlights the compactness, smoothness of the films, and good adhesion of films to the substrate. Film thickness (d), evaluated from SEM cross-sections was confirmed by step profilometry. Results are listed in Table 1, with an uncertainty of about 5nm caused by the inherent set of electron-beam radius and angular precision of sample orienting during the measurements.

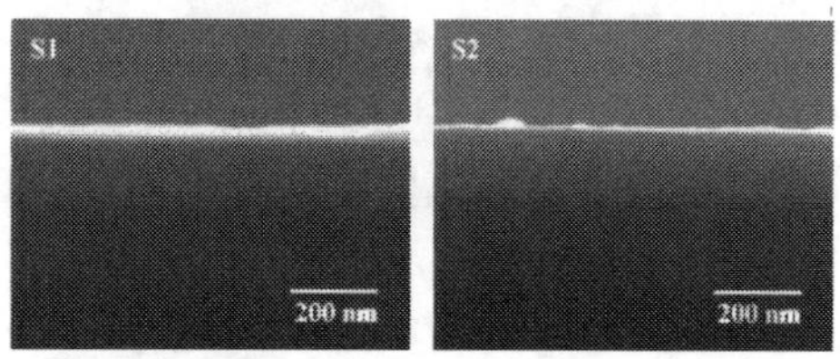

Fig. 1. SEM cross-sections of S1 and S2 films.

Figure 2 illustrates the AFM topographic 3D maps of samples S1-S4, which somehow look equivalent. Very low values for arithmetic mean roughness (R_a) were obtained, indicating extremely smooth surface conditions for these samples. Detailed results are referred in Table 1.

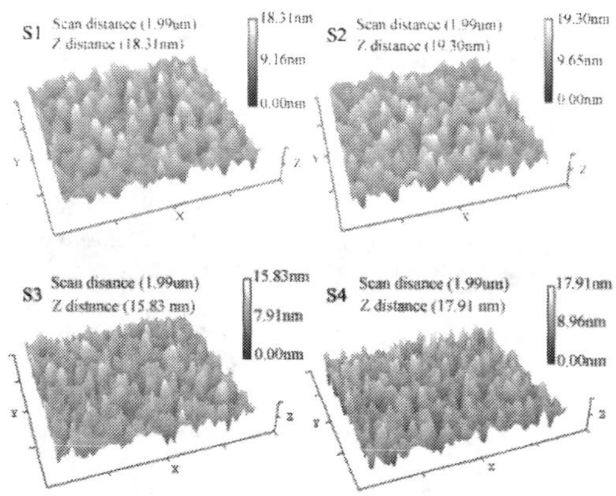

Fig. 2. AFM topographic 3D maps of samples.

The amorphous nature of samples is shown by the smooth XRD patterns in Figure 3. The broad peaks at about 2θ= 24° indicate a lack of long-range ordering in WO_{3-x} lattices. The broad hump at about 70° is instead originated by scattering of Si (100) substrate.

By XPS investigation of the W core-line, spectral peaks of W^{6+} deriving from WO_3 phase are detected (spin orbit components $4f_{7/2}$ and $4f_{5/2}$), as well as peaks that can be assigned to the $4f_{7/2}$ and $4f_{5/2}$ of $W^{+(6-x)}$ ions that derive from sub-stoichiometric WO_{3-x} phases and oxygen vacancies. Oxygen deficiency and related defects bring intrinsic n-type doping and increase electrical conductance of WO_{3-x} phases, both crystalline [14,15] and amorphous [17,27]. Therefore, we expected films to be somewhat conductive. Hot probe measurement on *as-deposited* samples proved the *n*-type semiconducting nature of WO_{3-x} samples. Values of sheet resistance R_s obtained by 4-probe method are listed in Table 1.

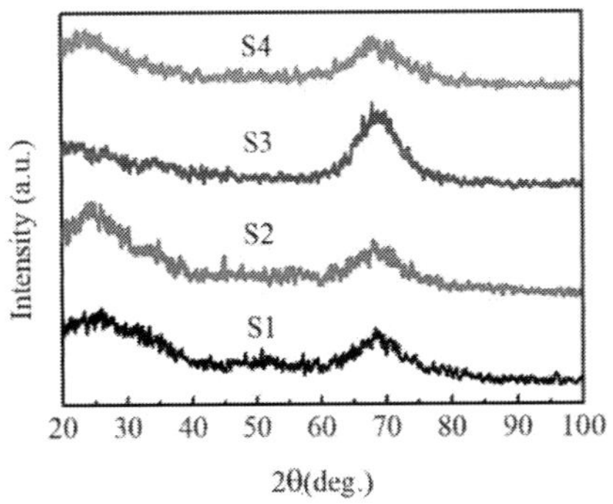

Fig. 3. XRD plots showing the amorphous structure of films.

By considering electrical conductivity at RT to be mainly contributed by band electrons, the *n*-type character of WO_{3-x} can be correlated to the position of the Fermi level E_F and to the value of work function (Φ) as measured by KP. As Φ increases, the distance in energy of the Fermi level from the bottom of conduction band (E_c-E_F) also increases and the density of free electrons N exponentially decreases, causing resistivity ρ to increase, as $\rho = (N\mu e)^{-1}$, where μ is the charge mobility, and e is the elementary charge. This is evident for the thicker set of samples, that by annealing both Φ and ρ increase. However, the KP analysis is not in consistency with the results of electrical conductivity for thicker samples using the previous theory. Instead, samples by annealing show a mild enhancement of electrical conductivity when a higher Φ is measured comparing to the *as-deposited* ones.

Table 1. Morphological, electrical, and optical properties of samples S1-S4.

Sample	S1	S2	S3	S4
Annealing (°C)	No	300	No	300
d (nm)	162±5	159±5	43±5	58±5
R_a (nm)	1.83	1.82	1.59	1.80
ρ (Ω·cm)	9.47E-2	5.10E-2	1.26	525.48
R_s (Ω·sq^{-1})	5.85E+3	3.21E+3	2.93E+5	9.06E+7
Φ (eV)	4.27	4.65	4.22	4.57
E_g (eV)	2.57	2.84	2.49	3.05
α @1550nm (10^5 cm^{-1})	0.89	0.76	1.18	0.34
n @1550nm	--	--	2.99	1.99
κ @1550nm	--	--	0.46	3.42E-9

The slight decrease in sheet resistance values with annealing can be explained in terms of a reduction in the density of grain boundaries and extended defects, which increases the scattering time for charges and charge mobility, thus compensating for the decrease of N.

Optical absorption coefficient α (cm^{-1}), as retrieved from the absorbance (A) spectra as $\alpha = 2.303A/d$ is reported in Figure 4. Post-growth annealing induces a decrease in absorption. The phenomenon is especially evident for the thinner film S4, for which the absorption coefficient is greatly reduced extensively along the whole examined spectral range. In particular, the absorption coefficient for S4 steadily decreases in the NIR and eventually reaches a value that is smaller than those of the other samples. Optical bandgap E_g (eV) values were retrieved by implementing Tauc plot analysis based on the theory of Davis and Mott [28],

$$(\alpha h\nu)^{1/\beta} = B(h\nu - E_g) \quad (2)$$

where $h\nu$ is the photon energy, B is a constant, and β is the index accounting for the nature of optical band-to-band transitions. The Tauc plot, as shown in the inset of Figure 4, agrees with the results reported elsewhere [5], both in terms of optical bandgap values for sub-stoichiometric WO_{3-x} and of the indirect nature of interband transitions.

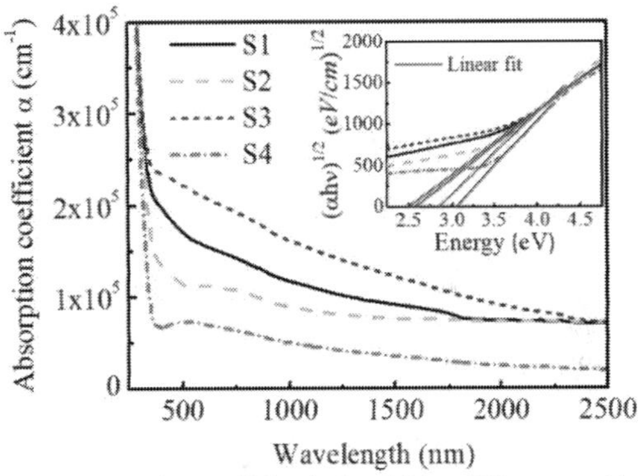

Fig. 4. Absorption coefficient in the visible to NIR range, with the Tauc plot analysis in the inset.

Figure 5 shows the dispersion of real and imaginary parts of the refractive index ($n(\lambda)$, $\kappa(\lambda)$) by VASE. It is interesting that the *as-deposited* sample S3 shows a metal-like anomalous dispersion within the examined spectral range. Instead, annealed sample S4 shows the normal chromatic dispersion, which is typical of dielectric materials.

Figure 6 compares Haacke's FOM calculated as a function of optical wavelength, for several TCFs and for WO_{3-x} films in present work. Generally, the good performance of TCFs is limited to around 1.5μm, above which the satisfactory FOM becomes invalid. In present work, however, tungsten oxide shows continuous and stable enhancement of FOM along wavelength up to 2.5μm. Though the performance within visible range is poor, tungsten oxide outrides the other TCFs when wavelength comes to above 2μm. This implies that tungsten oxide can be modulated as a promising transparent-conductive oxide specialized for NIR range.

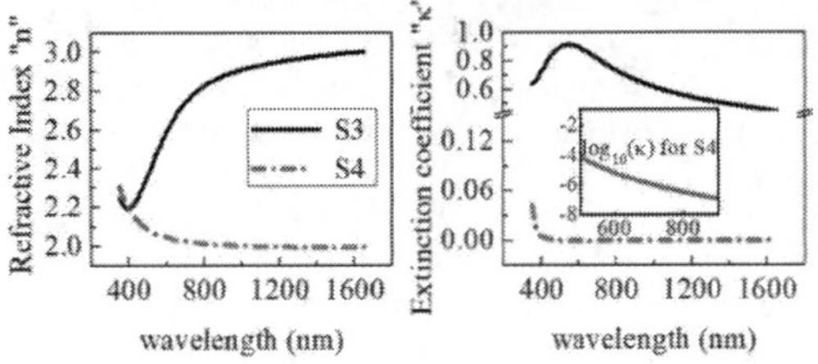

Fig. 5. Spectral dispersion of refractive index (n) and extinction coefficient (κ) from VASE. In the right an inset is added to show the $\log_{10}(\kappa)$.

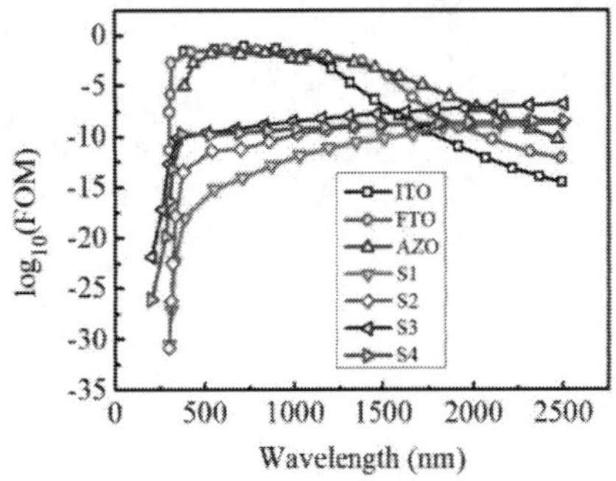

Fig. 6. Figure of merit for different transparent-conductive films, plotted by retrieving transmittance data and sheet resistance from: ITO [29], AZO [29], FTO [30].

4 Conclusions

Smooth and compact amorphous tungsten oxide thin films are fabricated by non-reactive RF-sputtering from a WO_3 target, as optical layers of tailorable dielectric properties. *As-deposited* WO_{3-x} films show good optical absorption and electrical conduction. Post-growth treatment at 300°C anneals out oxygen vacancies, leading to better transparency in the NIR. Effects of annealing on electrical resistivity, however, differs for thicker (about 160 nm) and thinner films (about 60 nm). Electrical resistivity increases by annealing for thicker samples, while for thinner counterparts, electrical resistivity decreases by annealing. Figure of merit, that evaluates the transparent-conductive performance of WO_{3-x} films, has been analyzed and compared with some popular and commercial transparent-conductive films. Result shows the shortage of WO_{3-x} as transparent-conductive films in visible range, but very good performance in NIR range.

Acknowledgement

Authors acknowledge co-funding of research by European fund - FESR, PON Ricerca e Innovazione 2014-2020.

References

1. B. Y. Oh et al., J. Appl. Phys. **99**, 124505 (2006).
2. H. Kim et al., J. Appl. Phys. **86**, 6451 (1999).
3. F. U. Hamelmann, J. Phys. Conf. Ser. **559**, 012016 (2014).
4. C. M.Lampert, Sol. Energy Mater. **11**, 1 (1984).
5. D. S. Ginley, C. Bright, and G. Editors, MRS Bull. **25**, 15 (2000).
6. B. G. Lewis and D. C. Paine, MRS Bull. **25**, 22 (2000).
7. L. Yu, C. Shearer, and J. Shapter, Chem. Rev. **116**, 13413 (2016).
8. T. Minami, Thin Solid Films **516**, 5822 (2008).
9. A. E. Rakhshani, Y. Makdisi, and H. A. Ramazaniyan, J. Appl. Phys. **83**, 1049 (1998).
10. W. Cao et al., J. Photonics Energy **4**, 040990 (2014).
11. K. Ellmer, J. Phys. D. Appl. Phys. **33**, (2000).
12. A. Al Mohammad and M. Gillet, Thin Solid Films **408**, 302 (2002).
13. E. Salje and K. Viswanathan, Acta Crystallogr. Sect. A **31**, 356 (1975).
14. D. B. Migas et al., J. Appl. Phys. **108**, 093713 (2010).
15. D. B. Migas, V. L. Shaposhnikov, and V. E. Borisenko, J. Appl. Phys. **108**, 093714 (2010).
16. F. Wang, C. Di Valentin, and G. Pacchioni, Phys. Rev. B - Condens. Matter Mater. Phys. **84**, 1 (2011).
17. F. Mehmood et al., J. Appl. Phys. **120**, (2016).
18. A. Vinante et al., Phys. Rev. Lett. **125**, 100404 (2020).
19. A. Chiasera et al., Opt. Mater. (Amst). **87**, 107 (2019).
20. C. Y. Kim and S. Park, Asian J. Chem. **25**, 5874 (2013).
21. A. Axelevitch and G. Golan, Facta Univ. - Ser. Electron. Energ. **26**, 187 (2013).
22. L. J. van der Pauw, Philips Res. Reports **13**, 1 (1958).
23. J. Tauc, A. Menth, and D. L. Wood, Phys. Rev. Lett. **25**, 749 (1970).
24. K. Ellmer, Nat. Photonics **6**, 809 (2012).
25. V. K. Jain and A. P. Kulshreshtha, Sol. Energy Mater. **4**, 151 (1981).
26. G. Haacke, J. Appl. Phys. **47**, 4086 (1976).
27. H. Simchi et al., J. Alloys Compd. **617**, 609 (2014).
28. E. A. Davis and N. F. Mott, Philos. Mag. A J. Theor. Exp. Appl. Phys. **22**, 903 (1970).
29. D. Langley et al., Nanotechnology **24**, (2013).
30. N. Noor and I. P. Parkin, J. Mater. Chem. C **1**, 984 (2013).

 https://doi.org/10.1051/epjconf/202125506001

Efficient watt-level 0.775 µm SHG and 3.3 µm DFG in annealed and reverse proton exchanged diced PPLN ridge waveguides

Sergiy Suntsov[1,*], *Kore* Hasse[1], and *Detlef* Kip[1]

[1]Faculty of Electrical Engineering, Helmut Schmidt University, 22043 Hamburg, Germany

Abstract. We report on the fabrication of ridge waveguides formed in congruent periodically poled lithium niobate (PPLN). The waveguides were fabricated by periodic poling, proton exchange, subsequent annealing and reverse proton exchange followed by diamond blade dicing. Up to 1 W of single-pass second-harmonic generation at 775 nm has been realized in 50 mm long ridge waveguides with internal conversion efficiency of 70%. Furthermore, difference frequency generation at 3.3 µm of two pump waves at 1.05 µm and 1.55 µm has been obtained in similarly fabricated PPLN ridge waveguides with an appropriate poling period.

1 Introduction

Wide transparency window, excellent electro-optical, acousto-optical and nonlinear properties combined with low-loss waveguides make lithium niobate a material of choice for a wide variety of photonics applications. Due to large $\chi^{(2)}$ nonlinear coefficients and the possibility of controlled domain inversion on a micrometer scale, PPLN waveguide nonlinear frequency converters are used in optical telecommunications, for generation of squeezed states and entangled photon-pairs, laser cooling of atoms, and supercontinuum generation. Regarding frequency doubling of 1550 nm light, impressive peak conversion efficiencies of up to 4600 %/W have been reported [1, 2]. On the other hand, watt-level second-harmonic generation (SHG) outputs, have been demonstrated in PPLN waveguides with relatively low efficiencies of (50 – 100) %/W [3, 4]. PPLN waveguides both with channel and ridge geometries, have been also used for generation of mid-infrared (MIR) light in the (3.2 – 4) µm wavelength range through difference frequency generation (DFG) of the two pump waves with wavelengths around 1050 nm and 1550 nm. In different waveguide types, DFG efficiencies up to 70 %/W in the up to 8 cm long PPLN waveguides have been demonstrated [5, 6].

In this work, we report on the first annealed/reverse proton exchanged, diamond blade diced PPLN ridge waveguides with 510 %/W (25 %/W·cm^2) conversion efficiency. In high pump power regime, up to 1 W of SHG at 775 nm at 70 % pump power conversion is obtained [7]. Furthermore, DFG in the mid-infrared range around 3.3 µm is demonstrated in similarly fabricated PPLN ridge waveguides under 1.05 µm and 1.55 µm pumping.

2 Sample fabrication

The PPLN ridge waveguide fabrication process comprises four main steps: electric field poling of a $LiNbO_3$ substrate, proton exchange with subsequent annealing (APE), reverse proton exchange (RPE) and ridge waveguide definition. As a sample substrate, 50 mm long pieces of an optical-grade z-cut $LiNbO_3$ wafer were taken. First, electric field poling with a quasi-phase-matching (QPM) period of 18.2 µm for SHG and 26.6 µm for DFG over 45 mm length was conducted. The samples were proton exchanged for 100 minutes (SHG) or 5 hours (DFG) in pure benzoic acid melt at 200 °C, and then annealed in air for 18 hours at 350 °C. These particular proton exchange times were chosen in order to obtain (after annealing and RPE steps) single mode waveguides at 1.55 µm and up to 4 µm wavelengths for SHG and DFG samples, respectively. The in-depth refractive index (RI) profiles at 532 nm of the obtained planar waveguides for SHG after annealing were reconstructed based on the effective RIs of guided TM modes measured with a prism coupler.

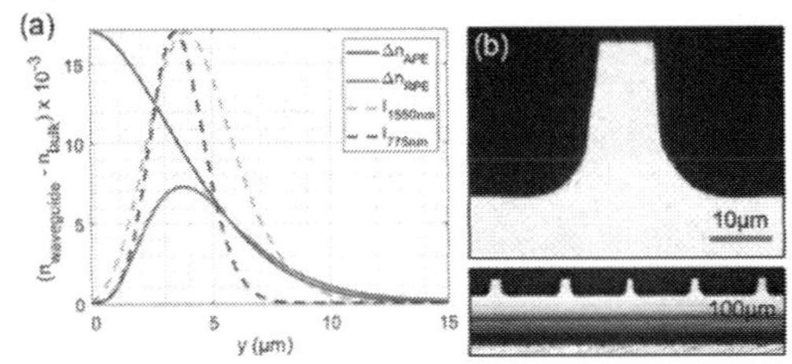

Fig. 1. (a) In-depth RI of measured APE (blue) and calculated RPE (red) planar waveguides at 532 nm. Dashed curves are simulated intensities of the fundamental modes after RPE at 1550 nm (yellow) and at 775 nm (violet). (b) Micro-graphs of the sample's facet at 500x and 2000x magnification.

* Corresponding author: suntsov@hsu-hh.de

The half-bell-shaped RI of the SHG sample is shown in Fig. 1 (a) with a blue curve and has a maximum increase of 0.017 at the surface with 4.0 μm width at 50 % height. The corresponding parameters of the fabricated DFG sample were found to be 0.038 and 3.5 μm. A subsequent RPE step was conducted in a Li-rich eutectic melt of $LiNO_3$:KNO_3:$NaNO_3$ for (5 – 7) hours at 330 °C. The RI profile after the RPE step is also shown in Fig. 1 (a). The simulated in-depth mode intensities of the RPE planar waveguide for SHG are shown with dashed lines in Fig. 1 (a). After RPE the guided modes are pushed away from the surface layer with damaged nonlinearity, and at the same time the more symmetric RI profiles lead to a significantly improved modes' overlap. Both effects are advantageous for efficient SHG. In the last step, ridges with widths (7 – 18) μm and a height of 25 μm were cut using a precision wafer saw (Disco DAD322), see micrographs of Fig. 1 (b).

3 Measurements setup

A schematic of the experimental setup for SHG measurements in our RPE PPLN ridge waveguides is presented in Fig. 2. The low power sub-milliwatt output of a fiber-coupled tunable laser source (New Focus Venturi TLB 6600 H CL) was amplified by a polarization maintained Er-doped fiber amplifier (EDFA, Keopsys CEFA-C-PM-HP-37) with an output power of up to 4.5 W. A collimator and an aspheric lens combination was used to inject the pump light into the waveguides. To facilitate coupling optimization, the in-coupling optics was mounted on a 3-axis piezo positioning system (Thorlabs Nanomax MAX341) with an auto-align option for tracking the output SH maximum (or throughput pump power maximum, when needed). The waveguide chip was placed inside an oven (Covesion PV50) for operation at elevated temperature. The second harmonic and the residual pump exiting the sample were collected and collimated with another aspheric lens. A beam sampler (BS) on a flip mount was used to reflect a small part of the output onto an infrared CCD camera while an appropriate filter (F, Thorlabs FGL1000M or Thorlabs FGB25) was inserted to separately record intensities of the pump and SH waveguide modes. The SH and the pump beams were then split in a short-pass dichroic mirror with 950 nm cut-off wavelength (DM, Thorlabs DMSP950) and their powers measured with calibrated Si and Ge high-sensitivity power sensors, respectively. To prevent sensors' saturation at higher power levels, neutral density filters as well as cross-check with a thermal power meter head were used. Furthermore, since the dichroic mirror has 2.5 % reflectance at 775 nm, and Ge sensor has non-negligible sensitivity at this wavelength, a long-pass filter (F1, Thorlabs FGL1000M) was used in front of the Ge power sensor to block the reflected second harmonic light. By determining the SHG and the residual pump powers at the waveguide's output, we accounted for the losses caused by the downstream optical components. To record the SHG and residual pump tuning curves, a swept-wavelength option of our laser and a digital oscilloscope were used.

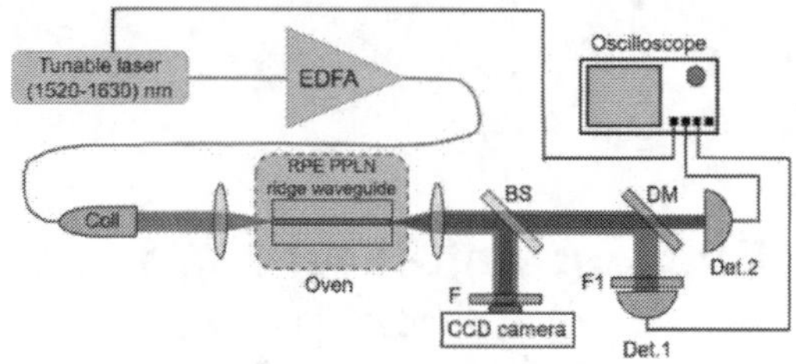

Fig. 2. Experimental setup for SHG characterization of RPE PPLN ridge waveguides. EDFA: erbium-doped fiber amplifier, Coll: collimator, BS: beam sampler, F, F1: filters, DM: short-pass dichroic mirror, Det.1 and Det.2: germanium and silicon power meters, respectively.

The setup for the DFG experiments (not shown) has an additional pump laser source at around 1050 nm (Sacher Lasertechnik Lion) with an output power of up to 400 mW. The two pump beams were combined in a dichroic mirror (Thorlabs DMSP1180). On the output side, the residual pumps were filtered out with a germanium filter, and the power of the generated MIR wave was measured with a mid-infrared detector (Hamamatsu P4631). Also, to record the intensity of the MIR waveguide modes an appropriate infrared camera (DataRay WinCamD-IR-BB) was used.

4 Experimental results

4.1 Second-harmonic generation

To reduce photorefractive damage (PRD) effects, during SHG experiment the sample's temperature was maintained at 120 °C. The recorded intensities of the pump and SH waveguide modes are shown in Fig. 3 (a). The measured low-power SHG tuning curve and the respective theoretical one, plotted in Fig. 3 (b), are in good agreement with a FWHM of 0.22 nm. We define the peak SHG conversion efficiency as η = 100 %$\cdot P_{SH}/(P_{p\text{-}in})^2$, where P_{SH} and $P_{p\text{-}in}$ are the exiting SH power and the pump power coupled into waveguide, respectively. From the parabola fit to the measured low-power dependence of P_{SH} on $P_{p\text{-}in}$ at QPM wavelength 1551.4 nm (inset of Fig. 3 (b)) a conversion efficiency of η = 510 %/W was extracted.

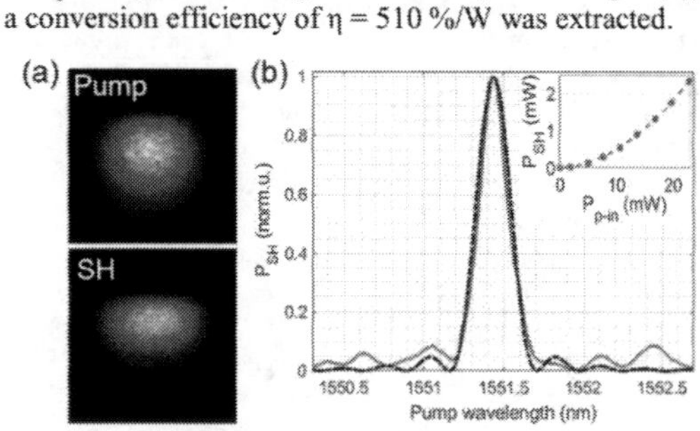

Fig. 3. (a) Recorded pump and SH mode intensities of the 10 μm wide RPE PPLN ridge waveguide (b) Measured (red solid) and calculated (black dashed) SHG tuning curve. Inset: low power P_{SH} vs. $P_{p\text{-}in}$ (circles) with a parabola fit (dashed).

P_{SH} and pump conversion efficiency into SH wave, defined as η_{SHG} = 100 %$\cdot P_{SH}/P_{p\text{-}in}$, vs. $P_{p\text{-}in}$ are plotted in

https://doi.org/10.1051/epjconf/202125506001

Fig. 4 (a). After the initial parabolic increase in the low pump depletion regime, a quasi-linear increase of P_{SH} was observed, and at $P_{p\text{-}in} = 1.2$ W, the maximum $\eta_{SHG} \approx 70$ % was reached with 0.85 W of the SH output.

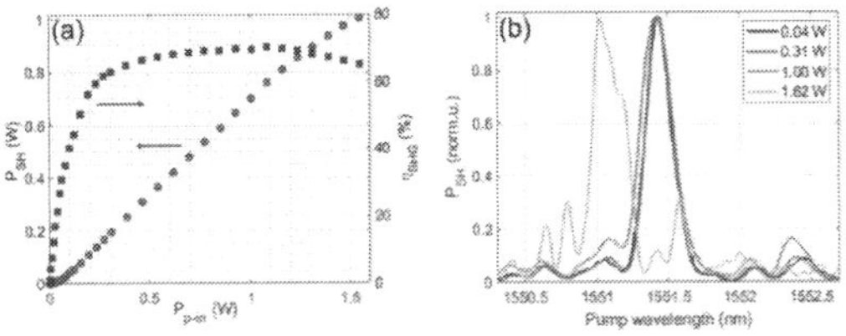

Fig. 4. (a) P_{SH} (circles) and η_{SHG} (squares) versus $P_{p\text{-}in}$. (b) SHG tuning curves for different coupled pump powers $P_{p\text{-}in}$.

With further $P_{p\text{-}in}$ increase, sub-linear P_{SH} grows and η_{SHG} decline were observed. The explanation for this behavior can be found in Fig. 4 (b) where the measured SHG tuning curves at different $P_{p\text{-}in}$ are shown. The observed blue shift of the spectral peak at $P_{p\text{-}in} > 1$ W clearly indicates PRD caused by the growing SH wave.

3.2 Difference frequency generation

Due to 3 times longer proton exchange time, ridge waveguides of our DFG PPLN sample supported the fundamental mode at up to 4 μm wavelength. To generate MIR light at a wavelength around 3.3 μm, a fixed pump at 1050 nm (Pump1) and a tunable pump in the (1520 – 1630) nm range (Pump2) were used. Figure 4 (a) shows the recorded modes of a 13 μm wide waveguide. For this ridge, the QPM was obtained at Pump2 wavelength of 1540 nm with the generated DFG wave at 3.3 μm.

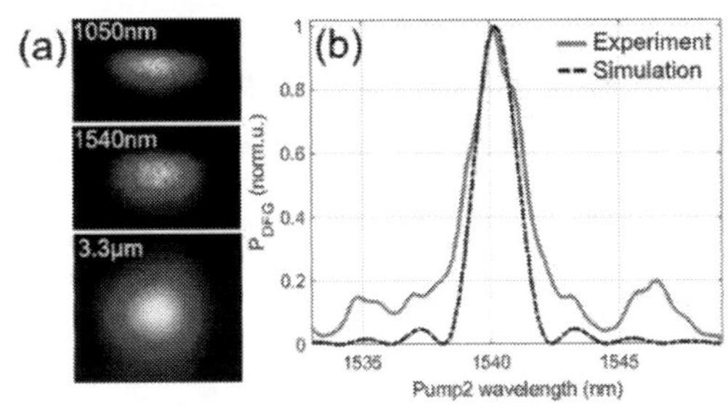

Fig. 4. (a) Experimentally measured Pump1, Pump2 and DFG mode intensities of the 13μm wide RPE PPLN ridge waveguide (b) Measured and simulated DFG tuning curves.

The measured and simulated DFG QPM tuning curves are presented in Fig. 4 (b). There is a clear deviation from the ideal $Sinc^2$ behavior caused by the imperfections in the fabrication process. Also, the experimental QPM peak is broader than the simulated one. The peak DFG conversion efficiency was found to be 3.2 %/W, which is an order of magnitude lower than the theoretical prediction. These discrepancies could be attributed to much higher propagation losses at 3.3 μm caused by OH^- absorption in proton-exchanged $LiNbO_3$ material and are currently under investigation.

4 Conclusion

To conclude, we have fabricated PPLN ridge waveguides using annealed and reverse proton exchange with subsequent diamond blade dicing. We reached the highest SHG power of 1 W at 775.7 nm and the internal conversion efficiency of 70 % reported to date for this type of PPLN waveguides. The maximum attainable SHG power is limited by PRD. Thus, PRD resistant substrates are needed for even higher power operation. In the RPE PPLN waveguides for mid-infrared DFG at 3.3 μm relatively low conversion efficiency was obtained that could be improved by shifting the DFG wavelength toward 4 μm, further away from the OH^- absorption peak.

Funding by Deutsche Forschungsgemeinschaft (DFG) (Ki482/17-1) is gratefully acknowledged.

References

1. T. Umeki, O. Tadanaga, and M. Asobe, IEEE J. Quantum Electron. **46**(8), 1206–1213 (2010).
2. R. Kou, S. Kurimura, K. Kikuchi, A. Terasaki, H. Nakajima, K. Kondou, J. Ichikawa, Opt. Express **19**(12), 11867 (2011)
3. C.-W. Hsu, R. Lai, C.-S. Hsu, Y.-T. Huang, K. Wu, M.-H. Chou, Proc. SPIE **10902**, 12 (2019)
4. L. G. Carpenter, S. A. Berry, A. C. Gray, J. C. Gates, P. G. R. Smith, C. B. E. Gawith, Opt. Express **28**, 21382–21390 (2020)
5. W. Denzer, G. Hancock, A. Hutchinson, M. Munday, R. Peverall, and G. A. D. Ritchie, Appl. Phys. B **86**, 437–442 (2007)
6. K.-D. F. Büchter, H. Herrmann, C. Langrock, M. M. Fejer, and W. Sohler, Opt. Lett. **34**, 470–472 (2009)
7. S. Suntsov, C. E. Rüter, D. Brüske, D. Kip, Opt. Express **29**, 11386 (2021)

 https://doi.org/10.1051/epjconf/202125506002

Indistinguishable Photons from a Single Molecule under Pulsed Excitation

Pietro Lombardi[1,2,*], *Maja* Colautti[1,2], *Rocco* Duquennoy[3], *Ghulam* Murtaza[1,2], *Prosenjit* Majumder[1,2], and *Costanza* Toninelli[1,2,**]

[1]National Institute of Optics (CNR-INO), Largo Enrico Fermi 6, Firenze 50125, Italy
[2]European Laboratory for Non-Linear Spectroscopy (LENS), Via Nello Carrara 1, Sesto F.no 50019, Italy
[3]Dipartimento di Fisica e Astronomia, Universita degli Studi di Firenze, Via G. Sansone 1, Sesto F.no 50019, Italy

Abstract. Quantum light sources are crucial for the future of quantum photonic technologies and, among them, single photons on-demand are key resources in quantum communications and information processing. Ideal quantum emitters providing indistinguishable photons in a clocked manner, negligible decoherence and spectral diffusion, and with potential for scalability are today still a major challenge. We report on photostable and indistinguishable single photon emission from dibenzoterrylene molecules isolated in anthracene nanocrystals (DBT:Ac NCs) at 3K. The visibility of two-photon interference is preserved even when they are separated more than thirty times the excited-state lifetime, or ten fluorescence cycles. One of the advantages of organic molecules is the low-cost mass production of nominally identical emitters, that also allow for on-chip integration. These aspects combined with high spectral stability and coherence make them promising for applications and future quantum technologies.

1 Introduction

The generation and control of optical qubits has paramount importance for the realization of quantum communication[1] and computing protocols[2]. Cyber-security in our information-driven society is currently based on the assumption of the technical difficulty in factorizing large integers and will crumble after the construction of quantum computers. Single photons are natural carrier[3] of data in both free space and guiding media. In the quantum-protected version, two communication parties share entanglement or secret keys in the form of qubits which cannot be cloned by any eavesdropper[4]. In the past decades different quantum emitters such as quantum dots[5] and colour centers in diamond[6] were investigated for the generation of single photon qubits. Among them, individual Dibenzoterrylene molecules in a host anthracene crystal show promising performances both at cryogenic[7] [8] and room temperature[9]. In this contribution we fully characterize the emission from DBT:Ac NCs, in terms of the emitted photon coherence properties. In particular, two particle Interference (TPI), observed in the Hong-Ou-Mandel (HOM) effect, is analysed in order to determine the photon wave-packet overlap, required to realize all-optical gates and many communication protocols. We demonstrate high and stable visibility of two-photon interference in the HOM configuration, both under CW pumping and upon triggered excitation[10].

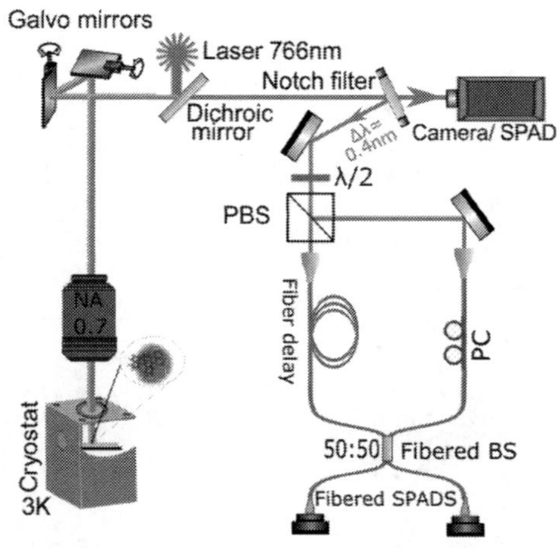

Figure 1. Experimental Setup: A Picoquant laser is used, which can be operated both in pulsed and CW mode. DBT:Ac NCs are kept inside a cryostat at 3K. An epi-fluorescence microscope with 0.7-N.A. objective is employed for excitation and emission collection. Notch Filter with 0.4-nm window filters the ZPL from the phonon side band (PSB). Different possible delay lines and a fibered beam splitter (BS) are used in the interferometer. EMCCD Camera and a free-space-coupled single photon avalanche diode (SPAD) are used to anaylize the ZPL and the PSB, whereas Fibered SPADs and a time to digital converter allow for reconstructing the histogram of the relative photon arrival times.

2 Experimental Setup and Result Analysis

Single molecules of Dibenzoterrylene (DBT) are an interesting choice for single photon emitters. DBT belongs to polyaromatic hydrocarbons (PAH). Embedded in small anthrecene crystals[11], they exhibit near-unity quantum yield, neglible blinking and Fourier-limited zero phonon lines (ZPLs) in the near infrared at low temperature. DBT:AC nanocrystals are placed inside a Cryostat working at 3K. Cryogenically cooled NCs are excited with tunable CW lasers or pulsed laser with adjustable repetition rate operating at a central wavelength of 766nm, with fluorescence collection of the ZPL at 783.5nm. The experimental setup is shown in Fig.1. Fluorescence light is collected with a 0.7-

*e-mail: pietroernesto.lombardi@ino.cnr.it
**e-mail: toninelli@lens.unifi.it

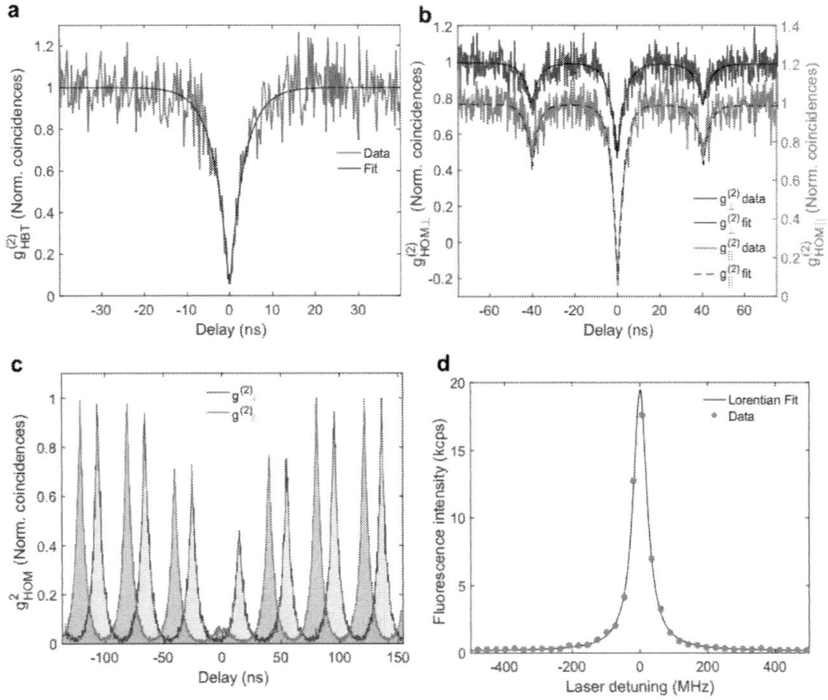

Figure 2. Optical characterization of molecular emission (a)Normalized second-order autocorrelation function of photon coincidences at both Fibered SPADS in continuous wave pumping (b,c) Hong-ou-Mandel Experiment: Histogram of the relative photon arrival time for parallel (red) and orthogonal (blue) polarizations in (b) CW and (c) pulsed excitation. (d) Linewidth measurement by scanning resonant laser across ZPL, exhibiting a lorentzian profile with $FWHM = 55.1 \pm 0.5\ MHz$.

Numerical-Aperture objective and isolated from the phonon sideband (PSB) with a 0.4-nm-wide notch filter in reflection. A 50/50 polarizing beam splitter separates the stream of photons into two arms. A fiber delay line of different possible length is added in one arm, whereas a fiber polarization controller (FPC) is included in the other arm. In this way, two non-correlated single-photon streams with different arrival times are obtained. A fiber beam splitter is used to recombine the two paths and coincidences are measured between the two output ports. Fiber polarization controllers are used to set parallel or orthogonal polarization in the two arms (switching between indistinguishable and distinguishable photons, respectively). Two fiber-coupled Single Photon Avalanche diodes (Fiber-SPADs) are employed for detection and the time histogram of coincidences is recorded by means of a time-correlated single photon counting module. DBT molecules are characterized both under CW and pulsed operation. Pulsed excitation is needed to have photons on demand. This optical setup works as Hanbury–Brown and Twiss (HBT) configuration when one arm (delay line) is closed. In Fig.2(a), the second order correlation function measured for an isolated DBT molecule is reported, showing high purity, with $g^{(2)}(0) = 0.03 \pm 0.02$. Instead, HOM interference measurement is shown in Fig. 2(b&c) for both indistinguishable(red) and distinguishable(blue) cases. The obtained Visibility is $V_{CW} = \frac{g_{\perp}^{(2)}(0) - g_{\parallel}^{(2)}(0)}{g_{\perp}^{(2)}(0)} = 0.89 \pm 0.05$ under CW excitation, with 10-min integration time and 250-coincidence as normalisation factor. The visibility under pulsed excitation is is calculated as a function of the peak areas within $\Delta t = 26ns$ which encompasses 96% area of photon wave packet around zero time delay in parallel $(A_{\parallel})$ and in the orthogonal $(A_{\perp})$ scenario as follows: $\frac{A_{\perp} - A_{\parallel}}{A_{\perp}}$. This yields to $V_P = 0.78 \pm 0.04$ with an integration time of 45 min. The pulsed laser has a tuneable repetition rate, which is adjusted in order to match the period between successive emission with the delay line. The laser repetition rate is finally adjusted to $24.79MHz = 1/\Delta t (\Delta t = 40.3 \pm 0.2)$, based on the delay estimation obtained from the HOM measurement in CW (shown in Fig. 2(b)). Importantly, the parameters extracted from the best fit to the experimental data are consistent with independent measurements. In particular, the linewidth is measured by collecting the fluoresence intensity in the PSB while scanning the resonant laser across ZPL, yielding a $FWHM = 55.1 \pm 0.5\ MHz$ (see Fig.1(d)), while the excited-state lifetime is estimated from the fluorescence decay, yielding $\tau = 4.01 \pm 0.01\ ns$. Non-classical interference of photons separated by different fluorescence cycles is shown in Fig. 3. The interference of two photons separated by first fluorescence cycle is clearly visible in Fig.3(a) corresponding to a repetition rate of 24.79MHz. Different laser repetition rates are required in order to obtain the temporal overlap of further photons. Quantum inference with photons arriving after the fifth and tenth fluorescence cycles are pre-

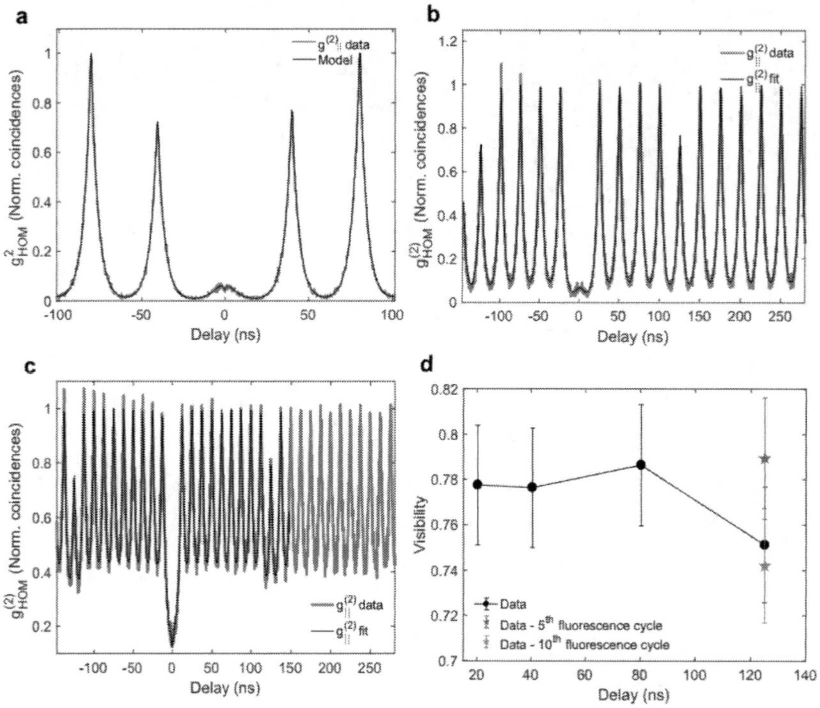

Figure 3. HOM Visibility: Quantum Interference Measurement of photons separated by (a)1st (b) 5th (c)10th fluorescence cycles with 24.79MHz, 40MHz and 80MHz repetition rate, respectively. (d) HOM visibility as a function of Delay line with longest delay of 125ns)

sented in Fig.3(b&c), corresponding to 40MHz and 80MHz repetition rate, respectively. The longest delay time of 12 5ns is achieved with a 25m-long fiber cable which is more than thirty times the excited state lifetime. Fig. 3(d) shows that visibility is only reduced with such long delay times. The overall visibility is limited by residual dephasing and could hence be increased to 96% by lowering the temperature to 1.5K. The finite extinction ratio of polarization optics, the spatial mode mismatch and PSB leaking through the notch filter are also accountable for lowering visibility. Overall spectral diffusion seems to be negligible, making this single photon source particularly interesting.

3 Conclusion

In conclusion, this paper reports indistinguishable single photons from an isolated single DBT molecule. This source also exhibits consistent HOM visibility for photon separated up to 30 times of excited state lifetime. Such type of molecular quantum emitters are easy to fabricate in hybrid system, have photostable emission and high degree of coherence.

References

[1] N. Gisin, R. Thew, Nature Photonics **1**, 165 (2007)
[2] E. Knill, R. Laflamme, G.J. Milburn, Nature **409**, 46 (2001)
[3] R. Uppu, F.T. Pedersen, et al, Science Advances **6**, eabc8268 (2020)
[4] W.K. Wootters, W.H. Zurek, Nature **299**, 802 (1982)
[5] X. Ding, Y. He, et al, Physical Review Letters **116**, 020401 (2016)
[6] C. Kurtsiefer, S. Mayer, P. Zarda, H. Weinfurter, Physical Review Letters **85**, 290 (2000)
[7] A.A.L. Nicolet, C. Hofmann, M.A. Kol'chenko, B. Kozankiewicz, M. Orrit, ChemPhysChem **8**, 1215 (2007)
[8] J.B. Trebbia, P. Tamarat, B. Lounis, Phys. Rev. A **82**, 063803 (2010)
[9] C. Toninelli, K. Early, J. Bremi, A. Renn, S. Götzinger, V. Sandoghdar, Optics Express **18**, 6577 (2010)
[10] P. Lombardi, M. Colautti, R. Duquennoy, G. Murtaza, P. Majumder, C. Toninelli, Applied Physics Letters **118**, 204002 (2021)
[11] S. Pazzagli, P. Lombardi, D. Martella, M. Colautti, B. Tiribilli, F.S. Cataliotti, C. Toninelli, ACS Nano **12**, 4295 (2018)

Stimulated Raman Microscopy Implemented by Three Femtosecond Laser Sources

Rajeev Ranjan[1,2*], *Maria* Antonietta Ferrara[2], Luigi Sirleto[2]

[1]CHT @Erzelli, Nanoscopy, Istituto Italiano Di Tecnologia, Genova, Italy
[2]National Research Council (CNR), Institute of Applied Sciences and Intelligent Systems, Napoli, Italy, Napoli, Italy

Abstract. In this work, the implementation of a femtosecond Stimulated Raman Scattering microscope, equipped with three femtosecond laser sources: a Titanium-Sapphire (Ti:Sa), an optical parametric oscillator (OPO), and a second harmonic generator (SHG); is presented. Our microscope is designed so that it can cover all the regions of Raman spectra, taking advantage of two possible laser combinations. The first, Ti:Sa and OPO laser beams, which cover the C-H region (>2800 cm^{-1}) in stimulated Raman gain (SRG) modality, whereas the second, Ti:Sa and SHG laser beams, covering the C-H region and the fingerprint region in stimulated Raman losses (SRL) modality. The successful realization of the microscope is demonstrated, reporting images of polystyrene beads using both SRL and SRG modalities.

1. Introduction

In spite of the fact that the fluorescence microscopy represents the fundamental pillars for biological imaging, it shows significant limitations: (i) some molecules and/or biological structures cannot be labeled without changing their biological function, such as lipids; (ii) many labels are cytotoxic, which leads to interference with the biological functionality. For these reasons, label-free imaging, obtained by microscopy techniques, is highly desirable, affording high chemical selectivity of unlabeled living cells.

In the last decades, nonlinear optics microscopy has become an attractive tool in both biological and biomedical research to overcome limitations due to fluorescence microscopy. By exploiting nonlinear optical effects, novel microscopy techniques have been developed, which afford high chemical selectivity of unlabeled living cells and implement real-time three-dimensional imaging with high spatial resolution and sensitivity. In multimodal imaging, various nonlinear optics microscopies (CARS, SRS, SHG, THG, and SFG) are combined in a single microscope to image cells or tissue by simultaneous image acquisition. The significant advantage is that the same excitation source can be used for all the nonlinear optics microscopy techniques in this combination of methods [1, 2].

Among novel label-free nonlinear microscopy techniques, Stimulated Raman scattering (SRS) techniques play an essential role. Over the past ten years, SRS has been investigated in nanophotonic [3-6] as well as in biophotonics as an analytical, label-free, non-invasive technique with unique cellular and tissue imaging capabilities [7-10]. SRS technique is sensitive to the same molecular vibrations, probed in spontaneous Raman spectroscopy, but it exhibits a nonlinear dependence on the incoming light fields and produces coherent radiation. In SRS microscopy, two collinear laser beams, a high-power pump laser and a low-power Stokes laser with different frequencies ($\omega_L > \omega_S$) are focused on a sample to generate an SRS signal. When their difference resonantly excites the vibrational frequency of the molecular bonds of interest, energy is transferred from the pump beam to the probe beam, observable in an increase in probe signal intensity (Stimulated Raman Gain, SRG) and a decrease of the pump signal intensity (Stimulated Raman Loss, SRL) [3,10].

In bio-photonics applications, label-free SRS microscopy allows the image of several molecular species, targeting their intrinsic chemical bonds at the congested cellular fingerprint region (<1800 cm^{-1}) or at the high-frequency C-H region (>2800 cm^{-1}). Moreover, recently, a bio-orthogonal chemical imaging platform has emerged by coupling SRS microscopy with tiny and Raman-active vibrational probes (e.g., alkynes and stable isotopes). Spectroscopically, these Raman active vibration probes exhibit Raman peaks at the cell-silent region where no other peaks from endogenous molecules exist, demonstrating sensitivity, specificity, and biocompatibility for imaging small biomolecules [11, 14].

In the reference [15], nonlinear vibrational bioimaging was demonstrated by taking advantage of femtosecond stimulated Raman loss. It is worth noting that a 12 times more significant SRL loss signal by femtosecond pulse

*Corresponding author: rajeev.ranjan@iit.it

excitation was measured with respect to picosecond pulse. Definitely, due to the low sensitivity of Raman-active vibrational probes, femtosecond stimulated Raman microscopy appears as an unavoidable perspective. In addition, interesting biological applications based on femtosecond SRS microscopy have been reported in the literature, extending in the silent region and taking advantage of bio-orthogonal chemical imaging platform [16].

We note that typical and widespread commercial laser sources combinations, such as Chameleon compact OPO (Coherent, Inc.) and Ti:Sa, although tailored for multimodal imaging, their minimum Raman shift is 2500 cm^{-1}. This means that by adopting this laser combination, only the CH-OH region of Raman spectra can be explored, while the silent and fingerprint regions are out of emission range [17-22]. Therefore, they cannot accomplish the demand of biorthogonal platforms based on femtosecond stimulated Raman microscopy.

2. Method and Results

In this work, the implementation of a femtosecond SRS microscope, covering all the regions of Raman spectra, is described. The microscope is equipped with three femtosecond laser sources: a Ti:Sapphire (Ti:Sa), a synchronized optical parametric oscillator (SOPO), and a frequency converter for ultrafast lasers, i.e., a second harmonic generator (SHG) optimized for the SOPO, which allows doubling the OPO energy radiation.

This SRS microscope is used to take advantage of two laser combinations. Ti:Sa and OPO laser combination cover the C-H region in SRG modality[17-22]. The latter Ti:Sa and SHG laser combination provide the extension of the microscope to the silent region (<1800 cm−1) and fingerprint region (1800–2800 cm−1) in the SRL modality [23].

In our experimental setup, a flip mirror (FM) placed after the OPO exit is used in order to switch the Ti:Sa-OPO and Ti:Sa-SHG laser combinations. When the FM is inserted in the optical path, OPO radiation is coupled into SHG, so Ti:Sa-SHG laser combination becomes available, whereas when the FM is flipped, the OPO can go on undisturbed, and Ti:Sa-OPO laser combination becomes available (see Figure 1). Considering the Ti:Sa laser excitation within the wavelength range of 740–880 nm, the corresponding OPO emission range is 1000–1600 nm, while the SHG emission range is 500–800 nm. When the Ti:Sa is used as pump at 830 nm, the largest range of Raman shift is obtained. Indeed, in this case, all the regions of Raman spectra can be covered (from 451 cm−1 to 3195 cm−1) using the Ti:Sa-SHG laser combination[23].

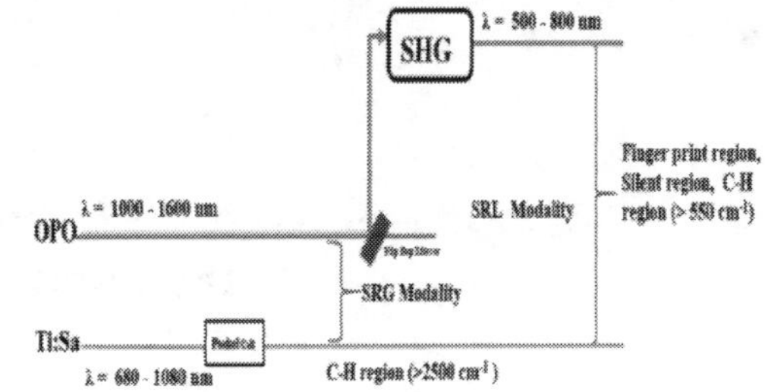

Figure 1. Generation system with combination Ti:Sa-OPO (SRG modality) and Ti:Sa-SHG (SRL modality).

In order to demonstrate its successful realization SRG and SRL images of polystyrene beads with a diameter of 3 μm are reported in figures 2 and 3, respectively, along with transmission image (TI). All the images are single recordings of 512 px × 512 px, and their acquisition

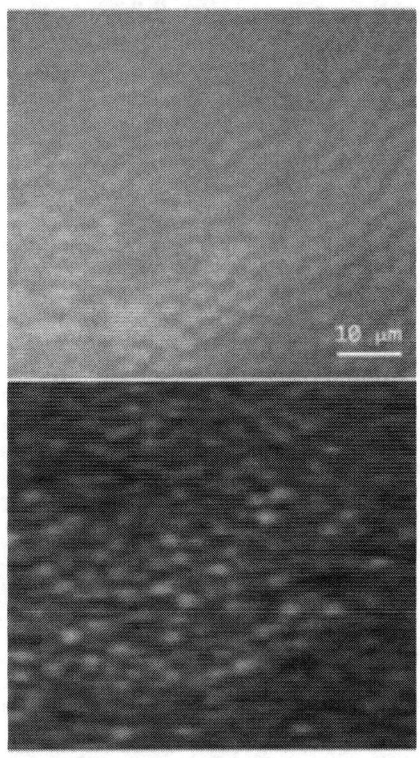

Figure 2. On the top transmission image, the bottom SRG image of polystyrene beads sample with a diameter of 3 μm.

time is 16 s. The Lock-in amplifier (LIA) time constant was set to 100 s with an 18 dB/oct slope and a sensitivity of 10 V. For SRG images, the pump signal generated by Ti:Sa was set at 808 nm, while the probe signal generated by OPO was set at 1072 nm. For both of them, the average focused power was about 10 mW. For SRL imaging, the pump signal generated by the SHG unit was set at 650 nm, while the probe signal generated by Ti:Sa was set at 811 nm, for Ti:Sa, the average focused power was 8 mW, while for SHG, the average focused power was of 6 mW.

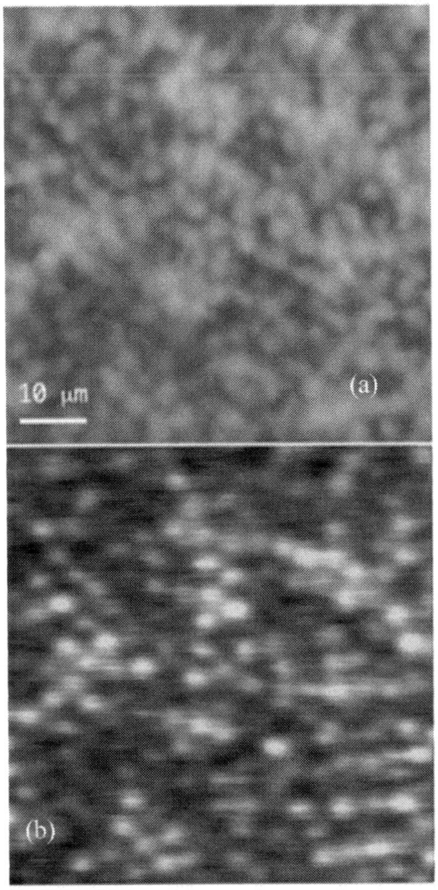

Figure 3 (a) Transmission image and (b) SRL image of a polystyrene beads sample with a diameter of 3 µm.

CONCLUSIONS

In conclusions, we have demonstrated the implementation of a femtosecond stimulated Raman microscope, extending its working range in fingerprint and in silent regions. Our microscope has been designed in order to allow the coexistence of SRL and SRG optical modalities without modifying optical path, i.e. without adding, removing or moving components.

ACKNOWLEDGEMENTS

The authors would like to thank M. Indolfi and V. Tufano (ISASI-CNR) for their technical assistance.

References

1. Streets, A. M.; Li, A.; Chen, T.; Huang, Y. Imaging without fluorescence: nonlinear optical microscopy for quantitative cellular imaging. *Anal. Chem.* **2014**, *86*, 8506–8513.
2. Min, W.; Freudiger, C.W.; Lu, S.; Xie, X.S. Coherent nonlinear optical imaging: beyond fluorescence microscopy. *Annu. Rev. Phys. Chem.* **2011**, *62*, 501-530.
3. Sirleto, L.; Ferrara, M.A.; Nikitin, T.; Novikov, S.; Khriachtchev, L. Giant Raman gain in silicon nanocrystals. Nat. Commun. **2012**, 3, 1–6. doi: 10.1038/ncomms2188;
4. Sirleto, L.; Vergara, A.; Ferrara, M.A. Advances in stimulated Raman scattering in nanostructures. Adv. Opt. Photon. **2017**, 9, 169–217. doi: 10.1364/AOP.9.000169
5. M.A. Ferrara, L.Sirleto, G.Nicotra, C.Spinella, I.Rendina. Enhanced gain coefficient in Raman amplifier based on silicon nanocomposites. 2011 vol.9, pp.1-7. doi:10.1016/j.photonics.2010.07.007.
6. Sirleto, L.; Aronne, A.; Gioffrè, M.; Fanelli, E.; Righini, G.C.; Pernice, P.; Vergara, A. Compositional and thermal treatment effects on Raman gain and bandwidth in nanostructured silica-based glasses. Opt. Mater. 2013, 36, 408–413. http://dx.doi.org/10.1016/j.optmat.2013.10.001
7. Freudiger, C.W.; Min,W.; Saar, B.G.; Lu, S.; Holtom, G.R.; He, C.; Tsai, J.C.; Kang, J.X.; Xie, X.S. Label-Free Biomedical Imaging with High Sensitivity by Stimulated Raman Scattering Microscopy. Science 2008, 322, 1857–1861. [CrossRef] [PubMed]
8. Cheng, J.-X.; Xie, X.S. Vibrational spectroscopic imaging of living systems: An emerging platform for biology and medicine. Science 2015, 350, aaa8870. [CrossRef]
9. Camp, C.H., Jr.; Cicerone, M.T. Chemically sensitive bioimaging with coherent Raman scattering. Nat. Photon 2015, 9, 295–305. [CrossRef]
10. Zumbusch, A.; Langbein,W.; Borri, P. Nonlinear vibrational microscopy applied to lipid biology. Prog. Lipid Res. 2013, 52, 615–632. [CrossRef]
11. Wei, L.; Hu, F.; Shen, Y.; Chen, Z.; Yu, Y.; Lin, C.; Wang, M. C.; Min, W. Live-cell imaging with alkyne tagged small biomolecules by stimulated Raman Scattering. Nat. Methods, 11, 410–412 (2014).
12. Wei, Y Shen, F Xu, F Hu, JK Harrington, KL Targoff, *et al.*: Imaging complex protein metabolism in live organisms by stimulated Raman scattering microscopy with isotope labeling, *ACS Chemical Biology*, vol. 10, no. 3, pp. 901-908, 2015.
13. Fanghao Hu, Michael R. Lamprecht, LuWei, Barclay Morrison Wei Min. Bioorthogonal chemical imaging of metabolic activities in live mammalian hippocampal tissues with stimulated Raman

scattering. Scientific Reports | 6:39660 | DOI: 10.1038/srep39660.

14. Lu Wei, Fanghao Hu, Zhixing Chen, Yihui Shen, Luyuan Zhang, and Wei Min, Live-Cell Bioorthogonal Chemical Imaging: Stimulated Raman Scattering Microscopy of Vibrational Probes. *Acc. Chem. Res.*,**2016**,*49*(8), pp 1494–1502. **DOI:** 10.1021/acs.accounts.6b00210
15. D. Zhang, M. N. Slipchenko, and J. X. Cheng, *J. Phys. Chem. Lett.*, vol. 2, no. 11, pp. 1248-1253, 2011.
16. C. Zhang, K.-C. Huang, B. Rajwa, J. Li, S. Yang, H. Lin, C.-S. Liao, G. Eakins, S. Kuang, V. Patsekin,J. P. Robinson, and J.-X. Cheng: Stimulated Raman scattering flow cytometry for label-free single-particle analysis, Optica, vol. 4, no. 1, 2017
17. R. Ranjan, M. A. Ferrara, A. Filograna, C. Valente and L. Sirleto, “Femtosecond Stimulated Raman Microscopy: Home-built realization and a case study of Biological imaging”, Journal of Instrumentation 14, P09008 (2019).
18. R. Ranjan, M. Indolfi, M. A. Ferrara, L. Sirleto, “Implementation of a Nonlinear Microscope Based on Stimulated Raman Scattering”, J. Vis. Exp. 149, e59614 (2019).
19. M. A. Ferrara, A. Filograna, R. Ranjan, D. Corda, C. Valente, L. Sirleto, “Three-dimensional label-free imaging throughout adipocyte differentiation by stimulated Raman microscopy”, PLoS ONE 14, e0216811 (2019).
20. A. D'Arco, M. A. Ferrara, M. Indolfi, V. Tufano and L. Sirleto, “Label-free imaging of small lipid droplets by femtosecond-stimulated Raman scattering microscopy”, Journal of Nonlinear Optical Physics & Materials 26, 1750052 (2017).
21. A. D'Arco, N. Brancati, M. A. Ferrara, M. Indolfi, M. Frucci, L. Sirleto, “Subcellular chemical and morphological analysis by stimulated Raman scattering microscopy and image analysis techniques”, Biomedical Optics Express 7, p. 1853-1864 (2016).
22. Sirleto, L.; Ranjan, R.; Ferrara, M.A., Analysis of Pulses Bandwidth and Spectral Resolution in Femtosecond Stimulated Raman Scattering Microscopy. Appl. Sci. 2021, 11, 3903.
23. R. Ranjan, A. D’Arco, M. A. Ferrara, M. Indolfi, M. Larobina and L. Sirleto, “Integration of stimulated Raman gain and stimulated Raman losses detection modes in a single nonlinear microscope”, Optics Express 26, pp. 26317-26326 (2018).

BIC in waveguide arrays

Vladimír Kuzmiak[1,*] and *Jiří* Petráček[2]

[1]Institute of Photonics and Electronics AS CR, Chaberská 57, 182 00, Praha 8, Czech Republic
[2]Institute of Physical Engineering, Faculty of Mechanical Engineering, Brno University of Technology, Technická 2896/2, 61 609, Brno, Czech Republic

Abstract. We propose a simple theoretical model based on the coupled-mode theory which allows to calculate the spectral properties and transmittance of the one-dimensional waveguide structures. The model was verified on the common coupled-waveguide array in which the existence of the symmetry-protected bound state in the continuum (BIC) was confirmed experimentally by Plotnik et al. [Phys. Rev. Lett. 107, 28-31 (2011)]. The method can be extended to topologically nontrivial lattices to explore the properties of the BICs protected by time-reversal symmetry.

1 Introduction

Bound state in the continuum (BIC) represents a fascinating exception to the conventional wisdom as despite it lies within the continuum and coexists with the extended state, it remains confined and does not radiates away; for recent reviews see, e.g., [1]-[3]. BICs exist in a wide range of material system due to various confinement mechanisms which are fundamentally different from the discrete levels of the conventional bound states, which lie outside the continuum, for example the bound state of an atom, the guided modes of an optical fiber below the light line and the defect states in a bandgap. In this contribution, we focus on BICs that may occur in coupled photonic lattices. We present a simple theoretical model based on the coupled-mode theory (CMT) which allows to calculate the spectral properties and transmittance of the one-dimensional waveguide structures. The model was applied to a perturbed coupled-waveguide array in which the existence of the symmetry-protected BICs was confirmed experimentally by Plotnik et al. [4]. We examine the properties of the bound modes induced by the perturbation and show how the perturbation affects the transmittance of the structure. The approach can be extended to topologically nontrivial lattices to explore the conditions under which the robust topological BIC states protected by time-reversal symmetry exist.

2 Theoretical model

We consider a 1D waveguide array – see Fig. 1 - that supports a continuum of extended states propagating along x-axis, which are even along perpendicular direction (y), along which the two defects are attached above and below this array. The evolution of the light in the coupled-waveguide structure depicted in Fig. 1 is described within

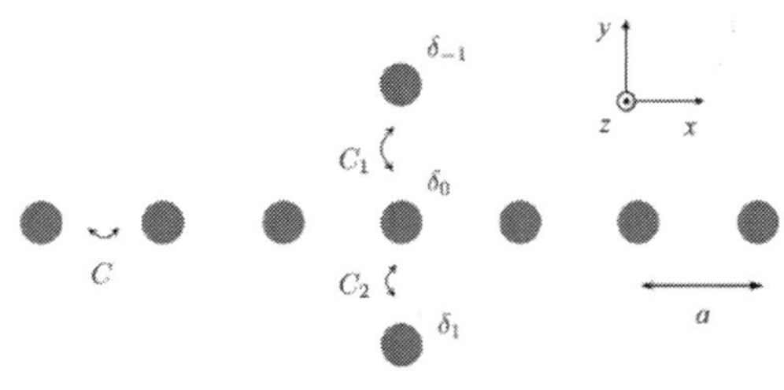

Figure 1. A coupled-waveguide system consisting of a 1D array and two additional waveguides above and below the array.

CMT by a set of coupled equations for the slowly varying modal amplitudes ψ_m on the m−th waveguide ($m \in Z$) in the periodic array along x-direction which reads as

$$
\begin{aligned}
\mathrm{i}\frac{\mathrm{d}\psi_m}{\mathrm{d}z} &= C\left(\psi_{m-1}+\psi_{m+1}\right),\ m \neq 0 && (1)\\
\mathrm{i}\frac{\mathrm{d}\psi_0}{\mathrm{d}z} &= \delta_0\psi_0 + C\left(\psi_{-1}+\psi_1\right) + \\
&+ C_1\varphi_{-1} + C_2\varphi_1 && (2)\\
\mathrm{i}\frac{\mathrm{d}\varphi_{-1}}{\mathrm{d}z} &= \delta_{-1}\varphi_{-1} + C_1\psi_0 && (3)\\
\mathrm{i}\frac{\mathrm{d}\varphi_1}{\mathrm{d}z} &= \delta_1\varphi_1 + C_2\psi_0, && (4)
\end{aligned}
$$

where C is the coupling coefficient in the array, $\varphi_{\pm 1}$ are the modal amplitudes on the waveguides attached below and above the array, $\delta_{\pm 1}$ are the variations of their propagation constants with respect to the propagation constant in the unperturbed array, δ_0 is the variation of the propagation constant in the 0−th waveguide in the periodic array and $C_{1,2}$ are the coupling coefficients between upper and lower additional waveguides attached to 0−th waveguide in the periodic array.

*e-mail: kuzmiak@ufe.cz

Instead of solving the whole set of the Eqs. (1)-(4), we first consider separately a one-dimensional periodic array of waveguides in the x-direction and a finite array in the vertical direction along y-axis, consisting of the 0-th waveguide in the periodic array and two additional waveguides above and below the array. By analysing the solutions of both subsystems we prove that asymmetric state belonging to the discrete spectrum of the vertical array is the solution of the set of the Eqs. (1)-(4), provided the system is vertically symmetric $\delta_1 = \delta_{-1}$. Due to the symmetry mismatch between this asymmetric state and the continuum, the asymmetric mode is decoupled from the Bloch modes of the periodic array and thus it forms BIC, provided its propagation constant lies in the continuum. To prove its existence numerically, we employ the set of the Eqs. (1)-(4) to derive the analytical expressions for the transmittance of the whole system and demonstrate the appearance of the peaks of the transmittance which arise when vertical asymmetry is introduced by imposing the condition $\delta_{-1} = -\delta_1$.

3 Results

In order to inspect the spectral properties of the waveguide array depicted in Fig. 1, we at first solve the Eqs. (1)-(2) for an unperturbed periodic array, i.e., we assume vanishing coupling coefficients $C_{1,2} = 0$ and perfect periodicity $\delta_0 = 0$. We seek the solution in the form of the plane wave $\psi_m = A\exp(-\mathrm{i}k_x am - \mathrm{i}\varepsilon z)$, where ε stands for change of the propagation constant (which corresponds to the energy in the quantum theory) and a is the period of the lattice and we obtain the following dispersion relation:

$$\varepsilon = 2C\cos(k_x a). \qquad (5)$$

By using the latter equation we evaluate the real and imaginary components of the Bloch wavenumber k_x, which are shown in Fig. 2.

In the second step, we find the states in the vertical waveguide, i.e., we solve the Eqs. (2)-(4) when $C = 0$, where we also neglect the variations of the propagation constants associated with attached waveguides, i.e., $\delta_{\pm 1} = 0$ and $\delta_0 = 0$. The discrete spectrum has the form $(\varphi_{-1}\ \psi_0\ \varphi_1)^T = (A_{-1}\ A_0\ A_1)^T\exp(-\mathrm{i}\varepsilon z)$, which corresponds to three different modes, namely two symmetric (S-modes), which exist when $A_0 \neq 0$

$$\varepsilon_{S_{1,2}} = \pm\sqrt{C_1^2 + C_2^2} \qquad (6)$$

$$A_{-1} = \frac{C_1}{\varepsilon_{S_1}} A_0 \qquad (7)$$

$$A_1 = \frac{C_2}{\varepsilon_{S_2}} A_0, \qquad (8)$$

and one asymmetric (A-mode), which exists when $A_0 = 0$

$$\varepsilon_A = 0 \qquad (9)$$

$$C_1 A_{-1} = -C_2 A_1. \qquad (10)$$

The eigenvalues of the symmetric modes are indicated by full circles in Fig. 2, the eigenvalue of the asymmetric mode is denoted by full square. The eigenvalues belonging to the continuum with the ε/C corresponding to ε_A and $\varepsilon_{S_{1,2}}$ which form the pairs of the counter-propagating waves with $k_x a = \pm\pi/2$ and $k_x a = \pm\pi/4$; $k_x a = \pm 3\pi/4$ are indicated by crosses and open circles, respectively in Fig. 2 (the upper panel). We note, that the values $\varepsilon_{S_{1,2}}$ of the symmetric modes can lie within or outside the continuum depending on the ratio $C_{1,2}/C$, while the asymmetric A-mode lies always within the continuum.

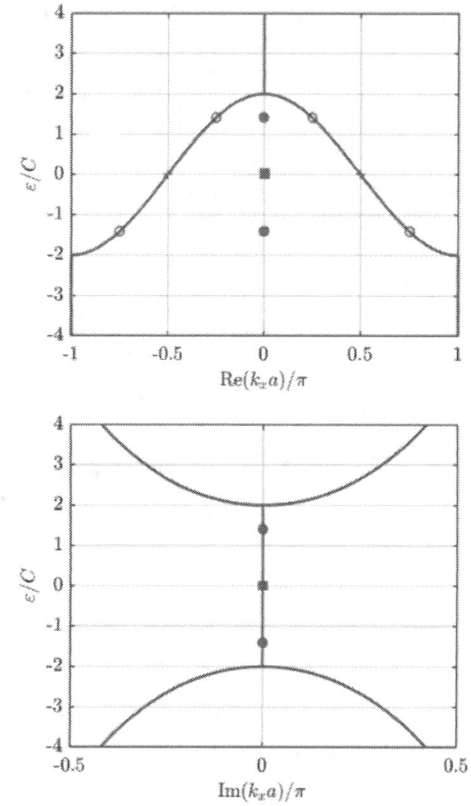

Figure 2. The real and imaginary parts of the band structure of 1D coupled waveguide array shown in Fig. 1. The positions of the states in the vertical waveguide are also indicated: S-modes (full circles) and A-mode (full square). The parameters are $C_1 = C_2 = C$ and $\delta_{-1} = \delta_1 = 0$.

In the third step, we consider the whole coupled system shown in Fig. 1, which is described by the Eqs. (1)-(4). Now, for the moment, we assume the trivial vertical symmetry $\delta_{-1} = \delta_1 = 0$. The localized A-mode given by the Eqs. (9)-(10) is also the solution of the whole system as $\psi_m = 0$, i.e., the A-mode cannot couple to propagating modes of the periodic array and thus it is a true BIC. Generalization of the result for the case, when $\delta_{\pm 1} \neq 0$, is straightforward: the localized A-mode can occur only if $\delta_{-1} = \delta_1$; instead of the Eq. (9), its energy is given by $\varepsilon_A = \delta_{-1} = \delta_1$ and the Eq. (10) remains

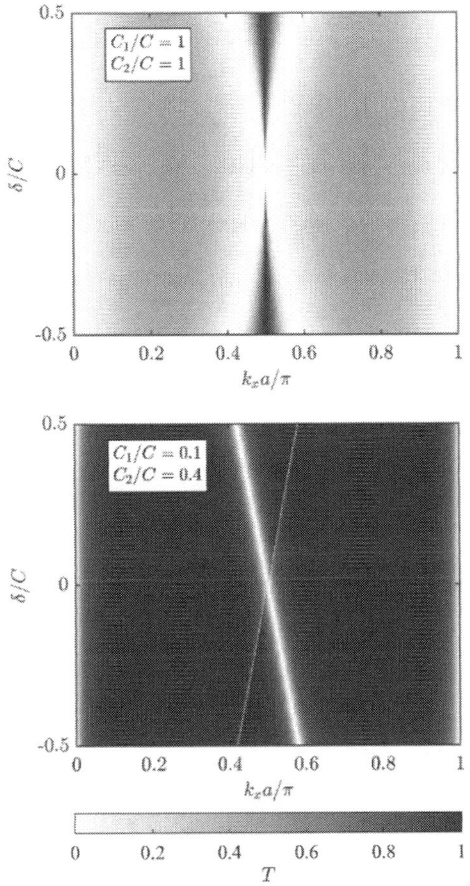

Figure 3. The transmittance T of the coupled-array system shown in Fig. 1 as a function of the Bloch wavenumber k_x and the asymmetry parameter $\delta = -\delta_{-1} = \delta_1$. The coupling coefficients C_1 and C_2 are indicated in the insets.

valid. Therefore, the asymmetric A-mode is a BIC provided $-2C < \delta_{-1} = \delta_1 < 2C$.

To examine the effect of a BIC on the characteristics of the structure we have calculated its transmittance and reflectance. We follow the approach described in Ref. [5] and assume the incident and reflected plane wave

$$\psi_m = e^{-ik_xam-i\varepsilon z} + re^{ik_xam-i\varepsilon z}, \; m \leq 0 \tag{11}$$

and the transmitted plane wave

$$\psi_m = te^{-ik_xam-i\varepsilon z}, \; m \geq 0. \tag{12}$$

After substituting the Eqs. (11) and (12) into the coupled equations (1)-(4) one obtains the analytical expressions for the amplitude transmittance t and reflectance r as

$$t = r + 1 = \frac{2C\sin(k_xa)}{2C\sin(k_xa) + i\mu}, \tag{13}$$

where

$$\mu = \delta_0 + \frac{C_1^2}{\varepsilon - \delta_{-1}} + \frac{C_2^2}{\varepsilon - \delta_1}. \tag{14}$$

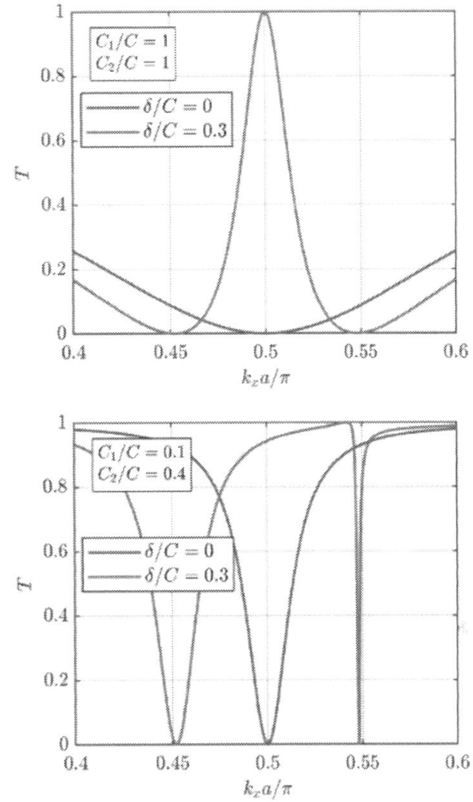

Figure 4. The transmittance spectra T for the parameters of Fig. 3 corresponding to the system with preserved (blue line) and broken (red line) symmetry.

The relation for the transmission given by the Eq. (13) for the coupled array system coincides with Fano scattering function [6] which reads as

$$\sigma(\epsilon) = \frac{(\epsilon + q)^2}{\epsilon^2 + 1}, \tag{15}$$

where ϵ is dimensionless energy and q is the shape parameter which determines the asymmetry of the profile. As a matter of fact, the intensity transmittance T defined as $T = |t|^2$ can be written in the form

$$T = \frac{\alpha_k^2}{\alpha_k^2 + 1}, \tag{16}$$

where $\alpha_k = 2C\sin(k_xa)/\mu$ plays the role of the dimensionless energy ϵ.

One can see that the transmittance T shown in Figs. 3 and 4 vanishes when $2C\sin(k_xa) = 0$, which happens at $k_xa = 0$ and $k_xa = \pm\pi$ corresponding to the edges of the continuum band, where the group velocity tends to zero. The transmittance also vanishes at $\varepsilon = 0$ due to diverging parameter $|\mu| \to \infty$ when $\delta = 0$ in the Eq. (13), see also the Eq. (14), and corresponds to BIC.

Breaking the vertical symmetry yields coupling of the A-state to the continuum and mediates the leakage of light into periodic array. It follows from our numerical experiments that the parameter δ_0 does not affect significantly BIC-induced features in the transmittance. Therefore, we set $\delta_0 = 0$ hereafter. Similarly, it is a "strength" of the asymmetry between δ_1 and δ_{-1} rather than actual values of the parameters that has the most prominent effect on the mentioned features. Therefore, in the following we focus on the class of asymmetric structures where vertical asymmetry is introduced through the parameter δ which is defined as $\delta = \delta_1 = -\delta_{-1}$. When the vertical symmetry is broken, μ becomes finite, coupling of the asymmetric mode to the continuum becomes non-negligible and the existence of the quasi-BIC state manifests itself as the peak $T = 1$ in the transmittance. The peak can be assigned to the constructive interference of the two counter-propagating modes with $k_x a = \pm\pi/2$ in the case $C_1 = C_2$ (see the upper panels in Figs. 3 and 4, $k_x a = \pi/2$) and excited localized A-mode given by the Eqs. (9) and (10). When $C_1 \neq C_2$, position of the peak moves with δ (see the lower panels in Figs. 3 and 4). The shape of the resonances is affected by the ratio C_1/C_2: when $C_1 = C_2$, the peak is Lorentzian and corresponds to the diverging asymmetry parameter $|q| \to 0$ – see upper panels of Figs. 3 and 4, while for $C_1 \neq C_2$, for example $C_1 = 0.1C$ and $C_2 = 0.4C$, the resonance possesses asymmetric Fano-like line shape – see lower panels of Figs. 3 and 4. The peak is always accompanied by two minima, at which the whole system becomes totally reflective, both of them arise due to the vanishing denominators in the second and third term in the Eq. (14) leading to an infinite scattering potential at the poles $\varepsilon = \delta_{\pm 1}$, i.e., $|\mu| \to \infty$ in the Eq. (13).

Within the continuum the ratio $C_{1,2}/C$ determines the fraction of the energy density transferred through the periodic chain of the waveguides. When $C = C_1 = C_2$ the transmittance has maxima $T = 0.5$ at $k_x a = \pm m\pi/4$ and $k_x a = \pm 3m\pi/4$, while for smaller values of the coupling coefficients the structure becomes totally transparent as it is shown in lower panels of Figs. 3 and 4.

4 Conclusions

We developed a theoretical model which allows to characterize spectral properties of the 1D photonic lattices consisting of a periodic chain of waveguides with two additional waveguides attached above and beneath the array and to analyze various kinds of BICs with broken vertical symmetry. The method can be extended to topologically nontrivial lattices to explore properties of the optical BIC pseudospin states.

Acknowledgments

We acknowledge financial support by the Czech Science Foundation (CSF) through the project 1900062S.

References

[1] C. W Hsu, B. Zhen, A. D. Stone, J. D. Joannopoulos, and M. Soljačić, Nat. Rev. Mater. **1**, 16048 (2016).

[2] S.I. Azzam and A. V. Kildishev, Adv. Optical Mater. **31**, 38-45 (2020).

[3] K. Koshelev, A. Bogdanov and Y. Kivshar, Optics and Photonics News, **9**, 2001469 (2020).

[4] Y. Plotnik, O. Peleg, F. Dreisow, M. Heinrich, S. Nolte, A. Szameit, M. Segev, Phys. Rev. Lett. **107**, 28-31 (2011).

[5] Weimann, S., Xu, Y., Keil, R., Miroshnichenko, A. E., Tünnermann, A., Nolte, S., Y. Kivshar, Y. S., Phys. Rev. Lett. **111**, 1-5 (2013).

[6] U. Fano, Phys. Rev. **124**, 1866 (1966).

Coupling of waveguide mode and graphene plasmons

Jiří Petráček[1,*], *Jiří* Čtyroký[2], *Vladimír* Kuzmiak[2], *Pavel* Kwiecien[3], *Ivan* Richter[3]

[1]Brno University of Technology, Faculty of Mechanical Engineering, Institute of Physical Engineering, Technická 2896/2, 61669 Brno, Czech Republic
[2]CAS Institute of Photonics and Electronics, Department of Fiber Lasers and Nonlinear Optics, Chaberská 57, 18251 Prague, Czech Republic
[3]Czech Technical University in Prague, Faculty of Nuclear Sciences and Physical Engineering, Department of Physical Electronics, Břehová 7, 11519 Prague 1, Czech Republic

Abstract. Photonic waveguides with graphene layers have been recently studied for their potential as fast and low-power electro-optic modulators with small footprints. We show that in the optical wavelength range of 1.55 μm, surface plasmons supported by the graphene layer with the chemical potential exceeding ~0.5 eV can couple with the waveguide mode and affect its propagation. This effect might be possibly utilized in technical applications as a very low-power amplitude modulation, temperature sensing, etc.

1 Introduction

Unusual electrical and optical properties of novel two-dimensional materials like graphene have revolutionized many branches of science and technology. Strong dependence of the surface conductivity of graphene on the chemical potential μ_C, controlled by either doping or applied voltage, makes it possible to modify the effective index of modes propagating in an optical waveguide with a graphene sheet overlay. For the optical communication wavelength range of 1550 nm, a graphene layer with the chemical potential μ_C below about 0.5 eV introduces a very strong optical attenuation, while for $\mu_C > 0.5$ eV the attenuation is low when the real part of the mode effective index is changed. In principle, a graphene layer can thus be utilized for both amplitude and phase electro-optic modulation [1-7].

Interestingly, the phase modulation can be accompanied with coupling of the waveguide mode with surface plasmon modes supported by the graphene stripe ("graphene modes"), even at the telecommunication optical wavelength band around 1550 nm [8].

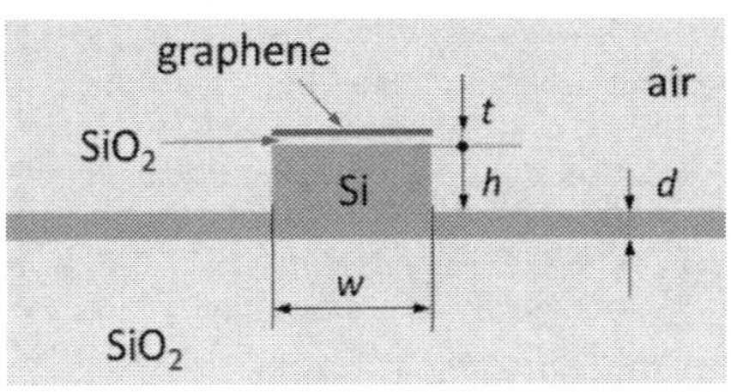

Fig. 1. Si rib waveguide modulator with a graphene layer.

In this paper, we consider a simplified example of a graphene-on-silicon modulator (see Fig. 1), inspired by the design described in [4], and demonstrate the coupling of the fundamental waveguide mode with graphene plasmon modes. We approximately describe the plasmon modes as the superposition two graphene plasmons reflected at the edges of the graphene stripe. Based on our recent comparison of various approaches to numerical modelling of light propagation in a silicon waveguide with graphene overlay [9], we employ a simplified coupled mode theory (CMT) [8] and calculate the complex propagation constant of the quasi-TE mode propagating in the waveguide modulator. We compare the result with a full-wave numerical simulation using the commercial software packet COMSOL Multiphysics. We show that despite the rather crude simplifications used in our implementation of the CMT, the similarity of both results is convincing. Furthermore, the coupling may strongly affect the modulator properties.

2 Model

The waveguide structure - the Si rib waveguide modulator with a graphene layer - is shown in Fig. 1. For the calculation we used the following parameters: $h = 220$ nm, $w = 450$ nm, and $d = 50$ nm. Graphene layer is deposited only on the top of the rib waveguide, separated from silicon with a thin SiO_2 layer, $t = 10$ nm. The operating wavelength is 1550 nm.

The optical properties of a graphene monolayer are determined by its complex surface conductivity σ_s, which can be described with an approximate expression presented in [1, 10]. It is well-known that, in the range of $\mu_C > 0.5$ eV, the positive imaginary part of the surface conductivity strongly prevails (we use the convention $\exp(-i\omega t)$ for time-harmonic quantities) which is the condition allowing propagation of a surface plasmon at

* Corresponding author: petracek@fme.vutbr.cz

the interfaces of a graphene layer, considered as an infinitely thin layer with a finite surface conductivity σ_s, sandwiched between two dielectrics.

The surface plasmon propagating along the waveguide axis cannot couple with the guided mode of the silicon waveguide because of a huge (1-2 orders of magnitude) mismatch of their mode effective indices (the effective index of the quasi-TE mode of the bare silicon waveguide calculated with COMSOL is 2.3754). However, the graphene stripe on top of the silicon ridge waveguide, shown in Fig. 1, supports a number of higher order graphene or "nanoribbon" modes [11-14] with smaller propagation constants, and some of them can match with that of the mode of the silicon waveguide. We use an approximate approach for determination of their propagation constants and field distributions. We assumme that similarly to a mode of a planar dielectric waveguide, the graphene modes result from the interference of two plasmons that are propagating under some (complex) angle with respect to the waveguide axis and are reflecting from the stripe edges. The dispersion equation for such modes can be written in the form of the transverse resonance condition (in the x direction in Fig. 2) [8]. Then, reflection of a surface plasmon from the edge of a graphene stripe is assumed to be total, and is modelled with a perfectly magnetic wall. This choice allows for a very simple evaluation of the effective indices and the electromagnetic field distributions of the graphene modes which are quite close to those obtained by using more rigorous COMSOL simulations. It should be noted that some lowest-order modes of the graphene stripe (including central and edge "ribbon plasmons" [14]) are out of scope of this approach. However, these modes cannot couple with the mode of a silicon waveguide due to strong mismatch of their propagation constants.

Coupling of a mode of a silicon waveguide with graphene modes can be considered as mutual coupling among modes of several parallel waveguides. One of the waveguides is the bare silicon waveguide, the other "waveguides" correspond to individual plasmonic modes supported by the graphene stripe. Here, we applied the principles of the complex CMT [15, 16] and used it for determination of modes of the composed structure. Our implementation [8] is based on several simplifications; namely, we ignore mutual coupling among various graphene modes and assume symmetrical matrix of coupling coefficients. Such procedure helps significantly simplify the calculation while keeping all aspects important for our analysis.

3 Results

Figure 2 shows the distribution of the horizontal electric field intensity component of the fundamental quasi-TE mode of the composed structure, calculated with COMSOL. It will be shown in the following text that the coupling is a resonant effect, therefore for Fig. 2, we used the resonance value of $\mu_C = 1.61$ eV where the modal loss is maximized, see Fig. 3 (c) (the fundamental quasi-TE mode is labeled as 'Si-Gr' in Fig. 3). The "decoration" of the mode field with the field distribution of the graphene mode is apparent.

Although the dominant electric field component of the plasmons is vertical, for the given structural parameters, there is curiously negligible coupling among the graphene plasmon modes and the TM-polarized waveguide mode. The reason stems from the nature of the coupling mechanism: only electric field components parallel with the graphene layer can contribute to the coupling; apparently, there is no graphene conductivity in the vertical direction.

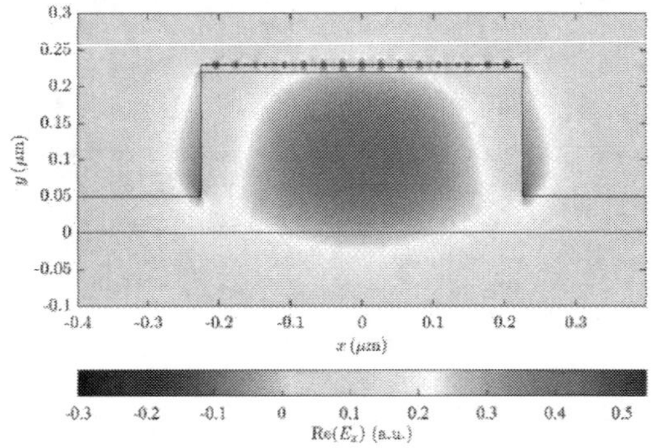

Fig. 2. Distribution of the horizontal (E_x) component of the fundamental quasi-TE mode. The chemical potential of graphene is $\mu_C = 1.61$ eV.

To demonstrate the coupling, we calculated the (complex) mode effective index N_{eff} for various modes with respect to the change of the chemical potential μ_C as follows:

(i) Using COMSOL, we first rigorously simulated the bare Si waveguide (with no graphene) and calculated its fundamental quasi-TE mode. Then we considered a graphene layer as a perturbation of the bare waveguide and employed a perturbation technique [9] to determine N_{eff} for the composed structure. The result is labeled as 'Si' (red line) in Figs. 3 (a), (b), and (c). Note that the perturbation method does not take into account the coupling.

(ii) We used the simplified theory and analytically calculated graphene modes propagating on the graphene stripe on top of the silicon ridge (silica substrate neglected). We considered symmetric graphene modes only (the mirror symmetry of the component E_x with respect to the $x = 0$ plane; as in Fig. 2), which can couple with the fundamental quasi-TE mode of the bare waveguide. Re(N_{eff}) for the resulting modes are shown in Fig. 3 (a) and marked as 'Gr' (yellow line in Fig. 3 (a) only). In general, for graphene modes, Re(N_{eff}) quickly decreases with increasing μ_C; therefore, only the relevant parts of the dependencies are shown in Fig. 3 (a). The modes exhibit extremely strong modal attenuation that for fixed Re(N_{eff}) decreases with μ_C; for the graphene modes with Re(N_{eff}) displayed in Fig. 3 (a), we observed the decreasing trend with values of Im(N_{eff}) in the range $2 < \mathrm{Im}(N_{\mathrm{eff}}) < 5$ (not shown in Fig. 3).

(iii) We used CMT that considers coupling between the waveguide mode (determined in (i)) and graphene modes (determined in (ii)) and calculated the relevant mode of the composed structure; the result is shown in Figs. 3 (b), (c) and marked as 'CMT' (yellow lines in Figs. 3 (b), (c) only).

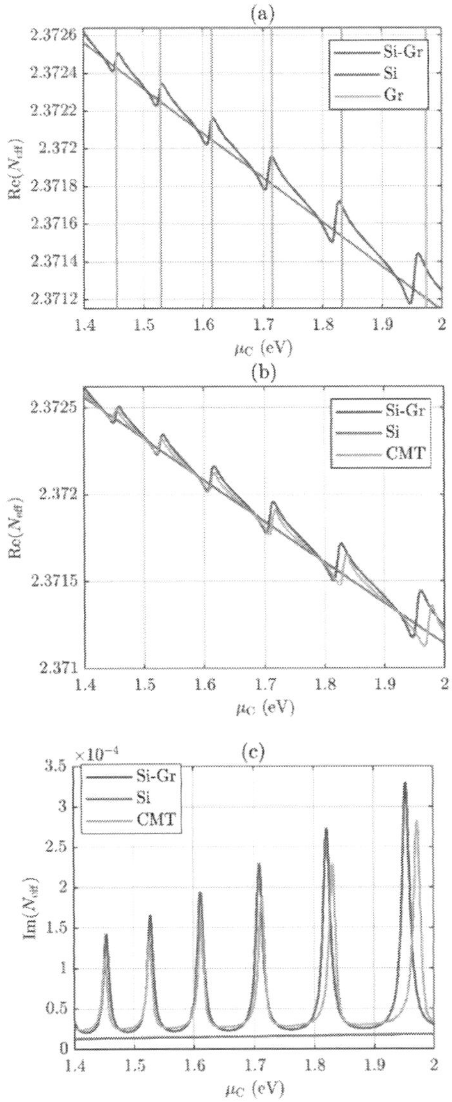

Fig. 3. (a), (b) Real and (c) imaginary parts of the mode effective index N_{eff} of various modes as functions of the graphene chemical potential μ_C. N_{eff} for the fundamental quasi-TE mode of the composed structure is marked as follows: 'Si-Gr' – rigorous calculation with COMSOL, 'CMT' – CMT calculation, 'Si' – the perturbation approach. N_{eff} for graphene modes is marked as 'Gr'. Only symmetric graphene modes (with the symmetry as the mode in Fig. 2) are shown.

(iv) Finally, we found the modes of the composed structure rigorously with COMSOL; the result for the fundamental quasi-TE mode is shown in Figs. 3 (a), (b), (c) and marked as 'Si-Gr' (blue line). In addition to the fundamental mode, we also display other modes of the composed structure in Fig. 4.

The graphs in Fig. 3 show that the coupling of the silicon waveguide mode with graphene modes has a typical resonant character; see 'Si-Gr' dependencies that reasonably agree with CMT and thus also provide convincing physical arguments for the presence of the coupling (the perturbation method 'Si' does not take into account the coupling). When the chemical potential is gradually changing, just one of the 'Gr' modes in a time comes into resonance with 'Si' mode, i.e., they are phase matched, see Fig. 3 (a). Consequently, at the resonance both the phase and amplitude of the waveguide mode are affected by the coupling. For $\mu_C > 1$ eV the peaks of Im(N_{eff}) caused by the coupling are of the order of 10^{-4}, which corresponds to a rather strong modal attenuation of the order of several dB/mm.

Next, in Fig. 4, we can see a simulated behavior of both real and imaginary parts of the effective indices, with respect to the change of the chemical potential, for the symmetric modes of the composed structure. We display only 'guided' modes with Re(N_{eff}) above the cutoff, Re(N_{eff}) > $N_{eff,c}$ = 1.6172; $N_{eff,c}$ being the effective index of the mode in the lateral waveguide (i.e. the waveguide with the core thickness d) in Fig. 1. In addition to the fundamental mode (marked as 'Si-Gr'), we observe a rich spectrum of modes that correspond to graphene modes in Fig 3 (a). The 'graphene' modes in Fig. 4 exhibit a strong attenuation that significantly increases when Re(N_{eff}) approaches value of $N_{eff,c}$ and decreases with increasing μ_C, provided Re(N_{eff}) is kept constant.

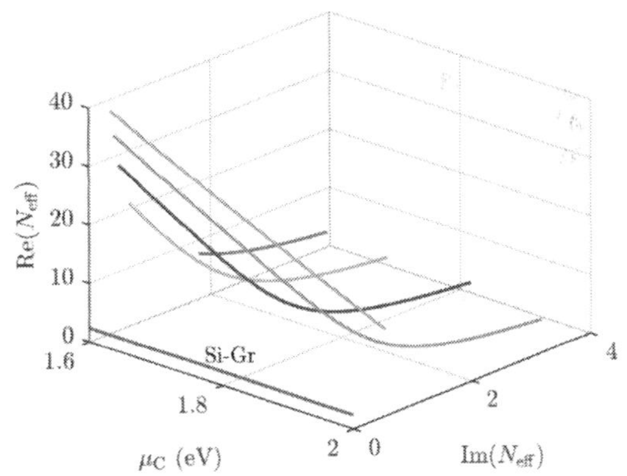

Fig. 4. Modal structure of the Si rib waveguide modulator with a graphene layer with chemical potential μ_C (rigorous calculation with COMSOL). The curve labeled as 'Si-Gr' describes the fundamental quasi-TE mode (the same mode as 'Si-Gr' dependencies in Fig. 3). Only symmetric modes are shown.

4 Conclusions

Properties of graphene plasmons and graphene modes have been previously studied mostly in THz and infrared frequency ranges. On the other hand, graphene layers on silicon waveguides have been recently very often used in the design and construction of photonic devices for modulation, switching, etc. These devices typically operate within the telecommunication band around 1550 nm, where the surface plasmon propagation on graphene layers has attracted much less attention. In this communication we show that in the range of the chemical potential of graphene above 0.5 eV, surface plasmons supported by graphene stripes deposited on the top of a silicon waveguides rather strongly affect their guiding properties due to the coupling of graphene plasmon modes with the mode propagating in the waveguide. This effect has been independently studied by both the approximate method based on the CMT, and by "rigorous" numerical simulations using COMSOL. Although the accuracy of the approximate method is not high, it offers a deep insight into the process and contributes to the understanding of the details of the coupling mechanism. The effect has not necessarily been considered harmful for the operation of silicon photonic devices, rather it may be employed in design of specific devices such as low-power modulators and sensors. Although we have focused with our analysis at the near infrared telecom wavelength range, we believe that this effect takes place not only in the near- to mid- IR silicon transparency window, but also in a 'standard' THz spectral range where the silicon waveguides are being used, too [6, 17, 18].

This work was financially supported by the Czech Science Foundation under Project 1900062S and by the Project Center of Advanced Applied Sciences – CZ.02.1.01/0.0/0.0/16_019/0000778 - European Structural and Investments Funds — Operational Programme Research, Development, and Education.

References

1. V. Sorianello, M. Midrio, M. Romagnoli, Design optimization of single and double layer graphene phase modulators in SOI, Optics Express, **23**, 6480 (2015)
2. R. Amin, et al., Attojoule-efficient graphene optical modulators, Applied Optics, **57**, D130 (2018)
3. G. Kovacevic, C. Phare, S. Y. Set, M. Lipson, S. Yamashita, Ultra-high-speed graphene optical modulator design based on tight field confinement in a slot waveguide, Applied Physics Express, **11**, 065102 (2018)
4. V. Sorianello, et al., Graphene–silicon phase modulators with gigahertz bandwidth, Nature Photonics, **120**, 40 (2018)
5. V. Sorianello, G. Contestabile, M. Romagnoli, Graphene on silicon modulators, Journal of Lightwave Technology, **38**, 2782 (2020)
6. M. Mittendorff, S. S. Li, T. E. Murphy, Graphene-based waveguide-integrated terahertz modulator, ACS Photonics, **4**, 316 (2017)
7. Y. S. Tao, H. W. Shu, M. Jin, X. J. Wang, L. J. Zhou, W. W. Zou, Numerical investigation of the linearity of graphene-based silicon waveguide modulator, Optics Express, **27**, 9013 (2019)
8. J. Čtyroký, J. Petráček, V. Kuzmiak, P. Kwiecien, I. Richter, Silicon waveguides with graphene: coupling of waveguide mode to surface plasmons, Journal of Optics, **22**, 095801 (2020)
9. J. Čtyroký, J. Petráček, P. Kwiecien, I. Richter, V. Kuzmiak, Graphene on an optical waveguide: comparison of simulation approaches, Optical and Quantum Electronics, **52**, 149 (2020)
10. F. H. L. Koppens, D. E. Chang, F. J. G. de Abajo, Graphene plasmonics: a platform for strong light-matter interactions, Nano Letters, **11**, 3370 (2011)
11. J. Christensen, A. Manjavacas, S. Thongrattanasiri, F. H. L. Koppens, F. J. G. de Abajo, Graphene plasmon waveguiding and hybridization in individual and paired nanoribbons, ACS Nano, **6**, 431 (2012)
12. N. Zhao, Z. X. Zhao, I. A. D. Williamson, S. Boutami, B. Zhao, S. Fan, High reflection from a one-dimensional array of graphene nanoribbons, ACS Photonics, **6**, 339 (2019)
13. K. O. Wedel, N. A. Mortensen, K. S. Thygesen, M. Wubs, Edge-dependent reflection and inherited fine structure of higher-order plasmons in graphene nanoribbons Phys. Rev. B, **99**, 045411 (2019)
14. H. W. Hou, J. H. Teng, T. Palacios, S. Chua, Edge plasmons and cut-off behavior of graphene nano-ribbon waveguides, Optics Communications, **370**, 226 (2016)
15. W. P. Huang, J. W. Mu, Complex coupled-mode theory for optical waveguides, Optics Express, **17**, 19134 (2009)
16. D. Marcuse, Light Transmission Optics, 2nd ed. (New York: Van Nostrand Reinhold), 1982
17. N. Vermeulen, J. L. Cheng, J. E. Sipe, H. Thienpont, Foundry-compatible SOI waveguides with a graphene top layer for wideband wavelength conversion, SPIE Proceedings **9891**, 98911B (2016)
18. A. Locatelli, G. E. Town, C. De Angelis, Graphene-based terahertz waveguide modulators, IEEE Transactions Terahertz Science Technology, **5**, 351 (2015)

EPJ Web of Conferences **255**, 07003 (2021) https://doi.org/10.1051/epjconf/202125507003
EOSAM 2021

Second and third harmonic generation from gold nanolayers: experiment versus theory

L. Rodríguez-Suné[1,*], J. Trull[1], N. Akozbek[2], D. De Ceglia[3], M. A. Vincenti[4], M. Scalora[5], and C. Cojocaru[1]

[1]Department of Physics, Universitat Politècnica de Catalunya, Rambla Sant Nebridi 22, 08222 Terrassa, Spain
[2]AEgis Technologies Inc., 401 Jan Davis Dr., Huntsville, Alabama 35806, USA
[3]Department of Information Engineering – University of Padova, Via Gradenigo 6/a, 35131 Padova, Italy
[4]Department of Information Engineering – University of Brescia, Via Branze 38, 25123 Brescia, Italy
[5]Aviation and Missile Center, US Army CCDC, Redsone Arsenal, AL 35898-5000 USA

Abstract. The use of semiconductors, metals, or ordinary dielectrics in the process of fabrication of nano-devices is at the front edge of nowadays technology, exploiting the properties of light propagation and localization at nanometric scale in new and surprising ways. Understanding accurately how light interacts with these materials at the nanoscale is crucial if one is to properly engineer nano-devices. When the nanoscale is reached, light-matter interactions display new phenomena where conventional approximations may not always be applicable and they should be either revised or voided. In this work, we measure the efficiency of second and third harmonic generation from gold nanolayers. The experimental results are compared with numerical simulations based on a detailed microscopic hydrodynamic model that considers different effects playing a role in the nonlinear response, not usually considered by more generic models. The agreement between experimental and theoretical results proves the importance of all these contributions.

1 Introduction

More effort to properly engineer nano-antenas, filters, and other devices whose geometrical features approach atomic size, has been put in the recent years. For this reason, it is very important to understand how light interacts at the nanoscale with metals, semiconductors, or ordinary dielectrics. At this scale, light-matter interaction can display completely new phenomena and conventional approximations should be revised. This is the case, for instance, of two well-known nonlinear processes such as second and third harmonic generation (SHG and THG). These phenomena have been studied extensively in different optical materials. Usually, high conversion efficiencies are sought, what requires thick nonlinear materials with high nonlinearities, phase-matching conditions and low material absorption. Under these circumstances, the leading nonlinear polarization term corresponds to the bulk contribution described through the second and third order nonlinear susceptibility tensors ($\chi^{(2)}$ and $\chi^{(3)}$). However, when the nanometric size is reached, SH and TH efficiencies may decrease and phase matching conditions and absorption may no longer play a primary or significant role. Moreover, the effective $\chi^{(2)}$ and $\chi^{(3)}$ may not coincide with their bulk, local counterparts and may depend on the type of nonlinearities that are triggered. In addition, contributions to the nonlinear polarization arising from electric quadrupole-like and magnetic sources should also be taken into account.

Most of the models used to explain harmonic generation from metals at the nanoscale rely on assigning effective surface and volume $\chi^{(2)}$ for SHG, and an effective volume $\chi^{(3)}$ to describe THG, which generally lack a detailed, microscopic, dynamical description of light propagation and light-matter interactions. With this, most theoretical predictions appear to accurately describe the general shape of the angular dependence of the SH signal, but fail in describing the observed amplitude.

In this work, we report experimental measurements on SHG and THG from 20nm and 70nm-thick Au nanolayers. These measurements are compared with numerical simulations based on a microscopic hydrodynamic model which accounts for surface, magnetic and bulk nonlinearities arising from free and bound charges, preserving linear and nonlinear dispersion, nonlocal effects due to pressure and viscosity, and an intensity dependent free electron density. This model is adapted and applied anew based on previous work reported in references [1-4].

* Corresponding author: laura.rodriguez.sune@upc.edu

2 Results

2.1. Experiments

Measurements of SH signals from the 20nm-thick Au nanolayer have been conducted in transmission and in reflection while the SHG from the 70nm-thick Au nanolayers has been studied in reflection. In both cases, incident pulses tuned at 800nm and 1064nm have been used. The process of THG from the 20nm-thick layer was conducted either in transmission and reflection using incident pulses tuned at 1064nm. For this purpose, we have developed an experimental set-up, shown in Fig. 1, capable of analysing the angular dependence of the harmonic signals in both transmission and reflection configurations. First of all, a half-wave plate is used to control the polarization of the incident field. Then, we filter out any possible SH or TH coming from different optical components placed before the sample. We use a lens to focus the beam on the sample plane. A focal length f=200mm was used for the SHG experiments to obtain beam intensities in the range of 1-2 GW/cm^2. A focal length f=100mm was employed for the THG measurements in order to achieve higher beam intensities. The sample is mounted on a rotary support which allows us to take measurements as a function of the angle of incidence. Just after the sample, the fundamental field is attenuated by means of a filter in order to avoid any potential SHG or THG from the surfaces of the optical elements placed along the set-up after the sample. After that, a lens with focal length f=100mm is used to collimate the beam, and a polarizer helps us to select the SH or TH polarization. Then, we use a prism and a blocking edge to separate and obscure the remaining fundamental field radiation from the SH or TH path. Finally, the harmonic signals are detected by means of a photomultiplier tube, on which we place a narrow-band spectral filter having 20nm band pass transmission around either the SH or the TH wavelength. This hole set-up is mounted on a rotary platform which allows us to take measurements in transmission and in reflection.

A calibration procedure has been performed to estimate the efficiency of a given process as the ratio between the SH or TH intensity generated in transmission or reflection, and the total peak pump pulse intensity just before the sample.

2.2 Theory

To understand the experimental results, we have performed numerical simulations reproducing the experimental situation. These simulations are based on a theoretical model that embraces full-scale, time-domain coupling of matter to the macroscopic Maxwell's equations. Our approach consists in formulating a microscopic, hydrodynamic model to understand linear and nonlinear optical properties of metals by accounting for competing surface, magnetic and bulk nonlinearities arising from both free and bound electrons, preserving linear and nonlinear material dispersion, nonlocal effects due to pressure and viscosity, and an intensity dependent free electron density to which we refer as hot electron contribution. When applying Newtonian dynamics to free and bound electrons, we obtain the simultaneous material equations of motion Eq. 1 and 2. Here, $\mathbf{P}_{b,j}$ and $\mathbf{P}_f$ are bound and free electron polarizations, respectively. Eq. 1 describes the behaviour of bound electrons. The j counter indicates multiple bound electron spices. In a metal like Au, one free (Drude) and two bound (Lorentz) electron spices generally suffice to describe the local, linear dielectric function down to a wavelength of ~200nm. Each Lorentz spices is characterized by a third order, isotropic nonlinearity $\mathbf{P}_{b,j}^{NL} = -\tilde{\beta}(\mathbf{P}_{b,j} \cdot \mathbf{P}_{bj})\mathbf{P}_{bj}$, where the coefficient $\tilde{\beta}$ may be derived for a nonlinear, two-dimensional oscillator model, and taking into account typical bound electron densities, lattice constant and

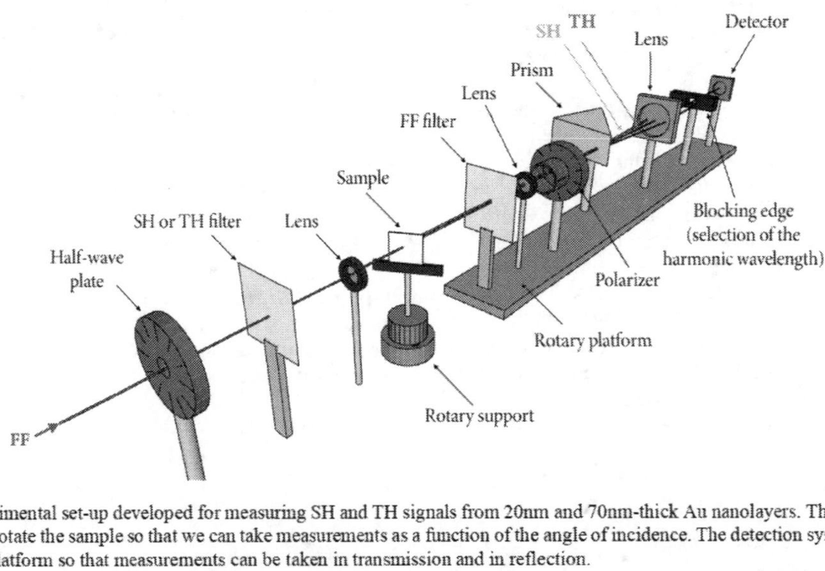

Fig. 1. Experimental set-up developed for measuring SH and TH signals from 20nm and 70nm-thick Au nanolayers. The rotary support allows us to rotate the sample so that we can take measurements as a function of the angle of incidence. The detection system is mounted on a rotary platform so that measurements can be taken in transmission and in reflection.

$$\ddot{\mathbf{P}}_{b,j} + \tilde{\gamma}_{b,j}\dot{\mathbf{P}}_{b,j} + \tilde{\omega}_{0,b,j}\mathbf{P}_{b,j} + \mathbf{P}^{NL}_{b,j} = \frac{n_{0,b}e^2\lambda_0^2}{m_b^* c^2}\mathbf{E} + \frac{e\lambda_0}{m_b^* c^2}\mathbf{E}(\nabla\cdot\mathbf{P}_{b,j}) + \frac{e\lambda_0}{m_b^* c^2}\dot{\mathbf{P}}_{b,j}\times\mathbf{H} \quad (1)$$

$$\ddot{\mathbf{P}}_f + \tilde{\gamma}_f\dot{\mathbf{P}}_f = \frac{n_{0,f}e^2\lambda_0^2}{m_{0,f}^* c^2}\mathbf{E} - \frac{e\lambda_0}{m_{0,f}^* c^2}\mathbf{E}(\nabla\cdot\mathbf{P}_f) + \tilde{\Lambda}(\mathbf{E}\cdot\mathbf{E})\mathbf{E} + \frac{e\lambda_0}{m_{0,f}^* c^2}\dot{\mathbf{P}}_f\times\mathbf{H} + \frac{3E_F}{5m_{0,f}^* c^2}\left(\nabla(\nabla\cdot\mathbf{P}_f) + \frac{1}{2}\nabla^2\mathbf{P}_f\right) +$$
$$-\frac{1}{n_{0,f}e\lambda_0}\left[(\nabla\cdot\dot{\mathbf{P}}_f)\dot{\mathbf{P}}_f + (\dot{\mathbf{P}}_f\cdot\nabla)\dot{\mathbf{P}}_f\right] \quad (2)$$

resonance frequencies for this material, it has a value of $\tilde{\beta} \approx 10^{-8}$. This parameter governs all third order effects triggered by the background crystal, i.e. bound electrons, including self-phase modulation, nonlinear absorption, and THG power conversion efficiencies. Eq. 2 determines the action of free electrons. As it can be seen, apart from the usual linear driving term there are different free electron contributions that give rise to harmonic generation. A quadrupole-like Coulomb term $-\frac{e\lambda_0}{m_{0,f}^* c^2}\mathbf{E}(\nabla\cdot\mathbf{P}_f)$, a magnetic Lorentz term $\frac{e\lambda_0}{m_{0,f}^* c^2}\dot{\mathbf{P}}_f\times\mathbf{H}$, convection terms $-\frac{1}{n_{0,f}e\lambda_0}\left[(\nabla\cdot\dot{\mathbf{P}}_f)\dot{\mathbf{P}}_f + (\dot{\mathbf{P}}_f\cdot\nabla)\dot{\mathbf{P}}_f\right]$, nonlocal effects represented by pressure and viscosity $\frac{3E_F}{5m_{0,f}^* c^2}\left(\nabla(\nabla\cdot\mathbf{P}_f) + \frac{1}{2}\nabla^2\mathbf{P}_f\right)$, and the hot electrons contribution $\tilde{\Lambda}(\mathbf{E}\cdot\mathbf{E})\mathbf{E}$. In this work, we highlight and discuss for the first time the relative roles bound and hot electrons play in THG and end up with the conclusion that the generated TH signal is mostly triggered by hot electron dynamics.

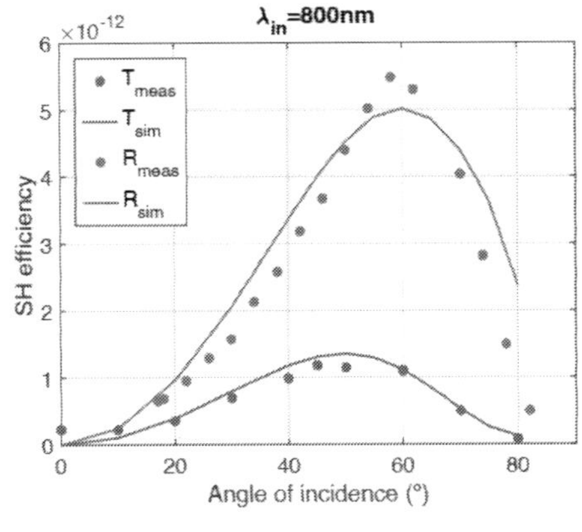

Fig. 2. Measured (full circles) and predicted (solid curves) SHG efficiencies as a function of the angle of incidence from the 20nm-thick Au layer when the incident pulse is tuned at 800nm. Transmission and reflection are plotted in blue and red, respectively.

In this section, we present some of the experimental results that we have obtained, which are compared with their corresponding numerical predictions.

In Fig. 2 we show results of SHG from the 20nm-thick Au layer when the sample is illuminated at 800nm. Experiments are depicted using full circles and simulations using solid curves. Transmitted (blue) and reflected (red) efficiencies are plotted as a function of the angle of incidence. As it can be seen, the maximum of the reflected curve is a little bit shifted towards a larger angle of incidence with respect to the transmitted curve. This phenomenon can be appreciated in both experimental and predicted results. Moreover, the ratio between the measured reflected and transmitted efficiencies is also reproduced in the theoretical case.

In Fig. 3 we present the reflection measurements (full circles) of the SH signal generated by the 70nm-thick Au layer when it is illuminated at 1064nm. As it can be seen, the maximum of the curve is achieved at a large angle of incidence, around 70°, fact that is reproduced in the predicted results.

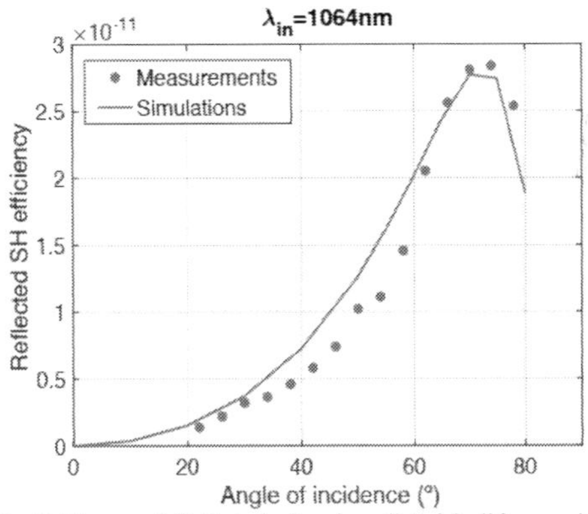

Fig. 3. Measured (full circles) and predicted (solid curves) SHG transmitted efficiencies as a function of the angle of incidence from the 70nm-thick Au layer when the incident pulse is tuned at 1064nm.

In Fig. 4, we show the results of the transmitted TH from the 20nm-thick Au layer as a function of the angle of incidence when the input pulse is tuned at 1064nm. In blue, we plot the transmitted TM-polarized TH for a TM-polarized incident pump, and in red we represent the transmitted TE-polarized TH generated when the incident field is TE-polarized. The simulations are performed by taking both bound and hot electrons into account. By only introducing the bound electrons contribution, efficiencies of order 10^{-11} where obtained, clearly inadequate to explain our measurements. Instead, what is required to reproduce the conversion efficiencies that we observe is the introduction of the hot electrons contribution.

In order to see how important the bound and hot electrons contribution to the TH signal can be, we have plotted in Fig. 4 predictions of the reflected TH efficiency from a 20nm-thick Au layer as a function of the input wavelength for three different scenarios: only taking bound electrons into account (black), only taking hot electrons into account (red) and taking both contributions into account (blue). It can be seen that, separately, each type of third order nonlinearity yields quantitatively and qualitatively a similar response, with a TH peak for pump wavelength of ~600nm. However, their combined response redshifts the TH peak. This prediction runs counter to intuition because an increased free electron density should blueshift the plasma frequency, with expectedly similar outcome for the TH peak. It is obvious that the two components interfere and conspire to instead redshift the peak, an effect that encapsulated a cautionary tale for any experimental result that may be cavalierly extrapolated without the benefit of proper assumptions and theoretical support.

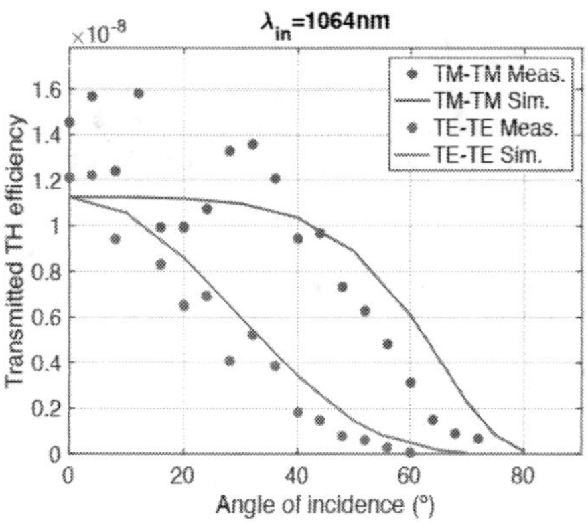

Fig. 4. Measured (full circles) and predicted (solid curves) transmitted THG efficiencies as a function of the angle of incidence from the 20nm-thick Au layer when the incident field is tuned at 1064nm. A TM- (blue) and TE-polarized (red) TH field is detected when the fundamental field is TM- or TE-polarized, respectively.

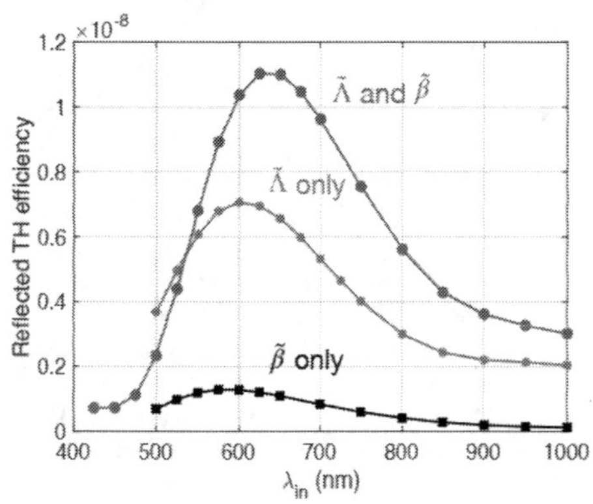

Fig. 5. Predictions of the reflected TH efficiency as a function of the pump wavelength obtained under three different scenarios: taking only bound electrons into account (black), takin only hot electrons into account (red), and taking both bound and hot electrons into account (blue).

3 Conclusions

In conclusion, we have reported measurements on SHG and THG from 20nm and 70nm-thick Au layers. The SH field generated by the 20nm-thick sample measured as a function of the angle of incidence carries information about combined surface and volume currents excited on and inside the sample. The angular dependence of the reflected SH signal from the 70nm-thick Au layer supports mostly surface currents. These measurements are compared with numerical simulations based on a hydrodynamic approach to model light-matter interactions that make no assumptions about effective surface or volume nonlinearities. Instead, we rely on temporal and spatial derivatives and mere knowledge of the effective electron mass to determine the relative magnitudes of surface and volume contributions. With this approach we find remarkable agreement with experimental observations. The THG measurements and simulations are reported for TM and TE polarizations of the generated field from the 20nm-thick layer in transmission and as a function of the angle of incidence. The generated TH signal is attributed mostly to hot electron dynamics.

References

1. M. Scalora, M. A. Vincenti, D. de Ceglia, V. Roppo, M. Centini, N. Akozbek and M. J. Bloemer, Phys. Rev. A **82**, 043828 (2010).
2. M. Scalora, M. A. Vincenti, D. de Ceglia, C. M. Cojocaru, M. Grande and J. W. Haus, J. Opt. Soc. Am. B **32**, 2129-2138 (2010).
3. L. Rodríguez-Suné, M. Scalora, A. S. Johnson, C. Cojocaru, N. Akozbek, Z. J. Coppens, D. Perez-Salinas, S. Wall and J. Trull, APL Photonics **5**, 010801 (2020).
4. M. Scalora, J. Trull, D. de Ceglia, M. A. Vincenti, N. Akozbek, C. Coppens, L. Rodríguez-Suné and C. Cojocaru, Phys. Rev. A **101**, 053828 (2020).

 https://doi.org/10.1051/epjconf/202125508001

Multiplexed fiber-optic Fabry-Pérot cavities for refractive index and temperature sensing fabricated using diamond-blade dicing

Ivonne Pfalzgraf[1], *Sergiy* Suntsov[1], *Kore* Hasse[1], and *Detlef* Kip[1,*]

[1]Faculty of Electrical Engineering, Helmut Schmidt University, 22043 Hamburg, Germany

Abstract. We report on multiplexing several Fabry-Pérot (FP) cavities in single-mode optical fibers for high-precision spatially resolved sensing of refractive indices (RI) of liquids. Resonators are fabricated by cutting small slots into fibers using a diamond-blade dicing saw and additional coating with thin Ta_2O_5 layers to increase cavity reflectance. Temperature compensation of RI measurements is achieved either by evaluating the reflection signals resulting from the solid core parts between different open-cavity sensor elements, or by using a thin Si inlay glued into one of the open cavities. The multiplexing performance and accuracy of the fabricated sensors with up to four open cavities were tested on sucrose solutions over a range of temperatures.

1 Introduction

Fiber-optic sensors have gained much attention in diverse areas due to their excellent characteristics including small size, high sensitivity, biocompatibility, and immunity to electromagnetic interference [1-4]. Among them, fiber sensors based on FP interference have shown very high sensitivity for measuring different physical parameters like RI, pressure, or temperature. At the same time, there is an interesting option for multiplexing of several sensors in a single fiber. Cutting open cavities into an optical fiber and through its core is perhaps the easiest realization of FP resonators. Here blade diced cavities [4-6] constitutes a simple alternative to other methods, such as chemical etching, laser micromachining, ion beam milling, or fiber splicing.

A possible application of multiplexed FP sensors at the fiber tip is the detection of the spatial distribution inside a liquid under test. Furthermore, FP cavities can be filled with molecularly imprinted polymers or coated with functional layers sensitive to trace substances, allowing the realization of an ultra-compact lab-on-a-fiber.

In this work, we present multiplexing of four open-cavity FP RI sensors in one optical fiber, clearly exceeding our recent results described in [6]. Individual sensors are formed by dicing through the core of a standard step-index single-mode fiber using a precision diamond-blade wafer saw. Either the solid fiber parts between open cavities or an additional Si inlay glued into one of the cavities can be used for temperature sensing [2]. To test the sensor's performance and the ability to sense refractive index of solutions with high precision at different spatial locations, we perform temperature dependent measurements in sucrose solutions with precisely adjusted refractive index.

2 Fabrication Method and Setup

For cutting the optical fiber is glued into a quartz ferrule to increase mechanical stability. Slots of different width are cut through the core of the optical fibers using a Disco DAD322 wafer saw to realize open FP micro-cavities. To achieve smooth surfaces, resin-based blades with very fine grit are used. In order to increase the mechanical stability of the whole sensor structure with its many cuts, a surrounding quartz glass ferrule is used. Due to blade conditioning with a special dresser board, all cuts have a nearly rectangular shape. For an optional Si inlay a rectangular plate is prepared with the dicing saw, placed inside one of the open cavities and glued with NOA61. The reflectivity of open FP cavities can be increased using dielectric layers on both sidewalls. For this we evaporated single high-index Ta_2O_5 layers ($n = 2.07$ at 1550 nm) on

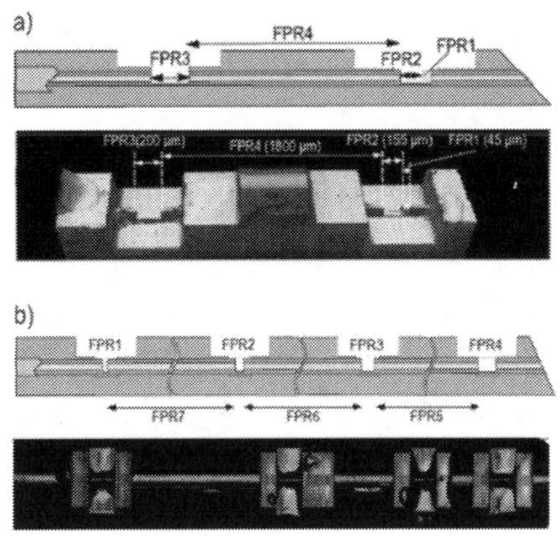

Fig. 1. Examples of two FP fiber sensors (top: schematic; bottom: confocal microscope image): (a) two RI sensors with additional Si inlay, (b) four different open-cavity RI sensors.

[*] Corresponding author: kip@hsu-hh.de

both facets of an open cavity while monitoring the optical reflection spectra inside an electron-beam coater. In Fig. 1 two examples of fabricated sensors are shown.

For the experimental setup an interrogation system (Micron Optics Hyperion Si155) operating in the wavelength range of (1460 – 1620) nm with an accuracy of 1 pm and a computer system for data read-out is used. The refractive index of the medium under test is changed by using sucrose solutions with calibrated RI in a temperature stabilized water bath (Julabo FP45) having an accuracy of 0.01°C.

In order to extract small changes of the optical length L_O of the individual cavities induced by temperature T and/or RI changes, phase tracking of the essential fast Fourier transform (FFT) components is used. Temperature and RI changes are calculated from the measured total phase changes $\Delta\Phi$ using appropriate calibration functions:

$$\Delta n = \frac{\lambda}{4\pi L_R}\Delta\phi = \Delta n_F(T) + \Delta n_L(T) = \Delta n_m + f_1$$

where Δn_m is the RI change related to the composition of the water based liquid (e.g. the concentration of sucrose in water), Δn_F is the temperature dependent fluid's RI change and Δn_L is a contribution caused by the sensor's thermal expansion ΔL_R of the physical cavity length L_R. The two temperature dependent parts can be combined in a temperature calibration function f_1.

Calibration of the temperature-sensing solid core regions is performed using

$$\Delta T_m = \frac{\lambda}{4\pi(dn/dT)L_R}\Delta\phi\, f_2 = \Delta T\, f_2$$

where ΔT_m is the temperature measured by the fiber parts, dn/dT is the thermo-optic coefficient of silica, and f_2 is another calibration function.

3 Experimental Results

The sensing region of the first sensor from Fig. 1 comprises four adjacent FPRs, namely the two open cavities, the solid fiber part in between and the Si inlay. To determine the sensitivity of the open cavities, the sensor was first immersed in solutions with different RI. Figure 2 (a) shows the measured phase change and calculated change in RI for the two open cavities, resulting in a sensitivity of 1153 nm/RIU for FPR2 and 1168 nm/RIU for FPR3. Here the sensor is calibrated in order to measure value Δn_m without cross-sensitivity to temperature. The response of the Si inlay and of the fiber piece between FPR2 and FPR3 to temperature changes is presented in Fig. 2 (b). The temperature is measured with a spectral sensitivity of 79 pm/°C for the Si inlay and a lower value of 13 pm/°C for the fiber piece. Here the high sensitivity of the Si inlay results from the large thermo-optic coefficient 1.7×10^{-4} K^{-1} of Si. Overall, this sensor allows for spatially resolved temperature-compensated RI sensing with an accuracy of 2×10^{-5} RIU.

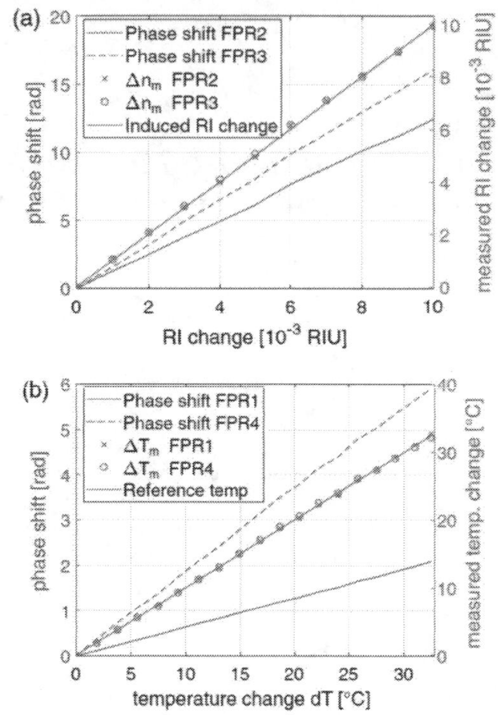

Fig. 2. (a) Response of the open cavities to a change in RI, induced with sucrose solutions. (b) Response of Si inlay and fiber piece for an induced change in temperature.

In the second example, we present a fiber sensor where we explored a higher number of accurately evaluable open cavities on a single fiber with additional enhancement of their reflectivity using dielectric coatings. The sensor is measuring simultaneously refractive index distributions using four different open cavities distributed along the fiber. As before, additional FP resonators formed by solid fiber parts in between the open cavities are used for temperature measurement. After appropriate calibration, this allows for automatic compensation of crosstalk between temperature- and composition-induced fluids' refractive index changes. Starting with FP resonator FPR4 (1st coating) located towards the tilted tip of the fiber (used to avoid reflections) and afterwards FPR3 (2nd coating) were additionally coated with thin Ta_2O_5 layers, while the reflection spectra of the sensor were simultaneously recorded during evaporation. While the optical thickness of the coating for sensor FPR4 was chosen to match a quarter-wave layer. The thicknesses for the coatings of FPR3 were chosen to be significantly lower and are about 1/6.5 of the thickness of that for FPR4. In this way, the response (i.e. strength of corresponding FFT peak) of each open cavity was optimized in order to improve the accuracy of RI measurements for both FPR3 and FPR4. The obtained experimental FFT spectra of Fig. 3 are in good quantitative agreement with the simulated ones. The slight increase in the reflectivity of FPR2 is here probably due to insufficient shielding of this sensor in the second coating step.

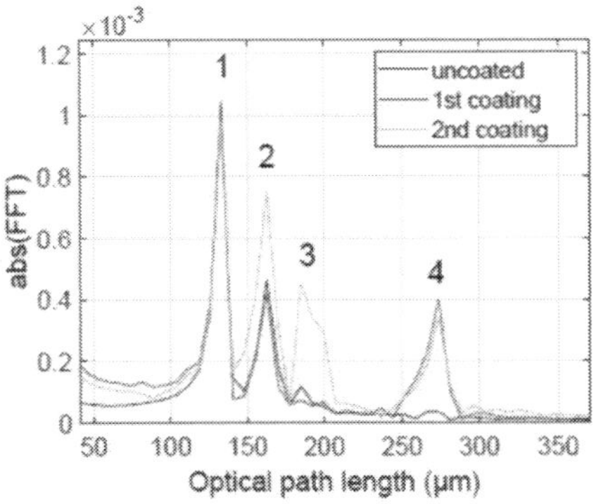

Fig. 3. Calculated FFT of the measured reflection spectra of the sensor with different configurations of coatings while immersed in DI water: no coating (blue), coating on the side walls of FPR4 (red) and additional FPR3 (green).

The RI sensing properties of this multiplexed sensor are presented in Fig. 4. Obviously, the accuracy of RI measurements is significantly enhanced by additional reflective coatings (cf. Fig. 4(b)) when compared to the initially uncoated ones (cf. Fig. 4(a)). The achieved sensitivity of ~1185 nm/RIU for RI sensing is similar for all four sensors. Again, calibration of the sensor allows for full compensation of any temperature cross-sensitivity.

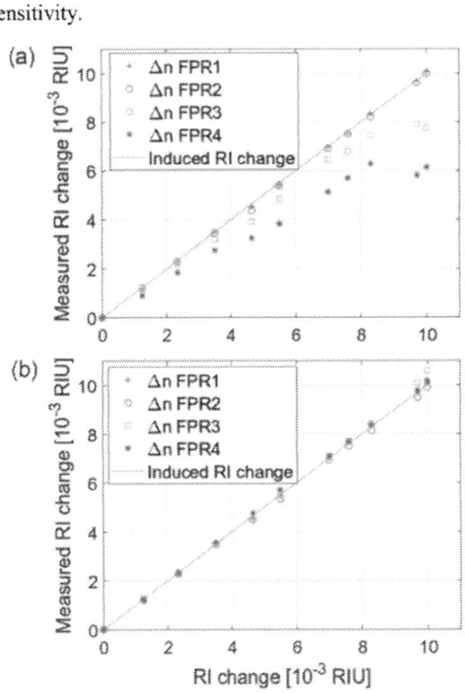

Fig. 4. Measurement of RI of the same liquid using the sensor from Fig. 1 (b) with four independent FP sensors with (a) uncoated FPRs) and (b) coated sensors FPR3 and FPR4.

Next, we characterize this sensor in terms of its temperature measurement sensitivity using the solid core fiber sections. For this, in Fig. 5 we determined the response (phase change) to induced temperature (0.5 °C steps) changes when the senor is immersed in DI water. Sensitivities of 19.4 pm/°C (FPR5), 19.33 pm/°C (FPR6) and 20 pm/°C (FPR7) were achieved with an accuracy of 0.01/°C. This sensor thus shows high average sensitivities of 1160 nm/RIU and 20pm/°C for RI and temperature, respectively, with good suppression of cross-talk among measured parameters and sensor elements, as required for biomedical applications.

Finally, with the performed calibration of the different sensing regions for either RI or temperature changes, we test the cross-sensitivity of the open cavities FPR1 to

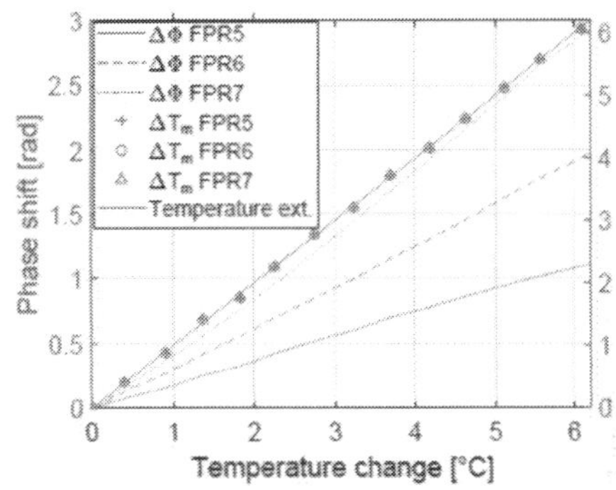

Fig. 5. Response of the solid fiber sections for an induced change in temperature.

FPR4 with respect to temperature. For this, the sensor is place in DI water while the temperature is changed. From this measurement shown in Fig. 6, we determine the accuracy, i.e. the deviation of the values Δn_m measured in this way from the expected value $\Delta n_m = 0$ is in the range of 6×10^{-6} RIU.

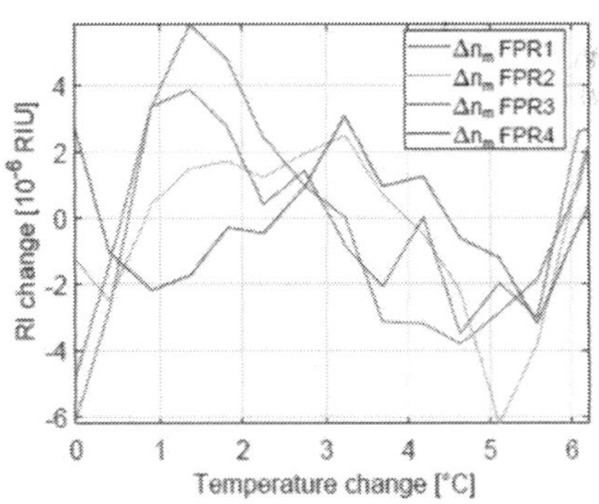

Fig. 6. Measurement of corrected RI change Δnm after temperature calibration of all open-cavity sensors.

4 Conclusion

In summary, we have demonstrated a very compact sensor at the tip of a single-mode optical fiber, with which four open FP cavities can be read out simultaneously with high accuracy. At the same time, real-time temperature compensation is possible, thus avoiding any cross-talk induced by inherent temperature changes during a measurement.

Such multiplexed fiber sensors can be effectively fabricated by diamond-blade dicing into single-mode fibers and have great potential to be used as lab-on-a-fiber. Additional reflective coatings allow for an improved accuracy, while the solid fiber parts can be applied to compensate for any cross-talk by temperature changes. Due to its compact dimensions, this novel fiber sensor may suit ideally to monitor both RI and temperature changes in small volume samples like in bio-medical applications.

Funding by Deutsche Forschungsgemeinschaft (DFG) (KI482/19-1) is gratefully acknowledged.

References

1. H. Joe, H. Yun, S. Jo, B. Min, Mater. Struct. **5**, 173 (2018)
2. B. Lee, Opt. Fiber Technol. **9**, 57 (2003)
3. S. Suntsov, C. E. Rüter, D. Kip, Appl. Opt. **58**, 2076 (2019)
4. H. Bae, D. Yun, H. Liu, D. A. Olson, M. Yu, J. Lightwave Technol. **32**, 1585 (2014)
5. S. Suntsov, C. E. Rüter, T. Schipkowski, D. Kip, Appl. Opt. **56**, 9139 (2017)
6. I. Pfalzgraf, S. Suntsov, D. Kip, Appl. Opt. **60**, 1428 (2021)

Simulation of the optical properties of gold nanoparticles on sodium alginate

Caterina Summonte[1,*], *Piera* Maccagnani[1], *Alberto* Maurizi[1,2], *Giulio* Pizzochero[1], and *Gabriele* Bolognini[1]

[1]CNR-IMM, u.o.s Bologna, via Gobetti 101, 40129 Bologna Italy
[2]Dipartimento di Scienze Statistiche "Paolo Fortunati" Università di Bologna, Italy

Abstract. In this contribution, we report on the simulation of optical reflectance and transmittance (R&T) taken on a set of gold nanoparticles thin film, deposited on sodium alginate by magnetron sputtering. The gold layer is very thin, so that the films are not continuous and the material is arranged in nanostructured layers. R&T spectra are simulated using the Generalized Transfer Matrix method applied to the film-on-substrate model. The gold NP films are simulated using the Drude-Lorentz model, by taking into account that the optical function of nanostructured gold exhibits increased collision frequency and reduced relaxation time. Moreover, the signal of localized surface plasmon, evident in the spectra, is simulated by introducing a dedicated modified Lorentz oscillator. The experimental results are well reproduced by the applied model. All trends (amplitude and energy position of the plasmon oscillator, film thickness, relaxation time) are correlated with the deposition parameters. The procedure represents a useful tool in the characterisation of such nanoparticles thin films.

1 Introduction

Fabrication of thin conductive films on polymers is an important research field for the next generation of devices, where flexibility and lightweight are two main properties for the fabrication of wearable electronics and soft robotics.

Sodium alginate (SA) is a biopolymer extracted from marine algae, which is presently rising interest for applications in green electronics; its electrical functionalization by deposition of gold nanoparticles has already been proposed [1, 2].

Considerable attention has been payed so far to the description of the optical properties of gold nanoparticles (NP) [3] for their numerous applications in optics an electronics, also due to the possibility of introducing controlled enhanced absorption in specific devices and techniques, with applications in sensing, microscopy, nanophotonics, light harvesting, biomedicine, and many others [4–9]. Details on morphology, dispersion of the nanoparticle shape and size, and how these occurrences reflect on the optical spectra, are all argument of interest. The ability of performing quantitative simulation of experimental optical spectra is therefore an important tool to detect the details of the growing material at the nanoscale. In this paper, we report the optical properties of a set of several gold NP thin films on SA in the range 250-1100 nm, identify specific features, perform quantitative simulations, and show the trend of optical parameters.

2 Experimentals

Gold NP are deposited by magnetron sputtering at room temperature, using either 20 or 30 W Ar plasma power and deposition time spanning from to 45 to 160 s, on free standing SA membranes of several tens of micron thickness, prepared by solution casting from 4% wt Alginic Acid Sodium Salt (Sigma-Aldrich) solution in water. The resulting films appear as an aggregate of NP, increasingly in contact to one another as thickness increases. SEM images of selected samples are reported in figure 1.

Optical spectra were obtained by a UV-visible Avantes spectrophotometer equipped with an integrating sphere, which was used because of the inherent roughness of the samples. Given the high reflectivity of gold, the T spectra were also corrected by accounting for the light back-reflected into the sphere. If this correction is neglected, erroneous negative absorption is measured.

3 Optical simulation

Reflectance and transmittance (R&T) spectra are simulated using the Generalized Transfer Matrix (GTM) method [10] applied to the film-on-substrate model. The dielectric function (DF) of the SA material used as a substrate for this specific set of samples was separately determined by R&T spectroscopy. The material shows some absorption in the UV and up to 400 nm; for longer wavelengths it is transparent with a refractive index of about 1.5 [11]. The DF of gold is described by a sum of three modified (see below) Lorentz oscillators with the addition of a Drude term, to account for both interband transitions and free carrier absorption. Starting from this formalism, the gold NP films are simulated by allowing the relaxation energy in the Drude term to vary as a free parameter [3, 12]. Moreover, an additional oscillator, described by four free parameters, was added to account for a strong absorption signal visible for thicknesses below about 7 nm, attributed

*e-mail: summonte@bo.imm.cnr.it

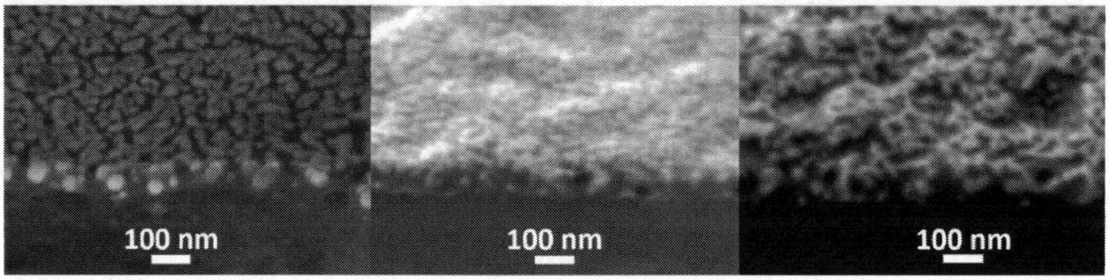

Figure 1. SEM images of selected gold/SA films. Thickness from simulation, from left to right: 4, 5, and 6 nm.

to the onset of the localised surface plasmon (LSP) that sets up in presence of gold NP [3]. The final form for the imaginary part of the dielectric constant of gold NP is therefore:

$$\varepsilon_2(E) = \sum_i \frac{(A_i - 1)E_{0i}{}^2\Gamma_{Bi}E}{(E_{0i}{}^2 - E^2)^2 + (\Gamma_{Bi}E)^2} + \frac{E_P{}^2E_T}{E(E^2 + E_T{}^2)} \quad (1)$$

In equation (1) the sum spans over four oscillators, three of which refer to bulk gold, and the fourth oscillator (LSP oscillator in the following) describes the LSP contribution. The second term on the right represents the Drude term, where E_P is the plasma frequency, and $E_T = h\nu_c$, where ν_c is the collision frequency and h is the Planck constant, is the relaxation energy. All oscillators, modified with respect to the three-parameter Lorentz oscillator, include an energy dependent broadening parameter Γ_{Bi}, which is given by:

$$\Gamma_{Bi} = \Gamma_i e^{-\alpha_i\left(\frac{E-E_{0i}}{\Gamma_i}\right)^2} \quad (2)$$

where α_i is the Gaussian broadening parameter, which has been introduced to limit the contribution of the oscillator to energies in the vicinity of the oscillator energy [13].

The parameters for bulk gold were obtained by fitting the gold DF in [14], and are reported in table 1. For the plasma frequency, $E_P = 8.78 \pm 0.02\,\text{eV}$ is obtained, similar to literature data [7–9, 15–17]. For E_{TB} , where "B" stands for "bulk", the value $0.073 \pm 0.005\,\text{eV}$ is obtained. This value is in agreement with several results in the literature [4, 7, 9, 18] although not in agreement with the static collision frequency obtained from electrical measurements [19], because of its dependence on frequency [20]. In the simulation of the DF of gold NP, E_T is left as a free parameter, to account for the increased ν_c in that case [3, 4, 7, 12].

Table 1. Amplitude, energy, broadening, and energy dependend broadening parameter of Lorentz oscillators for bulk gold, obtained using equation (1) and equation (2) fitted over the experimental data in [14]. Moreover, $E_P = 8.78 \pm 0.02\,\text{eV}$ and $E_{TB} = 0.073 \pm 0.005\,\text{eV}$.

Osc.	A	E (eV)	Γ (eV)	α
1	2.79 ± 0.04	6.25 ± 0.01	3.81 ± 0.05	2.00 (fixed)
2	3.18	3.97	1.80	1.48 ± 0.05
3	2.21	2.84	0.88	0.78

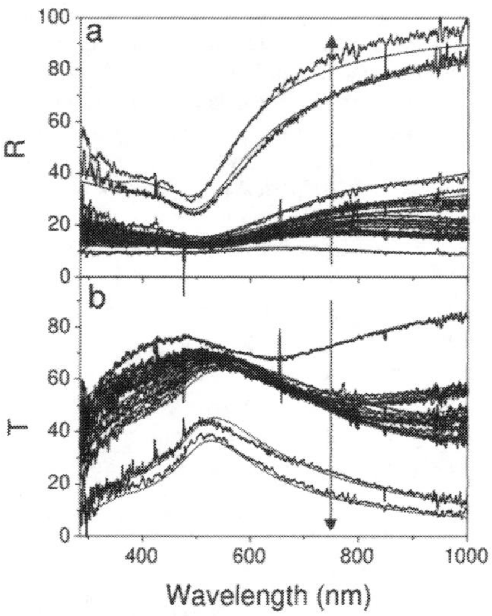

Figure 2. Reflectance (a) and Transmittance (b) of the whole set of samples. Black: experimental data. Color: simulation. The arrows indicate the direction of increasing overall deposited material.

In all cases the real part of the dielectric constat is obtained from the imaginary part by means of Kramers-Kronig transformation. All calculations are performed using the code GTB [13].

The total number of parameters in the simulation of the DF of gold NP is therefore 5: four for the modified LSP oscillator, plus the relaxation energy E_T. Film and substrate thicknesses were also left as free parameters in the simulation.

4 Results and discussion

The entire set of R&T spectra, both experimental and simulated, is reported in figure 2. The arrows indicate the direction of increasing overall material (increasing deposition time, or plasma power). The sample with highest

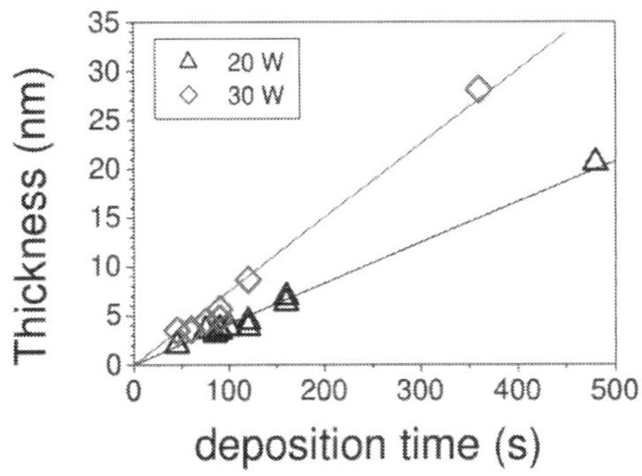

Figure 3. Thickness of all samples, obtained from the simulation.

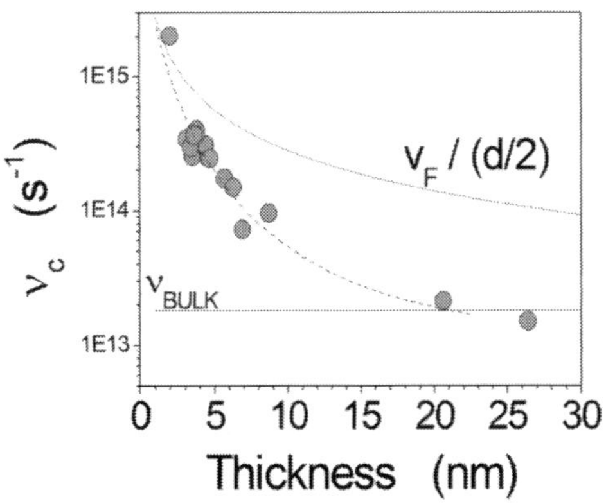

Figure 5. Collision frequency versus sample thickness. The value for bulk gold obtained from the simulation is also indicated. The continuous line represents the condition of mean free path equal to half the film thickness. The dashed line is a guide for the eye.

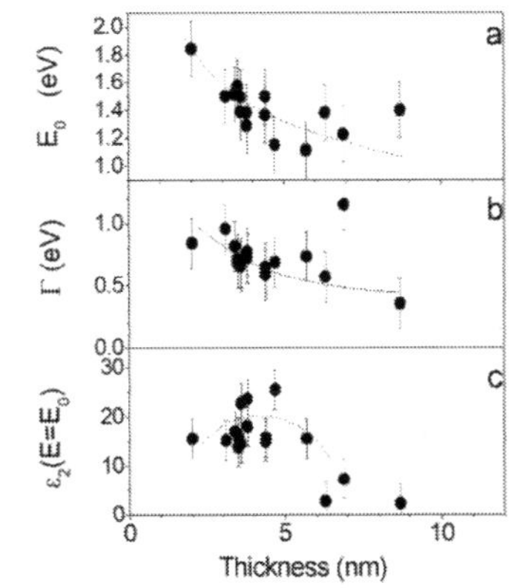

Figure 4. Energy of the oscillator describing the LSP (a); broadening of the oscillator (b); peak value of the associated contribution to the imaginary part of the dielectric constant (c). The lines are a guide to the eye.

R and lowest T refers to a 26 nm thick Au sample, which resulted to be best simulated using the optical constant of bulk gold, namely, E_T practically unaltered and the amplitude of the fourth oscillator vanished.

The good quality of the simulation confirms the choice of the optical model. The samples exhibit a broad minimum in T at about 650 nm for the thinnest sample, then shifting to the red for thicker samples. Less evident is the corresponding maximum in R, because it is superimposed to the increase associated to FC absorption. Both occurrences are due to the presence of the additional absorption signal, which can be attributed to the onset of a localized surface plasmon [5, 6].

The thickness of Au films obtained from the simulations is reported in figure 3. The growth of Au layer follows a linear process, with higher slope for higher plasma power. In the simulation the gold layer was simulated by a continuous film; an effective medium comprising voids has not been considered. This was done because in this range of thickness, thickness and void fraction are correlated parameters and the determination of both of them is not possible. In other words, the obtained thickness is the "equivalent" thickness, *i.e.*, the thickness of a dense film containing the same amount of material.

For a better insight on the material properties, we have a look to the features of the LPS oscillator (figure 4). We note that the LSP is located between 1.2 and 1.8 eV, increasing for decreasing thickness (figure 4a). Such energies are lower than what expected for monodispersed, isolated NP's, for which a value about 2 eV, decreasing only slightly for increasing NP diameter [12], also depending on shape, dielectric environmentent, and separation [3, 4, 8, 9], is expected. We attribute the result to the specific NP arrangement, with no fully isolated structures, yet not continuous, still showing resonance effects. Besides all non-idealities, according to [6] the shift of the LSP signal to the red (900 nm and beyond) is favoured by a high deposition rate that hinders the formation of isolated NP's.

We also observe (figure 4c) that the peak intensity increases for thicknesses about 5 nm, then it decreases, to fully disappear for about 10 nm. Since the LSP signal is expected to persist to much higher diameters [12], the result indicates that for this thickness, under our experimental conditions a continuous film is formed.

An indication on NP aggregation comes from the relaxation energy, which is markedly altered with respect to the "bulk" value. In case of isolated NP's, the collision frequency is related to NP radius as $1/r$ [12], possibly through a fitting factor [4, 7]. In figure 5 we report the collision frequency determined from E_T for our samples. The "bulk" value $\nu_{BULK} = E_{TB}/h$, is indicated. The curve v_F/d, where $v_F = 1.4x10^{15}$ nm/s is the Fermi velocity for gold [19] and d is the film thickness, is also indicated. This would be the curve expected for isolated NP with diameter $2r$ equal to film thickness. The figure shows that only for the very initial stages of the deposition is the collision frequency controlled by film thickness; for higher thicknesses, ν_c rapidly decreases and reaches the "bulk" value for about 20 nm, thus indicating that in-plane aggregation and rapid interconnection of the deposited material takes place. Similar conclusions were previously obtained by means of electrical measurements [2].

By comparing the results of figure 4 and figure 5, we observe that the LSP signal is strictly related to the non-continuous morphology of the films; it decreases and eventually vanishes as soon as sufficient material is deposited so that a continuous film is obtained, as detected by optical measurements, and in agreement with previously obtained electrical results [2]. This is in agreement with the observation that such band is possibly associated to the presence of islands and holes in low quality films [20]. In fact our material markedly differs from the ideal case, due to an inhomogenoeus, granular organization, where, although self arrangement on the substrate surface occurs, physical separation is not guaranteed, and the material is to be regarded as an aggregation of particles somewhat in contact among one another, evolving to a continuous film as the total amount of material increases. In this context, the observed remarkable broadening of the LSP oscillator (figure 4b) it is likely to reflect the convolution of a family of signals that arise from such an articulated structure, and to the multidispersed shape and size of NP's.

5 Summary and conclusions

The optical properties of very thin films of gold NP deposited by sputtering on SA substrates have been studied by means of simulation of optical properties. For low thickness, we observe the presence of an absorption band, which reaches the maximum intensity at about 5 nm, associated to a peak energy of 1.3 eV. For higher thickness, both intensity and peak energy decrease. The band eventually vanishes for thicknesses larger than 10 nm. The band is attributed to the onset of the local surface plasmon in conditions of highly disordered material and limited separation of the gold NP's.

For low thickness, we observe a high collision frequency which is to be expected given the low mean free path of such an articulated material. For increasing thickness the collision frequency rapidly decreases, which indicates aggregation of the deposited NP's. The results indicate that on this kind of substrate the initial stages of the deposition result in the formation of non-continuous material aggregates. Continuous film are formed for 20 nm deposition. The analysis is suitable to be applied to the study of the formation of the material on different substrates of different morphology and nature.

References

[1] C. Barone, M. Bertoldo, R. Capelli, F. Dinelli, P. Maccagnani, N. Martucciello, C. Mauro, S. Pagano, Nanomaterials **11**(3), 565 (2021)

[2] P. Maccagnani, M. Bertoldo, F. Dinelli, M. Murgia, C. Summonte, L. Ortolani, G. Pizzochero, R. Verucchi, C. Collini, and R. Capelli, Adv. Sust. Syst. **3**, 1900001 (2019)

[3] C. F. Bohren and D.R.Huffman, *Absorption and Scattering of Light by Small Particles* (1998)

[4] S. Berciaud, L. Cognet, P. Tamarat, and B. Lounis, Nano Letters **5**, 515–518 (2005)

[5] A. Axelevitch, B. Gorenstein and G. Golan, Physics Procedia **32**, 1–13 (2012)

[6] V.N. Rai, A.K. Srivastava, C. Mukherjee, and S.K. Deb, Appl. Opt. **51**, 2606 (2012)

[7] A. Derkachova, K. Kolwas and I. Demchenko, Plasmonics **11**, 941–951 (2016)

[8] D. Sikdar and A.A. Kornyshev, Scientific Reports **6**, 33712 (2016)

[9] S. Karimi, A. Moshaii, S. Abbasian, and M. Nikkhah, Plasmonics **14**, 851–860 (2019)

[10] E. Centurioni, Appl. Opt. **44** 7532–7539 (2005)

[11] M. Cocchi *et al.*, ACS Sust Chem Eng. (submitted)

[12] W. Haiss, N.T.K. Thanh, J. Aveyard and D.G. Fernig, Anal. Chem. **79**, 4215–4221 (2007)

[13] M. Allegrezza, F. Gaspari, M. Canino, M. Bellettato, A. Desalvo and C. Summonte, Thin Solid Films **556**, 105–111 (2014)

[14] P. Johnson and R. Christy, Phys Rev B **6**, 4370–4379 (1972)

[15] B.R. Cooper, H. Ehrenreich and H.R. Philipp, Physical Review **138**, A494 (1965)

[16] R.L. Olmon, B. Slovick, T.W. Johnson, D. Shelton, S.-H. Oh, G.D. Boreman, and M.B. Raschke, Phys. Rev. B **86**, 235147 (2012)

[17] P.G. Etchegoin, E.C. Le Ru and M. Meyer, J. Chem. Phys. **125**, 164705 (2006)

[18] C. Novo, DI Gomez, J. Perez-Juste, Z. Zhang, H.Petrova, M. Reismann, P. Mulvaney and G. V. Hartland, Phys. Chem. Chem. Phys. **8**, 3540–3546 (2006)

[19] N.W. Ashcroft and N.D. Mermin, *Solid State Physics* (Saunders College, Pennsylvania, 1976)

[20] M.L. Theye, Phys. Rev. B **2**, 3060 (1970)

EPJ Web of Conferences 255, 09001 (2021) https://doi.org/10.1051/epjconf/202125509001
EOSAM 2021

Highly symmetric aperiodic structures -INVITED

Uwe Grimm[1,*]
[1]School of Mathematics and Statistics, The Open University, Walton Hall, Milton Keynes MK7 6AA, United Kingdom

Abstract. The symmetries of periodic structures are severely constrained by the crystallographic restriction. In particular, in two and three spatial dimensions, only rotational axes of order 1, 2, 3, 4 or 6 are possible. Aperiodic tilings can provide perfectly ordered structures with arbitrary symmetry properties. Random tilings can retain part of the aperiodic order as well the rotational symmetry. They offer a more flexible approach to obtain homogeneous structures with high rotational symmetry, and might be of particular interest for applications. Some key examples and their diffraction are discussed.

1 Introduction

Motivated by the discovery of quasicrystals in intermetallic alloys by Dan Shechtman in 1982 [1], the investigation of aperiodic tilings and points sets has become an increasingly active area of mathematical research. One of the key fundamental questions from crystallography is how to properly define the concept of *order*, which is challenging, as can be seen from the ongoing discussion concerning the definition of a crystal; see [2] for background. While quasicrystals are now considered *aperiodic crystals* displaying aperiodic order, there is no clear demarcation line; in particular, there are many potential manifestations of partial order that may or may not be included.

The main tool to describe order in structures is linked to the experimental tool that is used to determine the structure of crystals, which is diffraction. Mathematical diffraction is a measure that is obtained by Fourier transform of the appropriately normalised pair correlation measure, and corresponds to the experimental diffraction in the kinematic far-field (Fraunhofer) limit; see [3] for details. Pure point diffraction is seen as the hallmark of an ordered structure, while continuous components, and in particular absolutely continuous components, are considered a sign of disorder. Traditionally, singular continuous components did not play a major role, but they have become more prominent in the discussion of aperiodic structures, because they are prevalent in substitution (or inflation) based structures displaying some self-similarity properties. A related approach to quantify order, which has recently become popular, is the concept of hyperuniformity [4], which is related to a scaling property of the diffraction measure; see [5] and references therein.

Aperiodic tilings and point sets not only serve as models for the atomic structure of quasicrystals, but are becoming increasingly important as metamaterials. The possibility to realise higher symmetries than in the severely constrained setting of periodic structures makes them particularly interesting for applications where isotropic properties are desirable [6]. With the recent advances in additive manufacturing, this opens up potential applications for metamaterials at any length scales, for applications exploiting their optical, phononic or mechanical properties.

In this brief note, we shall discuss some paradigmatic planar structures, focussing on their symmetry and diffraction. For mathematical and crystallographic background on the diffraction of aperiodic structures we refer to [7–9] and references therein.

2 Cut and project and random tilings

Cut and project sets (or model sets in the mathematical terminology) are obtained by projecting part of a higher-dimensional periodic lattice. While there is are crystallographic restrictions for any dimensions, the number of possible symmetries increases with dimension. In particular, four-dimensional lattices account for axes with fivefold, eightfold, tenfold or twelvefold symmetries, and six-dimensional lattices can accommodate icosahedral symmetry as observed in physical quasicrystals.

Due to the underlying higher-dimensional periodic structure, these systems are generally well understood, provided some conditions on the projected subset of the lattice are satisfied. In particular, all cut and project sets satisfying these conditions are pure point diffractive, and the diffraction is concentrated on the projection of the dual (or reciprocal) lattice. As an example, the Ammann–Beenker tiling is shown in figure 1. It can be obtained as a projection from the four-dimensional hypercubic lattice $\mathbb{Z}^4$. The corresponding diffraction measure for the infinite tiling is a pure point measure with perfect eightfold symmetry which is supported on a dense set in the plane. Figure 1 shows the diffraction image obtained for the finite patch, which already closely resembles the diffraction of the infinite structure.

*e-mail: uwe.grimm@open.ac.uk

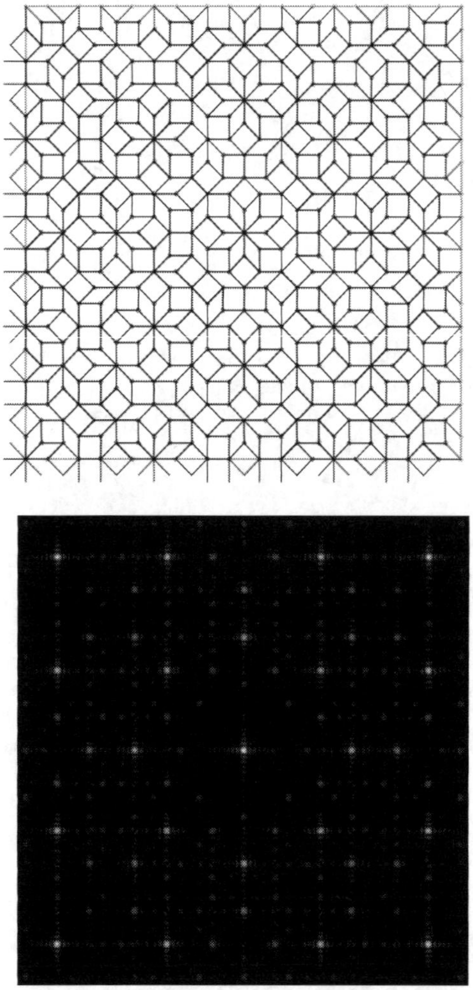

Figure 1. Finite patch of the Ammann–Beenker tiling (top) and its diffraction (bottom)

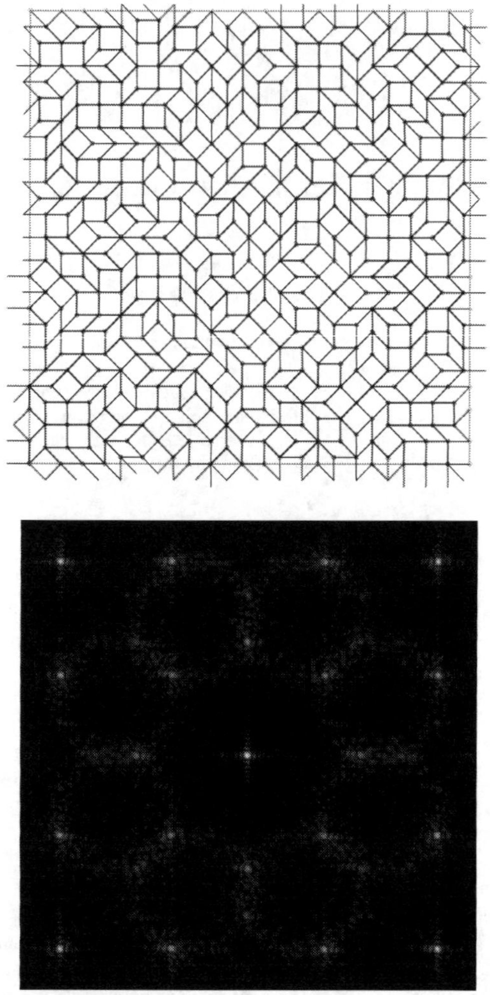

Figure 2. Finite patch of the Ammann–Beenker random tiling (top) and its diffraction (bottom)

One way to introduce some disorder into the structure is by considering a random tiling, or more precisely the ensemble of all tilings of the plane with the same set of tiles. It turns out that the proportion of the two tiles in the eightfold tiling are optimal in the sense that they maximise the entropy of the tiling. Figure 2 shows an example of a randomised perfect tilings obtained by performing local flips (which are ergodic) of the tiles around vertices with three neighbours. The corresponding diffraction image (of this finite patch) is also shown in figure 2.

Since the random tiling preserves the eightfold symmetry (any finite patch is equally likely to occur in any orientation), the diffraction of the random tiling still shows perfect eightfold symmetry. Rigorous mathematical proofs for the nature of the diffraction of a random tiling are lacking, but heuristic arguments point to a singular continuous diffraction in the planar case. Note that the random tiling can still be lifted into a subset of $\mathbb{Z}^4$, but this lift will, in general, no longer satisfy the conditions that guarantee pure point diffraction. Random tilings provide one way to obtain partially ordered structures with high rotational symmetries, and with a diffraction which is less strongly localised.

3 Inflation tilings

Inflation tilings are obtained by an iterative procedure consisting of a rescaling of the tiles followed by the dissection of the rescaled tiles into tiles of the original size. The resulting tilings possess self-similar properties. There are cut and project sets, such as the Ammann–Beenker tiling, that also possess an inflation symmetry, but in general cut and project sets do not possess an inflation symmetry, nor can general inflation tilings be lifted into a higher-dimensional lattice.

Planar inflation tilings can also realise any rotational symmetry, and, maybe surprisingly, can even possess an infinite rotational symmetry, making them completely isotropic. The paradigm for the latter is the pinwheel tiling which is obtained from the inflation rule of figure 3.

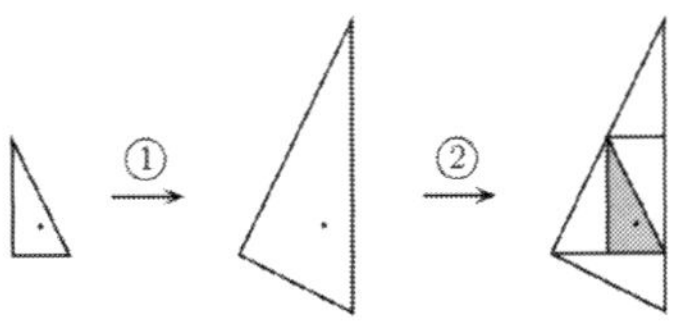

Figure 3. Inflation rule of the pinwheel tiling, consisting of a rescaling by a linear factor of $\sqrt{5}$ (step 1) followed by a dissection into five triangles of the original size (step 2). The rotation ensures that the shaded triangle has the same orientation as the original one, which ensures that iterating the rule can be used to define an infinite tiling; see [3] for details

The tile is a right triangle of side lengths 1, 2 and $\sqrt{5}$. Each inflation step produces edges in a new direction in the plane. A patch of the pinwheel tiling obtained by inflating an initial patch consisting of two triangles joined along their hypotenuses is shown in figure 4. Note that the two triangles are mapped onto each other by reflection, and that the reflected triangle is dissected by the reflected rule of that shown in figure 3, so that reflection symmetry is preserved.

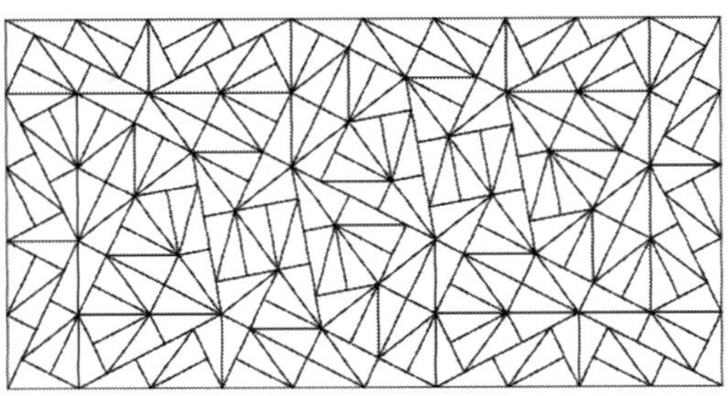

Figure 4. Patch of the pinwheel tiling obtained by three inflation steps from a couple of triangular tiles. It consists of $2 \times 5^3 = 250$ triangular tiles

The challenge with this tiling is that in each inflation step the number of tiles grows by a factor of five, while only one new direction is obtained. This means that any reasonably-sized finite patch will still be far from rotationally symmetric.

It has been proven that the diffraction measure for this tiling is fully rotationally symmetric, though it is still not clear whether it consists of sharp rings only (which amount to a singular continuous measure) or whether there is also an absolutely continuous component [10]. Due to the slow convergence to a rotationally symmetric structure, the investigation of finite patches does not provide sufficient evidence for or against the presence of absolutely continuous diffraction. However, recent results point towards weak hyperuniformity of the pinwheel tiling [11].

There are many other tilings with complete rotational symmetry, but the pinwheel tiling is arguable the simplest of these, using a single shape. In view of applications, which would necessarily use finite patches of such tilings, it is however less obvious whether these tilings really offer higher isotropy than tilings with finite rotational symmetry, such as the Ammann–Beenker tiling or a corresponding random tiling. For fivefold, eightfold, tenfold and twelvefold symmetry, the rotational symmetry can already be well realised in a rather small patch, as can be seen from the diffraction images in figures 1 and 2, while comparatively large patches of the pinwheel tiling may still be quite far from isotropic. The reason that these particular symmetries work well lies in the nature of the linear inflation factor for these cases, which is a quadratic Pisot–Vijayaraghavan (PV) number; see [3] for details. At the same time, these are symmetries that can be realised in cut and project sets obtained from four-dimensional lattices, as mentioned above. They are also the planar symmetries that have been observed to occur in quasicrystalline materials [12, 13], so are also preferred by nature; so far, no material with other non-crystallographic symmetries have been found.

4 Summary and Outlook

While the class of cut and project tilings is generally well understood, this is much less so for tilings obtained by an inflation procedure, or indeed for any structures that involve randomness, such as random tilings. There exists a vast realm of unexplored structures which possess a degree of (partial) order, and a proper classification of ordered structures is nowhere in sight, leaving lots of opportunities for mathematical as well as application-oriented research. Random tilings with non-crystallographic symmetries and stochastic long-range order offer a large class of promising structures that has not yet received as much attention as it deserved.

Acknowledgements. This work was supported by Engineering and Physical Sciences Research Council (EPSRC) via grants EP/S010335/1 and EP/V047108/1.

References

[1] D. Shechtman, I. Blech, D. Gratias and J.W. Cahn, Metallic phase with long-range orientational order and no translational symmetry, Phys. Rev. Lett. **53**, 1951–1953 (1984)

[2] U. Grimm, Aperiodic crystals and beyond, Acta Crystallogr. B **71**, 258–274 (2015)

[3] M. Baake and U. Grimm, *Aperiodic Order. Vol. 1: A Mathematical Invitation* (Cambridge University Press, Cambridge, 2013)

[4] S. Torquato and F.H. Stillinger, Local density fluctuations, hyperuniformity, and order metrics, Phys. Rev.

E **68**, 041113 (2003) and Phys. Rev. E **68**, 069901 (2003) (erratum)

[5] M. Baake and U. Grimm, Scaling of diffraction intensities near the origin: Some rigorous results, J. Stat. Mech.: Theory Exp. **2019**, 054003 (2019)

[6] L. Dal Negro, *Waves in Complex Media* (Cambridge University Press, Cambridge, 2021)

[7] M. Baake and U. Grimm, Kinematic diffraction from a mathematical viewpoint, Z. Kristallogr. **226**, 711–725 (2011)

[8] M. Baake and U. Grimm, Mathematical diffraction of aperiodic structures, Chem. Soc. Rev. **41**, 6821–6843 (2012)

[9] M. Baake and U. Grimm, Inflation versus projection sets in aperiodic systems: The role of the window in averaging and diffraction, Acta Crystallogr. A **76**, 559–570 (2020)

[10] M. Baake, D. Frettlöh and U. Grimm, A radial analogue of Poisson's summation formula with applications to powder diffraction and pinwheel patterns, J. Geom. Phys. **57**, 1331–1343 (2007)

[11] F. Sgrignuoli and L. Dal Negro, Hyperuniformity and wave localization in pinwheel scattering arrays, Phys. Rev. B **103**, 224202 (2021)

[12] W. Steurer, Twenty years of structure research on quasicrystals. Part I. Pentagonal, octagonal, decagonal and dodecagonal quasicrystals, Z. Kristallogr. **219**, 391–446 (2004)

[13] W. Steurer and S. Deloudi, *Crystallography of Quasicrystals: Concepts, Methods and Structures* (Springer, Heidelberg, 2009)

Light confinement in liquid crystals for optofluidic integrated microsystems -INVITED

Antonio d'Alessandro[1,*] *Anju M.* Kumaran[1]

[1]Sapienza University of Rome, Department of Information Engineering, Electronics and Telecommunications, 00184 Rome, Italy

Abstract. In this paper technology to make optical waveguides and microfluidic channels integrated on the same substrate will be reported to envisage novel micro-optofluidic chips. PolyDiMethylSiloxane (PDMS) is used to make microchannels to be filled with biological solutions. Liquid crystals (LC) are used to confine light to produce optical interaction with biological fluidic specimen. Optical waveguides base on PDMS channels filled with LC, named LC:PDMS waveguides, including both straight and bending channels are reported to design photonic devices. Electro-optic effect of LC allows to make tuneable optical waveguides to reconfigure the entire optofluidic microsystem which can include gold nanoparticles for photo-thermal therapies. Coplanar gold electrodes can switch LC molecules with applied voltage of about 2 V. Such electrode configuration can be used to make optical switches and wavelengths demultiplexers. A zero-gap directional coupler based on LC:PDMS waveguides has been designed to switch light from one waveguide to another with an extinction ratio of 16 dB by applying a voltage of just 1.62 V. A multimode interference demultiplexer has been also designed to demultiplex wavelengths at 980 nm and 1550 nm in two output waveguides with an extinction ratio better than 11 dB by applying about 7 V.

1 Introduction

Optofluidic channels have been successfully proposed by infiltrating liquid crystals (LC) in silicon grooves [1] and in PolyDiMethylSiloxane (PDMS) [2].

Both the variation in the diffraction pattern of an array of microfluidic channels acting as a switchable grating using ITO electrodes [3] and simple optical waveguides were experimentally demonstrated using LC infiltrated in PDMS, named LC:PDMS waveguides. PDMS channels have been made by using casting and molding technique, which is a low cost process to easily obtain well defined PDMS patterns [4].

PDMS has been demonstrated to be very effective and reliable for soft lithographic fabrication of many microfluidic and micro-optical devices [5]. PDMS with respect to other materials, such as silicon or glass, is the easy and cheap technological lithographic processing needed to make patterned geometries including bends. Flexibility is another interesting feature of PDMS enabling emerging applications which are impossible to achieve using conventional rigid or flexible technologies. Examples can be found in robotics health care, biomedical technologies, and consumer electronics.

Liquid crystals (LC) are effective materials to make both electrically [6] and optically [7-8] controlled optical waveguides to make switchable and reconfigurable waveguides to be used both at the visible and near infrared wavelengths [9]. In particular the efficient electro-optical effect with the additional advantage of negligible electrical current absorption and the nonlinear optical properties allow very low power consumption without significant energy dissipation. Such characteristics make LC-based photonic devices interesting potential candidates for both datacom and optofluidic applications [10]. In this paper we present design of both straight and bending optical waveguides based on LC:PDMS. Moreover some examples of electro-optically controlled devices based on LC:PDMS waveguides combined with coplanar gold electro-plated electrodes are shown. The fabrication process of both waveguides and gold electrodes will be reviewed. A study of LC orientation in PDMS channels based on Monte Carlo simulations can be used. Furthermore the design of an electro-optical switch consisting of a zero gap directional coupler and a multimode interferometer used as a wavelength demultiplexer based on LC:PDMS is presented.

2 LC: PDMS optical waveguides

PDMS channels are made by using casting and molding technique. Soft photolithography with SU-8 2005 negative photoresist by Micro Chem is used to create the mold of micro-channels. Waveguides are shaped in a PDMS bulk by pouring the liquid mixture on the SU-8 mold, placed in a holder to host 3 mm thick PDMS. Then the channels for NLC filling is shaped by pouring the PDMS on a cleaned silicon substrate, in order to provide a very smooth surface for PDMS. The SU8 mold is placed in the PDMS to shape the micro-channels. In a final step, a uniform PDMS film is obtained by spinning the liquid

* Corresponding author: antonio.dalessandro@uniroma1.it

 https://doi.org/10.1051/epjconf/202125510001

PDMS on a glass substrate. The PDMS samples, after vacuum treatment to avoid air entrapment in the channels, are baked in oven at 80°C for approximately 15 minutes. The sample with micro-channels is then simply peeled from the SU-8 mold and the PDMS in excess will be sharply cut away by using a razor blade. The complete assembled sample, like that shown in Figure 1 as an example, is finally baked at 80°C for 60 minutes. The NLC E7 is infiltrated in its isotropic phase at 80°C through a hole in a vacuum oven and then cooled down at room temperature. No alignment layer is used to orient the LC molecules. Light transmission through LC:PDMS waveguides has been simulated after calculating the refractive index profile of the LC related to its orientation. The distribution of the LC orientation inside the PDMS channel has been obtained by minimizing the Oseen-Frank free energy equation which includes both the elastic and the dielectric components related to an external applied field and solving the Poisson equation by using finite elements through COMSOL Multiphysics®, as already performed for other types of LC waveguides [11].

Fig. 1. A set of LC:PDMS waveguides.

The orientation of the LC corresponds to the minimum of the free energy which is given by:

$$F= F_{el} - F_{diel} - F_{opt} \quad (1)$$

which includes the elastic term F_{el}, the dielectric term F_{diel} and the optical term F_{opt} deriving by the electric field of the optical beam itself propagating in the LC.

The elastic energy according to the Oseen–Frank theory is expressed by [34]:

$$F_{el}=\frac{1}{2}\iiint [k_{11}(\nabla\cdot\mathbf{n})^2+k_{22}(\mathbf{n}\cdot\nabla\times\mathbf{n})^2+k_{33}(\mathbf{n}\times\nabla\times\mathbf{n})^2]\, dv \quad (2)$$

where k_{11}, k_{22} and k_{33} are the elastic constants corresponding to the splay, twist and bend respectively and dv is the elementary integration volume. The dielectric energy is given by:

$$F_{diel}=\frac{1}{2}\iiint \left[\epsilon_0\epsilon_{\perp,ac}|\mathbf{E}_{ac}|^2+\epsilon_0\Delta\epsilon_{ac}(\mathbf{E}_{ac}\cdot\mathbf{n})^2\right] dv \quad (3)$$

where ϵ_0 is dielectric permittivity in vacuum, $\epsilon_{\perp,ac}$ is the dielectric permittivity perpendicular to the optical axis, $\Delta\epsilon_{ac}$ is the dielectric anisotropy, and $\mathbf{E}_{ac}$ is the applied electric field at low frequency. The optical term of the energy is given by:

$$F_{opt}=\frac{1}{2}\iiint \left[\epsilon_0\epsilon_{\perp,opt}|\mathbf{E}_{opt}|^2+\epsilon_0\Delta\epsilon_{opt}(\mathbf{E}_{opt}\cdot\mathbf{n})^2\right] dv \quad (4)$$

where ϵ_0 is dielectric permittivity in vacuum, $\epsilon_{\perp,opt}$ is the dielectric permittivity perpendicular the optical axis, $\Delta\epsilon_{opt}$ is the dielectric anisotropy, and $\mathbf{E}_{opt}$ is the electric field at the optical frequencies. The minimization of F is achieved by solving the Euler - Lagrange equation. This approach is simpler than global or relaxation approaches, and therefore more adapted for large 3D problems of typical integrated optical structures. To solve the partial derivative equation, a finite element method is well suited since it allows the implementation of the weak form of the Euler - Lagrange equation. The minimization of F is coupled with the solution of the Poisson equation for the distribution of the electric potential Φ in the LC structure:

$$\nabla\cdot\left[\varepsilon_{\perp,ac}\nabla\Phi + \Delta\epsilon_{ac}(\nabla\Phi\cdot\mathbf{n})\mathbf{n}\right]=0 \quad (5)$$

The resulting output is the spatial distribution of the LC director orientation, from which the refractive index profile can be derived. Simulation of light propagation by using RSoft BeamPROP™ indicates a multimodal behaviour of the waveguide. Figure 2 shows the first order mode for TE light propagating along a straight LC:PDMS waveguide.

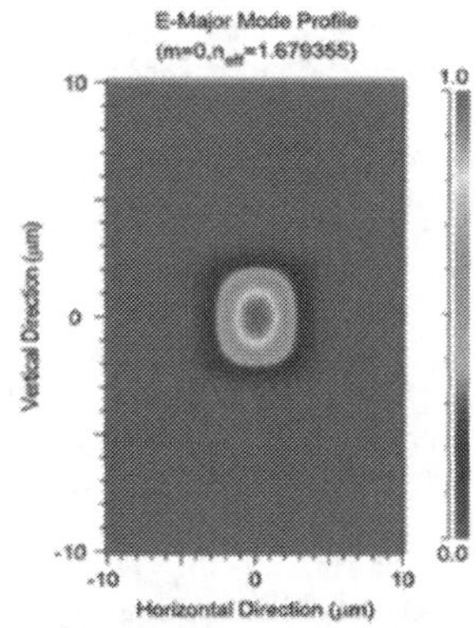

Fig. 2. Contour plots of the first TE-like mode propagating through an LC:PDMS waveguide.

A significant advantage in using LC:PDMS waveguides is that PDMS channels can be patterned to obtain bending paths. Bends are essential to make a large variety of optical circuits including ring resonators with compact dimensions. Figure 3 shows propagation of light though an S-bend with a bend radius of 1.3 mm over an extension of 10 mm.

Such a bend radius guarantees minimum optical losses as shown in Figure 4 in which optical loss in dB/cm and the normalised transmittance versus bend radius are plotted. The power loss is the total sum of the radiation and transmittance losses of the complete S-bend structure obtained from the BPM simulation. For a bend radius of both 1.3 and 1.4 mm a power loss of 0.622 dB/cm has been calculated. Losses increase for bend radii higher than 1.4 mm but saturating at a value of about 0.628 dB/cm. Consistently power transmission reaches a peak for a bend radius of 1.3-1.4 mm and decreases saturating at higher bend radii.

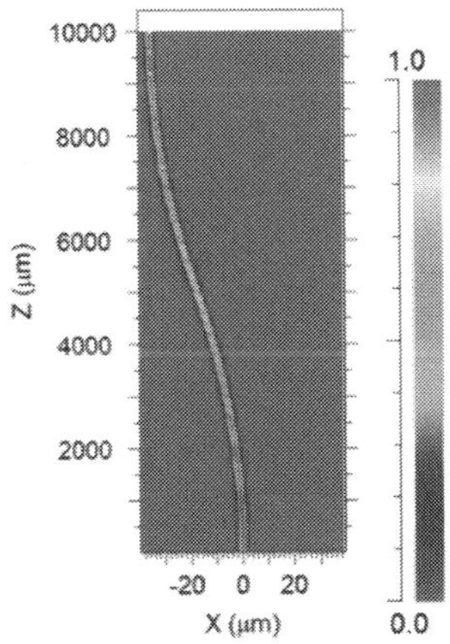

Fig. 3. Optical power propagation through an S-bend LC:PDMS waveguide.

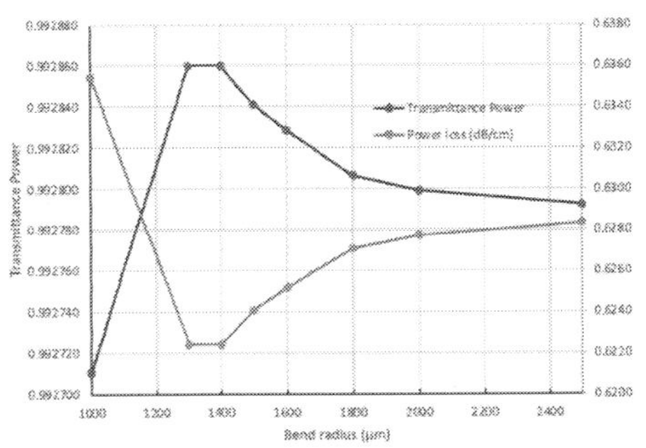

Fig. 4. Normalised optical transmittance and losses versus bend radius in a S-bend LC:PDMS optical waveguide.

Monte Carlo simulation method has been applied to investigate LC molecular orientation [12]. In particular, the Lebwhol and Lasher (LL) [13] model, the prototype potential for the mesoscale simulations of nematic LC, allows to simulate a high number of molecules, represented by three-dimensional headless spins, and then to reproduce or predict observables which can be obtained in real experiments. Some of these experiments are the microscopy polarized images which can give clear indications of the molecular organization inside a nematic LC sample [14].

The optical characterization to check the LC alignment in the LC:PDMS waveguides is performed by using a polarized microscope. Polarized light is transmitted through the channels revealing the homeotropic arrangement of the LC molecules inside the PDMS channels. Such orientation of the LC is proven by the black part in the central region for the waveguides indicating that polarization of input light is not changed when transmitted through the LC and is stopped by the crossed analyser. Light leaks through the edges of the waveguides because of the optical retardation induced by the LC molecules homeotropically aligned on the vertical sides of the PDMS channels.

The homeotropic alignment of the LC inside the PDMS channel creates a distribution of the LC refractive index which allows polarization independent propagation of light at both visible and near infrared wavelengths. Standard butt coupling technique is used to inject light beam into the LC:PDMS waveguides by using cleaved single-mode optical fibers. Both visible and near infrared light sources were used.

Figure 5 shows that it is possible to couple and transmit visible light. Near infrared wavelengths have been also injected and polarization independence of light propagation has been observed as expected due to the homeotropic alignment of LC molecules.

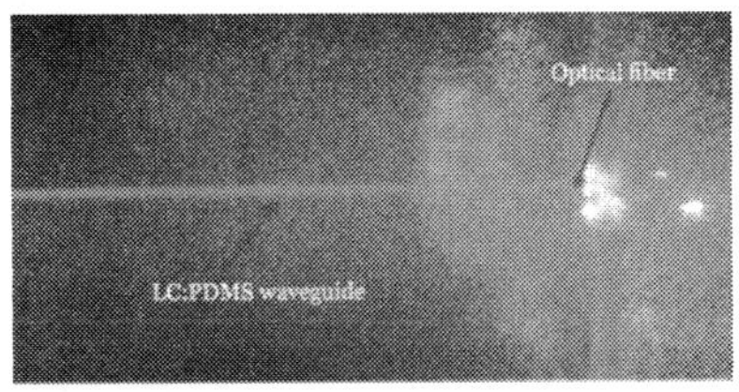

Fig. 5. Green light propagation in an LC:PDMS waveguide.

It is also possible to reorient LC molecules by applying an electric field between coplanar electrodes, as shown in Figure 6. The electrodes can be deposited by means of electroplating. It has been shown that a voltage as low as 1.8 V is enough to switch the LC molecules between the coplanar electrodes.

Such electrode configuration can be proposed to make electro-optical switchable LC:PDMS waveguide based devices.

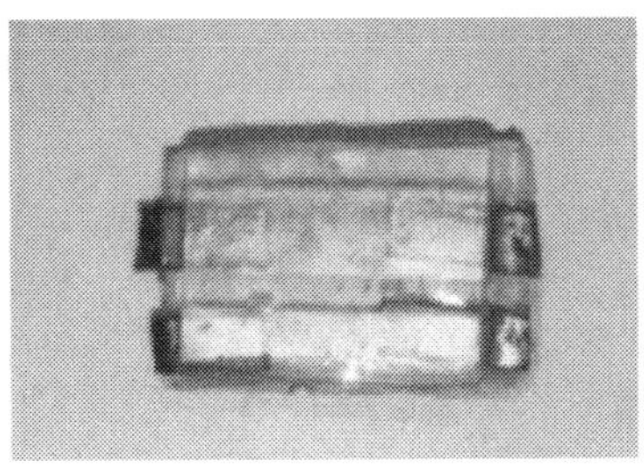

Fig. 6. Gold plated coplanar electrodes embedded in a two layers of PDMS [14].

3 Device applications and future developments

Photonic devices based on LC:PDMS waveguides have been demonstrated such as directional couplers, optical switches, multimode interferometers (MMI). In an electro-optical controlled directional coupler with zero gap, as shown in left-hand side of Figure 7, it is possible to switch light from one waveguide to a second one with an extinction ratio of 16 dB with an applied voltage of just 1.62 V. Light remains in the same waveguide with an extinction ratio of about 18 dB with a voltage of 1.76 V. A MMI based on LC:PDMS technology, as sketched in the right-hand side of Figure 7, has been also designed to demultiplex wavelengths at 980 nm and 1550 nm in two output waveguides with an extinction ratio better than 11 dB by applying about 7 V.

LC:PDMS based devices can be also used for biochip applications. One example consists of biochip for bacterial killing through the temperature increase caused by the heating of gold nanoparticles. Such a biochip can be made of PDMS in which microfluidic channels, gold nanorods (GNRs) and optical waveguides are combined and integrated.

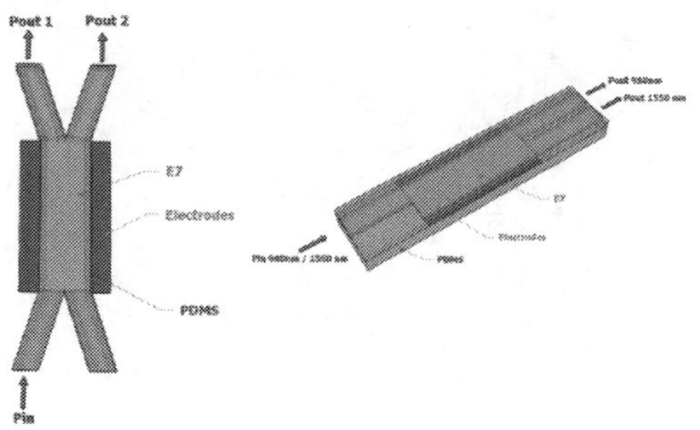

Fig. 7. (Left) Zero-gap directional coupler based and (right) tuneable MMI based on LC:PDMS waveguides [14].

References

1. D. Donisi, B. Bellini, B., R. Beccherelli, R. Asquini, G. Gilardi, M. Trotta, A. d'Alessandro, IEEE J. Quantum Electron. **46**, 762 (2010)
2. A. E. Vasdekis, J. Cuennet, D. Psaltis, Proc. SPIE **8475**, 847507 (2012)
3. L. De Sio, M. Romito, M. Giocondo, A. E. Vasdekis, A. De Luca, Lab on a chip **12**, 3760 (2012)
4. A. d'Alessandro, L. Martini, G. Gilardi, R. Beccherelli and R. Asquini, IEEE Photon. Technol. Lett. **27**, 1709 (2015)
5. D. A. Chang-Yen, R. K. Eich, B. K. Gale, J. Light. Technol. **23**, 2088 (2005)
6. D. Donisi, B. Bellini, R. Beccherelli, R. Asquini, G. Gilardi, M. Trotta, A. d'Alessandro, IEEE J. Quantum Electron. **46**, 762 (2010)
7. A. d'Alessandro, R. Asquini, M. Trotta, G. Gilardi, R. Beccherelli, I. C. Khoo, Appl. Phys. Lett. **97**, 093302 (2010)
8. J.F. Henninot, M. Debailleul, R. Asquini, A. d'Alessandro, M. Warenghem, J. Opt. A: Pure Appl. Opt. **6,** 315 (2004).
9. I. C. Khoo, *Liquid Crystals*, (Wiley, 2nd Ed., 2005)
10. D. C. Zografopoulos, R. Asquini, E. Kriezis, A. d'Alessandro, R. Beccherelli, Lab on a chip **12**, 3598 (2012)
11. A. d'Alessandro, L. Martini, L. Civita, R. Beccherelli, R. Asquini, Proc. SPIE **9384**, 93840L (2015)
12. P. Pasini, C. Chiccoli, C. Zannoni, *Advances in the computer simulations of liquid crystals*, (Kluwer, 2000)
13. P. A. Lebwohl, G. Lasher, Phys. Rev. A. 6, 426 (1972)
14. R. Asquini, C. Chiccoli, P. Pasini, L. Civita, A. d'Alessandro, Liq. Cryst. **45**, 1 (2018)

Novel optofluidic concepts enabled by topological microfluidics -INVITED

Anupam Sengupta[1]

[1]Physics of Living Matter, Department of Physics and Materials Science, University of Luxembourg, 162A, Avenue de la Faïencerie, 1511 Luxembourg City, Luxembourg

Abstract. The coupling between flow and director orientation of liquid crystals (LCs) has been long utilized to devise wide-ranging applications spanning modern displays, medical and environmental solutions, and bio-inspired designs and applications. LC-based optofluidic platforms offer a non-invasive handle to modulate light and material fields, both locally and dynamically. The flow-driven reorientation of the LC molecules can tailor distinct optical and mechanical responses in microfluidic confinements, and harness the coupling therein. Yet the synergy between traditional optofluidics with isotropic fluids and LC microfluidics remains at its infancy. Here, we discuss emerging optofluidic concepts based on *Topological Microfluidics*, leveraging microfluidic control of topological defects and defect landscapes. With a specific focus on the role of surface anchoring and microfluidic geometry, we present recent and ongoing works that harness flow-controlled director and defect configurations to modulate optical fields. The flow-induced optical attributes, and the corresponding feedback, is enhanced in the vicinity of the topological defects which geenerate distinct isotropic opto-material properties within an anisotropic matrix. By harnessing the rich interplay of confining geometry, anchoring and micro-scale nematodynamics, *topological microfluidics* offers a promising platform to ideate the next generation of optofluidic and optomechnical concepts.

1 Optofluidics

Optofluidics–the exciting interface of optics and microscale fluid dynamics–occupies a conspicuous niche in today's world of cross-disciplinary research. Despite its relatively recent introduction spanning little over a decade, optofluidic investigations, both fundamental and applied, have seen a steady growth, allowing it to pervade into various domains of contemporary research. From microfabrication and on-chip innovations, non-invasive actuation and optical control, to precise sensing platforms, optofluidics has contributed significantly to the fields of physics, biology, engineering, and analytical instrumentation [1, 2].

The steady expansion and acceptance of optofluidics as a mainstream technology was possible due to the compact, integrated and miniaturized devices, rapid and dynamic response times, and the reconfigurability of the optical systems. Its growing presence in practically all fields of fundamental and applied research suggests a promising future over the years to come. Engineering suitable optofluidic systems involves design and optimization of microfluidic geometries, selection of appropriate fluidic interfaces (or fluid-solid interfaces) and detection of signals as optical or electronic readouts. By varying one, or a combination therein, several ground breaking discoveries have been realized including nonlinear alteration of the refractive indices due to thermocapillary effects [3], adaptive optics for biomedical imaging via high-order wavefront modulation [4], and three-dimensional manipulation of sub-100-nm objects [5], allowing incorporation of optofluidics for medical sensing and diagnosis. More recently, nanofluidic waveguides [6], and liquid-core waveguide chips [7] were used for rapid detection of the SARS-CoV-2 virus and infection with single molecule sensitivity.

Optofluidic specs and performance have improved considerably due to the concomitant progress in the design and fabrication techniques, thanks to the advancement of microfabrication processes, including soft lithography based microfluidic fabrication, laser cutting and micro-milling, and 3D printing techniques [8]. Altering the fluidic properties through external fields (e.g., thermal effects) has further contributed to diversification of optofluidic applications [9]. Despite the phenomenal advancement, majority of the initial studies was carried out with isotropic fluids, leaving out many promising complex anisotropic fluids unexplored until recently [10, 11]. Liquid crystals (LCs)–owing to their anisotropic viscous, elastics, and crucially, optical properties [12]–is a key example of complex anisotropic fluid, which offer a multi-variate handle to precisely tune the specs and performance of optofluidic systems. In addition to the exotic and emergent dynamical attributes of anisotropic fluids [13, 14], LCs furnish topological defects due to spontaneous symmetry breaking [15, 16]: such defects offer strong prospects as labile and reconfigurable optofluidic elements [17, 18]. Conflicting

* Corresponding author: anupam.sengupta@uni.lu

boundary conditions, in combination with geometric constraints and appropriate external fields (for instance hydrodynamic), allow tailored generation of topological defects in microfluidic environments [19-21], thus paving the way for the next generation of dynamic, flow-stabilized, defect-based optofluidic platforms [22].

2 Liquid crystal optofluidics

LCs represent a rich system where an interplay between long-range order, spontaneous symmetry breaking and hydrodynamics elicits exotic effects distinct from those observed in isotropic fluids. The exotic properties are mediated by the anisotropic coupling in LC systems, which, traditionally, have been utilised to develop a range of liquid crystal based applications [23]. LCs offer an additional degree of freedom in manipulating the topology of the director field–the average molecular orientation of the LCs–by means of the anisotropic coupling of the director and an external field, for instance hydrodynamic field [14, 24]. Over the last decade, this intrinsic property of LCs has been leveraged toward developing the domain of *Topological Microfluidics*, that harnesses one or a combination of microfluidic geometry and architecture, surface anchoring conditions, and the strength and nature of the microscale flows (pressure-driven vs constant volume flows) to enable precise flow-governed director fields and programmable topological defect states [24-29]. This timely synergy between microfluidics and liquid crystals has led to basic and applied developments in LC-based lab-on-a-chip concepts, and demonstrated the unique role that flow-director coupling could play in triggering tractable emergent properties.

A local flow field can induce a change in the LC director field alignment, and, conversely, a local change of the LC orientation could trigger flows, also known as the backflow [30, 31]. The coupling between the flow and the LC director fields has been utilized to devise tunable optofluidic platforms that allow control of light intensity, polarization, and spectral composition within a microfluidic chip [32, 33]. Variation of the flow parameters leads to different LC orientational profiles: weak, medium, and strong, which can be further controlled by applying external fields including temperature or electromagnetic gradients [34, 35]. This forms the basis of tunable filters based on the micro-flow of nematic liquid crystals. When light passes through a flowing LC matrix, its wavelength-dependent birefringence can be tuned by modifying the flow, confinement or the surface anchoring conditions in the microfluidic chip, allowing for high spatio-temporal modulation of the light properties. Furthermore, upon integration of a dye-filled microchannel, one can measure the absorption spectrum of the dye, allowing for tunable optofluidic colour filters and spectroscope-on-a-chip.

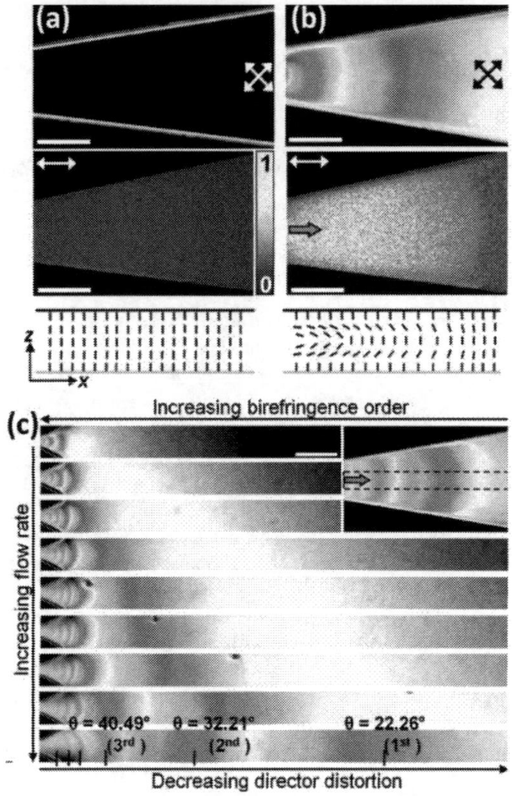

Fig. 1. Optofluidic velocimetry using nematic liquid crystal. (a) Polarized optical micrograph (top row) and fluorescence confocal polarized micrograph (middle row) of a channel filled with 5CB in absence of any flow. (b) At low flow rates, colourful birefringent domains appear, the corresponding fluorescent micrograph revealing the flow-induced director field. The laser is polarized along the flow direction. The bottom row schematically represents the molecular orientation in each case. Scale bar: 300 μm. (c) Polarized micrographs capture the evolution of the birefringent domains with gradual increase of the flow rate, as observed within the region marked in the inset. Adapted from Sengupta *et al.* [11].

In addition to flows, application of an external electric field provides a direct way of controlling the LC director, owing to the dipole moment of the LC system. Light-guided variation of the LC director and fluid actuation [36] are central to the modulation of the index of refraction, often leveraged in display technologies. However, the electric field by itself may trigger complex nonlinear responses. Taken together, the combined effect of hydrodynamic flow and an external electric field, could drive the emergence of a secondary axis in the anisotropic fluid. This was confirmed recently *in silico*, where a uniaxial NLC under equilibrium conditions produced an effective biaxiality under steady-state non-equilibrium conditions [13].

3 Topological defects as optical elements

Topological defects in liquid crystals have gained significant traction as optical vortex generators in recent years. The spin-orbit interaction of light, i.e., the coupling between the spin and the orbital angular momentum of light in anisotropic LCs has studied in thin films and droplets [37]. A key challenge therein lies in the generation of well-controlled orientation of the local optical axis at the microscale. Though micro-fabrication technology has been leveraged for devising high quality microscopic vortex beam generators, their passive characteristics limit the applicability. Liquid crystals offer highly efficient and tunable variants–as droplets or films embedding topological defects–realized by using electric field [38-40]. The operating wavelength and mode of such optical spin-orbit microscopic encoders can be controlled in real time using low-voltage electric fields. Complementing the field-induced topological defect generation in LC films and droplets, dispersions of NLCs in immiscible fluids and gels (like water or glycerin), have offered self-assembled alternatives [41]. These include microdroplets which have been utilized as tunable optical microresonators [42], 3D microlasers [43, 44] and Whispering-Gallery Mode (WGM) microlasers [45]. Such self-assembled fluidic structures have provided the impetus to employ NLC dispersions as basic photonic elements for soft-matter photonic circuits [46].

Despite our current understanding of topological defects in the context of *topological microfluidics*, and the unique potential of defects as optical elements, the two domains are yet to forge a meaningful synergy. Topological microfluidics has enabled programmable generation, stabilization, and guidance of defects and director topology, which taken together with droplet-based microfluidic techniques to generate dispersed phases like droplets and shells [47] offer an rich and exciting, albeit underexplored path ahead. Integrating microfluidic programming of the director and defect topology with optical manipulation offers a versatile potential for on-chip platform, with upstream opto-mechanical and opto-elastic components (actuators, valves and soft rails [26, 48]) that will enable downstream optofluidic manipulation and applications within the same microfluidic pipeline.

4 Toward novel optofluidic concepts enabled by topological microfluidics

Though the advent of *topological microfluidics* has propelled our understanding of the microscale LC flow behavior, the integration of optics–either to modulate the flow-driven LC director fields or to tune optical field via microflows–is still in its infancy. Optical fields have been extensively applied to modulate the fluidic properties of isotropic materials on-chip, yet replacing an isotropic fluid by anisotropic matrix remains grossly underexplored. Broadly, *topological microfluidics* will enable novel optofluidic and optomechanical concepts via two primary flow-mediated modes, as depicted in Figure 2: (i) non-singular LC director fields, *i.e.*, through continuous director fields and non-singular defect states; and (ii) singular LC director fields, *i.e.*, topological singular defects possessing different winding numbers. Both modes allow for dynamic programmability of the director and defect topology, thus offering a range of possibilities for optofluidic manipulations in the parameter space of microfluidic geometry, surface anchoring conditions, and the strength and nature of the imposed flow fields.

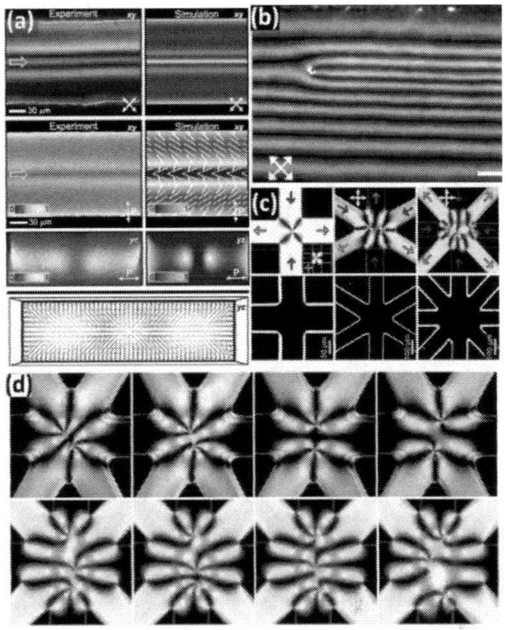

Fig. 2. Topological Microfluidics allows manipulation of director and defect topology toward novel optofluidic concepts. (a) Tunable complex flow-induced director field observed in microfluidic channels with homeotropic anchoring modulation, observed in both experiments and simulations. The topology of the director field can be modified by altering the microfluidic geometry, surface anchoring, and flow rates [34]. (b) Formation of a π-wall due to the nematic microflow in a channel with degenerate planar surface anchoring. The leading defect (bright spot) evolves spontaneously, which over time, reduces in length and stabilizes by aligning vertically within the channel [27]. (c) Microfluidic geometry can be harnessed to program the topology of nematic defects, shown here for generating effects of strength -1, -2 and -3 [19]. (d) Higher strength topological defects and lower multiples thereof (each with -1 strength) can be spatio-temporally manipulated by hydrodynamic flows [19]. The images are adapted from works of the author (Refs. [19,27,34]).

At a fundamental level, one could consider a fourth parameter, through the choice of the LC material. LC director field–in general–encounter short- and long-range elastic deformations in shear flows, depending on the order parameter and aligning (or tumbling) nature of the LCs. Four distinct regimes, spanning different

Deborah (*De*) and Ericksen (*Er*) numbers, are observed [14], of which the elastic-driven steady state corresponds to tumbling (or non-aligning) nematics and nematic elastomers, and the viscous-driven steady state corresponds to the flow-aligning behaviour at higher Er numbers. The two dimensionless numbers, *De* and *Er*, respectively capture the ratio of the viscous to short range elastic effect, and the ratio of the viscous effect to long-range elastic effect. Additionally, the intrinsic non-linearity in the system can be harnessed to drive the system into flow bifurcations, which promises novel parameter space for optofluidic applications. The synergy between *topological microfluidics* and optofluidics has already initiated, as evidenced by the recent progresses in several directions [49, 50]. For instance, the orientation of the nematic director is highly susceptible to electric or magnetic fields, thus the presence of additional fields would vastly expand the scope of LC-based optofluidic systems. Another novel direction would be the choice of chiral nematic phases for optofluidic applications: the chiral phase structure, known for versatile optical properties, can be integrated into the microfluidic pipeline for unique applications [50]. Finally, LC-based optofluidic platforms can be used to generate tailored light-environments at micron and sub-micron dimensions, with large dynamic range for tuning the intensity, wavelength and polarization of light waves. Such precision light fields which could be guided to a desired target, will allow simulated light conditions in different biological research systems (where specific light conditions are desired), for instance *in vitro* research on photosensitive microorganisms and plant cells.

This work was supported by the ATTRACT Investigator Grant (Grant no. A17/MS/11572821/MBRACE) and the FNR-CORE Grant (No. C19/MS/13719464/TOPOFLUME/Sengupta) of the Luxembourg National Research Fund.

References

1. D. Psaltis, S.R. Quake, C. Yang, Nature **442**, 381 (2006)
2. P. Minzioni, R. Osellame, C. Sada, S. Zhao, J. Optics **19**, 9 (2017)
3. S. Rubin, Y. Fainman, Phys. Rev. Lett **120**, 243904 (2018)
4. P. Rajaeipour, K. Banerjee, A. Dorn, H. Zappe, Ç. Ataman, Adv. Phot. **2**, 066005 (2020)
5. Y-C. Li, H-B. Xin, H-X. Lei, L-L. Liu, Y-Z. Li, Y. Zhang, B-J. Li, Light: Science & Applications **5**, e16176 (2016)
6. M. Makela, P.T. Lin, Anal Chem. **93,** 4154 (2021)
7. A. Stambaugh, J.W. Parks, M.A. Stott, G.G. Meena, A.R. Hawkins, H. Schmidt, PNAS **118**, e2103480118 (2021)
8. W. Zhang, W.C. Cheng, L. Dong, J.L. Di, J.L. Zhao, Proc. SPIE **115470F** (2020)
9. A.A. Kayani, K. Khoshmanesh, S.A. Ward, A. Mitchell, K. Kalantar-zadeh, Biomicrofluidics **6**, 031501 (2012)
10. A. E. Vasdekis, J. G. Cuennet, D. Psaltis, Proc. SPIE **847507** (2012)
11. A. Sengupta, S. Herminghaus, C. Bahr, Appl. Phys. Lett. **101**, 164101 (2012)
12. M. Kleman, O.D. Lavrentovich, Springer-Verlag New York, Inc., ISBN: 978-0-387-21759-8 (2003)
13. J-C. Eichler, R.A. Skutnik, A. Sengupta, M.G. Mazza, M.Schoen, Mol. Phys. **117** (2019)
14. P. Steffen, E. Stellamanns, A. Sengupta, Phys. Fluids **33**, 072005 (2021)
15. P. Oswald, P. Pieranski, CRC Press, ISBN: 9780429215742 (2005)
16. P. G. DeGennes, J. Prost, Oxford University Press, Inc., New York (1993)
17. K. Peddireddy, V.S.R. Jampani, S. Thutupalli, S. Herminghaus, C. Bahr, I. Muševič, Optics Express **21**, 30233 (2016)
18. M. Urbanski, C.G. Reyes, J.H. Noh, A. Sharma, Y. Geng, V.S.R. Jampani, J.P.F. Lagerwall, J. Phys.: Cond. Mat. **29**, 133003 (2015)
19. L. Giomi, Ž. Kos, M. Ravnik, A. Sengupta, Proc. Natl. Acad. Sci. U. S. A. **114**, E5771 (2017)
20. A. Sharma, I. L. H. Ong, A. Sengupta, Crystals **11**, 35 (2021)
21. S. Čopar, Ž. Kos, T. Emeršič, U. Tkalec, Nature Comm. **11**, 1 (2020)
22. A. Sengupta, Liquid Crystals Today **24**, 70 (2015)
23. A. Sengupta, S. Herminghaus, C. Bahr, Liquid Crystals Reviews **2**, 73 (2014)
24. A. Sengupta, Topological Microfluidics (Springer International Publishing, Switzerland, ISBN: 978-3-319–00857-8 (2013)
25. A. Sengupta, B. Schulz, E. Ouskova, C. Bahr, Microfluid. Nanofluid. **13**, 941 (2012)
26. A. Sengupta, C. Bahr, and S. Herminghaus, Soft Matter **9**, 7251 (2013)
27. A. Sengupta, U. Tkalec, C. Bahr, Soft Matter **7**, 6542 (2011)
28. T. Emeršič, R. Zhang, Ž. Kos, S. Čopar, N. Osterman, J.J. De Pablo, U. Tkalec, Science Advances **5**, eaav4283 (2019)
29. A. Sengupta, Int. J. Mol. Sci. **14**, 22826 (2013)
30. G. Tóth, C. Denniston, J.M. Yeomans, Phys. Rev. Lett. **88**, 105504 (2002)
31. M. Ravnik, J.M. Yeomans, Phys. Rev. Lett. **110**, 026001 (2013)
32. J.G. Cuennet, A.E. Vasdekis, D. Psaltis, Lab Chip **13**, 2721 (2013)
33. O. Wiese, D. Marenduzzo, O. Henrich, Soft Matter **12**, 9223 (2016)

34. A. Sengupta, U. Tkalec, M. Ravnik, J. Yeomans, C. Bahr, S. Herminghaus, Phys. Rev. Lett. **110**, 048303 (2013)
35. O. Henrich, D. Marenduzzo, J. Lintuvuori, Soft Matter **10**, 4580 (2014)
36. J-P. Delville et al., J. Opt. A: Pure Appl. Opt. **11** 034015 (2009)
37. E. Brasselet, C. Loussert, Proc. SPIE **820416** (2011)
38. E. Brasselet, N. Murazawa, H. Misawa, S. Juodkazis, Phys. Rev. Lett. **103**, 103903 (2009)
39. E. Brasselet, Phys. Rev. Lett. **108**, 087801 (2012)
40. E. Brasselet, C. Loussert, Optics Letters **36**, 719 (2011)
41. K. Peddireddy, V.S.R. Jampani, S. Herminghaus, C. Bahr, M. Vitek, I. Muševič, Proc. SPIE **9182** (2014)
42. M. Humar, M. Ravnik, S. Pajk, I. Muševič, Nat. Photonics **3**, 595 (2009)
43. M. Humar, I. Muševič, Opt. Express **18**, 26995 (2010)
44. D.J. Gardiner, S.M. Morris, P.J.W. Hands, C. Mowatt, R. Rutledge, T.D.Wilkinson, and H.J. Coles, Opt. Express **19**, 2432 (2011)
45. M. Humar, I. Muševič, Opt. Express **19**, 19836 (2011)
46. I. Muševič, Liquid Crystals Reviews **4** (2016)
47. M. Schwartz, G. Lenzini, Y. Geng, P.B. Rønne, P.Y.A. Ryan, J.P.F. Lagerwall, Adv. Mat. **30**, 1707382 (2018)
48. E.K. Fleischmann, H.L. Liang, N. Kapernaum, F. Giesselmann, J.P.F Lagerwall, R. Zental, Nature Commun. **3**, 1178 (2012)
49. Y.J. Na, T.Y. Yoon, S. Park, B. Lee, S.D. Lee, Chem Phys Chem. **11**, 101 (2010)
50. G. Tkachenko, E. Brasselet, Nature Commun. **5**, 3577 (2014)

Selective etching of 10 MHz repetition rate fs-laser inscribed tracks in YAG

Kore Hasse[1,2,3*], *Detlef* Kip[1], and *Christian* Kränkel[4]

[1]Faculty of Electrical Engineering, Helmut Schmidt University, 22043 Hamburg, Germany
[2]Institut für Laser-Physik, Universität Hamburg, Luruper Chaussee 149, 22761 Hamburg, Germany
[3]The Hamburg Center for Ultrafast Imaging, Luruper Chaussee 149, 22761 Hamburg, Germany
[4]Leibniz-Institut für Kristallzüchtung, Max-Born-Str. 2, 12489 Berlin, Germany

Abstract. We investigated fs-laser structuring of YAG crystals at high writing velocities up to 100 mm/s using a commercial 10 MHz fs-laser system supplied by Coherent Inc. and selective etching of these structures for fabrication of ultrahigh aspect ratio microchannels. Usage of a diluted acid mixture of 22% H_3PO_4 and 24% H_2SO_4 accelerated the etching process significantly to an etching parameter D of 11.2 µm²/s, which is three times higher than previously reported. Additionally, the selectivity of the etching process was increased by an order of magnitude.

1 Introduction

Selective etching of fs-laser inscribed structures has become a huge research field. Three-dimensional hollow micro- and sub-microstructures with aspect ratios of length to diameter in excess of 8000 [1] find applications in e.g. micro- and optofluidics, microsensors, photonics, and integrated lab-on-a-chip applications [2]. For glasses the process has been extensively studied and even finds industrial applications [3], but it is also applied for crystals such as sapphire [4], YAG [5], quartz [6], and CaF_2 [7].

The fabrication time for complex hollow structures is mainly limited by the selective etching rate, especially in crystals [8]. Nevertheless, also the fs-laser writing velocity can be a limiting factor [9] considering that lots of samples could be etched at the same time, while writing one single track requires one writing laser beam for each track and complex structures easily consist of a huge amount of written tracks.

For this reason, we systematically investigated the fs-laser structuring process using a 10 MHz-pulse repetition rate laser system. Due to the high repetition rate, a spatial pulse overlap is warranted also at very high writing velocities. In this way we reach up to 100 mm/s writing velocity, limited only by the maximum translation speed of the used translation stages. Furthermore, we investigated the selective etching behavior of the MHz-writing rate inscribed structures in YAG for the first time. Using the diluted acid mixture 22% H_3PO_4 24% H_2SO_4 we were able to accelerate the etching process compared to the undiluted mixture and to increase the etching parameter D by a factor of 3 to 11.2 µm²/s, while increasing also the selectivity by an order of magnitude. Using a precision dicing saw we excavated the channel walls to measure the channel wall roughness for the first time. A low roughness is crucial for photonic applications such as etched waveguide structures [1,10], gratings and lenses or resonators, what is known from waveguides fabricated using lithography [11].

2 Experimental procedure

For the fs-laser inscription we used a 10 MHz repetition rate *Fidelity HP High Energy* Yb-doped fiber laser oscillator supplied by Coherent Inc. The laser had a maximum output power of 10 W, 140 fs pulse duration, a center wavelength of 1040 nm, and a nearly diffraction limited beam quality of $M^2 < 1.3$. The beam was focused 364 µm deep into the polished surfaces of YAG crystals using focusing lenses of an *NA* of 0.50 and 0.55 (3.1 mm and 4.5 mm focal length, respectively). The laser beam polarization was chosen either parallel (π) or perpendicular (σ) to the writing direction (along the x-axis cf. Fig. 1 a)) which was perpendicular to the structuring laser beam (z-direction). Tracks were written along the [111] and [100] direction to investigate the influence of the crystal lattice orientation. For this purpose, the samples were translated on *Aerotech ABL1000* stages with velocities between 1 - 100 mm/s at average inscription laser powers of 0.54 - 1.85 W. More details on the fs-laser structuring setup can be found in [8].

We investigated the etching process using different etching agents at different temperatures including 43%H_3PO_4 48%H_2SO_4 at 83°C, 90°C, and 105°C and 22%H_3PO_4 24%H_2SO_4 at 83°C with respect to the characteristic etching parameter D and the selectivity S as defined in [8]. For the microscopic investigations, we used a *Keyence VHX7000* digital microscope.

To access the channel wall roughness, we cut slices containing one or two channels out of the samples with

* Corresponding author: hassek@hsu-hh.de

several channels inscribed, using a precision dicing saw *Disco D322,* first. The cuts were performed along the xz-plane. The channels were accessed from y-direction by increasing the cutting depth perpendicular to the xz-plane in 10 - 20 µm steps until the top of the channel was removed. In this way, we gained access to a direct measurement of the sidewall roughness with a laser-scanning microscope *Keyence VK-X.* The surface roughness was measured using the microscope software. Since the effort for this cutting procedure is very high, we limited the study of the channel wall roughness to selected processing parameters including the influence of the writing direction ([100] or [111]), the etching temperature, the focal length of the inscription beam as well as the etching agent concentration.

3 Results

Depending on the inscription pulse energy, we observed three regimes of material modification during inscription with 10 MHz repetition rate (cf. Fig. 1 b)). Pulse energies above 100 nJ lead to cracks, while those below 60 nJ did not lead to continuous track inscription. In the second regime the focus is split in vertical direction (cf. Fig. 1 b), regime 2). This vertical segmentation could result from self-focusing effects [1]. Third regime structures were strongly broadened by heat accumulation [12].

The most suitable regime (regime 1 in Fig. 1 b)) occurred at writing velocities of 30 to 100 mm/s and pulse energies between 60 and 100 nJ. These tracks were smooth and well confined, 15 to 40 µm in height and only 2 to 3 µm in width, similar to those observed at 1 kHz inscription frequency, and we regard this regime to be the only suitable scheme for selective etching of smooth hollow channels.

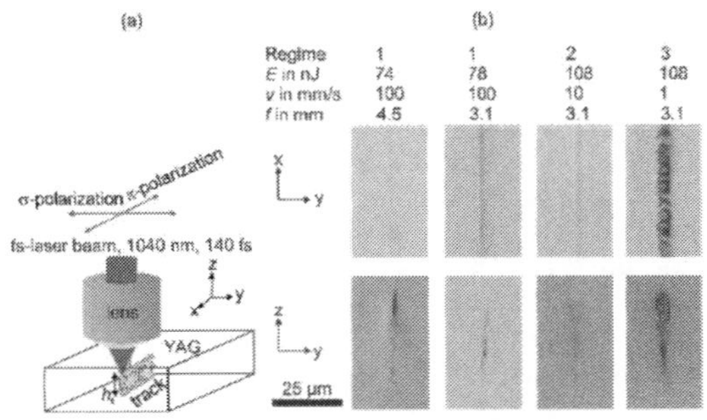

Fig. 1. Light microscope images of fs-laser written tracks inscribed in [111]-direction with π-polarization and the $f = 3.1$ mm lens, and σ-polarization with the $f = 4.5$ mm lens.

The etching within these tracks was highly selective for both 43%H_3PO_4 48%H_2SO_4 and 22%H_3PO_4 24%H_2SO_4 at 83°C and decreasing for higher temperatures. Due to this high selectivity, the diameter of the channels is mainly determined by the dimensions of the fs-laser inscribed material modification (cf. Fig. 2).

With regime 1 structures selectivity values in excess of 5000 were reached by etching tracks written in σ-polarization with a lens of 4.5 mm focal length and the diluted acid mixture at 83°C. The writing direction did not influence the selectivity significantly, neither did the pulse energy and writing velocity if chosen within the parameter range for regime 1. Focusing with a lens of 4.5 mm focal length though, leads to a higher selectivity and a larger etching parameter than using a lens of 3.1 mm focal length. With the 43%H_3PO_4 48%H_2SO_4 etchant, the selectivity was lower and no influence of polarization or writing direction was found.

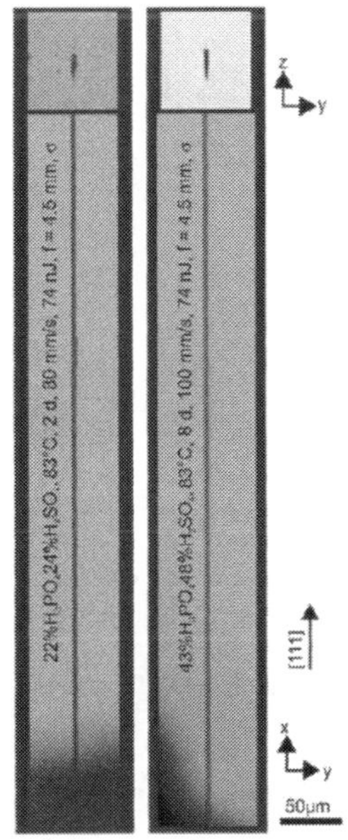

Fig. 2. Light microscope images of selectively etched channels. Writing direction and etching parameters are denoted in the image.

The etching depth d showed a dependence on time t like

$$d = (2 \cdot D \cdot t)^{1/2}. \quad (1)$$

The fitting parameter D characterizes the selective etching progress. Figure 3 a) shows the etched depth vs. time for both investigated etching agents at 83°C. Different than for tracks written at 1 kHz [8] no significant influence of the writing velocity on the etching velocity is observed (cf. Fig. 3 b)), enabling to use unprecedented inscription writing velocities of up to 100 mm/s. This is only a factor of 2 below the fastest fs-laser writing velocity in glasses [9] and only limited by the translation velocity of the stages used in this experiment.

The largest etching parameter value of $D = 11.2$ µm²/s was found for tracks inscribed with a focusing lens of 4.5 mm focal length, σ-polarization along the [111] direction and pulse energies between 60 and 100 nJ etching with 22%H_3PO_4 24%H_2SO_4 at 83°C (red curve in Fig. 3 a)). These also showed the highest selectivity. Extrapolation of the fit yields a total etching time of 13 days required for etching a 1 cm long

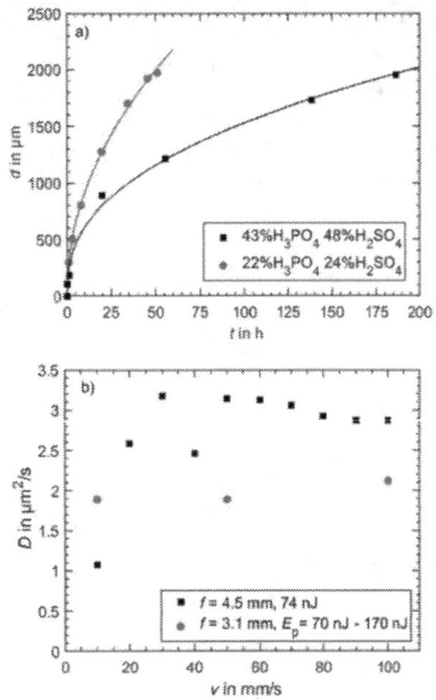

Fig. 3. a) Etched depth d vs. etching time t for both etching agents and first writing regime tracks and fitting curves. b) Fitting parameter D vs. writing velocity v for tracks etched in 43%H_3PO_4 48%H_2SO_4 at 83°C, the writing parameters are denoted in the figure.

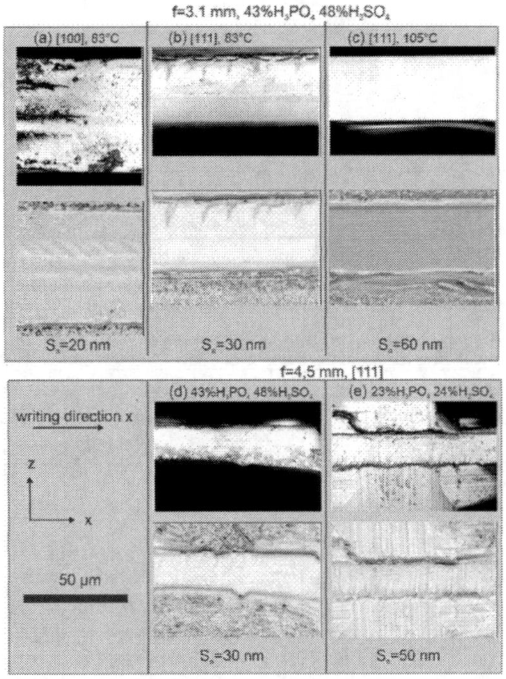

Fig. 4. Laser scanning microscope images of the opened selectively etched channels of regime 1 inscribed at 10 MHz. The different processing parameters are denoted in each picture.

channel from two sides. This is a factor of 3 shorter than under the previous fastest etching conditions (43%H_3PO_4 48%H_2SO_4 at 105°C, with a selectivity < 1000).

Fig. 4 shows microscope images of the channels opened to determine the side-wall roughness.

We did not observe any significant influence of the writing direction on the surface roughness of the channels. The channel cross section a) seems to be larger compared to b) - e) because of a rhombic cross section of the channel etched along [100] instead of the nearly rectangular channels obtained for [111] direction. We obtained an average surface roughness $S_a = 27 \pm 3$ nm for etching with 43%H_3PO_4 48%H_2SO_4 at 83°C, independent on the focusing conditions (cf. Fig. 4 b) and d)), though the height of the channels written with the lens of $f = 3.1$ mm is larger. Higher etching temperatures as well as a diluted acid increase the average surface roughness to $S_a = 55 \pm 4$ nm. Overall, the surface roughness is in the range of $\lambda/20$, what is typical for laser polish.

Nevertheless, compared to surface roughness values obtained using lithographic methods or chemical mechanical polishing ($S_a < 1$ nm [11]) the obtained roughness values are high. But, since the processing parameters were not yet investigated systematically with respect to reducing the surface roughness, it is likely that parameters yielding smoother channel surfaces exist. Overall, since the surface roughness is already comparable to laser polish, this is a promising result for future optical applications of selective etching.

4 Conclusion

For the first time we used a 10 MHz repetition rate laser system for fs-laser structuring of YAG. We achieved record-high writing velocities of up to 100 mm/s and identified inscription properties suitable for successive selective etching of the inscribed tracks. Using a diluted mixture of 22%H_3PO_4 24%H_2SO_4 at a temperature of 83°C three times improved etching parameters of up to $D = 11.2$ µm^2/s were achieved, enabling etching of 1 cm long microchannels in less than 14 days. We found the surface roughness of the channel walls to be below 60 nm. Our results represent a significant improvement towards future industrial applications of selective etching of fs-laser written structures in YAG concerning the enhanced processing speed. The obtained roughness parameters show the applicability for photonic applications.

This work was supported by the Cluster of Excellence 'Advanced Imaging of Matter' of the Deutsche Forschungsgemeinschaft (DFG) - EXC 2056 - project ID 390715994.

References

1. A. Ródenas, M. Gu, G. Corrielli, P. Paiè, S. John, A. K. Kar, R. Osellame, Nat. Photonics **13**, 105–109 (2018).
2. K. Sugioka, J. Xu, D. Wu, Y. Hanada, Z. Wang, Y. Cheng, K. Midorikawa, Lab Chip **14**, 3447 (2014).

3. J. Gottmann, M. Hermans, N. Repiev, J. Ortmann, Micromachines **8**(110), 1–10 (2017).
4. S. Juodkazis, H. Misawa, T. Ebisui, R. Waki, S. Matsuo, T. Okada, Adv. Mater. **18**(11), 1361–1364 (2006).
5. J. Siebenmorgen, K. Petermann, G. Huber, K. Rademaker, S. Nolte, A. Tünnermann, Appl. Phys. B **97**(2), 251–255 (2009).
6. S. Matsuo, Y. Tabuchi, T. Okada, S. Juodkazis, H. Misawa, Appl. Phys. A **84**(1–2), 99–102 (2006).
7. K. Hasse, D. Kip, C. Kränkel, Opt. Mater. Express **11**(5), 1546–1554 (2021).
8. K. Hasse, G. Huber, C. Kränkel, Opt. Mater. Express **9**(9), (2019).
9. M. Malinauskas, A. Žukauskas, S. Hasegawa, Y. Hayasaki, V. Mizeikis, R. Buividas, S. Juodkazis, Light Sci. Appl. **5**(8), 3–5 (2016).
10. J. Lv, B. Hong, Y. Tan, F. Chen, J. R. V. de Aldana, G. P. Wang, Photonics Res. **8**(3), 257 (2020).
11. J. Lin, F. Bo, Y. Cheng, J. Xu, Photonics Res. **8**(12), 1910 (2020).
12. S. M. Eaton, H. Zhang, P. R. Herman, F. Yoshino, L. Shah, J. Bovatsek, A. Y. Arai, Opt. Express **13**(12), 4708 (2005).

Dynamical imaging of local photovoltage at semiconductor surface by photo-assisted ultrafast scanning electron microscopy

Mohamed Zaghloul[1,3*], *Silvia M.* Pietralunga[2,3], *Gabriele* Irde[1,3], *Vittorio Sala*[1,3], *Giulio* Cerullo[1], *Hao* Chen[1,3], Giovanni Isella[4], *Guglielmo* Lanzani[1,3], *Maurizio* Zani[1], and *Alberto* Tagliaferri[1,3]

[1]Dipartimento di Fisica, Politecnico di Milano, Piazza Leonardo da Vinci, 32, 20133 Milano, Italy
[2]Institute for Photonics and Nanotechnologies (IFN)—National Research Council (CNR), Piazza L. da Vinci, 32, 20133 Milano, Italy
[3]CNST@PoliMi, Istituto Italiano di Tecnologia (IIT), Via Giovanni Pascoli 70/3, Milano, Italy
[4]LNESS-Dipartimento di Fisica, Politecnico di Milano, Piazza Leonardo da Vinci 32, 20133 Milano, Italy

Abstract. Photo-assisted Ultrafast Scanning Electron Microscopy (USEM) maps the dynamics of surface photovoltages and local electric fields in semiconducting samples. Photovoltages and their gradients close to surface affect the emission yield and the detection efficiency of secondary electrons (SE), leading to photo-excited SE 2D patterns. In this work, we present a method to characterize the evolution of the patterns up to ultrafast regime. These results reveal the role of surface states in affecting the external field dynamics at picoseconds. Moreover, we show that tiny changes in surface preparation express deeply different photo-excited voltage signals. We investigate the relation between the surface chemistry of Si and photo-induced SE contrast.

1 Introduction

The study of dynamical Surface Photovoltages (SPV) provides information on the photophysics and charge transport in semiconductors. Recently, techniques of photo-assisted Time-Resolved Scanning Electron Microscopy (TR-SEM) were introduced to provide dynamical 2D mapping of SPV [1,2,3,4]. TR-SEM interpretation is based on the assumption that photo-excited spatial modulations in work function induce secondary electron (SE) contrast patterns. These contrast modulations originate from the photoexcitation of local excess charge carrier distributions, which affect the configuration of surface states and the yield of SE emission. A less considered effect is that this change in charge distribution modifies the electric field outside the surface too, which in turn dynamically changes the SE collection in the SEM [1,2,4,5,6]. Both mechanisms, the changes in emission yield and in collection efficiency, contribute to the 2D dynamical SE patterns. Mapping of SPV by TR-SEM provides nanoscopic depth selectivity and lateral resolution, with a broad range of possible view fields and timescales.

Various TR-SEM setups were devised, from the quasi-steady-state temporal regime up to ultrafast regime. The latter is usually reached in laser-driven configurations (ultrafast scanning electron microscopy, USEM) based on pulsed electron beams and operated in pump-probe mode [4,7,8,9,10].

TR-SEM has certainly provided interesting insights into the photophysics of semiconductors. Nevertheless, published TR-SEM characterizations on nominally identical surfaces may still lead to controversial and even opposite results.

In the present work, we devise a method for analyzing photo-induced electric fields at semiconductor by 2D TR-SEM contrast pattern. Moreover, by operating USEM in ultra-high vacuum (UHV) conditions we performed in-situ chemical analysis of the surface by Auger micro-spectroscopy and correlate the dynamics of SPVs to the conditions of sample preparation. Our results suggest that, even in the case of nominally identical samples, tiny changes in preparation and in SEM environment in dramatically different TR-SEM dynamics.

2 Experimental

Our USEM apparatus is based on a modified UHV ($10^{-9}\div10^{-10}$ torr) SEM (PHI 660), equipped with a ZrO coated W Field-Effect electron tip. It is driven by a pulsed fiber laser (300 fs pulses @10 MHz repetition rate, $\lambda = 1030$ nm). The third harmonic (TH) beam ($\lambda = 343$ nm) is exploited to dynamically excite optical phenomena at semiconductor surfaces. The UHV-SEM is also equipped with an energy analyzer to perform Scanning Auger Microspectroscopy (SAM).

Sample tilting and lateral positioning of the electron detector introduce a symmetry breaking in the detection process that, combined with a proper numerical modeling of the SE collection process, enables us to single out the information on photo-excited electric fields.

We present here USEM analysis on highly *p*-doped (about 10^{19} acceptors/cm^3) Si(001) surfaces, prepared in air by initially sonicating the samples in acetone and ethanol for 20'' and then differently treating them in hydrofluoridric acid (HF), to remove native oxide with increasing effectiveness. After this preparation, samples are introduced into the USEM's UHV chamber.

In situ surface chemical composition at Si(001) is investigated by SAM in UHV at different chemical etching times. Auger spectra have been fitted by linear combinations of spectra taken on differently oxidized Si surfaces, to correlate USEM contrast with the surface oxygen concentration and get insight into the apparently contrasting results in the USEM literature, where opposite contrast was reported from the surfaces of silicon with similar doping [7,8].

3 Results and discussion

3.1 USEM dynamic imaging

Our USEM setup operates in pump-probe mode. The TH pulses from the laser source (*pump*) excites at time zero the sample surface over an elliptical spot (white dashed line in panels (a)-(c) of Fig.1, main axis of about 100 µm). Then an electron pulse (*probe*) shines the surface at a delayed time (positive when following the photon pulse) and probes the photo-excited effects. Panels (a)-(c) shows typical USEM 2D images, respectively detected at a large negative (-1 ns), at slightly positive (100 ps) and at large positive (1.3 ns) delay. The central panel (b), taken at delay close to zero, shows the largest SE contrast change at laser spot, and also in the external area around it.

In order to quantify photo-excited USEM contrast and its dynamics, a differential processing of the USEM images is operated as described hereafter. A multiple-pixel spatial averaging of the measured SE intensity is performed over a defined range, named hereafter *area of interest* (AOI). The AOI is a sub-portion of the width of the area where photo-excited effects are evident. A reference contrast value is also obtained by averaging SE intensity measured on another area, identical in shape to AOI and located in an unperturbed region far away from the laser spot. The USEM contrast is finally obtained as the difference between the AOI average and the reference contrast. By scanning the AOI over the image frame, 2D USEM contrast maps are obtained. Panel (e) of Fig. 1 shows a map acquired around zero delay on a non-HF etched sample surface (shown in Fig. 1.b). It is clear from Fig. 1.e that USEM contrast behaves differently on points at the left side (left area) and at the right side (right area) of the laser spot. Two regions of maximal contrast exist, positive on the left hand side of the laser spot (at the center of the image) and negative at the right hand side. The results for the red AOI at left and for the black AOI at right are plotted as a function of pump-probe delay in panel (d) of Fig. 1.

All of the temporal plots show a step-like evolution of the USEM contrast within 100 ps from zero delay, followed by a slower decay on a few ns timescale.

The state-of-the-art USEM literature focuses on the SE dynamics measured at the center of the laser spot and

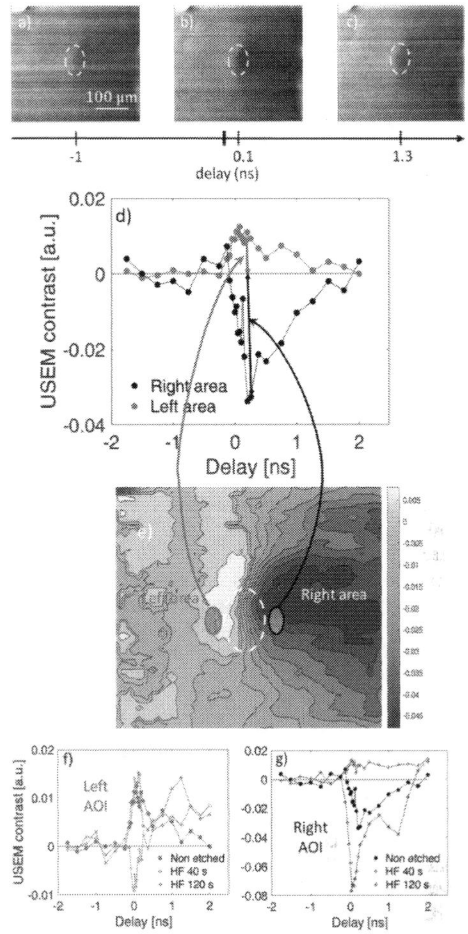

Fig. 1. Time evolution of USEM photoinduced contrast on highly *p*-doped Si, as a function of pump-probe delay. (a)-(c): 2D USEM contrast maps on a non-HF etched sample at three delay times, highlighting the bipolarity in sign of the contrast. (d) Time series of the USEM contrast, from the two AOI at left (red) and right (black) side of the laser spot (white), shown on the 2D USEM contrast step map in panel (e). (f): comparison of the SE contrast amplitude induced in differently HF treated samples in the left AOI; (g) the same for the right AOI.

relates it to the SE emission yield [3,7,8,10]. In our setup, a spatial bipolar modulation of the SE signal around the laser spot appears, which we attribute to the breaking of the cylindrical symmetry with respect to surface normal axis by the side SE detector. This modulation appears in the 2D patterns in Fig. 1, panels (a)-(c) (SE images) and panel (e) (USEM contrast map). We show that on highly *p*-doped Si surfaces, the USEM time series may have very different maximum amplitudes, and even switch in sign depending on SE emission point. This switching in sign cannot be explained by an spatial intensity modulation of SE emission yield within the laser spot itself. On the contrary, it mostly derives from the interplay between the gradients of the SPV and symmetry-breaking in detection. A positive SPV in the laser spot corresponds to a right-to-left electric field component parallel to the sample surface at the left side of the laser spot. As a consequence, the trajectories of SE emitted on the left side bend towards the detector positioned at the right, and their collection enhances. An opposite behavior appears at the right side, where the SE trajectories bend away and the collection reduces [1,2]. This switching in sign of the USEM contrast at the sides of the laser spot is therefore a fingerprint of the external fields at sample surface and has the potential to provide qualitative (sign) and quantitative (intensity) insight on SPVs. A similar effect was recently reported on GaAs surfaces [4].

The impact of the surface preparation on USEM contrast and on SPV is evident from Fig. 1, panels (f) and (g), where the result on three differently prepared highly *p*-doped Si(001) surfaces is compared. The first sample was just sonicated in acetone and ethanol, without HF etching (leaving the native oxide on the surface untouched); the second and third ones were subsequently etched in HF for 40 s and 120 s, respectively. On each surface, the inversion of the contrast amplitude with the displacement of AOI from the left hand (red) to the right hand side (black) of the laser spot is confirmed. As a very remarkable feature, we notice that the time series from the three surfaces show dramatic differences in amplitude and shape, with the 40" HF etched one presenting even a change in sign of the step with respect to the other two. This data reveals the sensitivity of the time varying photovoltages to the surface structure and to surface electronic states.

3.2 Auger electron spectroscopy analysis

Scanning Auger Microspectroscopy has been performed on samples after doing USEM measurements to analyze chemical differences in the highly *p*-doped Si(001) surfaces mentioned above and to relate them to the different dynamics shown by USEM signal as a function of surface preparation. Auger KLL lines for carbon, oxygen and silicon are reported in Fig. 2. The positive-to-negative peak intensity of those differential spectra is a monotonic function of the surface atomic concentration of the elements. Panel (a) shows carbon KLL line, featuring the negative peak around 269 eV. The C KLL intensity from the non-etched surface is a small fraction of the one from the two etched ones. Moreover, it is remarkably more intense from the surface HF etched for 120", than in the 40" HF etched one. Panel (b) shows the oxygen KLL negative peak around 510 eV. The O KLL more intense for non-etched with respect to the two etched surfaces, where it features a nearly equal height. The spectra of the Si KLL at about 1630 eV in panel (c) show almost the same strength in the non-etched and in the 40" HF etched surface, while its intensity is remarkably lower in the 120" HF etched one. In Fig. 2 it is noteworthy that the O KLL is less intense, while C KLL appears more intense after the etching process for both 40" and 120' This is an evidence of the removal of the native SiO2 layer by the HF etching, and at the same time of the increase in carbon

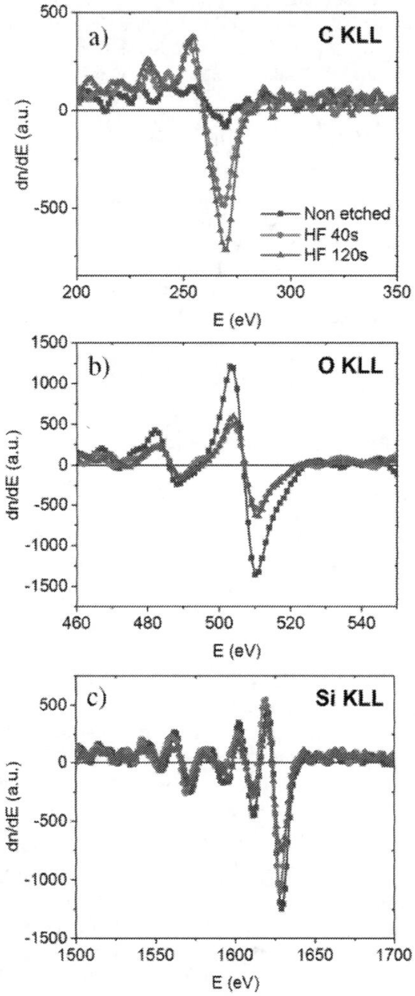

Fig. 2. Auger KLL spectra for highly *p*-doped Si surfaces (non-etched, 40" and 120" HF etching): (a) carbon, (b) oxygen and (c) silicon.

contamination of the surface during the post-etching time period until the sample is inserted in the vacuum chamber. This carbon contamination is more effective on the 120" HF etched than on the 40" HF etched surface. This is supported by the lower intensity of the Si KLL line in the former, which is tentatively attributed to the lower Si concentration in the carbon-rich over-layer at the surface. We report that carbon surface concentration is not influenced by the USEM measurements, as the Auger spectra do not show measurable changes on the e-beam exposed surface with respect to the as-inserted surface (not shown).

3.3 Correlation between USEM and SAM

The correlation between the USEM contrast reported in Fig 1.f and the atomic concentrations of carbon and oxygen at surface from the Auger spectra in Fig. 2 highlights the sensitivity of the USEM technique to SPVs, and in turn to the surface electronic structure as resulting from its chemical and structural state. USEM contrast 2D pattern in Fig 1.f reveals the same sign when measured on the non-etched surface and on the 120" HF etched one, while an opposite sign is evident on the 40" HF etched. Our interpretation of the influence of USEM contrast with duration time of chemical etching, is that the etching itself and the following exposure to air determine very different chemical and structural states of the Si(001) surface. This in turn deeply affects the value of the work function and of the charge carrier transport at surface that result in different photovoltages and even in their sign reversal.

4 Conclusion

The dynamics of photo-induced voltages at the (001) surface of highly *p*-doped silicon has been investigated by ultrafast scanning electron microscopy (USEM) and Scanning Auger Microspectroscopy (AES). A protocol to analyze differential USEM contrast maps was developed, to quantify the photo-induced contrast. USEM contrast is strongly influenced by chemical state resulting from the preparation of the Si(001) surface. Different HF etching times lead even to a switching in sign of the USEM contrast pattern. By correlating USEM and Auger electron microspectroscopy on differently prepared samples, we highlight the dramatic role of surface chemistry on photovoltages and charge dynamics at Si(001) surface.

References

1. G. Irde, S. M. Pietralunga, V. Sala, M. Zani, J.M. Ball, A.J. Barker, A. Petrozza, G. Lanzani, A. Tagliaferri, Micron **121**, 53 (2019)
2. S.M. Pietralunga, G. Irde, A.J.A.J. Barker, J.M. Ball, A. Petrozza, V. Sala, M. Zani, G. Lanzani, A. Tagliaferri, *Adv. Mater. Interfaces* **7**(16), 2000297 (2020)
3. Y. Li, U. Choudhry, J. Ranasinghe, A. Ackerman, B. Liao, *J. Phys. Chem. A* **124**(25), 5246–52 (2020)
4. M.W.H. Garming, M. Bolhuis, S. Conesa-Boj, P. Kruit, J.P. Hoogenboom, *J. Phys. Chem. Lett.* **11**(20), 8880–6 (2020)
5. W. Han, M. Zheng, A. Banerjee, Y.Z. Luo, L. Shen, A. Khursheed, *Sci. Rep.* **10**(1), 22144 (2020)
6. A. Khursheed, *Secondary Electron Energy Spectroscopy in the Scanning Electron Microscope*, (World Scientific, 2020)
7. E. Najafi, T. Scarborough, J. Tang, A.H. Zewail, Science **347**(6218), 164 (2015)
8. E. Najafi, V. Ivanov, A. Zewail, M. Bernardi, *Nat. Commun.* **8**(1), 15177 (2017)
9. M. Zani, V. Sala, G. Irde, S. M. Pietralunga, C. Manzoni, G. Cerullo, G. Lanzani, and A. Tagliaferri, *Ultramicroscopy*, **187**, 93 (2018)
10. E. Najafi, A. Jafari, *Appl. Phys. J.* **125**, 185303 (2019)

Efficient carrier-envelope phase tunable mid-infrared frequency combs based on CW-seeded optical parametric generation

Mikhail Roiz[1,*], *Jui-Yu* Lai[2], *Juho* Karhu[3] and *Markku* Vainio[1,4]

[1]Department of Chemistry, University of Helsinki, FI-00560, Helsinki, Finland
[2]HC Photonics Corp. Hsinchu Science Park, Hsinchu 30078, Taiwan
[3]Metrology Research Institute, Aalto University, Espoo, FI-00076, Finland
[4]Photonics Laboratory, Physics Unit, Tampere University, Tampere, FI-33101, Finland

Abstract. We present an efficient method for generating frequency combs in the mid-infrared (MIR) spectral range in bulk lithium niobate crystals as well as in waveguides. Our approach is simple and robust, since it is based on single-pass configuration of femtosecond Optical Parametric Generation (OPG) seeded by a telecom continuous-wave laser. Precise and fast tuning of the seed laser allows to rapidly change the offset frequency of the generated MIR comb independent of its repetition rate. The MIR comb offset frequency is inherently known, so its direct detection is not required. An additional degree of freedom that further complements the versatility of our light source is a unique way of repetition rate or mode spacing adjustments. This is achieved by our new concept of coherent multi-seeding that depending on the configuration allows for pulse train modulation and mode spacing, or repetition rate division by an integer number.

1 Introduction

Mid-infrared (MIR) Optical Frequency Comb (OFC) technology is a rapidly evolving scientific field since it has many crucial applications including frequency metrology, high-harmonic generation, microscopy and high-resolution spectroscopy. Especially in high-resolution spectroscopy, high-quality MIR OFCs are of great importance because they allow for parallel detection of tens of thousands absorption lines simultaneously with metrological precision [1]. Direct generation of MIR OFCs in Mode-Locked Lasers (MLL) is challenging, thus requiring parametric down-conversion techniques to be involved. However, the most common ways for MIR OFC generation based on parametric down-conversion – namely Optical Parametric Oscillation [2] and Difference Frequency Generation [3] – are technically quite demanding and complicated. In this conference contribution, we demonstrate an alternative way to generate MIR OFCs based on Optical Parametric Generation (OPG), which is remarkably simple and highly versatile.

2 Experimental setup

Our approach is general and can be used in any non-degenerate OPG system, however we have implemented the method using a readily available and popular nonlinear material - MgO doped periodically poled lithium niobate (MgO:PPLN). Here we demonstrate two CW-seeded OPG implementations - one is based on bulk crystals and the other on nonlinear waveguide technology in MgO:PPLN. The crystal/waveguide is pumped by a near-infrared MLL (250 MHz repetition rate, 100 fs pulse duration) centred at 1 μm wavelength, as schematically shown in Fig. 1 (a). The seeding is done in the signal region, which coincides with the convenient telecom

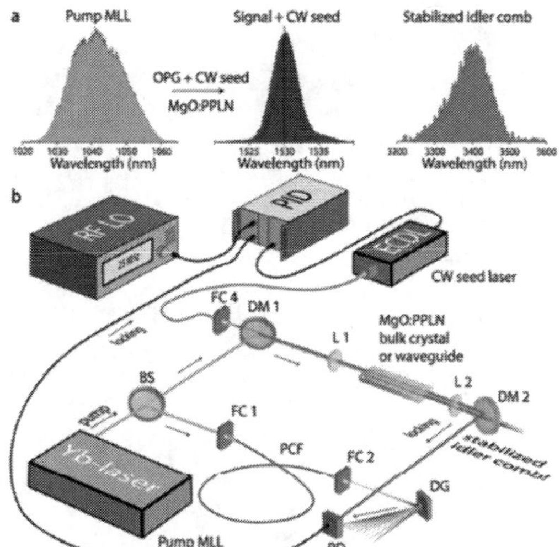

Fig. 1. (a) CW-seeded OPG concept (spectra are shown in the linear scale with arbitrary units for the case of bulk crystal). (b) Experimental setup. BS: beam splitter, FC: fiber collimator, DG: diffraction grating, DM: dichroic mirror, L: lens, PD: photodiode, PID: Proportional–Integral–Derivative controller, ECDL: External Cavity Diode Laser, PCF: Photonic Crystal Fiber.

* Corresponding author: mikhail.roiz@helsinki.fi

https://doi.org/10.1051/epjconf/202125511002

wavelengths. The idler comb typically has an instantaneous spectral coverage of 100 nm (full-width-at-half-maximum) for the bulk system and over 300 nm for the waveguides. The bulk system is realized using fanout crystal structure making the idler comb easily tunable across 3325 – 4000 nm (2380 – 3030 cm^{-1}). For more details on the bulk system as well as on waveguide see [4] and [5], respectively.

In this work we show that the CW seed laser can be phase-locked to a supercontinuum generated by the pump MLL, which guarantees full stabilization of the MIR comb [4,5]. When the signal comb of the OPG is seeded by a such phase-locked CW seed laser (see Fig. 1 (b)), it means that the signal comb shares the same Carrier-Envelope Offset (CEO) as the pump MLL, but with an additional frequency offset. This additional frequency offset is precisely defined by a conventional radio frequency (RF) local oscillator (RF LO in Fig. 1 (b)) used in the phase-locking process. Thus, by adjusting the RF LO frequency, one can continuously tune the signal and, most importantly, the idler CEO.

Table 1. Comparison of bulk and waveguide systems.

Parameter	Bulk	Waveguide
Threshold	8-16 nJ	25 pJ
Max. idler average power	700 mW	5.8 mW
Max. conversion efficiency	40 %	74 %
Pump-to-idler conversion	13 %	23 %
Instantaneous bandwidth	100 nm	> 300 nm
Min. integrated RIN	0.018 %	< 0.0047 %

3 Results

In Table 1 one can see a comparison of the main parameters of CW-seeded OPG in the bulk and waveguide systems. It is evident that the bulk system can be used for high power applications since it has a lot of output power: 700 mW of average power at 250 MHz repetition rate. However, the waveguide system has much lower threshold of just 25 pJ and a larger instantaneous optical bandwidth of over 300 nm (see Fig. 2 (a)). Additionally, in waveguides, we have achieved a remarkably high conversion efficiency of 74%, representing state-of-the-art for OPG systems [6].

Another important figure of merit is the Relative Intensity Noise (RIN). The importance of RIN in spectroscopy applications is hard to overestimate, since it directly defines detection limit or signal-to-noise ratio of the measurement. We show that in the OPG process RIN can be substantially reduced by CW seeding, which is evident from Fig. 2 (b). Remarkably, the integrated RIN in waveguides is much lower than in bulk systems because in waveguides one does not need to use high-power amplifiers (that induce additional intensity noise) for the pump laser. Consequently, the root-mean-square RIN of the MIR comb generated in a waveguide is smaller than 5×10^{-5} when integrated from 10 Hz to 2 MHz, where the noise is mostly below the detection limit.

Next we demonstrate how well the CEO of the two idler combs can be stabilized. We compare the two combs to one another using the same pump laser but with two different CW seed lasers locked to the supercontinuum as shown in Fig. 1 (b). The relative offset beat note can be seen in Fig. 2 (c) with sub-Hz linewidth, which is limited by the RF spectrum analyser. By performing frequency counting measurement of this beat note, we can conclude that the relative stability of the two combs is excellent with just 36 mHz standard deviation around the target frequency. The noise in this measurement averages out as white noise confirming that there are no relative drifts between the combs.

We would like to note that the bulk and waveguide implementations have excellent versatility in terms of offset frequency adjustments. The offset can be freely tuned, scanned or modulated with metrological precision [4, 5]. Our scheme does not have any cavities meaning that the signal/idler comb repetition rate can be freely adjusted by simply changing the pump laser repetition rate. Moreover, there is another useful degree of freedom that makes our light source even more versatile and flexible. Here we introduce a new concept of coherent CW multi-seeding that allows one to divide the repetition rate of the generated signal and idler combs by an integer fraction.

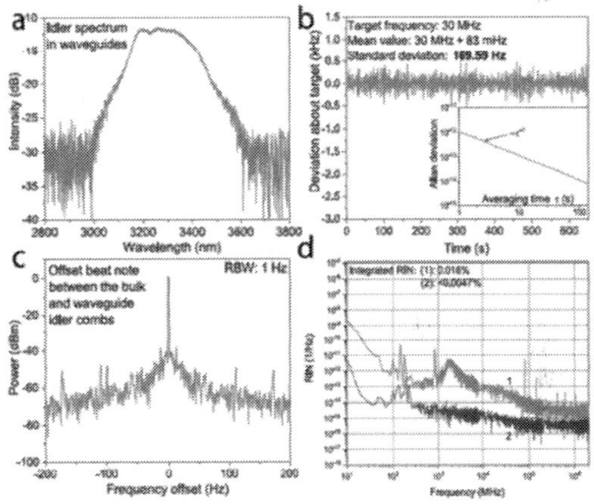

Fig. 2. (a) Typical idler spectrum in waveguides; (b) Relative Intensity Noise (RIN): (1) saturated OPG in bulk crystal with CW seeding; (2) same as (1) but in waveguides; (c) Offset beat note between the bulk and waveguide idler combs (pump laser is the same for both combs, see [5] for details); (d) Frequency counting experiment of the offset beat note in (c).

The term coherent CW multi-seeding refers to seeding of the OPG process with more than one mutually phase coherent CW lasers. In the simplest form of multi-seeding – namely dual-seeding – this approach requires two single mode CW lasers that are phase-locked to one another. The relative optical phases and frequencies of the seed lasers can be controlled by two independent RF LOs involved in the phase locking procedure. If we set the optical frequency difference to be exactly half of the pump laser repetition rate (250/2 MHz in our case), then the *CW-seeded OPG process will start up for both seed lasers simultaneously*. It leads to the generation

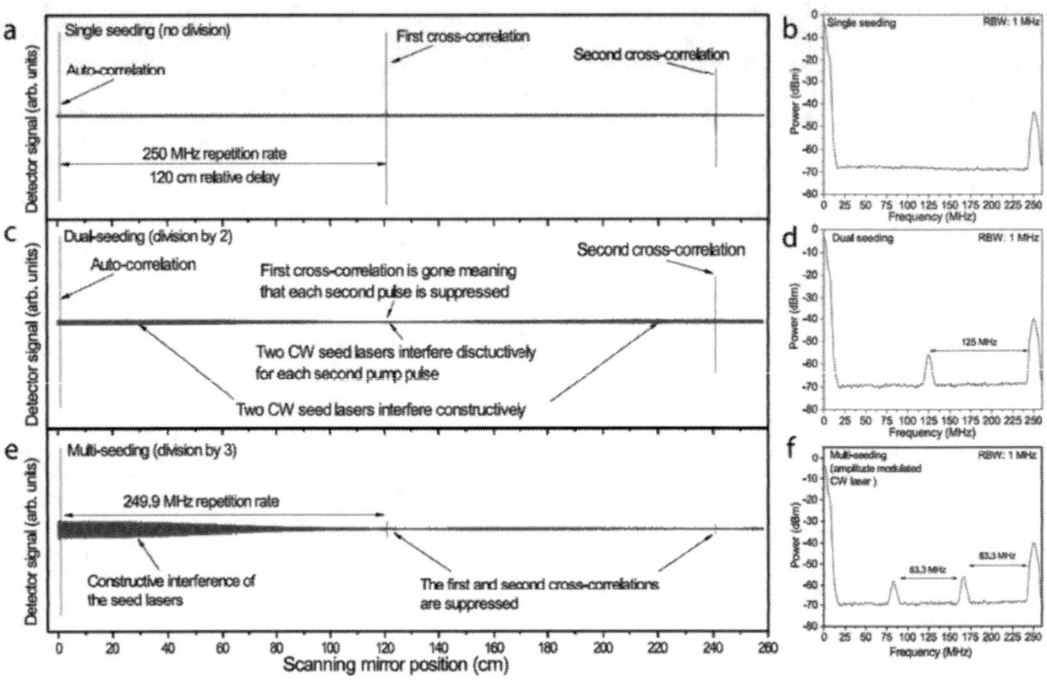

Fig. 3. (a) Interferogram in the case of single CW-seeded OPG; (b) RF spectrum of the signal/idler repetition rate for (a); (c) Interferogram for dual CW-seeded OPG corresponding to repetition rate division by a factor of 2; (d) RF spectrum of the signal/idler repetition rate for (c); (e) Interferogram for multi (one CW laser is amplitude modulated) CW-seeded OPG corresponding to repetition rate division by a factor of 3; (f) RF spectrum of the signal/idler repetition rate for (e).

of two signal/idler frequency combs within one laser pulse, optical frequencies of which are separated by exactly the frequency separation between the CW seed lasers. The dual-seeding configuration divides the repetition rate of the signal and idler combs in both frequency and time domain *only when the CW seed lasers are separated exactly by half of the pump repetition rate*. If the two seed lasers are separated by 83.3 MHz (250/3 MHz) pulse train modulation takes place (instead of pulse picking), where one of the pulses can be removed from the pulse train. This effect can be seen in the oscilloscope traces of Fig. 4, where we also exemplify the case of repetition rate division by a factor of 4. In order to efficiently divide the repetition rate in time domain one needs more than two seed lasers. Luckily, it can be easily done by using one CW laser with amplitude modulation. When a CW laser is amplitude modulated at the frequency of 83.3 MHz (division by 3), it generates additional sidebands (18 in our case) separated by 83.3 MHz serving as multi-seeding source. In Fig. 3e one can see how both the first and second cross-correlations are suppressed on the interferogram meaning that each second and third pulse is removed from the pulse train. In the frequency domain repetition rates of signal and idler combs are divided by 3 exactly the same way as it was done using dual-seeding approach (Fig. 3f).

Extreme repetition rate division – namely pulse picking at low frequencies – can be obtained by using a single seed laser, which is passed through a fast intensity modulator driven by short electrical pulses. The modulator can be used to switch the seed pulses on or off, hence gating the OPG process. This makes it possible to produce a repetition rate different from that of the pump (which is fixed at 250 MHz in all experiments), leading to repetition rate division by any integer. Alternatively, the seed laser pulses can be configured for arbitrary on/off modulation of the femtosecond pulses generated by the OPG process.

To summarize, in our conference contribution we demonstrate a simple concept for MIR frequency comb generation and provide its rigorous characterization. Our system makes it easy to stabilize and dynamically control the MIR comb offset frequency along with the free tuning of the repetition rate. Additionally, there is no need to directly detect the MIR comb offset frequency because its value is inherently known. We believe that the presented method is promising for a variety of applications in spectroscopy such as dual-comb spectroscopy, comb-assisted and Fourier-transform spectroscopy as well as in other scientific fields including microscopy and high-harmonic generation. Furthermore, our results in waveguides open up opportunities for efficient down-conversion of high repetition rate frequency combs (> 1 GHz), which is important for fast spectroscopic measurements, for example in reaction kinetics. This MIR comb generation method has an additional important advantage – one can perform repetition rate division directly in the OPG process using coherent multi-seeding approach. It helps to further improve the resolution of spectroscopic measurements by increasing the number of optical modes in the MIR comb (making it denser in frequency domain). In the pulse picking configuration one

* Corresponding author: mikhail.roiz@helsinki.fi

can selectively start the OPG process for any given pump pulse. Effectively, this filters out the pulse train and picks only those signal and idler pulses that are desired. This particular scheme can be applied in optical parametric chirped pulse amplifiers as a seed source since the seed sources need to have low repetition rate (<200 kHz) [8].

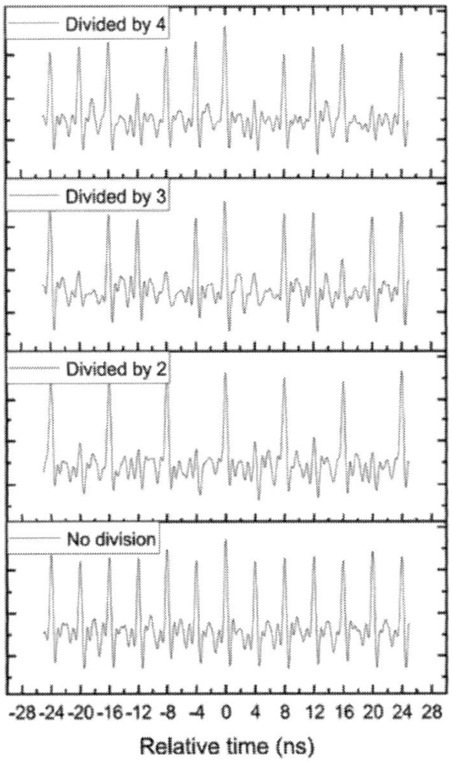

Fig. 4. Oscilloscope traces of the signal (same for idler) pulse train in different repetition rate division configurations in the dual-seeded case.

References

1. 1. A. V. Muraviev, V. O. Smolski, Z. E. Loparo, K. L. Vodopyanov, Nat. Photonics **12**, 209 (2018).
2. M. Vainio and L. Halonen, Phys. Chem. Chem. Phys. **18**, 4266 (2016).
3. G. Soboń, T. Martynkien, P. Mergo, L. Rutkowski, A. Foltynowicz, Opt. Lett. **42**, 1748 (2017).
4. M. Roiz, K. Kumar, J. Karhu, M. Vainio, APL Photonics **6**, 026103 (2021).
5. M. Roiz, J. Lai, J. Karhu, M. Vainio, Opt. Lett. **46**, 4037-4040 (2021).
6. M. Jankowski, N. Jornod, C. Langrock, B. Desiatov, A. Marandi, M. Lončar, M. M. Fejer, arXiv preprint arXiv:2104.07928 (2021).
7. J. Fan, W. Chen, C. Gu, Y. Song, L. Chai, C. Wang, M. Hu, Opt. Express **25**, 24594 (2017).
8. U. Elu, M. Baudisch, H. Pires, F. Tani, M. H. Frosz, F. Köttig, A. Ermolov, P. St.J. Russell, J. Biegert, Optica **4**, 1024-1029 (2017).

Interaction of Tungsten tips with Laguerre-Gaussian beams

Abhisek Sinha[1,*], *Debobrata* Rajak[2], *Shilpa* Rani[1], *Ram* Gopal[2], and *Vandana* Sharma[1,**]

[1]Indian Institute of Technology Hyderabad, India
[2]Tata Institute of Fundamental Research Hyderabad, India

Abstract. Interaction of femtosecond laser pulses with metallic tips have been studied extensively and they have proved to be a very good source of ultrashort electron pulses. We present our study of interaction of Laguerre-Gaussian (LG) laser modes with Tungsten tips. We report a change in the order of the interaction for LG beams and the difference in the order of interaction is attributed to ponderomotive shifts in the energy levels corresponding to the enhanced near field intensity supported by numerical simulations.

1 Introduction

Laguerre-Gaussian (LG) modes are a solution to the paraaxial Helmholtz equation with a cylindrical symmetry. The most interesting properties of LG beams are a result of their topological phase structure. The phase of an LG beam continually increases counter-clockwise, on a cross-section of the beam, along a closed loop from 0 to $2\pi l$, where l is an integer called the topological charge. Since the angle 0 is equal to $2\pi l$, a continuous distribution of phase is obtained resembling the topological structure of a mobius strip [1]. A phase singularity, also called a vortex is formed at the centre of the beam where the phase remains undefined. Due to this azimuthal phase variation, the wavefront appears twisted in shape. This twisting of the beam is identified as the OAM of the beam and the pitch of the twist determines the magnitude of the OAM which is related to the topological charge. It was shown that the projection of the OAM along the direction of propagation is equal to $l\hbar$ per photon averaged over the beam [2]. Thus, individual photons carry orbital angular momentum in addition to spin angular momentum in LG modes.

In practice, such beams are obtained from Gaussian modes using computer generated holograms [3, 4] or spiral phase plates [5, 6]. Other methods involve astigmatic mode conversions where high order Hermite-Gaussian (HG) beams are passed through a pair of cylindrical lenses and the Gouy phase shift thus introduced results in the conversion of the HG mode to an LG mode [7]. Lower order LG modes were generated in the cavity of solid state lasers using nanoscale mode selection elements [8]. The detection and characterization of the mode of the resultant beam can be done using interferometric techniques [9, 10] or by studying diffraction of the beam through different apertures [11, 12].

*e-mail: ph18resch01002@iith.ac.in
**e-mail: vsharma@phy.iith.ac.in

LG beams are central to various applications in diverse fields. In the field of communication, LG beams were used to encode multiple bits in a single photon [13] with an inherent security feature that does not require any mathematical or quantum mechanical encryption. LG beams are used in imaging applications to achieve super-resolution upto the order of $\lambda/25$ in Stimulated Emission Depletion (STED) microscopy [14, 15]. LG modes are well suited for Gravitational wave detectors as they are reported to reduce the influence of thermal noise in such systems [16, 17]. Other applications include micromachining [18] and optical tweezing [19].

Considerable amount of research has been done on the properties of LG modes and the transfer of OAM to matter at various scales. For instance, it was shown in the case of photoexcitation and photoionization of atoms by OAM beams, the selections rules can be different than those of plane waves due to transfer of OAM [20]. Properties of above threshold ionization spectra by OAM beams were calculated and predicted the existence of photoelectrons emitted in the direction of laser propagation [21]. In high harmonic generation, it was shown that the OAM was conserved in the process of generation of higher harmonics of the fundamental beam [22]. In metal nanoparticles, transfer of OAM to surface plasmons have been theoretically demonstrated and the excited plasmon mode was shown to be determined by the Total Angular Momentum (TAM) transferred in the excitation process [23]. For larger particles, the mechanical equivalence of spin and orbital angular momentum was established by the observation of cancellation (or addition) of spin angular momentum by orbital angular momentum to give total angular momentum [24].

In this report, we have presented our study of interaction of femtosecond LG beams with metallic tips. The electron emission and their dynamics from metallic tips illuminated by femtosecond laser pulses are well known [25, 26]. When light is focused on metallic tips of sizes smaller than the wavelength of the incident light, the field value near the apex of the tip rises several times the incident field [27]. This enhanced near field leads to emission of photoelectrons from the tip surface. In the case of vortex beams, this near field enhancement can be different and may lead to different emission properties. A comparison of the electron emission rates is presented and we show a change in the electron emission property for LG modes.

2 Experimental Methods

A schematic representation of the experimental setup is shown in figure (1). A Ti-Sapphire laser lasing 25 fs, 800 nm pulses at a repetition rate of 1 kHz was used as the source in the experiment. A combination of half-waveplate and polarizer was used to control the intensity of the laser beam. A spiral phase plate was used to obtain the desired LG mode of topological charge, $l = 1$. To detect the emitted electrons, a channeltron detector was placed parallel to the tip axis. The output signal from the detector was then fed to a Digital Oscillosocope triggered by a high speed photodiode. The position of the tip with respect to the laser beam was constantly monitored using a CCD camera.

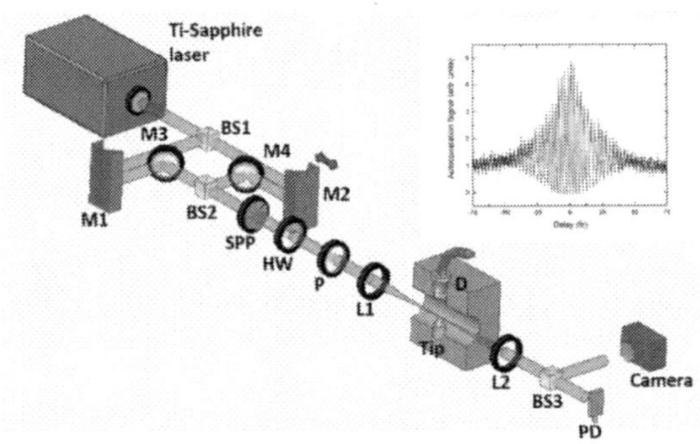

Figure 1: Schematic of the experimental setup used in the experiment. BS1, BS2, BS3: Beam Splitters; M1, M2: Movable mirrors; M3, M4: Mirrors; SPP: Spiral Phase Plate; HW: Half Waveplate; P: Polarizer; L1, L2: Focusing Lens; D: Channeltron Detector; PD: Photo diode. Inset: Interferometric Autocorrelation Trace (IAT) with the Tungsten tip as the nonlinear element.

3 Results

Inset in figure (1) shows the Interferometric Autocorrelation Trace (IAT) with the tip used as the nonlinear element. The emission process was fast enough for the measurement of 35 fs pulses ruling out thermal emission with characteristic timescales of 100 fs to 1 ps [28, 29]. Figure (2) shows the electron emission rates as a function of the average power for Gaussian beams. The data depicted in a log-log plot forms a straight line and the value of the slopes are 2 and 1.6 respectively for tip voltages 0V and 40V. For the case of an LG beam, figure (3), containing OAM ($l = 1$), the slopes of the log-log plot are nearly equal to 1 for both 0V and 40V.

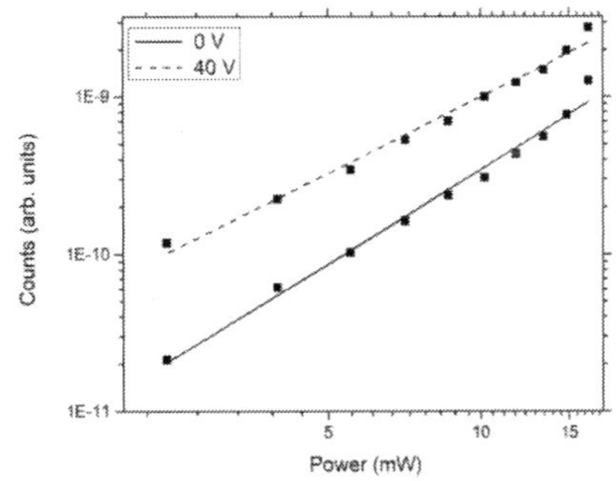

Figure 2: Gaussian beam when focused on the tip. In a log-log plot, the slope of the curve represents the expontiating index ($\Gamma = \alpha_n I^n$). The values of n are 2 (solid) and 1.6 (dash). Γ is the emission rate, I the intensity of the radiation, n the number of photons absorbed and α_n the cross-section of the interaction.

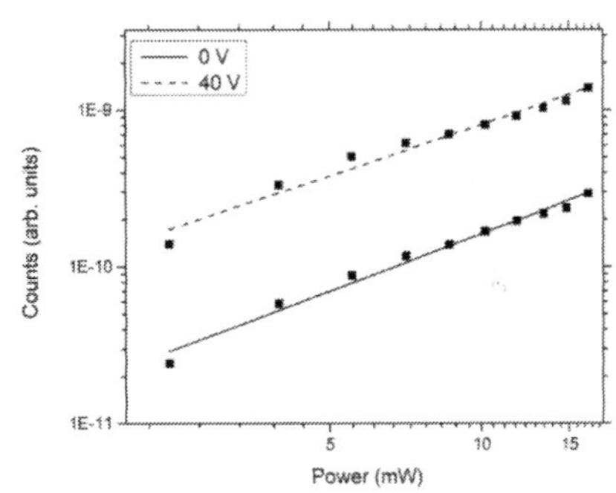

Figure 3: Counts variation with average laser power for OAM beams. Slope of the lines are close to 1 for both of the tip voltages.

4 Discussions

It is seen that in both Gaussian and LG beams, the yield of 40V is higher than 0V. This is because applying a potential on the tip decreases the height of the potential barrier giving rise to a higher ionization probability. The slope of the plot represents the number of photons absorbed in the process of ionization. As seen in numerical simulations, the intensity distribution indicates a higher field value at the tip apex for Gaussian beams. The ponderomotive energy due to this field distribution on the apex of the tip shifts the continuum [30] higher for Gaussian beams than LG beams. As a result of this shift in the continuum, the order of the interaction for the Gaussian beam is higher than that of the LG beam.

5 Conclusion

In this work we have explored the interaction of metallic tips with LG laser modes. The order of interaction for Gaussian beams was seen to be higher than LG beams. It is a result of a larger field enhancement factor for Gaussian beams near the apex of the tip that lifts the continuum higher so that more photons are required for ionization.

References

[1] Y. Shen, X. Wang, Z. Xie, C. Min, X. Fu, Q. Liu, M. Gong, X. Yuan, Light: Science & Applications **8**, 1 (2019)

[2] L. Allen, M.W. Beijersbergen, R. Spreeuw, J. Woerdman, Physical review A **45**, 8185 (1992)

[3] N. Heckenberg, R. McDuff, C. Smith, A. White, Optics letters **17**, 221 (1992)

[4] J. Arlt, K. Dholakia, L. Allen, M. Padgett, Journal of modern optics **45**, 1231 (1998)

[5] K. Sueda, G. Miyaji, N. Miyanaga, M. Nakatsuka, Optics express **12**, 3548 (2004)

[6] M. Beijersbergen, R. Coerwinkel, M. Kristensen, J. Woerdman, Optics communications **112**, 321 (1994)

[7] M.W. Beijersbergen, L. Allen, H. Van der Veen, J. Woerdman, Optics Communications **96**, 123 (1993)

[8] D. Lin, J. Daniel, W. Clarkson, Optics letters **39**, 3903 (2014)

[9] M. Harris, C. Hill, P. Tapster, J. Vaughan, Physical Review A **49**, 3119 (1994)

[10] M. Padgett, J. Arlt, N. Simpson, L. Allen, American Journal of Physics **64**, 77 (1996)

[11] J.P.C. Narag, N. Hermosa, Physical Review Applied **11**, 054025 (2019)

[12] L.E. de Araujo, M.E. Anderson, Optics letters **36**, 787 (2011)

[13] G. Gibson, J. Courtial, M.J. Padgett, M. Vasnetsov, V. Pas'ko, S.M. Barnett, S. Franke-Arnold, Optics express **12**, 5448 (2004)

[14] K.I. Willig, S.O. Rizzoli, V. Westphal, R. Jahn, S.W. Hell, Nature **440**, 935 (2006)

[15] S.W. Hell, J. Wichmann, Optics letters **19**, 780 (1994)

[16] M. Granata, C. Buy, R. Ward, M. Barsuglia, Physical review letters **105**, 231102 (2010)

[17] A. Allocca, A. Gatto, M. Tacca, R. Day, M. Barsuglia, G. Pillant, C. Buy, G. Vajente, Physical Review D **92**, 102002 (2015)

[18] K. Ladavac, D.G. Grier, Optics express **12**, 1144 (2004)

[19] H. He, M. Friese, N. Heckenberg, H. Rubinsztein-Dunlop, Physical review letters **75**, 826 (1995)

[20] A. Picón, J. Mompart, J.V. de Aldana, L. Plaja, G. Calvo, L. Roso, Optics express **18**, 3660 (2010)

[21] B. Böning, W. Paufler, S. Fritzsche, Physical Review A **98**, 023407 (2018)

[22] G. Gariepy, J. Leach, K.T. Kim, T.J. Hammond, E. Frumker, R.W. Boyd, P.B. Corkum, Physical review letters **113**, 153901 (2014)

[23] K. Sakai, K. Nomura, T. Yamamoto, K. Sasaki, Scientific reports **5**, 1 (2015)

[24] N. Simpson, K. Dholakia, L. Allen, M. Padgett, Optics letters **22**, 52 (1997)

[25] M. Krüger, C. Lemell, G. Wachter, J. Burgdörfer, P. Hommelhoff, Journal of Physics B: Atomic, Molecular and Optical Physics **51**, 172001 (2018)

[26] M.F. Ciappina, J.A. Pérez-Hernández, A.S. Landsman, W.A. Okell, S. Zherebtsov, B. Förg, J. Schötz, L. Seiffert, T. Fennel, T. Shaaran et al., Reports on Progress in Physics **80**, 054401 (2017)

[27] Y.C. Martin, H.F. Hamann, H.K. Wickramasinghe, Journal of applied physics **89**, 5774 (2001)

[28] P. Hommelhoff, C. Kealhofer, M.A. Kasevich, Physical review letters **97**, 247402 (2006)

[29] B. Barwick, C. Corder, J. Strohaber, N. Chandler-Smith, C. Uiterwaal, H. Batelaan, New Journal of Physics **9**, 142 (2007)

[30] M. Schenk, M. Krüger, P. Hommelhoff, Physical review letters **105**, 257601 (2010)

 https://doi.org/10.1051/epjconf/202125511004

Mid-infrared parametric wavelength conversion seeded with fiber optical parametric sources

Ronan A. Battle[1,*], *Anita M.* Chandran[1], *Timothy H.* Runcorn[1], *Arnaud* Mussot[2], *Alexandre* Kudlinski[2], *Robert T.* Murray[1], and *J. Roy* Taylor[1]

[1]Femtosecond Optics Group, Department of Physics, Imperial College London, Prince Consort Road, London SW7 2BW, UK
[2]Université de Lille, CNRS, UMR 8523-PhLAM—Physique des Lasers Atomes et Molécules, F-59000 Lille, France

Abstract. A new method of seeding χ^2 optical parametric converters with χ^3 fiber optical parametric sources is introduced. We demonstrate a tunable mid-infrared source operating at around 3 µm with the technique and discuss the potential of this architecture.

1 Introduction

The mid-infrared (MIR) spectral region contains strong characteristic molecular absorption features of great interest for a range of applications. Directly emitting MIR sources in this region are limited, but near-infrared (NIR) solid-state lasers can be used to generate MIR radiation through various $\chi^{(2)}$ parametric three-wave mixing (3WM) conversion techniques in nonlinear crystals [1]. Optical parametric amplification (OPA), and difference frequency generation (DFG) are examples of simple 3WM techniques which provide significant flexibility in output repetition rates, pulse durations, and wavelengths. However, a seed signal at the correct wavelength, synchronised with the pump laser, is also required, which can necessitate the use of expensive secondary lasers or complex nonlinear conversion stages. The pump and signal beams must also be precisely spatially overlapped in the crystal. These two requirements result in increased system complexity, and can lead to such sources being difficult to use by non-experts, restricting uptake in clinical and/or field applications.

In this contribution, we introduce and demonstrate a novel method of seeding χ^2 optical parametric converters with χ^3 fiber optical parametric sources, illustrated in Fig. 1(a). By pumping optical fibers in the low normal dispersion region, it is possible to generate widely spaced, high power spectral density Stokes/anti-Stokes pulses through four-wave mixing (4WM) [2]. These Stokes pulses are inherently temporally and spatially overlapped with the pump pulse at the exit of the fiber, provided the walk-off length is long compared to the length of fiber employed. The output of the fiber can therefore be focused directly into a nonlinear crystal to convert the pump and Stokes/anti-Stokes pulses into the MIR through 3WM. Consequently, the need for additional signal beam delay and combination optics - commonly required in OPA and DFG systems - is avoided.

The use of 4WM in PCF to generate the signal pulses required for 3WM has previously been explored by Yao and Knox [3]. However, they used pump pulses with a duration of 300 fs and a double zero dispersion wavelength (ZDW) PCF, neither of which are suitable for the architecture described here, owing to the large walk-off resulting from such short pump pulses, and the limited coupling efficiencies into the PCF. The single ZDW PCFs and longer pump pulse durations (35 ps) used in this work are key to enabling the direct focusing of the PCF output into the crystal. As a result, the architecture we demonstrate here could be fully fiber integrated [4], resulting in a compact, robust, and integrated pump system for the 3WM stage. This could ultimately lead to more user-friendly and field-deployable MIR sources, useful for a range of applications from surgery to remote sensing. We used the architecture to generate picosecond MIR pulses with pulse energies up to ~14 nJ and average powers up to 115 mW. The pulses were wavelength tunable around 3 µm, coinciding with the water absorption peak at 2.94 µm, making them well suited for biomedical laser ablation applications [5].

2 Experiment and Discussion

To demonstrate the technique, we first selected an optical fiber and nonlinear crystal with a common Stokes and signal wavelength, as highlighted in the 4WM and 3WM phase matching curves in Fig. 1(b) and Fig. 1(c). Pumping these nonlinear media at 1.064 µm results in a common Stokes and signal wavelength around 1.65 µm. Thus, the Stokes pulses generated in the 4WM stage can be used as the seed signal in the 3WM stage as illustrated in Fig. 1(a).

We employed an in-house developed ytterbium fiber master oscillator power amplifier (Yb:fiber MOPA) system shown in Fig. 2 as the pump source. The MOPA system consisted of a 50 MHz mode-locked fiber oscillator

*e-mail: ronan.battle15@imperial.ac.uk

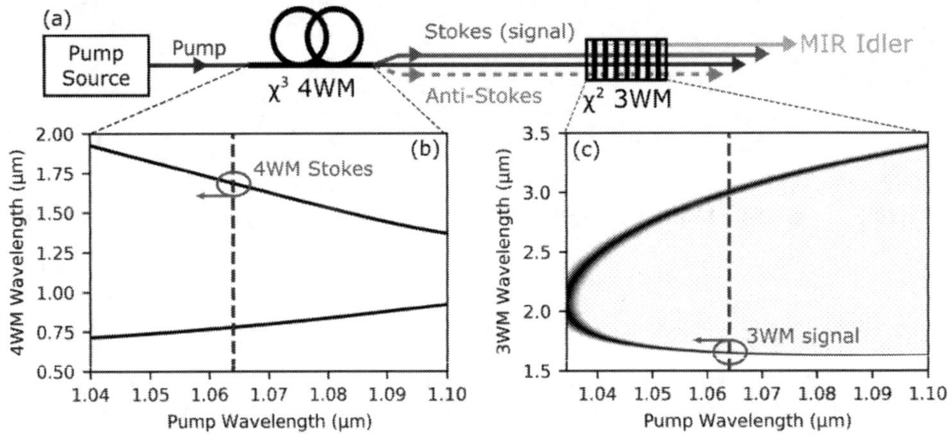

Figure 1. (a) Schematic of MIR generation via cascaded $\chi^{(3)}$ 4WM in fibre and $\chi^{(2)}$ 3WM in a bulk non-linear crystal. (b) Phase matching diagrams for the 4WM in PCF and (c) 3WM in PPLN (at 60°C). Circles highlight the common wavelength which is phase matched in both the 4WM and 3WM interactions.

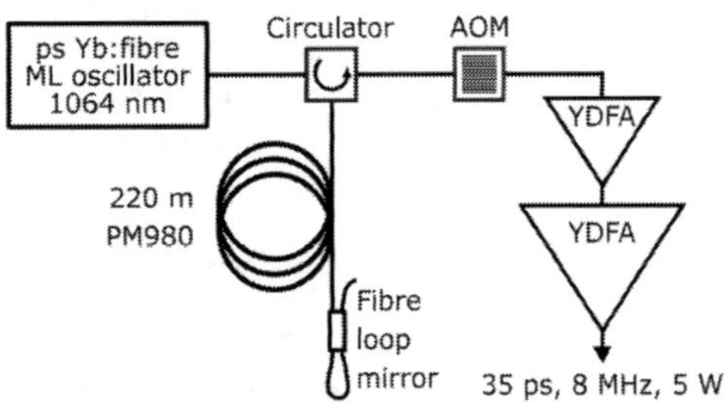

Figure 2. In-house developed ytterbium fiber master oscillator power amplifier system. ML, mode-locked; AOM, acousto-optic modulator; YDFA, ytterbium-doped fibre amplifier.

that produced 5 ps pulses, stretched to 35 ps in a double pass of 220 m of polarization maintaining fibre (PM980, Newport), picked down to ~8 MHz by an acousto-optic modulator (AOM) and amplified to 5 W in two ytterbium-doped fibre amplifiers (YDFA). The output spectrum of the MOPA, which has a 10-dB spectral width of 7.1 nm, is shown in Fig. 3(a). A long pass 1100 nm filter was used before the PCF to filter out light at 1120 nm resulting from stimulated Raman scattering in the final amplifier stage.

The custom polarization-maintaining photonic crystal fiber (PCF) used had zero dispersion wavelengths of 1.101 µm and 1.103 µm on the fast and slow axes respectively [6]. Pumping a 0.35 m length of PCF on the slow axis with 3.1 W of average power (coupled), we generated 1.65 µm Stokes and 0.785 µm anti-Stokes sidebands shown in Fig. 3(b). The average powers generated were 150 mW (Stokes) and 260 mW (anti-Stokes).

The output of the PCF was then focused directly into a periodically poled lithium niobate crystal doped with 5 mol.% MgO (PPLN, HC Photonics) to generate signal and idler light through 3WM. The 3x3x10 mm^3 crystal used had a poling period of 31.3 µm. The PPLN was held in a crystal oven to enable temperature tuning in the range 20-200°C. The maximum pump peak intensity in the crystal was estimated to be 1.7 GW/cm^2, calculated from a measured spot size of 35 µm at focus. The spatial overlap of the pump and Stokes pulses at the focus was confirmed using a scanning slit beam profiler. Due to the long walk-off length (2.0 m) between the pump and Stokes pulses relative to the short PCF length employed (0.35 m), the pulses were assumed to be well overlapped temporally for the 3WM stage.

As the Stokes seed generated in the 4WM stage was spectrally broader than the 3WM spectral acceptance bandwidth, tuning of the crystal temperature in the range 40-100°C resulted in tuning of the idler spectra as shown in Fig. 3(c). The associated variation in the generated idler and output signal powers are shown in Fig. 3(d). Moving away from the temperature at which the seed signal was phase matched resulted in no idler generation; no (spontaneous) optical parametric generation was observed. This confirmed that the pulses we generated in the 4WM stage were seeding the 3WM. A maximum 115 mW of generated 3WM idler alongside 189 mW of amplified 3WM signal was generated at a crystal temperature of 60°C. This corresponds to a pump conversion efficiency of 11% (amplified signal plus idler) in the 3WM stage.

The 3WM conversion efficiency was primarily limited by the broad bandwidth of the pump and signal relative to the spectral acceptance bandwidths of the crystal. Employing a narrower bandwidth pump/signal would allow an increase in conversion in the 3WM stage. This could be readily achieved by altering the pump laser parameters (bandwidth and pulse duration) and adjusting the length of the fiber in the 4WM stage.

In a system such as this, however, which employs cascaded nonlinear effects, various competing factors must be considered. In the 4WM stage, sufficient Stokes power must be generated to seed the subsequent 3WM. Ideally,

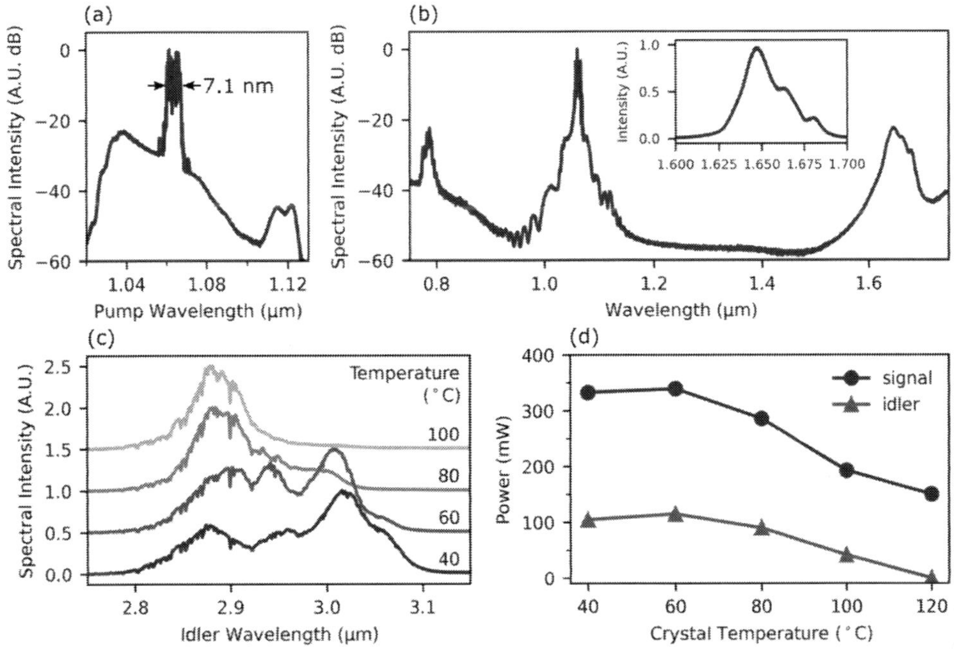

Figure 3. (a) MOPA pump laser spectrum centered at 1.064 µm, with bandwidth of 7.1 nm. (b) Spectrum at output of PCF showing 1.064 µm pump and 4WM sidebands. Inset: Stokes sideband on linear scale. (c) Idler spectral tuning with crystal temperature on linear scale (d) Generated signal and idler powers as a function of temperature.

the Stokes power generated would saturate the 3WM to achieve the greatest output power and stability in the generated MIR idler power [7]. Conversely, increased pump conversion to the Stokes pulses in the 4WM stage results in less residual pump power remaining to pump the 3WM stage, limiting the 3WM conversion. As discussed, the pump bandwidth must also not exceed the 3WM pump spectral acceptance bandwidth. Some self-phase modulation induced pump broadening is unavoidable in the 4WM stage, because the pump must propagate through normally dispersive fiber to generate the required 4WM sidebands. Ongoing work aims to balance these cascading nonlinearities to ascertain the performance limits of this architecture.

3 Conclusion

We have introduced and demonstrated a new method of seeding χ^2 parametric wavelength converters (3WM) directly with χ^3 fiber optical parametric sources (4WM). By exploiting the inherent spatial and temporal overlap of pulses generated by 4WM in PCF, we seed 3WM in PPLN to generate 115 mW at ~3 µm. Further investigations and simulations are ongoing to determine the limits of the conversion efficiency achievable using this architecture.

References

[1] V. Petrov, in *Progress in Quantum Electronics* (Elsevier Ltd, 2015), Vol. 42, pp. 1–106

[2] L. Lavoute, P.J. Mosley, S.A. Bateman, W.J. Wadsworth, Opt. Express **19**, 25337 (2011)

[3] Y. Yao, W.H. Knox, Opt. Express **20**, 25275 (2012)

[4] M. Baumgartl, M. Chemnitz, C. Jauregui, T. Meyer, B. Dietzek, J. Popp, J. Limpert, A. Tünnermann, Opt. Express **20**, 4484 (2012)

[5] S. Amini-Nik, D. Kraemer, M.L. Cowan, K. Gunaratne, P. Nadesan, B.A. Alman, R.J.D. Miller, PLoS ONE **5**, e13053 (2010)

[6] R.T. Murray, E.J.R. Kelleher, S.V. Popov, A. Mussot, A. Kudlinski, J.R. Taylor, Opt. Express **21**, 15826 (2013)

[7] F. Mörz, T. Steinle, A. Steinmann, H. Giessen, Opt. Express **23**, 23960 (2015)

Modeling femtosecond pulse propagation and high harmonics generation in hollow core fibers

Valer Tosa[1,*], *Anna Gabriella* Ciriolo[2], *Rebeca* Martinez Vazquez[2], *Caterina* Vozzi[2], and *Salvatore* Stagira[3]

[1]National Institute for R&D Isotopic and Molecular Technologies, 400293 Cluj Napoca, Romania
[2]Institute for Photonics and Nanotechnologies (IFN), National Research Council (CNR), Milano, Italy
[3]Politecnico di Milano, Physics Department, Milano, Italy

Abstract. One way to increase the well known low efficiency of high order harmonics generation is to place the gas medium in a hollow core waveguide. The numerical model used to obtain driving and harmonics field configuration along and across the waveguide was developed for an arbitrary gas density profile and arbitrary fiber diameter modulation. The model was tested against experimental measurements and excellent agreement was obtained for the fluorescence emission along the waveguide. We analyse the influence of the diameter modulation which result from the fabrication process on both pulse propagation and on harmonic generation.

1 Introduction and methods

High order harmonic generation (HHG) in gas, despite the low efficiency of the process, remains one of the main methods to obtain extreme ultra-violet (XUV) or soft X-ray attosecond pulses. Using guided structures to propagate the ultrafast pulse is one way to increase the HHG yield and thus is of great practical interest. Indeed, capillaries allow achieving both laser beam and gas confinement for a longer interaction length, thus leading to a significant increase of the generation yield. Moreover, capillaries can be filled with gas both in static and continuous-flow regimes, so that they can be used in combination with high-repetition-rate laser systems.

The integrated microfluidic glass device devoted to XUV generation by HHG in gas medium is composed [1] of a main micro-channel that works as a hollow-core waveguide (HCW) for the propagation of the femtosecond laser pulse. The interaction region is filled with the working gas using a micro-structured network of delivery channels fabricated in the same glass substrate. High-order harmonics of the fundamental field are generated within the waveguide by the well-known three step process happening in every optical cycle of the laser field: ionization, acceleration and recombination of the electron found in the outermost shell of the atom.

While the elementary three step process is well understood and described at quantum level, the macroscopic build up of the signal from the single radiating dipoles in the interaction volume is difficult to control experimentally and not easy to describe phenomenologically as it depends on many intensive and extensive parameters.

When building a model for HHG in macroscopic media [2] a first necessary step is solving the driving field propagation in the ionizing gas. We solved this problem by starting from the wave equation in time domain, then transforming it in the moving frame and in frequency domain. We then used the split-step method for advancing each spectral component along the propagation direction while assuming the field as a superposition of EH_{11} eigenmodes [3] to satisfy the boundary conditions in the hollow dielectric waveguide.

The second step is to estimate the response of the atom to the propagated field and this was calculated in the SFA approximation [4]. The time-dependent resulting polarization was eventually used as a source term in the wave equation for the harmonic field which was solved in a similar way as for the driving field. The far field was obtained taking into account the experimental configuration by solving the Hankel transform for spectral components of the harmonic field.

Our model allows for an arbitrary density distribution of the gas along the propagation direction. The density distribution was obtained in a separate investigation by solving the gas flow problem in the specific topology of the fabricated structure.

2 Testing the pulse propagation model

Solving the problem of pulse propagation in a HCW is the first essential step in modeling HHG. The method we used for the case of free space propagation [2,5] was adapted for fiber case as briefly described in the preceding section.

The numerical model takes into account a refractive index which is space-time dependent having contributions of dispersion from fiber and neutrals, as well as nonlinear contributions from Kerr and ionization effects. We note here that for all HHG experiments

* Corresponding author: tosa@itim-cj.ro

ionization is the necessary ingredient and it induces the dominant term in the refractive index build-up. The contribution of the nonlinear terms to pulse propagation was included in the second step of split step method.

Testing the model against experimental data was a necessary step in the model development. In the following we present simulation results of the model for conditions closed to a specific experimental situation [6]. In particular we consider a 5 cm long capillary with a 75 μm radius hole filled with argon at a pressure of 80 mbar in the central part of the capillary; the Ar pressure increases over 3 mm from 0 to 80 mbar at the fiber input. Laser pulses of 0.5 mJ energy, and 40 fs duration at 800 nm wavelength are focused with a 40 cm focal length element producing a Gaussian waist of 48 μm and an initial peak intensity of 3.8×10^{14} W/cm^2.

In the experiment [6] the plasma emission from the ionized Ar was recorded from the side of the waveguide as function of the propagation coordinate z (see for example Fig. 9 of [6]).

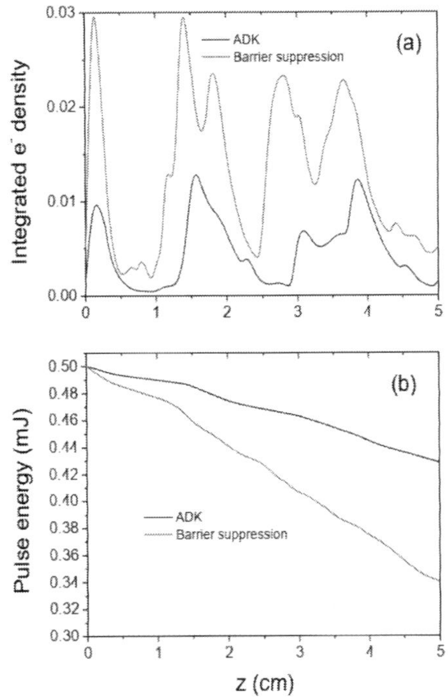

Fig.1 Total electron density (a) and pulse energy (b) as a function of z position along the HCW, for two ionization models, as detailed in the text.

In the modeling we calculated the total density of electrons (integrated across HCW area) and plotted it along the z coordinate, as seen in Fig. 1.a. The pulse energy is also shown in Fig. 1.b for the same conditions. The calculations have been done using two different models for the calculation of the Ar ionization rates: the classical ADK rates [7] and the static field ionization rates in the barrier-suppression regime [8]. The later yields ionization rates higher by almost one order of magnitude in the 10^{14} W/cm^2 range of intensities. This difference is reflected in both the integrated electron density and the pulse energy, higher ionization rates producing the obvious increase of the total electron yield but also the splitting of the density profile. Higher ionization rates also induce a larger pulse energy loss which is consumed for atoms ionization. The comparison with experimenal data favors the ADK rates which yield a closer agreement to the experiment both on plasma profile and on pulse energy dependence on propagation distance.

We also show in Fig. 2 the evolution of the energies contained in the first three modes, to demonstrate that almost 90% of the energy is initially in the fundamental mode.

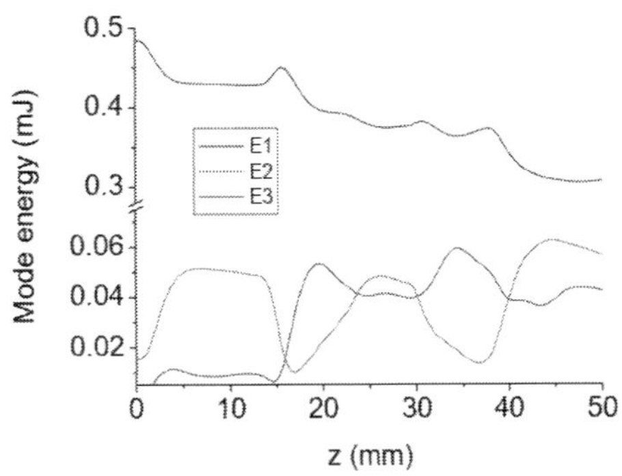

Fig.2 The energies contained in the first three modes

3 Laser pulse propagation in straight and modulated HCW

After the fabrication process [9] the hollow waveguide surface presents a ripple pattern, with a periodicity of about 5 μm, made of parallel grooves perpendicular to laser scanning direction. As the surface quality of the HCW walls influences the propagation of pulsed radiation, we here investigate the behaviour, in terms of driving field propagation and harmonics yield, between two devices characterized by the same internal structure, one with the typical surface roughness as resulting from the etching process and one smoothened by a thermal process.

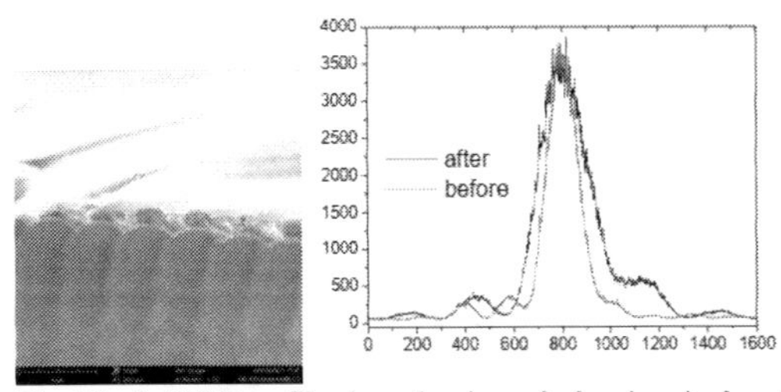

Fig. 3 Left: HCW profile along the channel, showing ripples of about 5 mm period. Right: Beam profile at fiber exit before and after thermal treatment, arbitrary units on both axes.

The experimental measurement of the beam profile at HCW exit shows clearly that the ripple pattern influences the pulse propagation. Shown in Fig. 3 is the radial profile of the beam at exit before an after thermal treatment which reduces the deepness of the modulation.

The width of the beam is larger after thermal treatment which suggests better conditions for propagation.

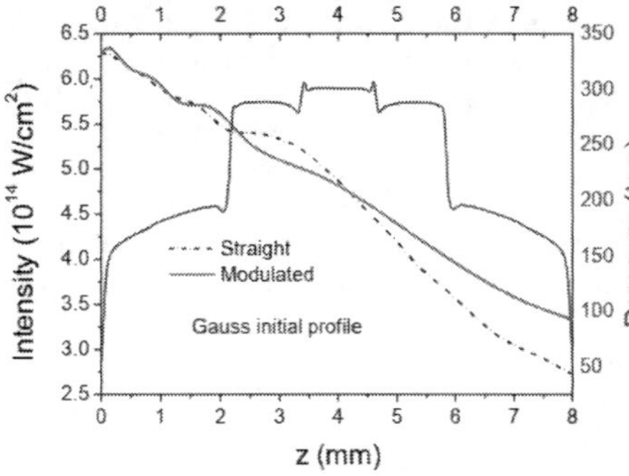

Fig. 4 On-axis peak intensity along the channel for the straight and a modulated HCW. The density profile of the He gas is also shown.

The modeling followed closely the experimental conditions: a pulse of 25 fs duration and 0.45 mJ energy propagates along a HCW of 140 μm diameter and 8 mm long, filled with He having a profile of density as shown in Fig. 4 (blue solid line). The on-axis peak intensity of the pulse along the channel is shown for both modulated (red solid line) and straight channel (black dash dot line). The initial radial beam profile was set to Gaussian with an optimized waist for coupling into the waveguide.

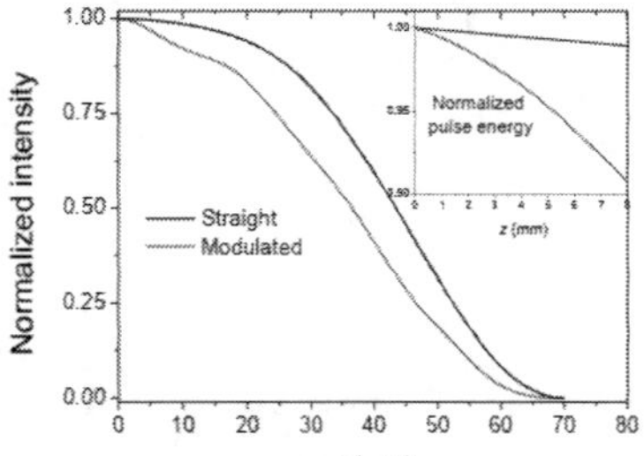

Fig. 5 Beam radial profile at HCW exit for the straight (black) and modulated (red) channel. The inset shows the evolution of the pulse eneergy along propagation direction.

The radial field profile at HCW exit is shown in Fig. 5 for the two cases. The straight case yields a larger waist (57.3 μm at $1/e^2$) than the modulated case (52.6 μm at $1/e^2$), in very good agreement with the experimental findings. The evolution of the pulse energy (shown in the inset of Fig. 5) confirms that the ripples enhance the losses in the channel during propagation.

4 Harmonic generation in straight and modulated HCW

As known, intensity variation of the driving field will induce phase variations for the induced dipole which is the source for the harmonic field. With this picture in mind we simulated the generation of harmonics in the two considered cases.

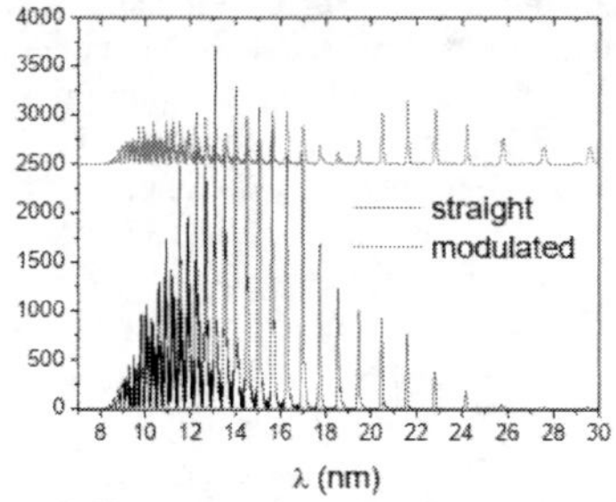

Fig. 6 Far field power spectra for the straight (black) and modulated (red) fiber. For better visibility the later is shifted on vertical axis, but (arbitrary) units are the same.

Shown in Fig. 6 is the far field power spectra for the straight (black line) and for the modulated HCW, the latter is shifted on vertical axis to see better the differences between them. As clearly seen, the modulated spectrum is about an order of magnitude weaker than the straight fiber spectrum. Also, from the analisys of the far field profiles of the harmonics (not shown here) we conclude that they are emited with a larger angle in the modulated fiber case.

In conclusion we investigated femtosecond pulse propagation and harmonic generation in HCW both in ideal conditions of surface quality and in modulated channel diameter caused by the fabrication process. Both fields are affected by the propagation, especially the harmonic field which suffer from the local variations of the intensity.

Acknowledgements: Funding from the European Union's Horizon 2020 research and innovation programme under grant agreement No 964588, and MIUR PRIN "aSTAR" 2017RKWTMY.

References

1. A. G. Ciriolo et al., J. Phys - Photonics **2**, 024005 (2020)
2. V. Tosa, H.T. Kim, I.J. Kim, and C.H. Nam Phys. Rev. A **71**, 063807 (2005)
3. E.A.J. Marcatili, R.A. Schmeltzer The Bell System Technical Journal **43**, 1783 (1964)
4. M. Lewenstein, Ph. Balcou, M.Y. Ivanov, A. L'Huillier, P. B. Corkum, Phys. Rev. A. **49** 2117 (1994)
5. V. Tosa, K. Kovacs, B. Major, E. Balogh, K. Varju, Quant. Electronics **46** 321(2016)
6. C.A. Froud, R.T Chapman, E.T.F. Rogers, M. Praeger, B Mills, J. Grant-Jacob, T.J. Butcher, S.L. Stebbings, A.M. de Paula, J.G. Frey, and W.S. Brocklesby . J. Opt. **A 11**, 054011 (2009)
7. V.S Popov, Physics Uspekhi **47**, 855 (2004)
8. X.M. Tong and C.D. Lin, J. Phys. B: At. Mol. Opt. Phys. **38**, 2593 (2005)
9. R. Martinez Vazquez, A. G. Ciriollo, G. Crippa, V. Tosa, F. Salla, M. Devetta, C. Vozzi, S. Stagira, R. Osellame, International Journal of Applied Glass Science, accepted 2021

EPJ Web of Conferences 255, 11006 (2021) https://doi.org/10.1051/epjconf/202125511006
EOSAM 2021

Near-single-cycle pulses generated through post-compression on FAB1 laser at ATTOLAB-Orme facility.

Jean-François Hergott[1], *Hugo* J. B. Marroux[1], *Rodrigo* Lopez-Martens[2], *Fabrice* Réau[1], *Fabien* Lepetit[1], *Olivier* Tcherbakoff[1], *Thierry* Auguste[1], *Lucie* Maeder[1], *Xiaowei* Chen[3], *Benoit* Bussière[3], *Pierre-Mary* Paul[3], *Pascal* D'Oliveira[1] and *Pascal* Salières[1]

[1]Université Paris-Saclay, CEA, CNRS, LIDYL, 91191, Gif-sur-Yvette, France
[2]Laboratoire d'Optique Appliquée, CNRS, Ecole Polytechnique, ENSTA Paris, Institut Polytechnique de Paris, 91120 Palaiseau, France
[3]Amplitude 2-4 rue du Bois Chaland CE 2926, 91029 Evry, France

Abstract. Generating high-energy few-cycle pulses is key in the study of light-matter interaction in the regime of high field physics. Attosecond science possess the necessary time resolution to study the underlying fundamental processes but requires repetitions rates on the order the kilohertz and stabilization of the Carrier-Envelope Phase. We present here a post-compression stage delivering 3.8fs pulses with 2.5mJ coupled to a Ti: Sa based 1 kHz TW-class laser which can deliver 17.8fs pulses with 350mrad shot to shot CEP noise. This is the first step towards high-energy few-cycle post-compression of the FAB laser at ATTOLAB-Orme.

The last decade has seen a lot of progress in attosecond science and high-field physics. Generating energetic, few-cycle laser pulses with a stabilized Carrier-Envelope Phase (CEP) constitutes the first step to access the ultra-fast dynamics underlying the interaction of matter with intense, ultrashort coherent light source [1-6]. High temporal resolution as well as high-field strengths with substantial repetition ratse (≥1 kHz to reduce acquisition time) are needed. Thus reducing the pulse duration to few-cycle pulses while keeping mJ range energies has been largely studied over the two past decades [7-9].

Post-compression systems exhibiting pulse durations lower than 4 fs with an energy yield above 2 mJ remain quite rare worldwide [10]. Here, we report on the generation of such intense few-cycle pulses by efficient spectral broadening in a stretched flexible HCF (Hollow Core Fiber) associated to the Attolab-Orme facility laser. The laser facility of Attolab-Orme is a Ti:Sapphire based system which can deliver up to 17 mJ, with pulses as short as 18 fs (0.9 TW) at 1 kHz with long-term CEP stability. This laser system is also capable of tuning the output central wavelength within a 90 nm range around 800 nm while maintaining a spectral bandwidth of 40 nm preserving a 60 fs pulse duration [11]. Amplitude Technologies has developed this laser in collaboration with CEA Saclay within their joint laboratory Impulse.

The full water-cooled 10 kHz front-end is based on an innovative double-crystal regenerative amplifier coupled to an in-line single pass amplifier delivering up to 120 nm ($1/e^2$) ultra-broad bandwidth laser pulses with more than 700 µJ [12]. This output is used to seed two different amplification stages. The thermal load is distributed over the two crystals to lower thermal lensing and ensure the regenerative cavity stability over a larger pump power range, keeping a very good beam quality and high efficiency. The front-end output is then split to seed two amplifiers operating respectively at 1 and 10 kHz.

(a)

(b)

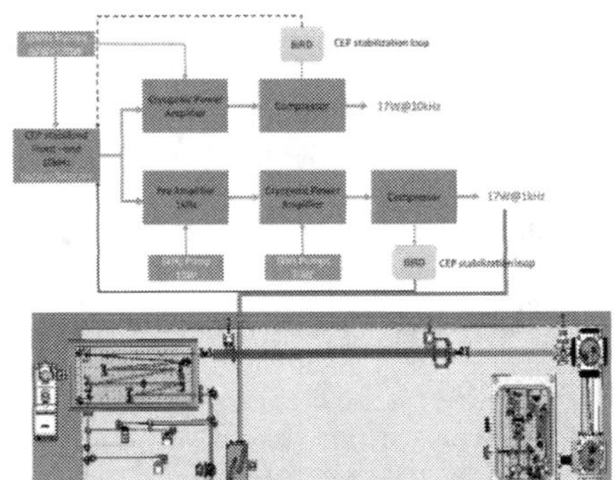

Fig 1. (a) General layout of CEP-stabilized, sub-20 fs, 1 kHz/10 kHz DUAL laser system. (b) 3D drawing of the HCF set-up with energy attenuator, coupling chamber including chirped mirrors and focusing astigmatism free telescope (in pink), output chamber with collimating telescope coupled to chirped mirror and wedges compressor (in yellow).

Routinely, the 10 kHz arm delivers laser pulses at 1,7 mJ and 23 fs while the 1 kHz laser pulses are amplified by two 'bow tie' multi-pass amplifier stages, leading to 17 mJ pulses at 23 fs duration after a grating-based compressor as shown on Fig. 1a. The pulse duration can respectively be lowered to 20 fs and 18 fs without energy loss. The difference in pulse duration between the two outputs is due to differences in their spectral phases'. The same DAZZLER operates the fine pulse compression for both arms so residual dispersion is then not perfectly

* Corresponding author: jean-francois.hergott@cea.fr

compensated simultaneously on each amplification line. We should precise that the compressors are located in the experimental rooms, separated from the laser. The last amplification stages are separated from the compressors by 3 meters. A pointing stabilization system corrects the slow drift of the laser during the free propagation to the compressors. This long propagation path associated to the large stretching ratio used before the amplification stages (5 ps/nm), necessary to limit the B-integral value, make CEP drift more sensitive to vibrations and thus more difficult to stabilize after amplification. However, the 10 and 1 kHz outputs show a shot-to-shot residual CEP noise that can respectively be as good as 250 and 350 mrad with an all analogical CEP stabilization loop [13].

The pulse post-compression experimental set-up used within ATTOLAB-Orme at CEA Saclay is based on the designs shown in references [8, 9]. The stretched fiber, developed by Laser Laboratorium Göttingen, is limited to 2 m length because of limited free space in the experimental room. A variable attenuator composed of a half wave plate and two reflective polarizers allows fine-tuning of the pulse energy from 150 µJ to 9 mJ within the full beam. An iris adapts the beam size in order to reach the focal spot size allowing an optimal coupling. The beam then enters then the first vacuum chamber (in pink in Fig 1b) containing a chirped mirrors pair compensating the dispersion of the entrance window and a quarter-wave plate to convert the polarization from linear to circular. An aberration free reflective focusing telescope composed of two spherical mirrors with a radius of curvature of - 2000 mm and 800 mm (5.9 m equivalent focal length) focuses the beam at the fiber entrance. The focal spot position is very sensitive to the distance between those two mirrors and has to be precisely controlled. Incidence angles on the two telescope' mirrors are tuned to minimized astigmatism resulting in a focus size of 340 µm ($1/e^2$) with 98 % circularity at the fiber entrance while the fiber' internal diameter is 530 µm. A second aberration free telescope placed 1.5 m after the fiber exit collimates the beam at 15 mm ($1/e^2$). The whole set-up fits on a 4.5 m long optical table as shown in Fig 1b. This length limits the maximum energy that we can couple in the fiber due to the damage threshold on the HCF surrounding mirrors, especially the first output mirror.

We use small energy leakages of a mirror before the focusing chamber and of the last mirror in the focusing chamber for imaging of the beam near and far field for daily alignment and pointing stabilization. A second beam stabilization system (TEM Aligna) placed after the laser compressor including all the focusing optics ensures for the perfect coupling of the laser with a RMS beam pointing fluctuation of 7.5 µm. A coupling efficiency in vacuum of 80 % with up to 6 mJ input and stable over 6 hours has been measured. To introduce spectral broadening, the output chamber is filled with He while the entrance chamber is continuously pumped. The circularly polarized light limits the degree of ionization of the Helium gas. Increasing the Helium pressure increases the spectral broadening up to saturation around 3000 mbar. The spectral width increases thus from 81 nm at $1/e^2$ for the fundamental pulse up to 310 nm for the broadened spectrum measured with an intensity calibrated spectrometer and shown by blue markers in Fig 2c. The coupling efficiency slightly reduces with increasing gas pressure from 80 % at low pressure to 70 % above 2000 mbars.

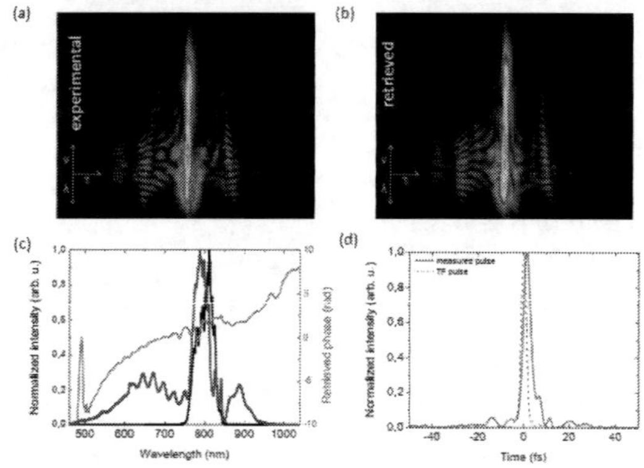

Fig 2. FROG measurement of the 2.5 mJ 1.5-cycle pulse. Panels (a) and (b) display respectively measured and retrieved FROG traces. Panel (c) displays initial (black line) and broadened (blue line) spectra with the retrieved spectral phase (red line). Panel (d) shows the retrieved temporal pulse shape (red line) and the calculated Fourier transform pulse (black dashed line).

The spectrally broadened pulse exits the gas-filled chamber through a 3 mm thick AR coated fused silica window after the polarization has been converted back to linear using an air-spaced achromatic quarter waveplate. The pulse then enters a compressor composed of a set of double-angled chirped mirror pairs allowing 12 reflections (PC70 from Ultrafast Innovations), corresponding to a dispersion compensation of approximately -480 fs^2. The compressor is associated to a motorized pair of fused silica wedges and to a 2-mm thick KDP plate, working either at air or under vacuum. KDP plates are used to compensate residual negative third-order dispersion because of their higher TOD/GDD ratio compared to fused silica. [14] Only the central part of the fiber output is reflected by the optics; the outer part of the beam containing around 20 % of the fiber output energy is cut by the limited size of the chirped mirrors. This limits the overall transmission of the collimating, steering and compression optics to approximately 50 % of the input energy. A commercial single-shot second harmonic FROG (Femtoeasy) is used to measure the pulse' spectral phase and duration. The experimental FROG trace is shown in Fig 2a along with the reconstructed trace in Fig 2b, which corresponds well to the experimental results (error of 16.10^{-3} on a 256x256 grid). The corresponding retrieved spectral phase is shown in Fig 2c leading to the temporal pulse profile in Fig 2d. Coupling a 5.5 mJ, 25 fs pulse in the differentially pumped stretched hollow core fiber, filled with 2800 mbar of He leads to a measured pulse duration of 3.8±0.3 fs with 2.55 mJ. The calculated Fourier Limited pulse duration is 3.1 fs. One has to note that we are approaching the measurement limits of the FROG device that we have in the laboratory. We are losing some signal below 500 nm in the doubling process during FROG measurement limiting thus the pulse

https://doi.org/10.1051/epjconf/202125511006

duration measurement, as it can be seen in Fig 2c. If in some conditions we were able to generate even larger spectra we were not able to measure smaller pulse duration. Another set of chirped mirrors more efficient in the 450 – 1000 nm spectral range should be necessary as well as an adapted pulse duration measurement tool to reach possible 3 fs pulse duration in the 3 mJ energy range [15].

Adjusting the gas pressure is a convenient way to adapt continuously the pulse duration by limiting the spectral broadening process. The He pressure has been varied from 0 mbar to 2800 mbar. The spectral width at $1/e^2$ increases from 81 nm and a pulse duration of 26 fs to more than 300 nm yielding the 3.8 fs pulse duration. The evolution of the spectral width and measured pulse duration are shown in Fig 3. One can also observe the central wavelength shifting from 801 nm to 752 nm. Fig 3 also shows the measurements performed using Ne as non-linear medium. A clear saturation is now observed above 1000 mbar as well in central wavelength shift and spectral broadening.

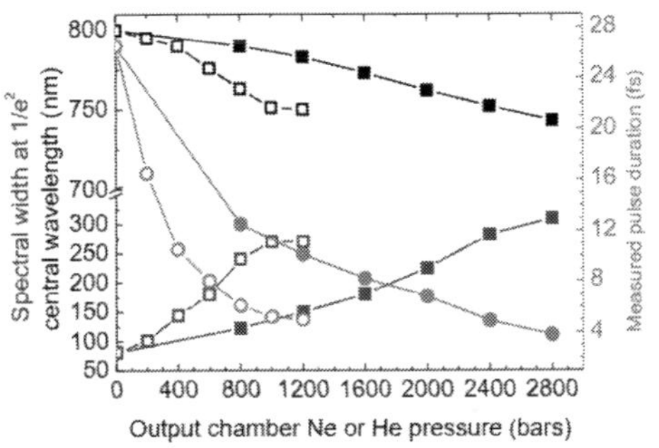

Fig 3. Spectral bandwidth, central wavelength and measured pulse duration evolution as a function of pressure in the case of Ne (open squares and circles respectively) and in the case of He (full squares and circles respectively).

The central wavelength spectral shift remains almost the same in Ne compared to He although it occurs at lower pressure for the former gas. The maximal spectral broadening is obtained for He, thus the shortest pulse duration is slightly higher in Ne than in He, 4.8 fs versus 3.8 fs, with wedge positions and KDP thickness re-optimization. It appears that it was more difficult to perfectly compensate for the residual dispersion in the Ne case. The central wavelength shift towards the shorter wavelengths is due to non-symmetric broadening; the blue wing is larger than the red wing which could be due to residual ionization despite the circular polarization independently of the gas. Numerical studies using a propagation code describing the different involved physical processes are under progress to better understand those results.

The 2.5 mJ 3.8 fs pulses obtained employing He in the HCF post-compression stage have been used for high-order harmonic generation despite a low quality of the CEP stability after the post-compression stage. The shot-to-shot CEP residual noise was around 450 mrad for approximately 200 consecutive shots but it was not possible to conserve this stabilisation over an extended period. Some work is still needed to efficiently cover the few-cycle pulse optical path and isolate the system from vibrations to preserve the CEP stabilization after the HCF stage as it has been demonstrated [8]. The beam was focused in a 1 cm long gas cell by a 1000 mm focal length silver coated spherical mirror for the HHG. A silica plate at grazing incidence was used to remove the majority of the driving field and direct the extreme ultraviolet (XUV) radiation towards a variable line-space grating for spectrally resolving the XUV pulse. The spectrum was collected using an assembly of MCP stack and phosphorous screen and imaged by an external CCD camera.

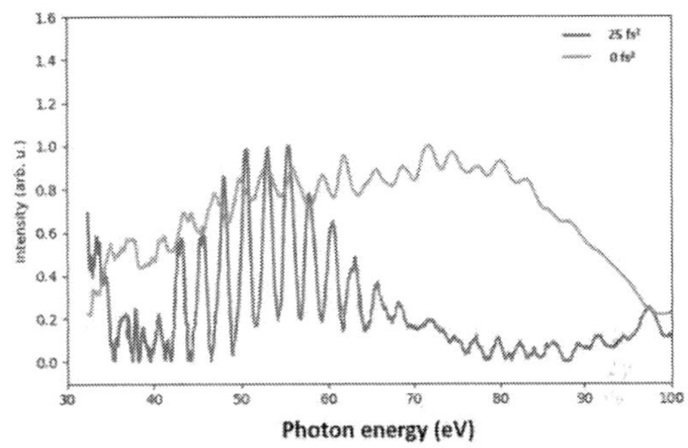

Fig 4. HHG spectra generated with post-compressed pulses with various amounts of GDD.

The spectra obtained for two different pulse compressions are shown in Fig 4. The compression was tuned by optimizing the amount of glass in the beam path using the wedges described previously. When the pulse is under compressed (i.e. positive GDD) the XUV spectra in blue shows clear harmonic peaks consistent with a multiple cycles driver field. Removing glass in the beam path compresses the driver field to 4 fs and generation of the XUV spectrum in orange. This spectrum is continuous with a slight 2ω modulation at lower energy indicating a possible isolated attosecond pulse.

This high-energy few-cycle post-compression of the FAB laser at ATTOLAB-Orme is the first step towards high-energy single attosecond pulse generation at ATTOLAB-Orme, paving the way for new atto-physics experiments. To our knowledge, only 2 other Ti: Sa systems deliver higher energies with sub-4 fs pulse duration obtained with hollow-core fiber compression technique. The generated harmonic spectra in Ne using more than 2 mJ are the direct proof of the quality of the few cycle pulses generated by our post-compression system. These pulses will soon be used in the generation and characterisation of high-intensity isolated attosecond pulses at ATTOLAB-Orme.

Acknowledgements

The authors want to acknowledge T. Nagy and P. Simon for fruitful discussions and information about the stretched hollow core fiber.

Authors' contributions

All authors contributed equally in compiling this paper. The authors read and approved the final manuscript

Fundings

The laser development received fundings from European Union grant H2020-MSCA-ITN-641789-MEDEA, Agence Nationale de la Recherche grant ANR11-EQPX0005-ATTOLAB and Conseil Régional Ile-de-France SESAME2012-ATTOLITE. The HCF set-up has received partial fundings from Laserlab-Europe EU-H2020 871124 (JRA PRISES) and from the Physics of Light and Matter (PhOM) Department of the Graduate School of Physics of Université Paris-Saclay.

References

1. M. Nisoli and G. Sansone, Prog. Quant. Electron. **33**, 17 (2009).
2. M. Uiberacker, T. Uphues, M. Schultze, A. J. Verhoef, V. Yakovlev, M. F. Kling, J. Rauschenberger, N. M. Kabachnik, H. Schröder, M. Lezius, K. L. Kompa, H. G. Muller, M. J. J. Vrakking, S. Hendel, U. Kleineberg, U. Heinzmann, M. Drescher and F. Krausz Nature **446**, 627 (2007).
3. M. I. Stockman, M. F. Kling, U. Kleineberg and F. Krausz, Nat. Photonics **1**, 539 (2007).
4. PB. Corkum, F. Krausz. Nat Phys.;**3** 381 (2007).
5. F. Lépine, MY. Ivano, MJJ. Vrakking, Nat Photon. , 195(2014)
6. M. Nisoli, P. Decleva, F. Calegari, et al. Chem Rev.;**117** 10760 (2017).
7. S. Bohmann, A. Suda, T. Kanai, S. Yamaguchi and K. Midorikawa Opt. Lett. **35** 1887 (2010).
8. M. Ouillé, A. Vernier, F. Böhle, M. Bocoum, A. Jullien, M. Lozano, J.-P. Rousseau, Z. Cheng, D. Gustas, A. Blumenstein, P. Simon, S. Haessler, J. Faure, T. Nagy and R. Lopez-Martens, Light Science & App. **9**, 47 (2020)
9. T. Nagy, M. Kretshmar, M. J.J. Vrakking and A. Rouzée Opt. Lett. **45** 3313 (2020)
10. T. Nagy, P. Simon and L. Veisz Advances in Physics: X **6**, 1845795 (2020)
11. A. Golinelli, X. Chen, E. Gontier, B. Bussière, P.M. Paul, O. Tcherbakoff, P. D'Oliveira, and J.-F. Hergott, Opt. Exp. **27** 13624 (2019)
12. A. Golinelli, X. Chen, E. Gontier, B. Bussière, O. Tcherbakoff, M. Natile, P. D'Oliveira, P.M. Paul and J.-F. Hergott, Opt. Lett. **42**, 2326 (2017)
13. C. Feng, J.-F. Hergott, P.-M. Paul, X. Chen, O. Tcherbakoff, M. Comte, O. Gobert, M. Reduzzi, F. Calegari, C. Manzoni, M. Nisoli, and G. Sansone, Opt. Exp. **21**, 25248 (2013).
14. M. Miranda, J. Penedones, C. Guo, A. Harth, M. Louisy, L. Neoricic, A. L'Huillier and C.L. Arnold, Jour. of the Opt; Soc; of Am. B **34**, 190 (2017)
15. F. Silva, B. Alonso, W. Holgado, R. Romero, J. San Roman, E. Conejero Jarque, H. Koop, V. Pervak, H. Crespo and I. J. Sola, Opt; Lett. **43**, 337 (2018)

Post-Compression of 9.2-μm Terawatt Laser Pulses to Femtoseconds

Igor V. Pogorelsky*, *Mikhail N.* Polyanskiy, *Marcus* Babzien and *Mark A.* Palmer

Accelerator Test Facility, Brookhaven National Laboratory, Upton, NY 11973, USA

Abstract. Contemporary CO_2 laser systems are capable to delivering picosecond pulses of the multi-terawatt peak power. Further pulse compression to a few cycles is possible using a post-compression in bulk materials with negative group velocity dispersion (GVD). We have experimentally demonstrated the post-compression of a long-wave infrared (9.2 μm) 150-GW peak power pulse from 1.85 ps to less than 500 fs using a combination of two optical materials with significantly different ratios of the nonlinear refractive index to the GVD coefficient. Such combination allows for optimization of the compression mechanism and promises a viable path to scaling peak powers to multi-terawatt levels.

1 Introduction

Recent years have seen a growing interest in applying long-wave infrared (LWIR) lasers (λ>4 μm) to strong-field physics research. This comes as the realization of benefits from the wavelength scaling for such applications as electron acceleration in plasma wake fields, ion beam generation from gas jets, intense gamma sources via inverse Compton scattering, energy transport through air, HHG, etc. The extension of experimental studies of these and other phenomena into the LWIR spectral domain can be enabled by CO_2 lasers capable to producing picosecond pulses at $\lambda\approx$10 μm and several joule energies [1, 2]. However, the continuing advancement of ultra-fast laser applications calls for down to few optical cycles, sub-picosecond LWIR pulses at the multi-terawatt peak power.

The duration of an ultra-short laser pulse is limited by the spectral bandwidth of the laser gain, which does not exceed ~1 THz for a CO_2 laser potentially allowing to attain 500 fs [3]. However, the present-day CO_2 laser systems are restricted to a few-picosecond minimum pulse duration due to the spectral narrowing of ultra-short seed pulses under a high amplification.

Sub-picosecond post-compression of picosecond pulses at the several terawatt peak power can be explored through their self-phase modulation (SPM) $\Delta\phi(t) = -\left(\frac{\omega_0}{c}\right) n_2 I(t) L_1$ on their propagation through an optical material of the thickness L_1 with the nonlinear refraction index n_2, where I is the laser intensity, ω_0 is the laser central frequency and c is the speed of light. This phase shift results in the frequency chirp $\Delta\omega(t) = \frac{d(\Delta\phi)}{dt} = -\left(\frac{\omega_0}{c}\right) n_2 L_1 \frac{dI}{dt}$. Followed by dispersive compression with diffraction gratings, a chirped mirror, or a bulk material of a length L_2 with a negative group velocity dispersion (GVD) this phase chirp can be compensated resulting in a pulse compression by $\Delta t \approx \Delta\omega(GVD)L_2$. This idea has been explored theoretically and experimentally mainly for the purpose of improving characteristics of near-IR solid state lasers [4, 5]. Some recent works address also LWIR lasers at $\lambda\approx$4 μm [6] and $\lambda\approx$10 μm with NaCl as the material for both SPM and GVD [7-9]. However, until now, this technique was never experimentally verified with CO_2 lasers.

We report here the first experimental demonstration on post-compression of 2-ps, terawatt-class CO_2 laser pulses to sub picoseconds using optical materials transparent at the λ=9.2 μm.

2 Experiment

We used in our experiment a 5-TW, 1.85-ps CO_2 CPA laser system operated at the BNL/ATF [2]. The central wavelength of this laser filled with a multi-atmosphere mixture of $C^{16,18}O_2$ isotopes is 9.2 μm. The experimental setup for a proof-of-principle demonstration of the post-compression is shown in Fig. 1.

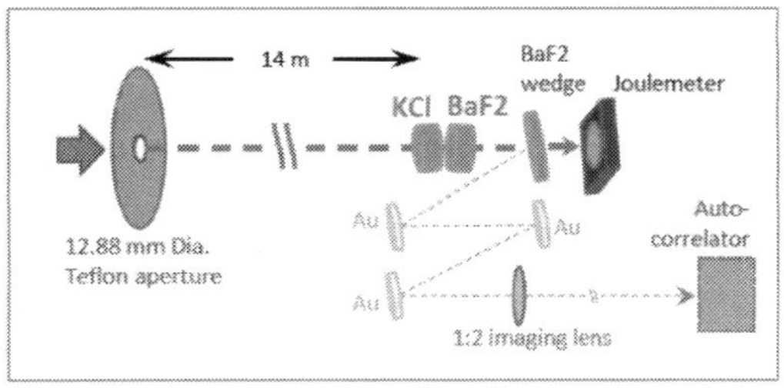

Fig. 1. Experimental setup for studying the post-compression.

A close to Gaussian beam of up to 500 mJ energy is produced in this scheme by sending the output beam from

* Corresponding author: igor@bnl.gov

a several-Joule laser amplifier of the 10 cm optical aperture through a 13 mm diameter Teflon aperture. The intensity distribution after 14 m of propagation is defined by far-field diffraction and practically does not depend on the profile of the input beam. The KCl and BaF_2 50 mm thick slabs were chosen for a SPM chirper and a GVD compressor, correspondingly, based on their n_2 and GVD characteristics shown in Table 1. Among the shown three materials, KCl has the highest absolute $n_2/(GVD)$ ratio, exceeding that of NaCl by roughly a factor of two and BaF_2 by more than a factor of eight. This led to the choice of KCl for the first element of the compressor, which is primarily responsible for the SPM part of the process, and BaF_2 for the downstream element, which provides the dispersive compensation of the chirp.

Table 1. Material properties at 9.2 µm.

	n_0	(GVD) (fs^2/mm)	n_2 (m^2/W)
NaCl	1.500	-1010	3.5×10^{-20}
KCl	1.459	-558	3.4×10^{-20}
BaF_2	1.412	-2340	1.7×10^{-20}

A BaF_2 lens was used for imaging the output surface of the BaF_2 slab on the 10×10 mm crystal of a single-shot autocorrelator (AC) with the ×2 optical magnification. Representative AC traces recorded at different laser energies are shown in Fig. 2 together with their horizontal intensity projections. Compression to sub-picosecond duration is observed for all pulses with input energies exceeding 200 mJ. The highest peak intensity is achieved at the 275 mJ joulemeter readout corresponding to ~450 mJ/cm^2 incident laser peak fluence. At higher energies, the AC signal gradually decreases due to nonlinear absorption in KCl and BaF_2 despite the continued shortening of the pulse.

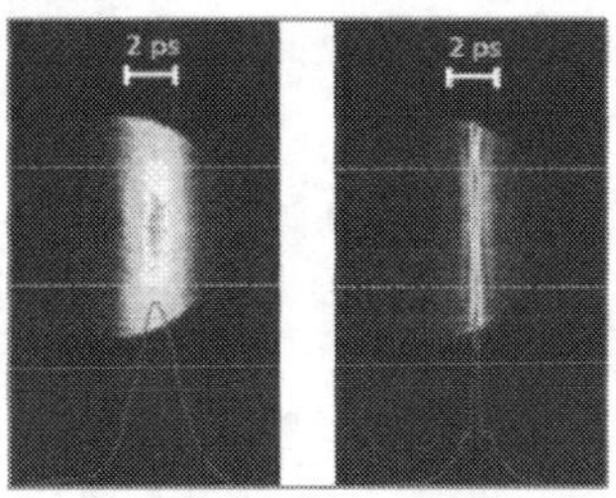

Fig. 2. AC images obtained at the 450 mJ/cm^2 incident laser peak fluence without post-compression (left) and with a two-element KCl+BaF_2 compressor (right).

Spectrum evolution has been recorded in a similar setup with a single-shot spectrometer of the 8 nm wavelength resolution installed in the position of the imaging lens in Fig. 1. Spectra of a pulse before and after the post-compression are shown in Fig. 3. The broadened spectrum is for the laser pulse with the energy matching the conditions of the optimal post-compression. The fine structure of the spectra is due to the rotational modulation of the gain spectrum of CO_2 laser amplifiers.

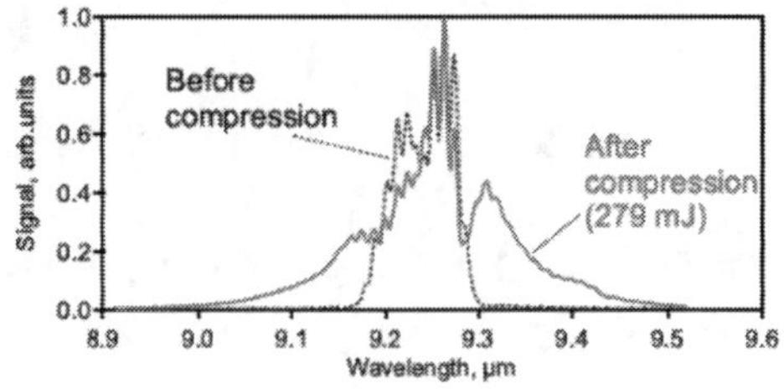

Fig. 3. A typical measured spectra of a laser pulse before and after compression at the energy corresponding to the optimal compression.

3 Discussions and prospects

The numerical modeling of the post-compression is done with the *co2amp* software [10]. For input parameters, it uses the results of a comprehensive modeling of a chirped-pulse amplification of a picosecond LWIR pulse in a chain of high-pressure, mixed-isotope CO_2 laser amplifiers followed by the propagation in air. A split-step algorithm is used for modeling the interaction of the laser pulse with the compressor materials. To compare the model predictions with experimental AC curves, we first model the dynamics of the pulse compression in the central portion of the beam of the 5 mm diameter as is seen by the AC chip. We have found that the model accurately reproduces experimental AC curves in a wide range of the laser pulse energies, including the optimal post-compression regime. The simulated output spectrum agrees reasonably well with the spectrum measured experimentally, yet again confirming the accuracy of the model.

The results of the pulse modeling at different stages of the post-compression (the input, after the first element, and at the output of the post-compressor) are shown in Fig. 4 by dotted black, solid blue, and solid red lines, respectively. The left portion of Fig. 4 corresponds to the 5 mm diameter central portion of the beam. Here, the output pulse profile agrees to a high accuracy with the deconvolution of the AC trace shown in Fig. 2 (right).

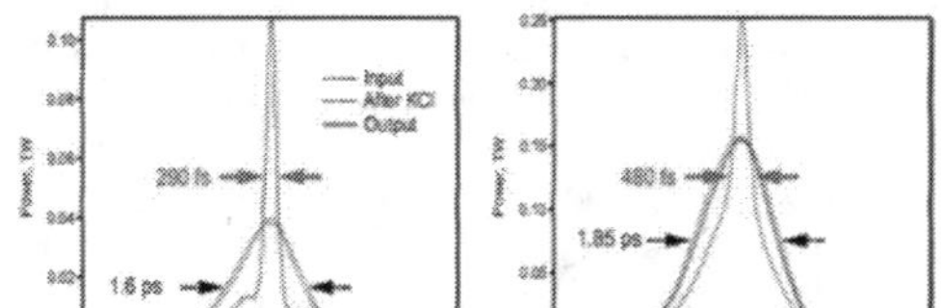

Fig. 4. Results of numerical modelling of the pulse compression in the KCl+BaF_2 configuration at the conditions corresponding to the optimum compression; left – input and output temporal profiles integrated over the 5 mm diameter central portion of a laser beam seen by the AC crystal; right – the same but integrated over the entire beam.

Note that the duration of the input pulse is shorter in the beam center than in the outer areas of the beam: 1.6 ps (FWHM) in the center vs. 1.85 ps beam-average. This is due to a pulse compression in air that occurs by the same manner as the compression in bulk solids (SPM

https://doi.org/10.1051/epjconf/202125511007

accompanied by negative GVD) but requires longer propagation due to orders-of-magnitude smaller n_2 and dispersion.

The simulated temporal profiles agree with deconvoluted experimental AC traces. This allows us to conclude that we achieved a post-compression of a 1.6-ps pulse to <300 fs in the center of a near-Gaussian beam with a 180% increase of the peak intensity.

At the same time, the simulations allow us to gain the understanding of the pulse evolution averaged over the entire beam. Figure 4 (right) shows that the entire 1.85-ps pulse gets compressed to 500 fs with the 60% of the power increase.

Our initial demonstration of delivering hundreds of millijoule on the femtosecond time scale surpass by far the results previously achieved in the LWIR spectral domain. The implemented here partial decoupling between the SPM spectral stretching and GVD pulse compression by using a combination of two materials with contrasting $n_2/(GVD)$ ratios allows matching the nonlinear spectrum broadening with the GVD phase delay compensation while keeping B-integral on each optical slab below the threshold for parasitic micro-filamentation and nonlinear absorption (B≤3).

Experiments will continue towards further improving the efficiency of the LWIR energy conversion to femtoseconds by optimizing the post-compressor staging. We expect that our continuing material search and optimization of the beam and the setup geometry will lead to a practical system for an effective femtosecond conversion of the entire CO_2 laser output at a multi-terawatt level as is required for advanced plasma accelerators and other strong-physics research and applications.

Acknowledgements

This research is funded by the US Department of Energy (DOE) Accelerator Stewardship Program grant, B& R #KA2601020. BNL Laboratory Directed Research and Development (LDRD) grant #20-010 AD/NPP.

References

1. D. Haberberger, S. Tochitsky and C. Joshi, "Opt. Express, **18**, 17865 (2010)
2. M. P. Polyanskiy, I. V. Pogorelsky, M. Babzien and M. A. Palmer, OSA Continuum, **3**, 459 (2020)
3. M. P. Polyanskiy, I. V. Pogorelsky, M. Babzien, R. Kupfer, N. Vafaei-Najafabadi and M. A. Palmer, Photonics, **8**, 101 (2021)
4. A. A. Voronin, A. M. Zheltikov, T. Ditmire, B. Rus and G. Korn, Opt. Commun., **291**, 299 (2013)
5. S. Hädrich, et al, Opt. Lett., **41**, 4332 (2016)
6. M. Hemmer, M. Baudisch, A. Thai, A. Couairon and J. Biegert, Opt. Express, **21**, 28095 (2013)
7. B. G. Bravy, V. M. Gordienko and V. T. Platonenko, Laser Phys. Lett., **11**, 065401 (2014)
8. J. Pigeon, S. Tochitsky and C. Joshi, AIP Conf. Proc., **1777**, 110005 (2016)
9. B. G. Bravy, V. M. Gordienko and V. T. Platonenko, J. of Russ. Laser Research, **37**, 465 (2016)
10. M. N. Polyanskiy, Appl. Opt. **54**, 5136 (2015)

Reconstruction of few-fs XUV pulses with a perturbative approach

Bruno Moio[1,2,*], *Fabio* Medeghini[1,2], *Gian Luca* Dolso[1], *Giacomo* Inzani[1], *Nicola* Di Palo[1,2], *Rocío* Borrego-Varillas[2], *Mauro* Nisoli[1,2], and *Matteo* Lucchini[1,2]

[1]Department of Physics, Politecnico di Milano, 20133 Milano, Italy
[2]Institute for Photonics and Nanotechnologies, IFN-CNR, 20133 Milano, Italy

Abstract. A precise temporal characterization of the pulses involved in pump-probe experiments is crucial for a proper investigation of the ultrafast dynamics in several physical systems. Indeed, it is required for the assessment of the dynamical properties under examination with sufficient temporal resolution. In the few-fs/attosecond domain, typical reconstruction procedures require time-consuming interative methods, which are also sensitive to the experimental noise and to the distortion of the measurement. We developed an approach, called Simplified Trace Reconstruction In the Perturbative regimE (STRIPE), which allows us for a precise characterization of the infrared (IR) and extreme ultraviolet (XUV) pulses, used in a pump-probe experiment. Our method is not based on a phase retrival algorithm, and for this it is typically much faster than the other ones currently known. Moreover, it allows for easily including in the reconstruction the experimental non-idealities that may affect the measurement, like possible distortion due to the measurement procedure itself.

1 Introduction

Recently, ultrashort extreme-ultraviolet (XUV) pulses, combined with few-cycle infrared (IR) radiation in a pump-probe scheme, open the possibility of studying dynamics in a wide range of systems, from molecules to solid stated materials [1, 2]. In addition, the possibility to select single harmonics from a much wider XUV spectrum [3] allows to selectively access single transitions, but maintaining at the sane time the possibility to tune the spectral range to investigate.

The use of single-harmonic extreme-ultraviolet radiation has been widely employed for the investigation of electron dynamics in molecules [4] with excellent simultaneous temporal and spectral resolution. The availability of few-fs single harmonic XUV pulses allowed to study the ultrafast conical-intersection dynamics in NO_2 [5] and the temporal evolution of ethylene cation with state-to-state resolution [6]. A proper analysis of such dynamics calls for an accurate characterization of the pulses. A commonly used method consists in measuring several photoelectron spectra, as produced by the XUV radiation in the presence of a delayed infrared pulse, and employing a phase retrival algorithm, based on iterative procedures, for charachterizing the pulses involved [7, 8]. These phase retrival algorithm are typically computationally expensive and require a very clean experimental measurement, since are sensitive the residual noise of the trace.

In this work, we propose a novel approach for the pulse temporal characterization, based on a simplified method called Simplified Trace Reconstruction In the Perturbative regimE (STRIPE). Our method is much faster than the retrieval algorithms currently used and can be applied directly to differential phototelectron traces, to remove the distorsions that may arise from background signals. Moreover, our method allows for a proper pulse characterization even if the signal to noise ratio is extremely poor. This makes the measurement and the reconstruction inherently robust with respect to noise. Finally, being more versatile than a standard iterative algorithm, STRIPE can be easily modified to explicitly include the system non-idealities due to the meaurement procedure itself, like electronic peak broadening.

2 The STRIPE model

In the strong field approximation (SFA) [9], the photoionazion process of a rare gas, in the presence of XUV and IR pulses can be described by the following formula (atomic units are used hereafter):

$$S(\mathbf{p},\tau) = \left| \int_{-\infty}^{+\infty} dt\, E_{XUV}(t+\tau)\, e^{i\phi(\mathbf{p},t)} e^{i\left(\frac{p^2}{2}+I_p\right)t} \right|^2 \quad (1)$$

where $E_{XUV}(t)$ is the XUV pulse, $\phi(\mathbf{p},t)$ is a phase term which accounts for the effect of the IR field on the photoelectron wavepacket of momentum $\mathbf{p}$, I_p is the atomic ionization potential and τ is the relative delay between the XUV and IR. We can now develop the simplified model by applying a series of approximations to the Eq. (1). In particular, under the central momentum approximation (CMA), the perturbative approximation and the slowly-varying envelope approximation (SVEA), we can reformulate the previous equation into the following expression

*e-mail: bruno.moio@polimi.it

https://doi.org/10.1051/epjconf/202125511008

[10]:

$$S(\omega,\tau) \simeq \left| \int_{-\infty}^{+\infty} dt\, E_{XUV}(t+\tau)\, e^{i\frac{p_c}{\omega_0^2}E_{IR}(t)}\, e^{i\omega t} \right|^2 \quad (2)$$

where we set $\omega = p^2/2 + I_p$. Equation (2) represents the model upon which STRIPE is based. Figure 1 shows

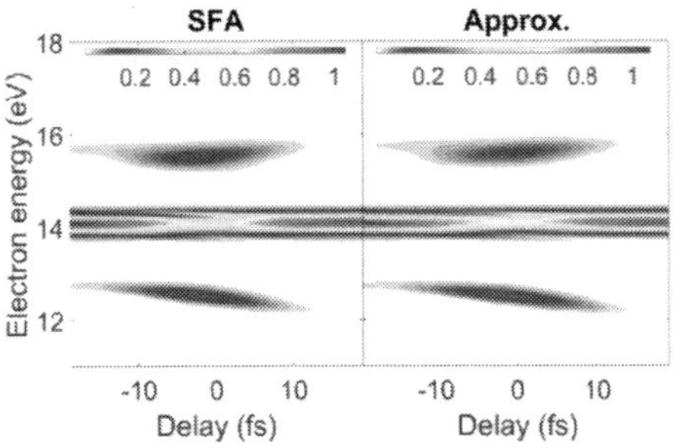

Figure 1. Comparison of a single-harmonic phototelectron trace, computed with the SFA formula (left) and the STRIPE model (right). The agreement is very good as the STRIPE model can describe all the features of the main band and the sidebands. However, the computational effort for computing the trace with the simplified model is significantly smaller than the one needed for the SFA calculations.

a comparison between a single-harmonic phototelectron trace, computed with the full SFA model (left panel) and one computed with the STRIPE formula (right panel). We considered an XUV and IR pulses, with durations of about 10 and 9 fs, respectively, a second order dispersion of 10 and 20 fs^2, a third order of 300 and 30 fs^3 and a fourth order of 2500 and -150 fs^4. The IR intensity is $5 \cdot 10^{11}$ W/cm^2.The parameters are deliberately set at the limits of validity of the approximated model, in order to test the procedure in the worst-case scenario.

The agreement between the two simulations is very good, as the STRIPE model can reproduce all the main features observed in the SFA trace. Its worth mentioning that the computational effort to compute the STRIPE trace is much smaller than the one needed for the SFA calculations. Indeed, the time to compute a single typical SFA trace amounts to about 30 seconds, against the 40 milliseconds needed for the computaton with the STRIPE model.

3 Trace reconstruction

If, on one side, it is clear that the approximated model produces traces that nicely compare to the SFA calculations, on the other, to prove the possibility to use the STRIPE model for the characterization of XUV and IR pulses in photoelectron experiments, we must apply it to some simulated traces, and see if it can recover the fields involved. We start by using STRIPE in a reconstruction of a clean trace, with noise, to first test the method. Then, we repeated the reconstruction also including noise, to prove the real applicability of the algorithm. Eventually, we tested the case of a real experimental trace.

3.1 Reconstruction with no noise

As a first attempt, we applied STRIPE for the reconstruction of pulses out of a simulated trace computed with SFA calculations, where no approximations are considered, and with no additional noise introduced. The reconstruction is done by a non-linear fitting procedure, employing the simplified model in Eq. (2). The trace on which we applied our algorithm is the one in the left panel of Fig. 1, whereas the pulses retrieved by STRIPE are shown in Fig. 2 (red dashed lines), together with the ones used in the SFA simulations (black lines). Both the XUV and IR

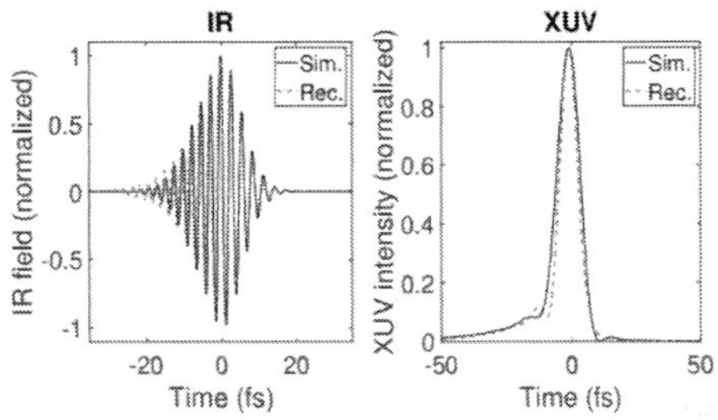

Figure 2. Result of the STRIPE reconstruction, applied to the trace simulated with SFA in Figure 1. The black solid lines show the simulated IR and XUV pulses, whereas the dashed red lines show the reconstructed ones.

pulses are correctly retrieved, with small deviations from the ones used in the simulation, despite the parameters are chosen to stretch as much as possible the approximations of the method. We must mention that an SHS experiment is not sensitive to the phase of the IR field. In this case, we chose the IR phase to match the one of the simulation, just to highlight the agreement between the simulated and retrieved pulse energy dispersion. In any case, this result proves that, despite the fact that the model upon which it is based is heavily approximated, STRIPE can serve as a useful and efficient tool for pulse characterization.

3.2 Reconstruction with white noise

Another important feature of our reconstruction method is a great robustness against the experimental noise. To prove this claim, we applied STRIPE to a trace, where an additional very large white noise is added. We choose an extreme noise level, much larger than in any reasonable experiment, to prove that our method is basically insensitive to any noise intensity. The noisy trace is shown in the top panel of Fig. 3, whereas the comparison between the simulated and reconstructed fields is displayed in the bottom left and right panels of the same figure. Also here we set the IR phase to match the simulated value, for a clearer visualization. The comparison shows that, despite the trace being completely distorted by the noise, with a signal-to-noise level ratio is unrealistically low, the information is not actually lost and STRIPE can access it, with little to no additional uncertainty, as one can see by comparing the fields if Figs. 2 and 3. Indeed, the precision

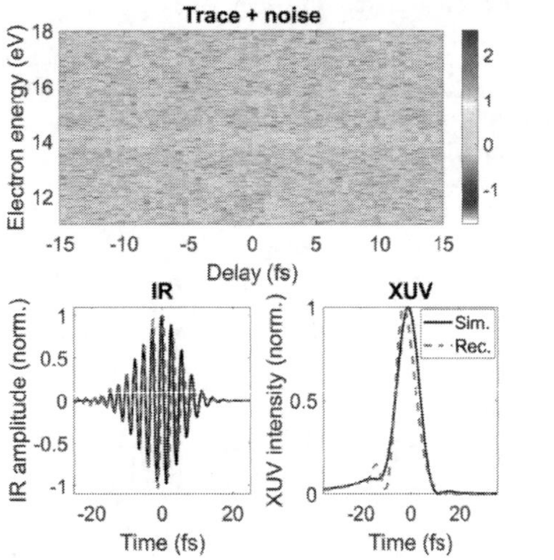

Figure 3. Results of the STRIPE reconstruction, with a large white noise applied to the SFA trace. The top panel shows the noisy simulated trace, whereas the bottom panels show a comparison of the simulated and reconstructed pulse. Despite the signal-to-noise ratio is very low, STRIPE can recover the pulses with almost the same accuracy.

of the field reconstruction is comparable to the case with no noise, showing that the trace distortion is not basically affecting the robustness of the algorithm.

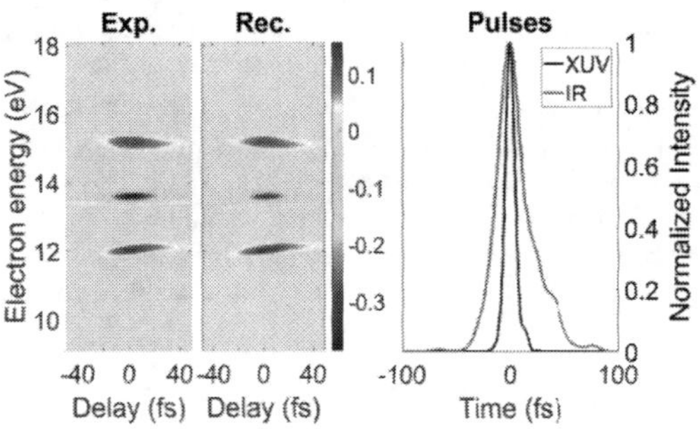

Figure 4. Experimental and reconstruced single-harmonic differential traces. Despite the simplified model, STRIPE can catch all the features of the experimetnal trace. On the right, we show the reconstructed XUV (black) and IR (red) pulses.

3.3 Reconstruction with experimental trace

To undoubtedly prove that our method can be applied to real-case measurements, we performed the reconstruction on an experimental trace. We carried out the retrieval on the differential trace, obtained by subtracting the XUV-only spectrum, to remove the effect of a background signal in the experiments, and to include in the fitting the artificial spectral broadening due to the electronic response of the acquisition system. Figure 4 shows a comparison between the experimental differential trace and the reconstruction one, and the reconstructed XUV (black line) and IR (red line) pulses. The reconstruction is very accurate, and all the features of the trace are correctly recovered. Moreover, the retrieval is not affected by the experimental noise. The extracted IR field duration of 26.9 fs compares very well to the one obtained with an independent FROG measurement, while for the XUV STRIPE gives a duration of 10.2 fs, in line with what expected from ray-tracing calculations.

4 Conclusions

We presented STRIPE, a novel method for the characterization of XUV and IR femtosecond pulses in a photoelectron experiment. Our method allows to reconstruct the pulses in a much faster way and with high degree of accuracy, despite all the approximation introduced. The reconstruction performed on a very noisy trace shows that STRIPE is intrinsically robust against any level of noise, and by applying it to an experimental trace affected by large non idealities, we proved this method to be capable of precise retrieval also in real, extreme cases. Indeed, the reconstructed pulses agree with what obtained with independent measurements. So far we tested our method only with single-harmonic spectrograms, but we already observed that the model in Eq. (2) can also be used to properly simulated photoelectron traces in a variety of settings (streaking, RABBITT, etc.), making our technique a valuable tool for a fast characterization of the pulses in a wide range of experiments.

References

[1] Krausz, F., and Misha I., Rev. Mod. Phys. **81**, 163 (2009)
[2] Calegari, F., et al., Science **346** 336-339 (2014)
[3] Poletto, L., et al., Opt. Lett. **32** 2897-2899 (2007)
[4] Eckstein, M., et al., J. Phys. Chem. Lett. **6**, 419-425 (2015)
[5] von Conta, A., et al., Nat. Commun. **9**, 1-10 (2018)
[6] Lucchini, M., et al. CLEO/Europe-EQEC, (2019)
[7] Lucchini, M., et al., Opt. Expr. **26**, 6771-6784 (2018)
[8] Murari, M., et al., Opt. Expr. **28**, 10210-10224 (2020)
[9] Kitzler, M. et al., Phys. Rev. Lett. **88**, 173904 (2002)
[10] Moio, B. et al., J. Phys. B: At. Mol. Opt. Phys. **54**, 154003 (2021)

A Spectral Resolution study in Femtosecond Stimulated Raman Scattering Microscopy

Rajeev Ranjan[1,2]*, *Maria Antonietta* Ferrara[2], and *Luigi* Sirleto[1]

[1]National Research Council (CNR), Institute of Applied Sciences and Intelligent Systems, Napoli, Italy
[2]CHT @Erzelli, Nanoscopy Istituto Italiano Di Tecnologia, Genova, Italy

Abstract. The simultaneous mapping and the specificity of different chemical species are desirable in several biological and biomedical applications. The stimulated Raman Scattering technique is a proven and well-established label-free method to map the distributions of individual species in a multi-component-based system due to the linear dependence of signals on concentration and its chemical selectivity. In this framework, spectral resolution, i.e., the ability to distinguish closely lying resonances, plays a fundamental role. Here in this work, the cross-correlation of Ti:Sa & OPO femtosecond laser beams in a stimulated Raman scattering microscope is measured. The separation between protein and lipid bands in the C-H region is important for biochemical research and is successfully classified.

1 Introduction

Over the past ten years, stimulated Raman scattering (SRS) has been investigated in nanophotonic [1–4] as well as in biophotonics as an analytical, label-free, non-invasive technique with unique cellular and tissue imaging capabilities [5–8]. Stimulated Raman Scattering (SRS) based spectroscopy is a quite popular and proven technique for the spectrochemical analysis of molecules. When the SRS technique is combined with microscopy, it brings a mass revolution in tiny molecules imaging and differentiation between specific molecules. It offers high chemical selectivity of unlabelled living cells and real-time 3-D imaging with high spatial resolution and sensitivity [5–8].
The C-H stretching (2800–3100 cm^{-1}) region of Raman spectra of biomolecules is the most used in SRS microscopy. Lipids and proteins can be investigated in this region, taking advantage of CH_2, near 2845 cm^{-1}, and CH_3, near 2930 cm^{-1}, Raman bands, respectively. The difference between the peaks of these two Raman bands and their width spectral shapes is about 100 cm^{-1}. This means that they are partially overlapped. Typically, SRS microscopy is implemented by using two Fourier transform-limited (FTL) tunable picosecond (*ps*) laser sources, which have a high spectral resolution (about 10 cm^{-1}). Still, the drawback is that *ps* pulses show a low peak intensity, thus needs high laser power for imaging [5–8]. When *fs* (femtosecond) laser sources are used, an improvement of about one order of magnitude of the SRS signal has been demonstrated with respect to ps pulses [9]. Still, a low spectral selectivity is obtained, and multi-band excitation can occur, not allowing, in principle, the separation of some bands of particular interest in biologies, such as those of lipids and proteins. To solve this issue, a number of methods for SRS multicolor imaging, based on broadband femtosecond pulses, have been developed [10-14].
In order to enhance spectral resolution in SRS microscopy based on fs laser pulses, an option is given by spectral focusing (SF). The equally chirping pump and Stokes beams with an energy spacing corresponding to the Raman line (IFD, $\Omega = \omega p - \omega s$) spectrally focuses the excitation energy into a single resonance. We note that the bandwidth $\delta\Omega$, in the limiting case, can simulate the *ps* SRS system. Nevertheless, due to a large number of parameters, the resulting spectral resolution of SF-SRS setups is often worse than theoretically predicted [14].
It is worth noting that in coherent Raman scattering, there is a trade-off between the best spectral resolution and the best ratio of image contrast and signal intensity, which is obtained when the spectral resolution and the width of the Raman lines under observation are almost the same [15]. This condition is satisfied in the fingerprint region in the case of excitation with both ps pulses, since they match the Raman linewidths (5–20 cm^{-1}), while in the case of the C-H region, broader bandwidth *fs* pulses can be considered as excitation for lipids and proteins Raman bands.
It is well known that, due to the effects of chromatic dispersion and nonlinearities, a pulse can gain a chirp, for example, through propagation in a transparent medium. Therefore, the optical elements embedded in an SRS optical circuitry microscope setup can also change the pulse width. In this paper, taking advantage of the small chirping introduced by simply propagating the beams through dispersive materials already present in the SRS microscope setup, we demonstrate that we can be distinguished protein and lipid bands in the C-H region.

*Corresponding author: rajeev.ranjan@iit.it

 https://doi.org/10.1051/epjconf/202125511009

2 Experimental Set-up

In our previous paper, we have successfully described the design and execution of femtosecond-based SRS microscopy, which is able to cover the C-H region (>2,500 cm^{-1}) [16-22]. In Figure 1, a femtosecond-SRS spectroscopy setup is integrated with a Nikon C2 confocal microscope, which comprises an inverted Nikon Ti-eclipse microscope and a scan head. This arrangement has been described in full in our earlier publications. A Ti:Sapphire (Ti:Sa—Chameleon Ultra II—pulse duration = 140 fs, repetition rate = 80 MHz, λ= 680–1080 nm) and a femtosecond optical parametric oscillator (OPO—Chameleon Compact OPO—pulse duration = 200 fs, repetition rate = 80 MHz, λ= 1000–1600 nm) are used in this configuration.

Our SRS microscope allows us to probe individual Raman bands at a time. By tuning the wavelength of the pump or Stokes beam with the matching Raman frequency of different samples at the required Raman band, we can obtain the images of the different Raman band matching molecules with high chemical contrast [16-23]. Thus, CH_2 (2,845 cm^{-1}) and CH_3 (2,940 cm^{-1}) stretching signals can be collected at one Raman shift at a time, leading in principle to map the distributions of the lipid and protein contents in the same image. We note that when *fs* pulses are used, the loss in spectral selectivity of SRS signal is not prejudicial in lipids imaging, so SRS images acquired at 2,845 cm^{-1} may be mainly attributed to the lipid. On the contrary, the chemical specificity of SRS images at 2,940 cm^{-1} is not particularly efficient. Since both lipids and proteins have significant Raman signals and their Raman bands are partially overlapped, SRS images acquired at 2,940 cm^{-1} contain both lipids and protein signals.

Considering that SRS is a two-pulse technique, its spectral resolution is not defined by the spectrum of the individual exciting pulses but by the spectrum of their temporal interference. Consequently, cross-correlation is fundamental for providing information about the entire system; in particular, the FWHM of pump and probe beams' cross-correlation allows us to evaluate the experimental spectral bandwidth [23].

Taking advantage of a non-linear process, when both pulses spatially and temporally overlapped, a cross-correlation pulse signal can be generated. Since the rate of two-photon absorption (TPA) is also proportional to the product of two intensity laser beams with different wavelengths, this process is suitable for cross-correlation measurements. Taking advantage of TPA, cross-correlation measuring is carried out, focusing two different beams into a detector. The average photocurrent is larger when the two beams are temporally aligned than when they are misaligned. The detected photocurrent usually has a non-zero background level because the two beams can produce TPA separately even when non-overlapped. In this phenomenon, the output signal is equal to the cross-correlation of the input signals.
Cross-correlators based on TPA has some significant advantages: (i) a higher sensitivity can be obtained with respect to second harmonic generation (SHG), due to the involvement of fields only at the original frequency, ω, owing to the TPA resonant second-order transition nature; (ii) extremely short pulses can be characterized since TPA can operate in a wide wavelength range not restricted by the narrow phase-matching bandwidth; (iii) in TPA non-linear signal multiplication and detection are combined into one process, leading to a simplification and a higher efficiency with respect to a two-step process of optical nonlinearity followed by linear detection.

A pulseCheck 50 (APE, Berlin) is used in our measurements, with a pulse width measurement range [10 fs-12 ps]. The pulse Check connects to the PC via USB, thanks to APE's standardized software interface, which can be remote-controlled or integrated into automated setups. An extra flip-flop mirror in our existing optical architecture is introduced (see figure 1) to establish cross-correlation measurements. The flip-flop mirror (FFP/AM) is installed in-between the mirror M4 and M5 so that each laser beam can be directly diverted into an autocorrelator. Once laser beams are directed into the autocorrelator, the beam is tuned to stabilize and maximize the intensity. After reaching the maximum intensity, the pulse is recorded, and the data analysis is performed. For the data analysis, we used the Gaussian fitting curve function *cftool* of MATLAB 2020.

For this measurement, a step-by-step acquisition method is utilized. Probe and pump beams are focused inside the TPA detector. The Ti:Sa is optically delayed by an optical delay line (Newport MOD MILS200CC) inserted between the Ti:Sa and the microscope. The delay line moves one step and stops, and when it stops during the process, the signal is acquired. The single or averaged signal decay is available in recorded files.

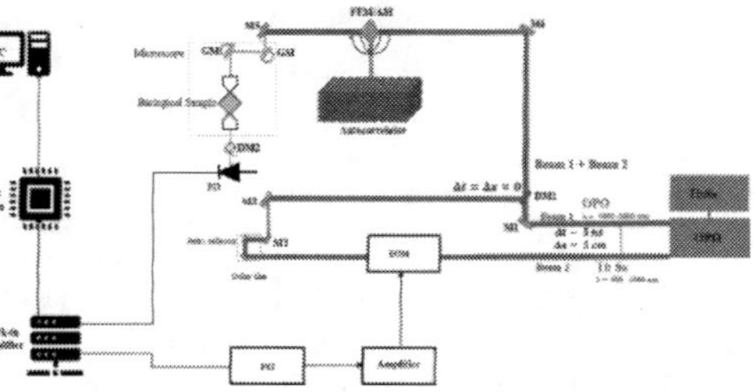

Fig. 1. Optical architecture for spectral characterization single beam

3 Results and Discussion

The measured FWHM (Full width at half maximum) of our Ti:Sa and OPO cross-correlation was 262 *fs* (see Figure 2) [22]. Thus, the obtained experimental spectral bandwidth, given by the FWHM of cross-correlation in the frequency domain, was 56 cm^{-1}.

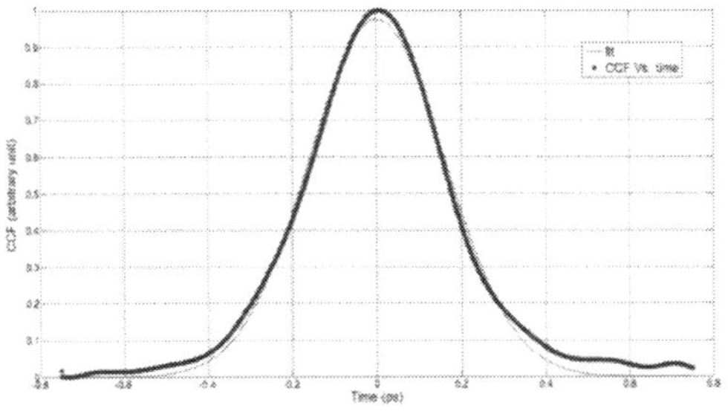

Fig. 2 Cross-correlation of Ti:Sa-OPO and its Gaussian fit

Since our experimental spectral bandwidth was 56 cm^{-1}, when we set the $\lambda_{Ti:Sa}$ at 808 nm and λ_{OPO} at 1049 to excite the 2845 cm^{-1} lipid band, but in-fact we are exciting from (FWHM) 2817 to 2873 cm^{-1}; in the same way, when we set the lasers beams of $\lambda_{Ti:Sa}$ at 803 and λ_{OPO} at 1052 to excite the protein band at 2940 cm^{-1}, then effectively we exciting the range 2912-2968 cm^{-1} (FWHM). In Figure 3, spectral bandwidth deriving by the chirping of the pulses (56 cm^{-1} – continuous lines) and two simulated Raman bands having the two peaks at a distance of 95 cm^{-1} with the width of about 100 cm^{-1} (in figure 3 blue and red) are illustrated. The overlap between two excited bandwidths is highlighted with yellowish shadow; however, they contribute to the Raman signal from this region can be neglected considering that the intensities are lower than FWHM values and, so, they can be regarded as under threshold. Thus, we can conclude that using 56 cm^{-1} chirped pulses, the 2845 cm^{-1} channel is essentially related to the lipid signal, and the 2940 cm^{-1} channel is principally due to protein content. Based on this explanation, we may conclude that our SRS microscope is suited for molecular imaging specificities such as lipids, proteins, and fatty acids in the C-H region.

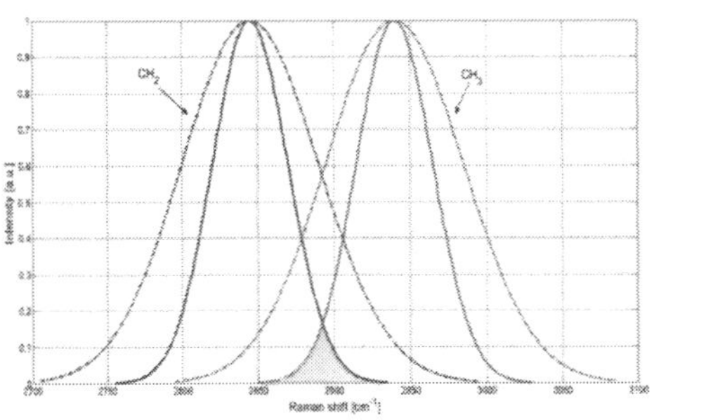

Fig. 3. Raman bands for CH_2 (2845 cm^{-1}, red circles) and CH_3 (2940 cm^{-1}, blue diamonds).

In SRS microscopy, the trade-off between high selectivity offered by picoseconds pulsed laser source and high SRS signal obtained by using femtoseconds pulses is still an open question, and it is widely investigated. In particular, there is interest in label free image both the lipids and proteins distribution for biological application. Unfortunately, the Raman bands of these two components are very close. Thus, ps pulses are required; however, the protein content could be low, leading to very weak Raman signals so that fs pulses would be appropriate.

In this paper, an alternative method is proposed to overcome the drawback of spectral focusing. Its basic idea is to avoid adding optical elements in the SRS microscopy optical setup and taking advantage of chirping, introduced by simply propagating the beams through dispersive materials already present in the SRS microscope. The pros are that no additional optical elements have to be introduced in the experimental setup, giving the significant advantages of a more cheap and straightforward design. The cons are that the spectral resolution is fixed. However, in our set up an experimental value of 56 cm^{-1} for spectral resolution is measured by cross-correlation techniques, and molecular specificity is demonstrated for lipids and proteins in the C-H region.

4 Acknowledgements

The authors would like to thank M. Indolfi and V. Tufano (ISASI-CNR) for their technical support.

References

1. Sirleto, L.; Ferrara, M.A.; Nikitin, T.; Novikov, S.; Khriachtchev, L. Giant Raman gain in silicon nanocrystals. Nat. Commun. **2012**, 3, 1–6. doi: 10.1038/ncomms2188;
2. Sirleto, L.; Vergara, A.; Ferrara, M.A. Advances in stimulated Raman scattering in nanostructures. Adv. Opt. Photon. **2017**, 9, 169–217. doi: 10.1364/AOP.9.000169
3. M.A. Ferrara, L.Sirleto, G.Nicotra, C.Spinella, I.Rendina. Enhanced gain coefficient in Raman amplifier based on silicon nanocomposites. 2011 vol.9, pp.1-7. doi:10.1016/j.photonics.2010.07.007.
4. Sirleto, L.; Aronne, A.; Gioffrè, M.; Fanelli, E.; Righini, G.C.; Pernice, P.; Vergara, A. Compositional and thermal treatment effects on Raman gain and bandwidth in nanostructured silica-based glasses. Opt. Mater. 2013, 36, 408–413. http://dx.doi.org/10.1016/j.optmat.2013.10.001
5. Freudiger, C.W.; Min,W.; Saar, B.G.; Lu, S.; Holtom, G.R.; He, C.; Tsai, J.C.; Kang, J.X.; Xie, X.S. Label-Free Biomedical Imaging with High Sensitivity by Stimulated Raman Scattering Microscopy. Science 2008, 322, 1857–1861. [CrossRef] [PubMed]
6. 6. Cheng, J.-X.; Xie, X.S. Vibrational spectroscopic imaging of living systems: An emerging platform for biology and medicine. Science 2015, 350, aaa8870. [CrossRef]
7. Camp, C.H., Jr.; Cicerone, M.T. Chemically sensitive bioimaging with coherent Raman scattering. Nat. Photon 2015, 9, 295–305. [CrossRef]

8. Zumbusch, A.; Langbein,W.; Borri, P. Nonlinear vibrational microscopy applied to lipid biology. Prog. Lipid Res. 2013, 52, 615–632. [CrossRef]
9. Zhang, D.; Slipchenko, M. N.; Cheng, J.-X. Highly sensitive vibrational imaging by femtosecond pulse stimulated Raman loss. J. Phys. Chem. Lett. 2011, 2, 1248–1253.
10. Fu, D.; Lu, F.K.; Zhang, X.; Freudiger, C.W.; Pernik, D.R.; Holtrom, G.; Xie, X.S. Quantitative chemical imaging with mul-tiplex stimulated Raman scattering microscopy. J. Am. Chem. Soc. 2012, 134, 3623–3626. [CrossRef]
11. Liao, C.-S.; Wang, P.; Wang, P.; Li, J.; Lee, H.J.; Eakins, G.; Cheng, J.-X. Spectrometer-free vibrational imaging by retrieving stimulated Raman signal from highly scattered photons. Sci. Adv. 2015, 1, e1500738. [CrossRef]
12. Fu, D.; Holtom, G.; Freudiger, C.; Zhang, X.; Xie, X.S. Hyperspectral Imaging with Stimulated Raman Scattering by Chirped Femtosecond Lasers. J. Phys. Chem. B 2013, 117, 4634–4640. [CrossRef]
13. Pegoraro, A.F.; Ridsdale, A.; Moffatt, D.J.; Jia, Y.W.; Pezacki, J.P.; Stolow, A. Optimally chirped multimodal CARS micros-copy based on a single Ti:sapphire oscillator. Opt. Express 2009, 17, 2984–2996. [CrossRef]
14. Rocha-Mendoza, I.; Langbein,W.; Borri, P. Coherent anti-Stokes Raman microspectroscopy using spectral focusing with glassdispersion. Appl. Phys. Lett. 2008, 93, 201103. [CrossRef]
15. Brückner, L.; Buckup, T.; Motzkus, M. Exploring the potential of tailored spectral focusing. J. Opt. Soc. Am. B 2016, 33, 1482–1491.[CrossRef]
16. D'Arco A., Brancati N., Ferrara M. A., Indolfi M., Frucci M., and Sirleto L., Biomed. Opt. Express, 2016, 7, (5), pp. 1853-1864, (2016).
17. Annalisa D'Arco, Maria Antonietta Ferrara, Maurizio Indolfi, Vitaliano Tufano and Luigi Sirleto., Journal of Nonlinear Optical Physics & Materials Vol. 26, No. 4, 1750052 ((2017).
18. R. Ranjan, A. D'arco, M. A. Ferrara, M. Indolfi, M. Larobina and L. Sirleto. Opt. Exp., Vol. 26, Issue 20, pp. 26317-26326, (2018).
19. Ranjan, R., Indolfi, M., Ferrara, M. A., Sirleto, L. J. Vis. Exp. (149), e59614, (2019).
20. Ferrara MA, Filograna A, Ranjan R, Corda D, Valente C, Sirleto L. PLOS ONE 14(5): e0216811, (2019).
21. R. Ranjan, M. A. Ferrara, A. Filograna, C. Valente and L. Sirleto. Journal of Instrumentation, vol. 14, P09008, (2019).
22. Sirleto, L.; Ranjan, R.; Ferrara, M.A. Appl. Sci., 11, 3903, (2021).
23. T.; Obara, Y.; Misawa, K. Single-beam phase-modulated stimulated Raman scattering microscopy with spectrally focused detection. J. Opt. Soc. Am. B 2017, 34, 1004. [CrossRef]

Ultrashort-pulse, terawatt, long-wave infrared lasers based on high-pressure CO_2 amplifiers

Mikhail N. Polyanskiy*, *Igor V.* Pogorelsky, *Marcus* Babzien, *Rotem* Kupfer, and *Mark A.* Palmer

Accelerator Test Facility, Brookhaven National Laboratory, Upton, New York 11973, USA

Abstract. We discuss the state of the art, the ongoing research and development, and the potential for achieving a supra-terawatt peak power in few-cycle pulses at a long-wave infrared wavelength with a laser system based on high-pressure, mixed-isotope CO_2 amplifiers.

1 Introduction

Extending the wavelength range covered by high-peak-power laser systems into the mid-wave- and long-wave infrared (MWIR and LWIR respectively) promises many advantages for high-energy physics and material research [1]. Some examples of applications that can benefit from increasing the laser wavelength are the laser wakefield acceleration of electrons, the laser acceleration of ions from gas targets, and the generation of high-energy, supercontinuum, attosecond pulses. Despite the progress in the field of infrared solid-state sources and frequency conversion techniques [2], generation of short LWIR pulses at energy levels above a sub-millijoule is presently only possible using molecular gas lasers. We present here the state of the art in the multi-terawatt, picosecond LWIR laser technology by the example of a laser system based on high-pressure, mixed-isotope CO_2 gas laser amplifiers operated at BNL and discuss the prospects for progressing towards few-cycle LWIR pulses of a supra-terawatt peak power.

2 Background

The carbon dioxide gas laser is historically one of the most used types of lasers due to its robustness, efficiency, and scalability. Pulsed atmospheric-pressure CO_2 lasers transversely excited by an electric discharge (TEA CO_2 lasers) can produce hundreds of joules of energy in long, typically a hundred of nanoseconds, pulses. The generation of short (<1 ns) pulses at such gas pressure is not possible due to a highly modulated rotational-vibrational structure of the gain spectrum shown in Fig. 1(a). A quasi-continuum gain spectrum shown in Fig. 1(b) can be achieved using an elevated gas pressure of ~10 bar for collisional broadening and admixing the ^{18}O containing isotopologues of CO_2 for increasing the density of the spectral lines. The ~1 THz bandwidth of the strongest branch of this spectrum, *9R* centred at 9.2 μm corresponds to the bandwidth of a transform-limited 500 fs pulse. However, minimum pulse duration of an amplified pulse is restricted to 2 ps due to the gain narrowing [3].

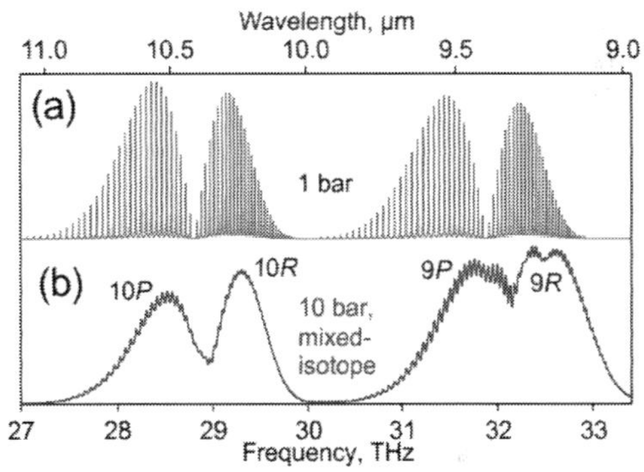

Fig. 1. Typical gain spectra of CO_2 lasers: (a) conventional atmospheric-pressure laser; (b) high-pressure, mixed-isotope $C^{16}O_2$ - $C^{16}O^{18}O$ - $C^{18}O_2$ laser.

3 State of the art in a short-pulse CO_2 laser technology

Achieving a 15 TW peak power in a 10.6 μm 3 ps (FWHM here and through the article) pulse using a CO_2 laser system based on a 180 *l* active volume amplifier producing one shot in 30 minutes was reported by UCLA [4]. A relatively moderate, 2.5 bar gas pressure in this amplifier resulted in a train of pulses with 18 ps period with the highest peak power attained in the middle of the train.

An LWIR system with a higher, 1 Hz, repetition rate, but of a lower, 1 TW, power based on much smaller, ~1 *l* volume, 12 bar CO_2 amplifiers is currently under development at the Naval Research Laboratory [5].

A presently operational single-pulse, multi-TW LWIR laser system at BNL's Accelerator Test Facility (ATF) features a sub-picosecond front end based on a solid-state

* Corresponding author: polyanskiy@bnl.gov

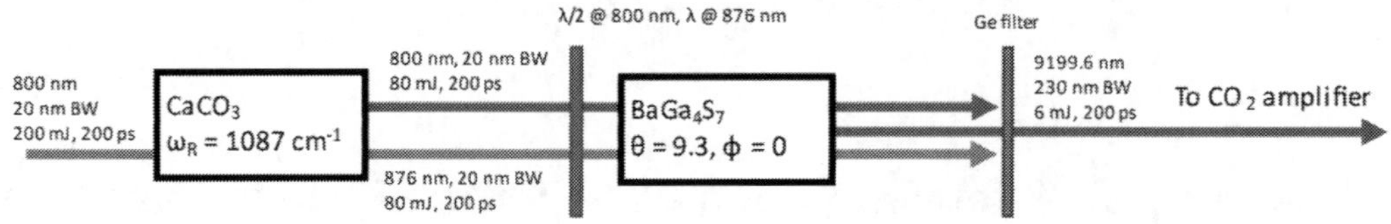

Fig. 2. Proposed seed generation scheme. A stretched 800 nm TiS laser pulse pumps a Calcite crystal to generate an 876 nm seed by stimulated Raman scattering. Polarization state of two NIR pulses is adjusted by a dual-wavelength waveplate. LWIR pulse is produced via type-I difference frequency generation in BGS and separated from the NIR pump by a Ge filter.

laser with an optical parametric amplifier that seeds a pulse into a series of two 8.5-10 bar, mixed-isotope CO_2 amplifiers with 0.16 *l* and 8.5 *l* active volumes. A 5 TW peak power in a single 2 ps pulse has been demonstrated with this system [6] operated at up to one shot per 20 seconds repetition rate. This laser system is configured for a chirped-pulse amplification (CPA) regime to avoid detrimental wave front distortions at the output amplifier window. Much higher, up to kHz, repetition rates might be possible from such laser systems [7].

4 Quest for few-cycle, multi-TW LWIR pulses

The near-term goal of the ongoing research at ATF is achieving a 10 TW LWIR peak power in sub-picosecond pulses. In long term, few-cycle pulses (optical cycle duration at 9.2 μm is 31 fs) with peak powers approaching 100 TW is a challenging but not an unrealistic goal. Numerical modelling and initial proof-of-principle experiments show that the ultimate reduction of the pulse duration to ~100 fs is possible by combining two techniques:

1) Amplification of a ~500 fs pulse using a combination of the *9P* and *9R* branches of the gain spectrum [8], and

2) Post-compression of the pulse [9].

4.1 Strong broad-band seed

To achieve amplification of a broad-bandwidth pulse to multi-joule energies while avoiding substantial gain narrowing, total amplification in the CO_2 amplifiers must be kept at a relatively low, $\sim 10^3$ level. Thus, the energy of the sub-picosecond seed pulse must be in the order of several millijoules. We are presently investigating a possibility for generating such a seed pulse using Raman conversion of a near-infrared pulse in a calcite ($CaCO_3$) crystal followed by difference frequency generation between the fundamental and Raman-shifted frequencies according to the scheme presented in Fig. 2.

Figure 3 shows the results of numerical modeling where a 300 fs, 10 mJ pulse at 10.3 μm central wavelength is amplified in the final ATF amplifier in CPA regime. The output pulse comprises a 500 fs leading peak containing 75% of the energy followed by a small secondary peak at an 1 ps delay. The four-fold shortening of the pulse compared to the present configuration results in the corresponding increase in the peak power from 5 TW to 20 TW.

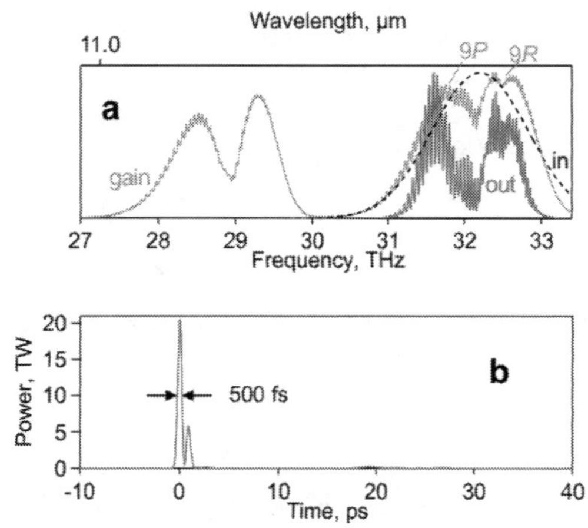

Fig. 3. Calculated spectrum (a, "out") and temporal pulse structure (b) of a 14 J pulse (20 J before compressor) with a 10 mJ, 300 fs pulse seeded into the final amplifier of the ATF's laser system. The spectrum of the seed pulse and a typical gain spectrum of a high-pressure mixed-isotope CO_2 amplifier are also shown in (a) as "in" and "gain," respectively.

4.2 Post-compression

The post-compression techniques for near-IR high-power laser pulses usually involve broadening of the pulse's spectrum via self-phase modulation (SPM) in a nonlinear material followed by the chirp compensation on a chirping mirror. Many relevant materials transparent to the LWIR exhibit a negative group velocity dispersion (GVD) allowing combining the SPM and the dispersive chirp compensation actions in one bulk optical slab [10].

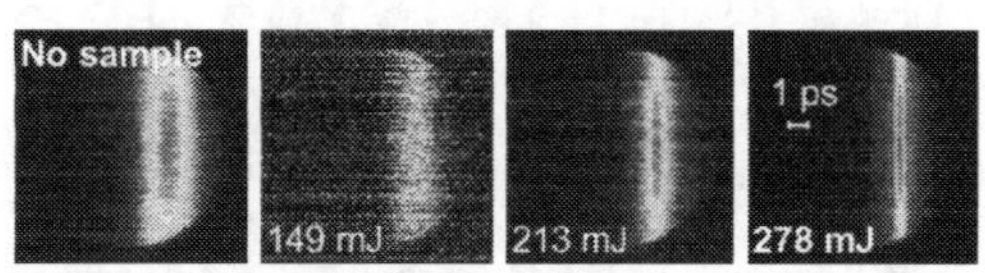

Fig. 4. Raw autocorrelation images of a pulse compressed in a KCl+BaF_2 (each 50 mm thick) post-compressor system at different energies.

We have recently proposed and experimentally demonstrated a modified version of this scheme where a

 https://doi.org/10.1051/epjconf/202125511010

combination of two materials, KCl and BaF_2, with substantially different ratios of nonlinear refractive index to GVD is used to partially decouple the SPM and the de-chirping actions and experimentally demonstrated compression of a 2 ps sub-TW pulse to <500 fs duration (see Fig. 4) [9].

For a 500 fs pulse (Fig. 3b) at the input to a post-compressor, a few-cycle pulse can be expected at the output if the thickness of the elements is properly optimized. Calculations shown in Fig. 5 for input energy fluence of 500 mJ/cm^2 in a flat-top beam and 15 mm thick KCl and BaF_2 plates indicate compression to 50 fs (less than two optical cycles at 9.3 μm) with the peak power exceeding 100 TW.

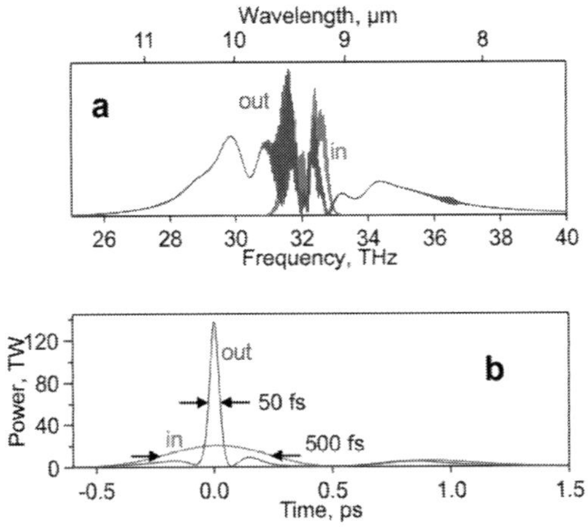

Fig. 5. Calculated spectrum (a) and temporal pulse structure (b) of a pulse before ("in") and after ("out") the post-compression in 15 mm thick KCl and BaF_2 plates. Input pulse corresponds to the 500 fs, 14 J output from the laser amplifier (Fig. 3 "out").

5 Conclusions

Both segments of the roadmap towards a few-cycle, supra-terawatt LWIR pulse, including an energetic sub-picosecond seed generator and a post-compressor, require further research. Once conclusive results are obtained for each of these techniques, they will be combined in one laser system for achieving the ultimate performance.

Note that the prospects towards >100 TW ultra-fast LWIR laser systems described here are based on the parameters of a relatively modestly sized ATF CO_2 laser. Up-scaling of the laser amplifier size and its output energy might allow to approach Petawatt power levels.

Funding

Department of Energy (DOE) Accelerator Stewardship Program grant, B&R #KA2601020. BNL Laboratory Directed Research and Development (LDRD) grant #20-010 AD/NPP.

References

1. D.J. Wilson, A.M. Summers, S. Zigo, B. Davis, S.-J. Robatjazi, J.A. Powell, D. Rolles, A. Rudenko, C.A. Trallero-Herrero, Sci. Rep. **9**, 6002 (2019)
2. K.L. Vodopyanov, *Laser based mid-infrared sources and applications* (Wiley, 2020)
3. M.N. Polyanskiy, I.V. Pogorelsky, V. Yakimenko, Optica 2, 7717 (2011)
4. D. Haberberger, S. Tochitsky, C. Joshi, Opt. Express **18**, 17865 (2011)
5. D.F. Gordon, V. Hasson, H. von Bergmann, Y.H. Chen, A. Schmitt-Sody, J.R. Penano, Proc. SPIE **9835**, 98350Z-1 (2016)
6. M.N. Polyanskiy, I.V. Pogorelsky, M. Babzien, M.A. Palmer, OSA Continuum **3**, 459 (2020)
7. I. V. Pogorelsky, M. Polyanskiy, T. Shaftan, Phys. Rev. Accel. Beams **23**, 120702 (2020)
8. M.N. Polyanskiy, I.V. Pogorelsky, M. Babzien, R. Kupfer, N. Vafaei-Najafabadi, M.A. Palmer, Photonics **8**, 101 (2021)
9. M.N. Polyanskiy, I.V. Pogorelsky, M. Babzien, R. Kupfer, K.L. Vodopyanov, M.A. Palmer, Post-compression of long-wave infrared 2 picosecond sub-terawatt pulses in bulk materials, submitted to Optics Express (2021)
10. B.G Bravy, V.M Gordienko, V.T. Platonenko, Laser Phys. Lett. **11**, 065401 (2014)

3D printing of long period gratings for curvature applications

Nuno F. Valente, *Lúcia* Bilro, and *Ricardo* Oliveira*

Instituto de Telecomunicações and Universidade de Aveiro, Campus Universitário de Santiago, 3810-193 Aveiro, Portugal

Abstract. In this paper we will discuss a new type of long period grating (LPG) that consists on an optical fibre glued on top of a 3D printed grooved plate. The LPG performance will be tested for the case when pressure is applied transversely to the grooved plate. Results showed the appearance of attenuation bands at specific wavelengths, that grow with increasing load. The evolution of the dip power at those resonances, as function of the curvature will be shown, demonstrating the potentialities of the sensor for low cost intensity detection schemes.

1 Introduction

Nowadays the possibility to change the guiding properties of optical fibres is being used in both sensor and communication applications. One way to achieve this effect is by using short and long period gratings (LPGs). These are structures capable to periodically modulate the refractive index of the fibre core. An LPG has typically a period between 100 μm and 1 mm. In this structure the fundamental core mode is coupled to certain cladding modes, causing attenuation bands to appear in the transmission spectrum at certain wavelengths defined by:

$$\lambda_i = (n_e - n_i)\Lambda \qquad (1)$$

where λ_i is the wavelength of the resonance band, n_e is the effective refractive index of the core mode, n_i is the refractive index of mode i^{th} cladding mode, and Λ is the grating period. The light transmission at the resonances could be expressed by the following expression [1]:

$$T = cos^2(k_{co-cl}^{ac} l) \qquad (2)$$

where l is the length of the grating and k_{co-cl}^{ac} is the *ac* component of the coupling coefficient between the core and the cladding modes.

In literature it is possible to find several types of LPGs and two major groups stand out: permanent gratings through laser engraving and mechanically induced through load. For the first, there are several methods to produce them, such as CO_2 laser [2], femtosecond laser [3], ion beam [4], and electric arc discharge [5]. For the later, there are also different procedures to fabricate them, such as: pressing a grooved plate [6,7] or laminated plate [8] onto an optical fibre, and recently, by surface bonding an optical fibre to a grooved plate, namely by torsion and shear-strain forces [9]. In this type of gratings, the refractive index of the fibre core is modulated according to the force applied and may be expressed by [9]:

$$\Delta n = -n_0 p_e \varepsilon_i = -n_0 p_e \frac{F}{E_i A_i} \qquad (3)$$

being, n_0, p_e, ε_i and F, the unperturbed refractive index, the effective photoelastic constant, the associated strain and the applied force, respectively. E and A are the effective Young's Modulus and radius of each region, respectively.

Taking this later example, in this work it will be demonstrated the capability to measure curvature through these fibre optic devices.

2 Methods

2.1 Sensor fabrication

The fabrication of the sensor followed the descriptions found in [9]. For that, the grooved plate was 3D printed using a digital light projector (DLP) 3D printer from Anycubic[10]. The plate has a total of 100 rectangular hollow regions, separated with a period of 650 μm, giving a total grooved plate length of 65 mm. The thickness of the plate was 5 mm and its width was 15 mm. The resin used for the 3D printing was a standard ultraviolet (UV) transparent resin from Anycubic. The computer aided design (CAD) image of the proposed grooved plate may be seen in Fig. 1(a).

The assembly of the fibre structure is done by positioning the optical fibre on top of the grooved plate followed by surface bonding it using the same UV resin that composes the grooved plate. When performing this step, some care must be taken, such as not placing too much glue between the optical fibre and the grooved plate, otherwise it may fill the hollow regions of the plate, taking away its periodicity. Another precaution that must be taken when performing this task is to keep the optical fibre completely stretched during the adhesion process, otherwise it may not be well fixed to the grooved plate. Once this step is completed, the fibre structure is ready to be mechanically characterized in order to induce the LPG.

* Corresponding author: oliveiraricas@av.it.pt

https://doi.org/10.1051/epjconf/202125512001

2.2 Curvature characterization

The mechanical apparatus used for the curvature characterization may be seen on Fig. 1 b) and c).

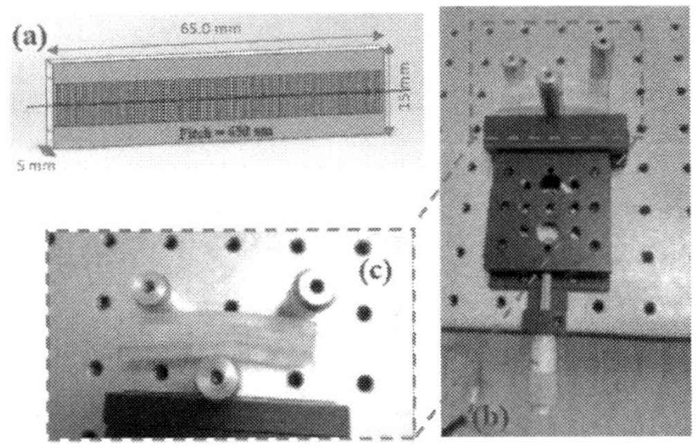

Fig. 1. (a) CAD image of the grooved plate. (b) Experimental setup for the curvature test. (c) Inset of the sensor under curvature.

The characterization setup is based on two mechanical posts tightened to an optical table, allowing to secure the fibre device at both ends. Then, a transversal force is applied at the middle region of the grooved plate, using a micrometer screw (10 μm resolution). The displacement was made in steps of 50 μm up to 300 μm. The curvature was calculated through the following expression [11]:

$$C = \frac{1}{R} = \frac{2h}{h^2 + D^2} \qquad (4)$$

where C is the curvature, R is the radius of curvature, h is the displacement imposed by the micrometer screw and D is the distance between the fixed posts (50.7 mm). Considering the values given for h and D, the curvature was made in steps of 0.04 m^{-1} up to a maximum value of 0.23 m^{-1}.

The curvature characterization was made in transmission, using a wideband light source (Fianium Whitelease model SC-400-2) and an optical spectrum analyzer (Advantest Q8384), is used to collect the grating spectra after each curvature step.

In order to see the influence of the uniaxial strain, we decide to measure the polarization dependent losses (PDL) of the fibre device. To accomplish this task, we used an optical network analyzer (Agilent Technologies 86038b).

3 Results

The spectra obtained for the different steps of curvature may be seen in Fig. 2.

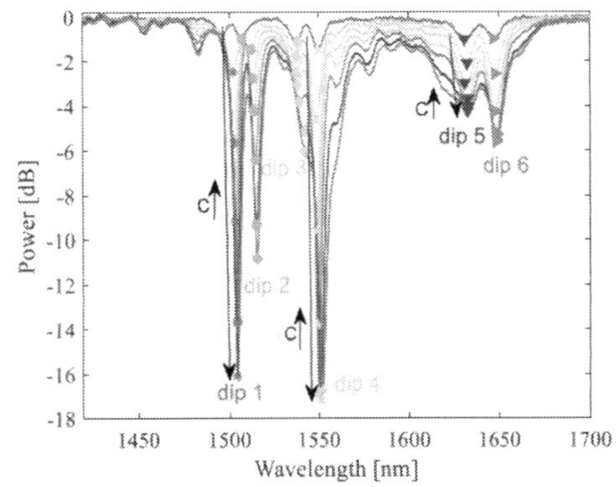

Fig. 2. Transmission spectra as function of the applied curvature.

The results presented on Fig. 2 show the appearance of six dip resonances that increase their attenuation with the increase of curvature value. By tracking the dip attenuation of the resonances that appear in Fig. 2 we were able to represent the evolution of the coupling strength as function of the applied curvature, as is shown in Fig. 3.

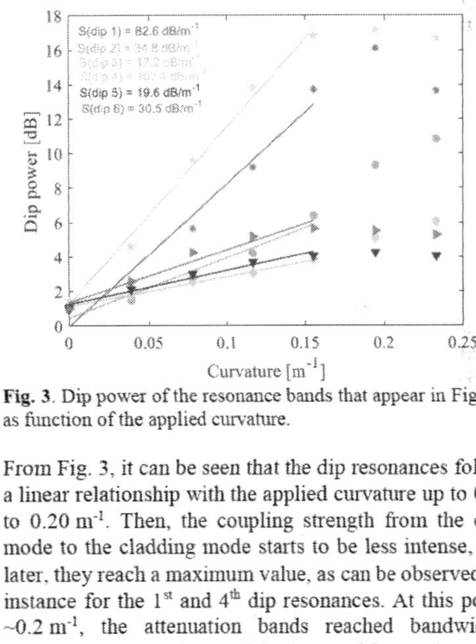

Fig. 3. Dip power of the resonance bands that appear in Fig. 2, as function of the applied curvature.

From Fig. 3, it can be seen that the dip resonances follow a linear relationship with the applied curvature up to 0.15 to 0.20 m^{-1}. Then, the coupling strength from the core mode to the cladding mode starts to be less intense, and later, they reach a maximum value, as can be observed for instance for the 1st and 4th dip resonances. At this point, ~0.2 m^{-1}, the attenuation bands reached bandwidths ≤ 4 nm and relatively small background losses (< 1 dB). After reaching the minimum attenuation value, ~16 dB to ~17 dB, respectively for the 1st and 4th dip resonances, the transmission starts to increase again, where the coupling is now reversed, being the cladding modes coupled to the core mode. The process continues up to maximum transmission (minimum attenuation). However, we stopped the process in order to prevent plastic deformation of the polymeric structure. Nevertheless, it is expected that the attenuation bands present an oscillation between maxima and a minima values, as indicated by the cosine function found in equation (2). In Fig. 3 it is also possible to observe that the sensitivities of the attenuation bands as function of curvature are 82.6 dB/m^{-1}, 34.8 dB/m^{-1}, 17.2 dB/m^{-1}, 102.4 dB/m^{-1}, 19.6 dB/m^{-1} and 30.5 dB/m^{-1} for the 1st, 2nd, 3rd, 4th, 5th and 6th dip resonances, respectively. These values were slightly lower than those found in literature (that is about 130.4 dB/m^{-1}) for other fibre optic structure based on multimode interference [12]. Despite this, we stress out that the sensitivities are still high compared to the state of the art and that the fabrication methodology used in this work is simpler and cost effective. Furthermore, we observed that the resonance bands have a wavelength shift of less than 1.5 nm during the linear tuning range. This low value could be an interesting opportunity to build a fibre device operating with low cost approaches, using a laser source cantered at one of the resonance bands and using a photodetector to monitor the power during time [9].

In order to compare the results obtained in Fig.2 with the ones predicted theoretically, we decide to run a simulation using the OptiGrating v4.2 software (Optiwave Corporation). The simulations were performed considering the first three linearly polarized modes (LP modes), an estimated refractive index of 1.44912 and 1.44400, and diameters of 8.2 µm and 125.0 µm, for the core and cladding, respectively. The grating period was set to 650 µm and the refractive index modulation was optimised in order to fit the spectral profile obtained experimentally. The results of this simulation, together with the ones obtained experimentally for a curvature of 0.19 m^{-1} are shown in Fig. 4

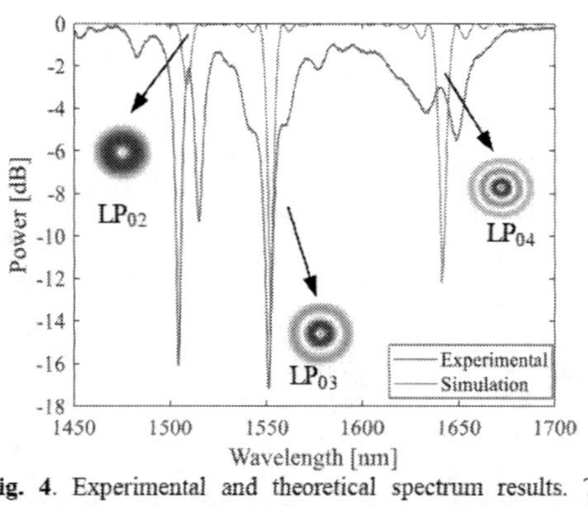

Fig. 4. Experimental and theoretical spectrum results. The images indicate the corresponding modes for which the coupling has occured.

As is seen in Fig. 4, the simulation results revealed three dip resoance bands located at ~1497 nm, ~1540 nm and ~1628 nm, that corresponde to the LP_{02}, LP_{03} and LP_{04} modes, respectively. Yet, each of these resonaces are split into two on the experimental results. However, we stress out that the result observed experimentaly is an indicative of the different refractive index modulation observed in both orthogonal polarizations. To deeply explore this effect we decide to measure the PDL of the fibre device. The results may be seen in Fig.

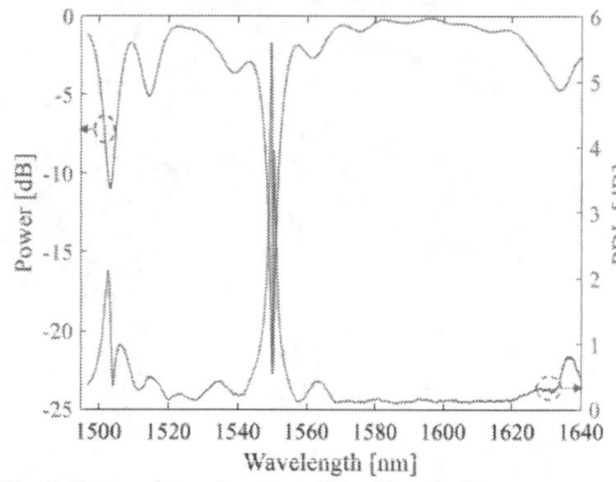

Fig. 5. PDL and insertion loss of the fibre device.

In the graph of 5 it is possible to observe the existence of a high PDL value, reaching values in the order of ~5.6 dB. This happens because asymmetric stresses are applied to the optical fibre bonded to the grooved plate, creating different refractive index variations in the two orthogonal axes of the fibre, which leads to the appearance of two attenuation bands in the transmission spectrum. This difference in modulation of the refractive index is due to the geometry of the fibre device, because, the fibre is surface bonded to the grooved plate and this restricts the fibre core to be positioned at 2.563 mm out of the central region (neutral region), considering half of the thickness of the grooved plate and half of the diameter of the optical fibre. Note that if the fibre is exactly at the middle region of the grooved plate, no strain would be induced at the core of the optical fibre. Because of that, an asymmetric strain distribution is created when the fibre is being curved, being one side of the fibre compressed while the other is being elongated. As a result, different refractive index modulations are built and thus, different coupling strengths could be found for the two orthogonal polarizations. This effect could in fact be detrimental in different applications and to solve this issue we propose the use of a polariser.

4 Conclusion

In this work, a low-cost LPG curvature sensor is proposed. The spectral results showed attenuation bands up to 17.2 dB, located at specific wavelengths that satisfy the resonance condition, presenting also low losses outside of the resonance bands (1.2 dB). The results revealed that the dip attenuation bands reach a linear sensitivity with increasing curvature, reaching values up to ~102.4 dB/m^{-1}, which is a high value compared to the state of the art. Besides that, the low cost and simplicity of the proposed fabrication process, makes this fibre sensor well suited for curvature sensing applications. Furthermore, it is worth to mention that the device could be used with simple detection schemes such as the ones based in intensity, by using a laser centred at one of the resonance bands and monitoring the power variation through a photodetector, allowing to reduce the cost and the complexity of the final device.

This work was funded by FCT-Fundação para a Ciência e Tecnologia and by the European Regional Development Fund (FEDER), through the Competitiveness and Internationalization Operational Programme (COMPETE 2020) of the Portugal 2020 framework and by national funds under the projects FOPEComSens (PTDC/EEI-TEL/1511/20), AQUATICsens (POCI-01-0145-FEDER-032057) and UIDB/50008/2020-UIDP/50008/2020.

References

[1] T. Erdogan, "Fiber grating spectra," *J. Light. Technol.*, vol. 15, no. 8, pp. 1277–1294, 1997, doi: 10.1109/50.618322.

[2] J. M. P. Coelho, C. Silva, M. Nespereira, M. Abreu, and J. Rebordão, "Writing of Long Period Fiber Gratings Using CO2 Laser Radiation," *Adv. Opt. Fiber Technol. Fundam. Opt. Phenom. Appl.*, vol. 32, pp. 287–314, Feb. 2015, doi: 10.5772/59153.

[3] B. Li, L. Jiang, S. Wang, H. L. Tsai, and H. Xiao, "Femtosecond laser fabrication of long period fiber gratings and applications in refractive index sensing," *Opt. Laser Technol.*, vol. 43, no. 8, pp. 1420–1423, 2011, doi: 10.1016/j.optlastec.2011.04.011.

[4] M. L. von Bibra, A. Roberts, and J. Canning, "Fabrication of long-period fiber gratings by use of focused ion-beam irradiation," *Opt. Lett.*, vol. 26, no. 11, pp. 765–767, 2001, doi: 10.1364/ol.26.000765.

[5] C. Colaço, P. Caldas, I. Del Villar, R. Chibante, and G. Rego, "Arc-induced long-period fiber gratings in the dispersion turning points," *J. Light. Technol.*, vol. 34, no. 19, pp. 4584–4590, 2016, doi: 10.1109/JLT.2016.2540678.

[6] R. Oliveira, L. M. Sousa, A. M. Rocha, R. Nogueira, and L. Bilro, "UV inscription and pressure induced long-period gratings through 3D printed amplitude masks," *Sensors*, vol. 21, no. 6, 2021, doi: 10.3390/s21061977.

[7] S. H. Shi, X. J. Zhou, Z. Y. Zhang, L. Lan, C. Yin, and Y. Liu, "Fiber-optic solution concentration sensor based on a pressure-induced long-period grating in a composite waveguide," *Chinese Phys. Lett.*, vol. 29, no. 9, 2012, doi: 10.1088/0256-307X/29/9/094206.

[8] I. Torres-Gómez, D. E. Ceballos-Herrera, and K. M. Salas-Alcantara, "Mechanically-induced long-period fiber gratings using laminated plates," *Sensors*, vol. 20, no. 9, 2020, doi: 10.3390/s20092582.

[9] R. Oliveira, R. Nogueira, and L. Bilro, "3D printed long period gratings and their applications as high sensitivity shear-strain and torsion sensors," *Opt. Express*, vol. 29, no. 12, pp. 17795–17814, 2021, doi: 10.1364/oe.427387.

[10] "ANYCUBIC 3D Printing." https://www.anycubic.com/ (accessed Jul. 30, 2021).

[11] Y. Liu, L. Zhang, J. A. R. Williams, and I. Bennion, "Optical Bend Sensor Based on Measurement of Resonance Mode Splitting of Long-Period Fiber Grating," *IEEE Photonics Technol. Lett.*, vol. 12, no. 5, pp. 531–533, 2000, doi: 10.1109/68.841276.

[12] Y. Gong, T. Zhao, Y. J. Rao, and Y. Wu, "All-fiber curvature sensor based on multimode interference," *IEEE Photonics Technol. Lett.*, vol. 23, no. 11, pp. 679–681, 2011, doi: 10.1109/LPT.2011.2123086.

Accommodative Lag by Open-field Autorefractor and Hartmann-Shack Aberrometer

Jessica Gomes, Kishor Sapkota, Patrícia Nogueira and Sandra Franco[1,*]

[1]Centre of Physics, University of Minho, Campus de Gualtar, 4710-057, Braga, Portugal

Abstract. The purpose of the present study was to compare an open-field autorefractor (AR) and an aberrometer for measuring the ocular accommodative lag. The measurements were. It was found higher accommodative lags when measured it with the AR specially for high accommodative stimuli. However, the differences in accommodative lag between the two instruments were not statistically significant (p >0,05) and were under the limits of agreement The results indicate that aberrometer may be used for measuring the accommodative lag and may be more efficient for measuring accommodative lag for high accommodative stimuli.

1 Introduction

Ocular accommodation is the process in which the crystalline lens changes its power to focus on the retina images of objects at different distances.[1]
One of the methods used to identify problems in the ocular accommodation is measuring the difference between the accommodative stimulus, which is presented to an eye, and the accommodative response of the subject, which can lead to a lag or lead of accommodation. This difference is typically between +050 D and +0.75 D. However, in the presence of an accommodative insufficiency, this value is exceeded; whereas, inferior values could indicate an accommodative excess.[1]
Accommodation response can be measured indirectly by objective methods, such as dynamic retinoscopy, and by autorefraction, aberrometry, and double-pass systems.
Dynamic retinoscopy is the method most used in clinical practice to measure the accommodative lag; however, as it is considered partially subjective [1], autorefraction and aberrometry are more used in research. [2,3] Moreover, some authors claim that subjective techniques overestimate the accommodative amplitude relative to objective methods. Whereas, autorefraction was considered a suitable instrument for measuring accommodation. [4]
Divers autorefractors were compared among themselves for measuring accommodative response, and showed no significative differences for stimulus lower than 5.00 D. [5] Furthermore, autorefraction is comparable to dynamic retinoscopy for measuring accommodative response [6], but not with minus lens technique for assessing accommodative amplitude [7].
Aberrometry has been used for determining accommodative response, and showed to be a reliable tool. [3,8]
Aberrometer and autorefractor were only compared for measuring refractive error, showing a spherical equivalent mean difference of 0.25 D. [9]
Given the importance of having precise methods to analyse the accommodative response and lag, the purpose of this study was to compare the accommodative lag measurements obtained with an open-field autorefractor and an aberrometer.

2 Methods

2.1. Subjects

Eight subjects (five females and three male) aged between 23 and 40 years-old were recruited from University of Minho.
The mean spherical equivalent was -0.16 ± 0.39 D with astigmatism less than 1D, and had no binocular and accommodative dysfunctions. Patients with any ocular pathology, previous ocular surgery or taking medication that could affect vision were excluded.
An informed consent was obtained from each subject after providing a verbal explanation of the nature and possible consequences of the study. The study was approved by the Ethics Subcommittee for the Life Sciences and Health of University of Minho.

2.2. Procedure

Accommodative response was measured with an open-field autorefractor (WAM-5500, Grand Seiko Co., Ltd., Japan) (figure 1.A) and an in-house Hartmann-Shack aberrometer (Thorlabs WF150-7AR) (figure 1.B). The values were first taken for the relaxed state, and then negative lenses (-1 D, -2 D, -3 D, -4 D and -5 D) were placed in front of the eye in a trial frame with a vertex

* Corresponding author: sfranco@fisica.uminho.pt

distance of 12mm to stimulate accommodation, while patient was asked to fix a target and keep it as clear as possible. The target was a high contrast black-on- white letter.
Accommodative lag was calculated from the difference between accommodative stimulus and response.

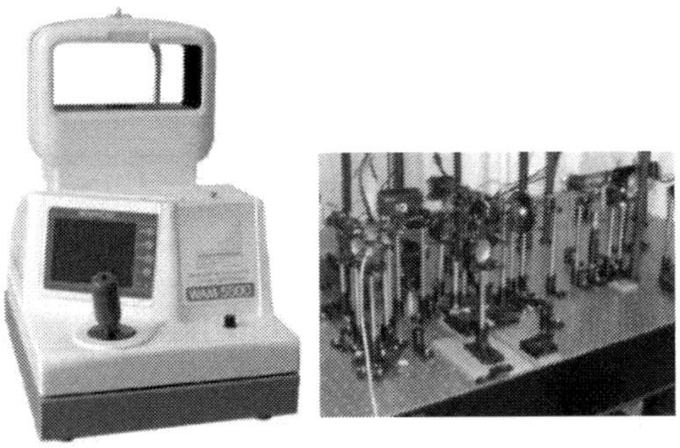

Fig. 1. Open-field utorefractor (WAM-5500, Grand Seiko Co., Ltd., Japan) on the left and Hartmann-Shack aberrometer (Thorlabs WF150-7AR) on the right.

The statistical analysis was performed with the software R 4.1.0. Shapiro-wilk test was used to test the normality of the data and paired t-test or Wilcoxon-test to compare the two related samples for parametric and non-parametric data, respectively. A p-value less than 0.05 was considered statistically significant.

3 Results

The values of accommodative lag/lead obtained from the open-field autorefractor and aberrometer for the different accommodative stimuli can be observed in the figure 2.

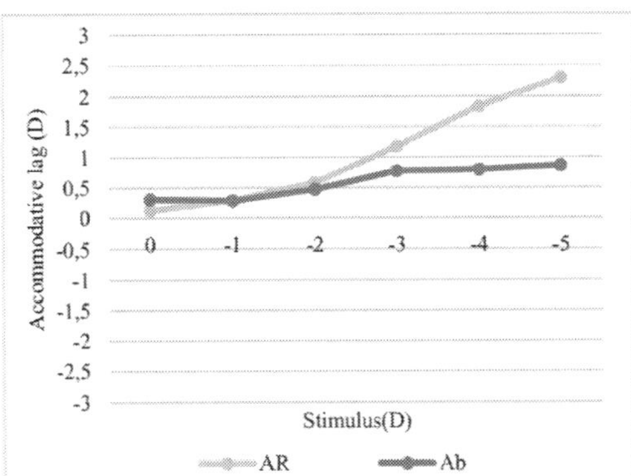

Fig. 2. The accommodative lag measured with open-field autorefractometer (AR) and Hartmann-Shack aberrometer (Ab) for the different lenses (0, -1, -2, -3, -4 and -5 D). Positive values represent accommodative lag and negative values accommodative lead.

The accommodative lag measured with AR presented values tendentially higher than with the aberrometer, especially for high accommodative stimuli (from the lens -3.00 D); However, there was no statistically significant difference for any accommodative stimulus ($p<0.05$).
The values obtained by the AR for the higher stimuli present values above the expected, while those obtained with the aberrometer are within the expected range.
The mean, standard deviation (SD), coefficient of agreement (COA), intra-class correlation coefficient of accommodative lag (ICC), limit of agreement (LA) and p-values of the differences are shown in the table 1.

Table 1. Mean, standard deviation (SD), coefficient of agreement (COA), intra-class correlation coefficient of accommodative lag (ICC), limit of agreement (LA) differences obtained with AR and aberrometer, for different accommodative stimuli.

Lens (D)	0	-1	-2	-3	-4	-5
Mean (D)	-0.183	0.006	0.11	0.406	1.039	1.433
SD (D)	0.766	1.699	1.355	1.227	1.745	2.421
COA (D)	1.501	3.330	2.656	2.405	3.420	4.744
ICC	0.486	0.478	0.454	0.508	0.446	0.430
LA↑ (D)	1.318	3.336	2.766	2.811	4.458	6.177
LA↓(D)	-1.683	-3.324	-2.546	-1.999	-2.381	-3.312
p-value	0.612	0.326	0.825	0.380	0.136	0.138

With Bland-Altman plots it was observed that all values were within the limits of agreement, except one, which were out of the limits with the lens -1.00 D and -2.00 D. Figure 3 shows Bland-Altman plots analysing the agreement of the measurements obtained with the two instruments for the relaxed state and for 3 D of accommodation.

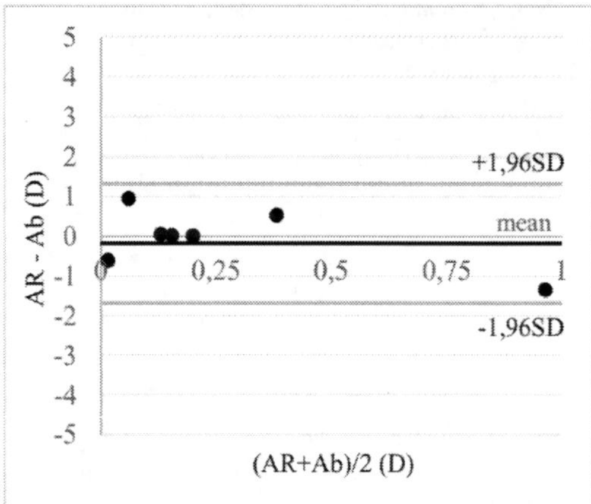

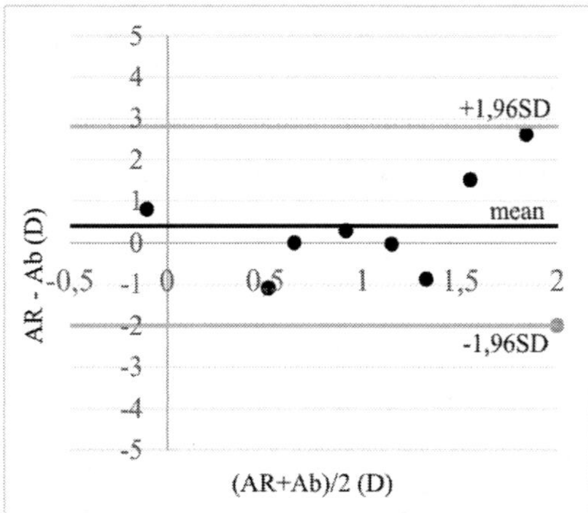

Fig. 3. Bland-Altman plot comparing the measurements of accommodative lag in the relaxed state (up) and with an accommodative stimulus of 3D (down) obtained with the open-field autorefractor (AR) and with the Shack-Hartmann aberrometer (Ab). In the x axis is represented the average of the measures of these two instruments and in the y axis the difference between them in diopters.

4 Conclusion

There are no significant differences between open-field autorefractor and Shack-Hartmann aberrometer for measuring accommodative lag until 5D, although the open-field autorefractor presented values tendentially higher than with Shack-Hartmann aberrometer, mainly for high accommodative stimuli.
These preliminary results indicate that Shack-Hartmann aberrometer appears to be more effective for measuring accommodative lag in high accommodative demands than the open-field autorefractor.
In the future, it is intended to increase the sample size.

5 Acknowledgments

This work was supported by the Portuguese Foundation for Science and Technology (FCT) in the framework of the Strategic Funding UID/FIS/04650/2019 and by the project PTDC/FIS-OTI/31486/2017 and POCI-01-0145-FEDER-031486.

References

[1] B. W. Borish's, *Clinical Refraction-E-Book*. 2nd ed. 2006.

[2] M. M. . McBrien NA, Clinical evaluation of the Canon Autoref R-1. Am J Optom Physiol Opt., **62**, no. 786Y92 (1985).

[3] R. E. T. Ang, J. A. S. Sarmiento, J. T. M. Remo, G. H. A. Martinez, and L. M. B. Canilao, Measurement of Accommodative Amplitude Using Wavefront Aberrometer. **June**, 3–10, (2015)

[4] D. Win-Hall. Open-field autorefractor measurements of accommodation in human subjects. OVS, Poster **44**, Scientific program (2004)

[5] M. Aldaba, S. Gómez-López, M. Vilaseca, J. Pujol, and M. Arjona. Comparing autorefractors for measurement of accommodation. Optom. Vis. Sci., **92**, 10, 1003–1010 (2015)

[6] P. S. Krishnacharya. Study on accommodation by autorefraction and dynamic refraction in children. J. Optom. **7**, 4, 193–202 (2014)

[7] Y. Chen *et al.* Comparison of three monocular methods for measuring accommodative stimulus–response curves. Clin. Exp. Optom., **100**, 2, 155–161 (2017)

[8] J. Tarrant, A. Roorda, and C. F. Wildsoet. Determining the accommodative response from wavefront aberrations. J. Vis. **10**, 5, 1–16 (2010)

[9] A. A. Martinez, A. Pandian, P. Sankaridurg, K. Rose, S. C. Huynh, and P. Mitchell. Comparison of aberrometer and autorefractor measures of refractive error in children. Optom. Vis. Sci., **83**, 11, 811–817 (2006)

 https://doi.org/10.1051/epjconf/202125512003

A comprehensive transmedia activity to commemorate the International Day of Light 2021 on social networks

Carlos Damián Rodríguez-Fernández[1,*], *Bastián* Carnero[1], *Mónica* Canabal[1], *Damián* Ínsua[1], *Alejandro* Doval[1], *Alicia* Muñoz[1], *Raúl* Sánchez[1], *Rocío* Liñares[1], *Xoán* González[1], and *M.T.* Flores-Arias[1]

[1]USC-OSA Student Chapter and USC-EPS Young Minds Section, Facultade de Óptica, Universidade de Santiago de Compostela, E-15782 Santiago de Compostela, Spain

Abstract. Since the advent of the COVID-19 pandemic, face-to-face outreach has been gradually replaced by dissemination through social networks. However, the use of social networks for outreaching is not easy since there is an important risk of biasing the communication by converting it into an exclusively one-way channel. In this contribution, we explain how a transmedia perspective can transform a regular activity, as a photo contest for the International Day of Light 2021, into an interactive, comprehensive and functional event in Twitter, Instagram and Twitch.

1 Introduction

Dissemination is a fundamental instrument to explain what scientific research is about, but also to raise awareness about the important role that science plays in our lives. In this context, a number of national and international scientific societies decided to support student chapters, groups of students devoted to dissemination. The USC-OSA/USC-EPS is a student chapter composed by physics students from the Universidade de Santiago de Compostela (USC) funded both by the Optical Society (OSA) and the European Physics Society (EPS), which has been active since 2013. Our outreach has been historically focused on raising awareness about the importance of light through face-to-face interactive activities with school and high school students [1, 2]. However, due to the advent of the COVID-19 pandemic, typical outreach activities were no longer allowed, so we decided to migrate our activity to social networks (SNs).

In this contribution, we present and analyze a comprehensive transmedia activity to commemorate the International Day of Light (IDL) 2021, entitled *The bright side of everyday life*. This activity involved the use of three distinct SNs: Twitter, Instagram and Twitch, each one selected to host a specific part of the event in accordance with their own distinguishing features.

2 Dissemination by SNs and transmediality

SNs are online environments that people use to interact with other people with similar interests and to develop social relationships. It is estimated that, currently, there are more than 4.20 billion of accounts in SNs worldwide, a figure which roughly corresponds to 54% of the global population [3]. During the pandemic, many outreachers have migrated their face-to-face outreach to SNs, which ensure access to a large number of people while guaranteeing the required social distance. However, using SNs for outreach is not easy because of the limitations of each platform and the inherent risk of biasing the communication by converting it into a unidirectional channel. These drawbacks can be avoided by opting for a transmedia approach [4]. It consists in involving people in an event that is held on different platforms at different times and at different stages. Note that it does not mean to do the same thing on different SNs but to do different actions on each platform as part of a whole. This strategy is very convenient since it allows taking the particular advantages of each platform. However, in order to make it work, the events held in the different platforms must have a common narrative, since only when all the pieces match together a transmedia activity is successfully addressed.

3 *The bright side of everyday life*

The bright side of everyday life is a transmedia activity devoted to commemorate the IDL 2021. The central narrative of the event is a photography contest which looks for the best picture of an optical phenomenon in daily life. From a regular perspective, this activity could be kept in locally spreading the contest, choosing a jury and announcing the winning picture. However, by using a transmedia approach, the photo contest becomes just one piece of a larger event which can be divided in three parts: active promotion, interactive photography contest and awarding by outreaching, each one held in a different SN.

*e-mail: usc.osa.sc@gmail.com

3.1 Active promotion: Twitter

Twitter was chosen as the main platform to actively promote our activity. Twitter is a microblogging SN where users can interact through 280 character-limited messages known as tweets, which also can feature pictures. These tweets can be concatenated to create threads, small segments of information that can be easily consumed and shared. This functionality was exploited to announce and update all the details in relation to the photography contest. Specifically, we created an appealing initial tweet together with a promotional poster (Fig. 1) which worked as a header of a thread that summarized the most relevant aspects of the contest. This thread was periodically quoted and updated to keep the contest up to date for our followers. A final mention must be made regarding language, which must be close and inclusive. This helps people to engage with the activity, which is important for a transmedia event.

Figure 1. . Promotional poster of *The bright side of everyday life* transmedia activity. This poster was uploaded to our Twitter profile as a header of a thread where we summarized the most relevant aspects of the contest.

3.2 Interactive photography contest: Instagram

Instagram was selected for holding this part of the event since it is the most relevant social network of those devoted to photo sharing and it counts with a large number of users. For the photography contest, we asked for pictures of optical phenomena that could be observed in everyday life together with a brief scientific description of them. In a first round, a jury constituted by professors graded the pictures according to their quality and to the accuracy of the given description. The top ten pictures were published on our profile and the winners were chosen by popular vote. The same day that the voting ended, the pictures were hidden to the public to announce the definitive winners in a specific award ceremony. The implemented two-round vote system has two main benefits. First, the jury vote ensures that the finalist pictures have a high scientific and artistic quality. Second, it generates engagement in our public since our followers are given an active role in the contest. This was especially clear for the authors of the finalist pictures, who competed between them to attract attention, often involving friends and relatives in the event. This dynamic is highly beneficial since it allows reaching people that were a priori not interested in optics, spreading the importance of light. In our case, the finalist photographs collected together thousands of votes and our number of followers increased by 300.

3.3 Awarding by outreaching: Twitch

The awarding ceremony was organized on Twitch, a cutting-edge SN focused on live streaming, a format which creates a very close feeling in the viewer. Instead of designing a standard award ceremony where the winners are directly announced, the occasion was used to go a step further and disseminate optics. During the live streaming, a group of five members of the chapter reviewed the ten finalist photos (Fig. 2a), analyzing, discussing and explaining the optical phenomena behind each picture in a relaxed talk format (Fig. 2b). The explanation of the photos was previously prepared in detail by the members of the group, so that the phenomena could be explained rigorously.

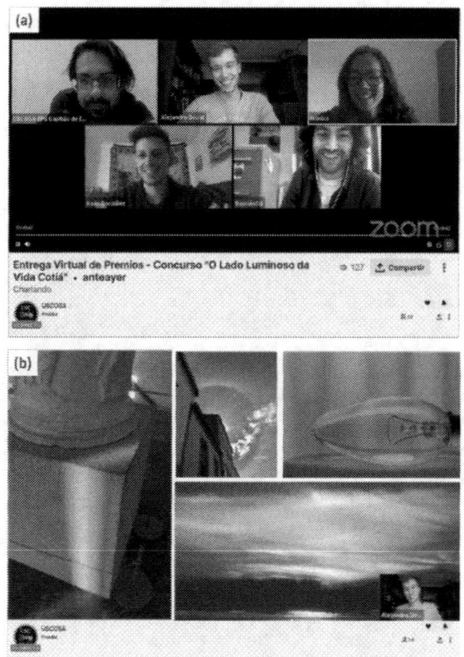

Figure 2. Awarding ceremony of *The bright side of everyday life* transmedia activity held in Twitch. a) Five members of the chapter participated in the ceremony where b) the analysis, discussion and explanation of the optical phenomena behind each picture was performed in a relaxed format.

To finish, the winner of the contest joined the streaming as a special guest to explain the principal aspects of her photography and to discuss the optics behind. The live award ceremony was followed by about 30 people, however, afterwards, it was watched in delay up to 150 times, which is a clear indicator of the interest generated by our outreach-oriented award ceremony.

 https://doi.org/10.1051/epjconf/202125512003

4 Conclusions

Transmediality goes beyond outreaching by using different social networks but creating an interactive event with a coherent narrative that flows through different platforms. This approach allows enhancing the engagement of the public by exploiting the specific strengths of each social network while getting rid of their cons. Applying this philosophy, we converted a standard photo contest into a comprehensive experience on Twitter, Instagram and Twitch, including a fluid interaction with our followers and a close but rigorously outreaching award ceremony.

References

[1] A.I. Gómez-Varela, A. Gargallo, H. González Núñez, T. Delgado-García, C. Almaguer-Gómez, F. Cambronero-López, M.T. Flores-Arias, *The USC-OSA Student Chapter: goals and benefits for the optics community*, in *12th Educ. Train. Opt. Photonics Conf.*, edited by M.F.P.C. Martins Costa, M. Zghal (2014), Vol. 9289, p. 92892T, ISBN 9781628413649, ISSN 1996756X

[2] A. Gargallo, A.I. Gómez-Varela, H. Gónzalez-Nuñez, T. Delgado, C. Almaguer, F. Cambronero, Á. García-Sánchez, D. Pallarés, M. Aymerich, Á.L. Aragón et al., J. Phys. Conf. Ser. **605**, 012040 (2015)

[3] Hootsuite & We Are Social, *Digital 2021 Global Digital Overview* (2021)

[4] L. Fleming, J. Media Lit. Educ. **52**, 370 (2013)

Evaluation of the best solution for the functionalization of photocatalytic, superhydrophobic, and self-cleaning properties on asphalt mixtures

Iran Rocha Segundo[1,7*], *Salmon* Landi Jr.[2], *Elisabete* Freitas[3], *Cátia* Afonso[4], *Manuel* F. M. Costa[5], *Behzad* Zahabizadeh[3], *Vítor* M.C.F. Cunha[3], *Graça* Soares[6], *Jorge* Santos[6], *Vasco* Teixeira[7], and *Joaquim* Carneiro[7]

[1]Civil Engineering Department, University of Minho, Azurém, Guimarães, Portugal
[2]Federal Institute Goiano, Rio Verde – GO, Brazil
[3]University of Minho, ISISE, Department of Civil Engineering, Guimarães, Portugal
[4]IPC - Institute for Polymers and Composites, University of Minho, Guimarães, Portugal
[5]Centre of Physics, University of Minho, Gualtar Campus, Braga, Portugal
[6]Textile Engineering Department, University of Minho, Azurém Campus, Guimarães, Portugal
[7]Centre of Physics, University of Minho, Azurém Campus, Guimarães, Portugal

Abstract. This research aims to develop asphalt mixtures with new capabilities, namely photocatalytic, superhydrophobic, and self-cleaning capabilities. Different solutions were prepared combining nano-TiO_2 and micro-PTFE on water, ethyl alcohol, and dimethyl ketone with different concentrations. The solutions were sprayed over the asphalt mixtures, and the photocatalytic and superhydrophobic capabilities were evaluated via decolourization of Rhodamine B dye aqueous solution under simulated solar irradiation and Water Contact Angle (WCA) measurements, respectively. Among the combinations analysed, the best solution was TiO_2-PTFE with ethyl alcohol (8 g/L), which provided a decolourization of about 60% after having spent 8 h in light exposure conditions and a superhydrophobic surface with WCA equal to 153°. Thus, this research work aims to contribute to advances in the field of clean technologies, assisting in the transition to the "Green Recovery", which is environmentally friendly, inclusive, and sustainable.

1 Introduction

Currently, several research works have contributed to advances in clean technologies, assisting in the transition to an innovative socio-economic model identified as "Green Recovery", which is environmentally friendly, inclusive, and sustainable. In this sense, the coatings composed of nano/microparticles applied over asphalt mixtures have been provided new capabilities, such as: photocatalytic, to photodegrade pollutants for air cleaning; superhydrophobic, to improve road safety; and self-cleaning, to avoid slipping problems [1,2]. In general, the photocatalytic capability is developed on asphalt mixtures by the application of semiconductor materials, mainly titanium dioxide (TiO_2), by using four significant methods: spraying coating, bitumen modification, volume incorporation, and spreading [3]. In addition to being actual recipients of substantial quantities of nanomaterials, these functionalized asphalt mixtures also act as essential tools for leveraging the emerging industrial segment dedicated to producing materials in the form of nanoparticles.

Road pavements are ideal infrastructures to reduce air pollution levels due to their enormous surface area and their proximity to the pollutant gases (mostly from vehicles), especially in urban areas [4]. When functionalized with semiconductor materials, the photocatalytic process on roads is initiated by the electron's excitation from the valence band to the conduction band via photons with proper energy. Then, charge carriers are formed (electrons e^- and holes h^+). They can react with oxygen and humidity, generating free radicals which can degrade, for example, NO_x and SO_x.

The functionalization process also endows the material self-cleaning properties, which are related to the degradation of organic compounds, such as oils and greases. This effect can mitigate the decrease of friction caused by these pollutants' contamination over the road surface [5].

On the other hand, the benefits of providing the superhydrophobic capability to asphalt mixtures are mainly to increase water repellence and prevent ice formation. Therefore, the decrease of friction caused by water/ice over the road surface can be reduced. In this specific case, the self-cleaning effect is based on water droplets' ability to carry dirt particles. For this purpose, the literature refers to the application of nano/micromaterials (with a grain size between 6 nm and 45 μm) over asphalt mixtures, namely: TiO_2, ZnO [6], modified-SiO_2 [7], modified fluorine-containing polymer

* Corresponding author: iran_gomes@hotmail.com

 https://doi.org/10.1051/epjconf/202125512004

with nano-CaO [8], Magnesium–Aluminum layered double hydroxides [9], and PTFE [10]. Usually, for this specific functionalization process, solutions (or dispersions) with these particles are prepared and sprayed over the asphalt mixture surface layer [6,10,11].

Thus, the main objective of this research work is to develop asphalt mixtures with photocatalytic, superhydrophobic, and self-cleaning capabilities. To achieve it, different types of solutions were prepared and sprayed over the asphalt mixture AC 10, namely with TiO_2 and Polytetrafluoroethylene (PTFE) using water, ethyl alcohol, and dimethyl ketone.

2 Materials and Methods

PTFE microparticles with the size of 1 µm were purchased from Merck (Algés, Portugal) while TiO_2 nanoparticles (80% anatase and 20% rutile, particle size is of about 23 to 28 nm) were acquired from Quimidroga (Barcelona, Spain).

First, different types of solutions were prepared with nano-TiO_2 and micro-PTFE with water, ethyl alcohol, and dimethyl ketone using two concentrations (4 and 8 g/L). It was not possible to prepare aqueous solutions with PTFE since the particles cannot be dispersed into this medium.

The method of application of the particles was by spraying the solutions over the surface of a conventional asphalt mixture already compacted. The solutions were sprayed over the asphalt mixture samples (8 mL/cm^2 at room temperature) with the dimensions of 25×25×15 mm^3 in order to choose the best solution. The asphalt mixture samples were named by an alphanumeric string, which starts by AC 10 to indicate the asphalt concrete, the particle used (TiO_2 and/or PTFE) and then the solvent used (W – water, ETH – ethyl alcohol, and CET – dimethyl ketone), respectively. If both particles, nano-TiO_2 and micro-PTFE, were combined, at the end of the string, the concentration of the solution was inserted, for example, AC 10 TiO2PTFE-ETH-8g/L. In this case, 4 g/L and 8 g/L represent 2 g/L and 4 g/L of each solute, respectively. In case of the solutions with just one type of particles, the concentration was 4 g/L.

The photocatalytic efficiency was evaluated via Rhodamine B (RhB) degradation following [12,13]. The asphalt mixture samples were immersed into 30 mL of RhB aqueous solution with a concentration of 5 ppm. The systems were placed inside a box containing a sunlight simulation lamp, located 25 cm above the samples, resulting in a power intensity of 11 W/m^2. First, the samples were kept in dark condition for 3 h and then they were irradiated during 8 h. In this way the adsorption and photocatalysis phenomena could be split and the photocatalytic efficiency could be better analysed. To avoid variations in concentration due to the evaporation of RhB solution, the systems were closed with a transparent plastic film with at least 90% of transmittance (between 292 and 900 nm). The maximum absorption (554 nm^{-1}) of the dye was evaluated as function of the time (using a Shimadzu 3101 PC). The photocatalytic efficiency (Φ) was calculated with the Equation 1, according to [5].

$$\Phi\ (\%) = \left(\frac{A_o - A}{A_o}\right) \times 100 \qquad (1)$$

Where A and A_0 represent the maximum absorbance of RhB solution for time "t" and 0 hour, respectively.

The Water Contact Angle (WCA) test was carried out to evaluate the wettability of the functionalized asphalt mixtures. Using the equipment OCA 15 plus Dataphysics, 3 readings of 5 µL water drops were carried out at 2 samples for 2 minutes, at room temperature and relative humidity, and the arithmetic mean was calculated. Surfaces with WCA higher than 150° present superhydrophobic capability [6,10,11,14].

3 Results and Discussions

Figure 1 shows the results of the photocatalytic essay. A global examination of these results reveals that the samples only with PTFE (AC 10 PTFE-ETH and AC 10 PTFE-CET) presented a photocatalytic efficiency 21% higher than AC 10. In addition, the samples with TiO_2 showed a photocatalytic efficiency at least 38% higher than the AC 10 at the end of the test, with the worst performing results for the sample AC 10 TiO2PTFE-CET-4g/L, which increased the efficiency from 34% to 47% (38% increase), when at least TiO_2 is used.

When ketone was used, the highest efficiencies were achieved with only TiO_2. Besides that, when the ethyl alcohol was used, the best treatment was AC 10 TiO2PTFE-ETH-8g/L, then AC 10 TiO2-ETH and AC 10 TiO2PTFE-ETH-4g/L. The effect of the solvent on the efficiency of samples with TiO_2 in decreasing order of performance was ketone followed by ethyl alcohol and water, with almost the same results.

For the combination of the particles, the increase of the photocatalytic efficiency was 5% for the ethyl alcohol and 15% for the ketone, when the concentration increases from 4g/L and 8 g/L. It can be seen that the best solutions achieved were TiO2-CET and TiO2-PTFE under an ethyl alcohol medium with a concentration of 8 g/L (AC 10 TiO2PTFE-ETH-8g/L).

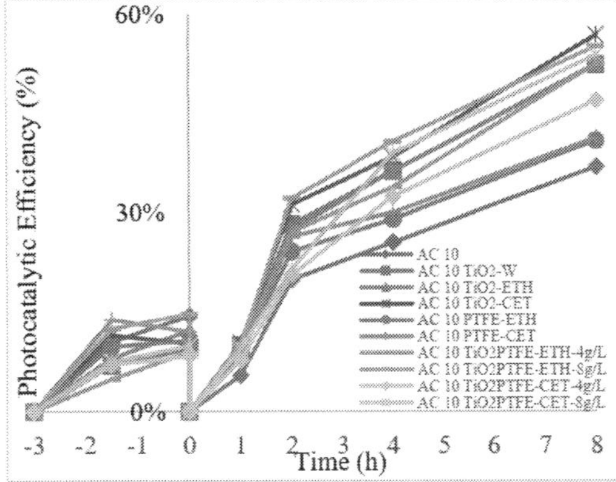

Fig. 1. Photocatalytic efficiency results.

Figure 2 shows the WCA results. All the treated samples increased WCA when compared to AC 10. The increase was at least 11% concerning the AC 10 TiO2-CET sample. All functionalized asphalt mixtures presented a superhydrophobic capability (WCA > 150°), except TiO2-W (145°) and TiO2-CET (130°), which are overhydrophobic surfaces (120° < WCA < 150°) [6,10,11,14]. Thus, the treatments TiO2-ETH, PTFE-ETH, PTFE-CET, TiO2PTFE-ETH, and TiO2PTFE-CET provided superhydrophobic capability.

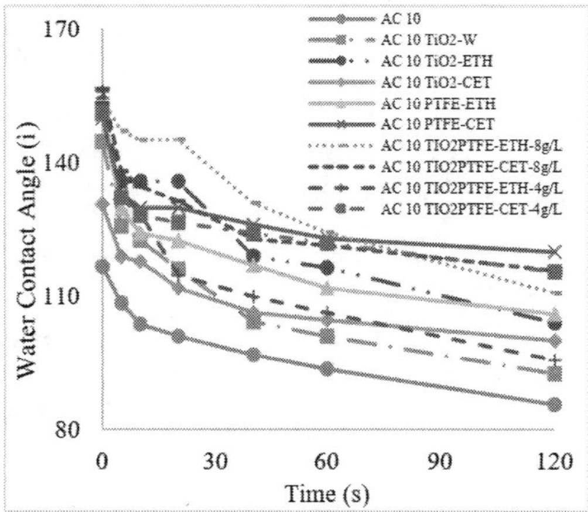

Fig. 2. Water Contact Angle results.

Regarding the effect of the solvents with TiO_2, ethyl alcohol led to the best WCA (153°), followed by water (145°). The worst of them was with ketone (130°), as already mentioned before. For the PTFE, both ketone and ethyl alcohol treatments presented similar initial results, but with time, the best of them was with ketone.

The increase in the concentration of TiO2-PTFE was more effective on the ethyl alcohol. For the whole testing, doubling the concentration leads to an increase of the WCA of 13% and 2% for ETH and CET, respectively.

When the particles were combined, the initial WCA was similar (155°), except for AC 10 TIO2-PTFE-CET-4g/L, that was slightly less (151°). Nevertheless, considering the whole testing duration, it is clear that samples with higher WCA were the AC 10 TIO2PTFE-ETH-8g/L. The average of WCA from 0 to 120 s increase 36% when compared to the reference sample. This same functionalized asphalt mixture showed a WCA at 20 s, and a better performance than all the other samples, that dropped the WCA to at least 131°, along this testing time. As explained before, the best results are represented by the highest initial WCAs.

For the same solvent, solutions with TiO_2 and PTFE and a concentration of 8 g/L delivered the highest WCA. While the results for the solutions with ketone were similar except for TiO_2-CET (the worst of them), for ethyl alcohol, the results were dispersed over time.

Thus, the treatment TIO2PTFE-ETH-8g/L was the best solution since it was the best considering the WCA and the second one regarding photocatalysis. The best one regarding the photocatalytic efficiency (AC TiO2-CET) presented the worst WCA for the treated samples. While the second best considering the WCA (AC 10 TiO2PTFE-ETH-4g/L) was the fifth performance concerning the photocatalysis. Thus, considering both tests, the best solution was TiO2PTFE-ETH with a concentration of 8 g/L.

4 Conclusions

This research aimed to identify the best solution to apply over asphalt mixtures to provide them with new capabilities, specifically photocatalytic, superhydrophobic, and self-cleaning. For this, different types of solutions were sprayed over a conventional asphalt mixture. Then, the functionalized asphalt mixture samples were submitted to dye degradation for photocatalytic efficiency analysis and Water Contact Angle (WCA) test.

The main conclusions were:

• The coatings with at least TiO_2 presented a photocatalytic efficiency at least 38% higher than the reference mixture without any treatment;
• The best solutions considering this aspect were TiO2-CET and TiO2PTFE-ETH-8 g/L;
• The treatments TiO2-ETH, PTFE-ETH, PTFE-CET, TiO2PTFE-ETH, and TiO2PTFE-CET provided superhydrophobic capability;
• The best treatment for WCA was TiO2PTFE-ETH 8 g/L;
• The best solution for both capabilities was TiO2PTFE-ETH 8 g/L.

It is important to emphasize that these functionalized asphalt mixtures can degrade gases like SO_x and NO_x, avoid road accidents by cleaning the dirt particles and degrade oils/greases over their surface, presenting benefits to road safety and the environment. In this sense, the present work showed that one of the potential destinations for the large-scale application of nanomaterials is the civil engineering, especially road construction. Furthermore, this destination assists in promoting the dynamism and commercial progress of nanomaterial and nanotechnology industries.

Acknowledgements

This work was supported by FCT (Fundação para a Ciência e a Tecnologia) under the frameworks of the Strategic Funding UIDB/04029/2020 and UIDB/04650/2020, and the project PTDC/FIS-MAC/6606/2020. The first author would like to acknowledge the FCT for the Ph.D. scholarship (SFRH/BD/137421/2018).

References

1. A. Nahvi, M. Sadoughi, A. Arabzadeh, and A. Sassani, J. Comput. Des. Eng. (2018)
2. M. Hassan, L. N. Mohammad, S. Asadi, H. Dylla, and S. Cooper, J. Mater. Civ. Eng. **25**, (2012)
3. I. Rocha Segundo, E. Freitas, S. Landi Jr, M. F. M. Costa, and J. O. Carneiro. Coatings **9** (2019).

4. C. Toro, B. T. Jobson, L. Haselbach, S. Shen, and S. H. Chung, Atmos. Environ. **139**, 37 (2016)
5. J. O. O. Carneiro, S. Azevedo, V. Teixeira, F. Fernandes, E. Freitas, H. Silva, and J. Oliveira, Constr. Build. Mater. **38**, 594 (2013)
6. I. Rocha Segundo, C. Ferreira, E. F. Freitas, J. O. Carneiro, F. Fernandes, S. L. Júnior, M. F. Costa, S. Landi Júnior, and M. F. Costa. Constr. Build. Mater. **166**, 36 (2018)
7. Y. Gao, L. Qu, B. He, K. Dai, Z. Fang, and R. Zhu. Constr. Build. Mater. **191**, 270 (2018)
8. J. H. O. Nascimento, P. Pereira, E. Freitas, and F. Fernandes. *7th International Conference on Maintenance and Rehabilitation of Pavements and Technological Control* (Auckland, 2012)
9. C. Peng, H. Zhang, Z. You, F. Xu, G. Jiang, S. Lv, and R. Zhang. Constr. Build. Mater. **189**, 227 (2018)
10. A. Arabzadeh, H. Ceylan, S. Kim, K. Gopalakrishnan, and A. Sassani. Transp. Res. Rec. J. Transp. Res. Board **2551**, 10 (2016)
11. I. Rocha Segundo, S. Landi, E. Freitas, V. Castelo Branco, M. F. M. Costa, and J. Carneiro. EPJ Web Conf. **238**, 12012 (2020)
12. W. Zhang, X. Xiao, X. Zeng, Y. Li, L. Zheng, and C. Wan, J. Alloys Compd. **685**, 774 (2016)
13. B. Madhukar, J. Wiener, J. Militky, S. Rwawiire, R. Mishra, K. I. Jacob, Y. Wang, B. M. Kale, J. Wiener, J. Militky, S. Rwawiire, R. Mishra, K. I. Jacob, and Y. Wang, Carbohydr. Polym. **150**, 107 (2016)
14. I. Rocha Segundo, E. Freitas, V. T. F. C. Branco, S. Landi, M. F. Costa, and J. O. Carneiro. Renew. Sustain. Energy Rev. **151**, 111552 (2021)

Experimental Mueller matrix polarimetry with full Poincaré beams and a CCD camera.

Juan Carlos Suárez-Bermejo[1], *J. Carlos* González de Sande[2], *Massimo* Santarsiero[3] and *Gemma* Piquero[4]

[1]Universidad Politécnica de Madrid, Materials Science Department, Av. de la Memoria 4, 28040 Madrid, Spain
[2]Universidad Politécnica de Madrid, ETSIS de Telecomunicación, Campus Sur, 28031 Madrid, Spain
[3]Università Roma Tre, Dipartimento di Ingegneria, V. Volterra 62, Rome 00146, Italy
[4]Universidad Complutense de Madrid, Departamento de Óptica, 28040 Madrid, Spain

Abstract. Recently, the use of full Poincaré beams for extracting the Mueller matrix of a sample has been proposed. These beams present all possible polarization states across their transverse section. By placing a CCD camera behind a simple polarization analyzer formed by a quarter wave phase plate and a linear polarizer, a polarization map of the beam cross section can be obtained. This polarization map is modified when a sample is inserted before the polarization state analyzer. Comparison of these two polarization maps allows to obtain the Mueller matrix of the sample. An overdetermined system of linear equations (thousands of equations) can be written from this comparison. Standard mathematical methods are used to find optimum solution of this overdetermined system of equations. Some experimental results will be presented to check the performance of the proposed polarimetric method.

1 Introduction

The use of non-uniformly totally polarized (NUTP) beams [1, 2] has been proposed as parallel polarization states generator (PSG) for Mueller polarimetry [3,4]. Experimental measurement of Mueller matrices has been done by using azimuthally [5] or spirally [6] polarized beams. In these cases, only linear polarization states are present across the beam section, so that only a partial Mueller matrix can be obtained, but this is enough for recovering the whole matrix for homogeneous and deterministic samples [5-8]. If four independent states of totally polarized light are present across the NUTP beam section [9], the complete Mueller matrix can be measured for any sample [10,11]. The so-called full Poincaré beams (FPBs) contain all possible polarization states across its transverse section [9,12,13] and are candidates to be used as parallel PSGs.

Two alternative schemes have been proposed to obtain the Mueller matrix of a sample with FPBs. One is based on the use of a commercial polarimeter for measuring the state of polarization at four different points across the beam transverse section, before and after the sample under study. The inversion of the polarimetric relation gives the Mueller matrix of the sample. This procedure has been carried out experimentally and good results has been obtained [10]. The second method is based on the use of a CCD camera and a polarization state analyser [14]. Intensity measurements are made behind a polarizer and a quarter wave phase plate that are placed at least in four different configurations. As the input beam is a FPB, the change in the state of polarization at each point of its transverse section can be measured and a large data set can be obtained. Numerical simulations using different strategies to obtain the Mueller matrix of the sample from these data have been recently presented. These strategies are the denoted as *center-point* strategy and *nearest-neighbor* strategy [14].

In this work, a different approach to the problem is proposed, which will be called the *Moore-Penrose (M-P) pseudo-inverse* strategy. To obtain information about the sample we will use not only four, but a larger number of points across the beam section and apply the M-P pseudo-inverse strategy to simulated images affected by Gaussian noise. The obtained results will be compared to those provided by previous procedures, in order to verify which one gives the most accurate results. The noise caused by inaccuracies in the measurement of the intensity captured by each pixel of the CCD camera will be carefully analysed. The theoretical results will be compared with the results obtained from experimental images in a particular case.

In Section 2, the general procedure for FPB Mueller matrix polarimetry is sketched. Numerical simulations are presented in Section 3 and some of them are compared to experimental results in Section 4. Finally, concluding remarks are outlined.

2 Mueller full Poincaré polarimetry

The scheme followed to acquire the raw data is shown in Fig. 1. A particular FPB is synthesized by means of a calcite crystal [1, 9] and is used as a PSG. The polarization pattern of this beam is characterized by means of a

* Corresponding author: juancarlos.suarez@upm.es

https://doi.org/10.1051/epjconf/202125512005

polarization state analyser (PSA), which consists of a polarizer and a quarter wave phase plate. The intensity profile for at least four different configurations of the PSA is measured by a CCD camera [see Fig. 1 a)]. From these intensity measurements, the Stokes parameters at each point of the input beam can be determined. Afterwards, a known sample is placed before the PSA and the intensity map acquisition is repeated [see Fig. 1 b)]. From this second set of intensity maps, the Stokes parameters at each point of the transverse section of the output beam can be obtained.

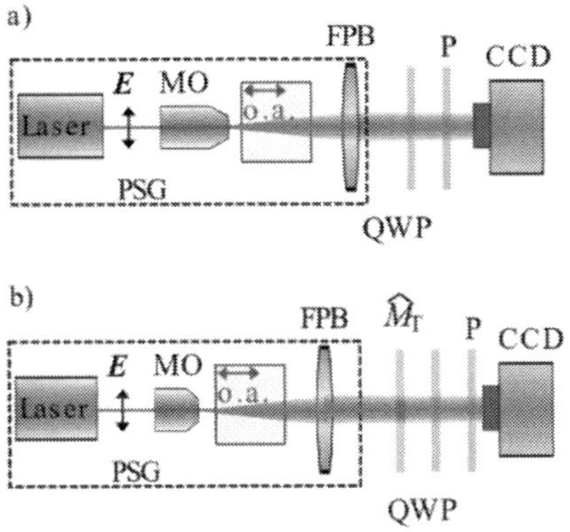

Fig. 1. Experimental scheme for Mueller FPB Polarimetry without (a) and with (b) a sample. E: MO: objective of microscope, o.a.: optical axis of a calcite crystal; L: lens QWP: quarter wave phase plate; P: polarizer.

The Stokes parameters of the input beam and those of the output beam are related by the Mueller matrix, $\widehat{M}_S$, of the sample by

$$\vec{S}^{out} = \widehat{M}_S \vec{S}^{in} . \quad (1)$$

We apply this relation to a wide area of the beam transverse section where the Poincaré sphere is mapped at least twice [9], so that an overdetermined system of equations is obtained. This system is inverted by means of the Moore-Penrose pseudo-inverse [15] to obtain the Mueller matrix of the sample.

An important source of error is the inaccuracy in the intensity measurements at each pixel of the CCD camera. Comparison of several images acquired under the same conditions has been performed to evaluate such source of error [16]. From this comparison, it has been deduced that the standard deviation of the measured intensity in each pixel of the CCD camera is about or below 0.01 times the mean value of the intensity [16].

3 Numerical simulations

Simulations for four different standard samples have been made: air (no sample introduced in the polarimetric system), a linear polarizer with horizontal transmission axis (LP0) and with its transmission axis at 60° with respect to the horizontal (LP60), and a quarter wave phase (QWP) retarder with its fast axis along the horizontal direction.

Theoretical images of CCD power distribution for six specific configurations [14] of the PSA are generated using a specifically dedicated code in MATLAB. Input images are independently generated and then passed through a function that adds Gaussian noise (characterized by the standard deviation ε of its probability density function). Input Stokes vectors for the four sample are derived from these noisy images. Regarding the output images, theoretical input Stokes vectors (without noise) are modified using the ideal Mueller matrix of each sample. These output Stokes vectors are used to obtain the output intensity images in the six considered states after the beam has passed through the sample. The power images are then sent to the modulus for adding Gaussian noise (randomly generated and, in consequence, independent from the noise in the input images). The Stokes vectors of the output beams at each pixel are then reconstructed from the output noisy images.

By using the noisy input and output sets of Stokes vectors, the Mueller matrix of the sample can be calculated by means of Eq. (1). This process is repeated many times. Then the difference between the theoretical Mueller matrix and those obtained from the simulations can be calculated. To give a measure of these differences with a sole number, the root mean square of such difference is calculated as

$$rms(\widehat{M}_S - \widehat{M}_T) = \frac{1}{4}\sqrt{\sum_{i,j=0}^{3}\left(m_{ij}^S - m_{ij}^T\right)^2}, \quad (2)$$

where m_{ij}^S are the values of the Mueller matrix elements obtained from the simulation and m_{ij}^T are the theoretical ones.

To obtain a quantitative estimate of the errors in the determination of Mueller matrix for a considered sample, three different strategies have been followed. In first place (*center-point* strategy), four points have been selected in such a way that their polarization states correspond to the vertices of a tetrahedron inscribed in the Poincaré sphere. Using the same points for the input and the output, and obtaining the Stokes vectors at these specific locations, it is possible to obtain the Mueller matrix of the sample including the noise in the simulated measurements. The process can be repeated up to 15 realizations maintaining one of the vertices fixed and rotating the other three around the axis s_1. Afterwards, it is possible to obtain an average Mueller matrix from all the realizations and compare every individual element with the theoretical matrix to have an estimation of the errors for every value of the parameter ε.

In the second approach (*nearest-neighbor* strategy) not only four points are selected in every realization to calculate the Mueller matrix inverting Eq. 1, but a wider set of points is used, all located around each of the center points considered in the previous case. A total of $N \times N$ nearest neighbor pixels are considered around every central point. Mueller matrices are obtained for all the possible four-point sets (input and output) that can be chosen from the $4 \times N^2$ pixels included in the analysis. An average value is then computed from all the valid cases (condition number of the matrix below 20), for the

15 analysed orientations of the tetrahedron in the Poincaré sphere.

Finally, the third strategy (*Moore-Penrose pseudo-inverse*) uses all points included in a circle around the center of the images, up to a certain radius, to construct an overdetermined system of linear equations that gives an estimate of the 16 elements of a Mueller matrix with the minimum mean square error value. The radius of the considered circle must be great enough to contain the center part of the beam where all possible polarization states can be found twice [9, 10]. Taking into account the CCD camera characteristics, this radius can be up to 600 pixels. The state of polarization, and hence the Stokes vector varies from pixel to pixel. Then, the calculations involve the individual values of around 1.13×10^6 pixels for every realization.

Gaussian noise has been introduced independently for every realization with a value for parameter ε=0.01, which is comparable to the standard deviation of the intensity measured values in each pixel of the CCD camera used in the experiment [16]. Figure 2 shows the root mean square of the difference for each realization of this simulation. In total, 500 realizations of the simulation have been done assuming the sample is air. It is found that the root mean square of the difference is around 1.14×10^{-3}, which is a quite low value [17]. It can be observed in Fig. 2 that the dispersion of these root mean squares is very small.

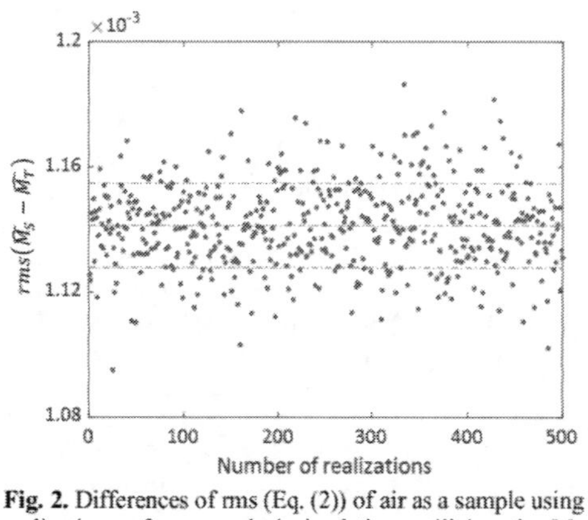

Fig. 2. Differences of rms (Eq. (2)) of air as a sample using 500 realizations of a numerical simulation, utilizing the Moore-Penrose pseudo-inverse strategy (ε=0.01).

The simulations have been repeated for three additional samples: a quarter wave plate with its fast axis along the horizontal direction, a linear polarizer with its transmission axis along the horizontal direction, and the same polarizer with its transmission axis forming a 60° angle relative to positive horizontal direction. Figure 3 summarized the results obtained from the 500 realizations of the simulation for the four samples that have been considered. Fig. 3 shows the average of the 500 computed root mean squares and their standard deviation for every sample.

An interesting issue is to determine if the differences of the sixteen elements of the Mueller matrix are evenly distributed or if there are some elements in the Mueller matrix that are prone to present higher errors than the mean.

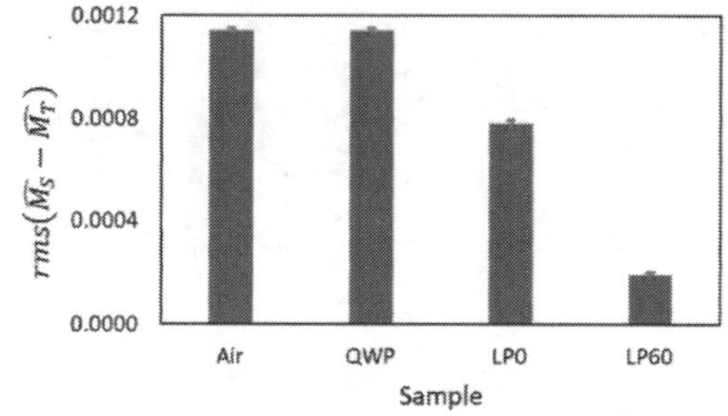

Fig. 3. Mean values and standard deviations of the root squared differences between the Mueller matrices obtained from numerical simulations and the ideal ones. For four samples: air, a linear polarizer with horizontal transmission axis (LP0) and with its transmission axis at 60° with respect to the horizontal (LP60), and a quarter wave phase (QWP) retarder with its fast axis along the horizontal direction.

To elucidate this question, the differences of each individual element between the theoretical Mueller matrix and the one obtained for each simulation have been evaluated. After that, they have been represented as a set of dots in the vertical of the matrix element indexes. Figure 4 shows this representation for the case of air and taking added Gaussian noise with standard deviation ε=0.02. Again, 500 realizations of the simulation have been done considering all the pixels around the center of the images up to a radius of 600 pixels. It can be observed that the absolute value of all differences are below 5×10^{-3}, except for one element. The biggest difference is found for m_{22} element for which the average value of the difference is 14×10^{-3}.

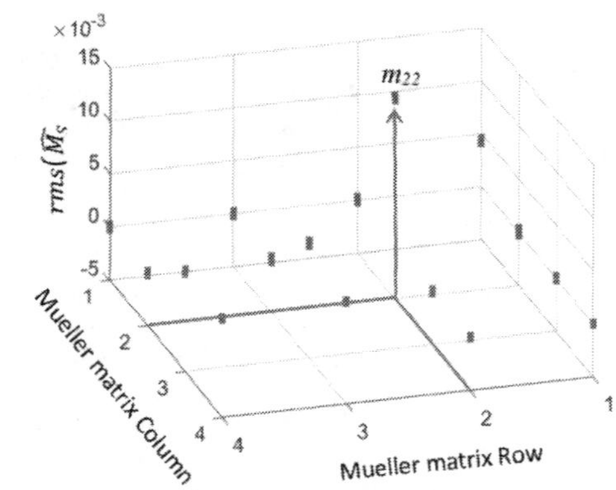

Fig. 4. Individual values of the differences of the 4×4 elements between Mueller matrix obtained for each realization of the simulation and the theoretical one. The case of air for 500 realizations of the simulation with a standard deviation of the added Gaussian noise ε=0.02 is shown.

4 Experimental results

As an example, just for comparative purposes, the measurements in air are shown here. The results for the rest of the samples will be published elsewhere. The experimental setup for generating the FPB can be found in ref. [9]. The obtained FPB is expanded to cover most of the CCD camera surface. Two different sets of imaged for six different configurations of the PSA [14] have been acquired. From these two sets of images, the air Mueller matrix has been obtained following the M-P pseudo-inverse strategy. The root mean square value of the difference between the measured Mueller matrix and the identity matrix is 4×10^{-3}. This result is presented as a red horizontal line in Fig. 5. This is compared to the rms levels found from simulations for the three described strategies when a different level of added noise is considered. In the numerical simulations, no matter what noise level is added to the images, the differences with the ideal Mueller matrix of the sample are lower when the Moore-Penrose pseudo-inverse strategy is employed to compute the matrix, as compared with the center-point and the nearest-neighbor strategies (see Fig. 5). However, the computational cost is higher for the calculations using the pseudo-inverse because it is necessary to solve a system of linear equations involving 1.13×10^6 pixels in the images.

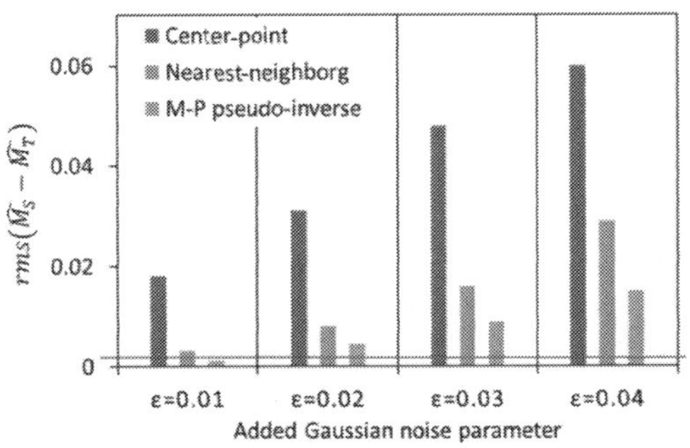

Fig. 5. Errors estimated in the numerical simulations for air, using three different strategies to compute the Mueller matrices, and experimental results for the sample.

It can be observed that the experimental result (red line in Fig. 5) is somewhere between the *rms* values obtained when the Moore-Penrose pseudo-inverse strategy is used in simulations where the standard deviation of the added Gaussian noise is 0.01 and 0.02. This level of Gaussian noise is in good agreement with the independently determined noise in the images obtained with our CCD camera and polarimetric system [16].

5 Conclusions

Full Poincaré Mueller Polarimetry can be carried out using different setups. Here a PSA consisting of a polarizer, a quarter wave phase plate and a CCD camera is proposed. With this setup three different strategies for processing the data from the camera can be applied. Two of them (*center-point* strategy and *nearest-neighbor* strategy) have already been proposed in a previous work [14]. Here a new strategy (Moore-Penrose pseudo-inverse strategy) is applied to the simulated imagen with added noise for different samples and compared with the previous results. Finally, for a particular sample the numerical results have also been compared with the results when the experimental images are considered.

Acknowledgements

This work has been partially supported by Spanish Ministerio de Economía y Competitividad under project PID2019-104268GB-C21.

References

1. G. Piquero and J. Vargas-Balbuena, Eur. J. Phys. **25**, 793 (2004).
2. G. Piquero. R. Martínez-Herrero, J.C.G. de Sande, M. Santarsiero, JOSA A **37**, 591 (2020).
3. S. Tripathi, K.C. Toussaint KC. Opt. Express, **17**, 21396 (2009).
4. F. Kenny, O. Rodríguez, D. Lara, C. Dainty, *Frontiers in Optics 2011/Laser Science XXVII*. Optical Society of America; (2011).
5. J.C.G. de Sande, G. Piquero, and M. Santarsiero, Opt. Commun. **410**, 961 (2018).
6. J.C.G. de Sande, M. Santarsiero, and G. Piquero, Opt. Laser Eng. **91**, 97 (2017).
7. R.A. Chipman, Polarimetry, in *Handbook of Optics*, ed, M. Bass, (McGraw-Hill Companies, 2010).
8. M. Swami, H. Patel, P. Gupta, Opt. Commun. **286**, 18 (2013).
9. G. Piquero. L. Monroy, M. Santarsiero, M. Alonzo, J.C. de Sande, J. Opt. **20**, 065602 (2018).
10. J. C. Suárez-Bermejo, J.C.G. de Sande, M. Santarsiero, G. Piquero, Opt. Laser Eng., **122**, 134 (2019).
11. J.C. Suárez-Bermejo, J.C.G. de Sande, M. Santarsiero, G. Piquero, *2019-Progress In Electromagnetics Research Symposium*, (PIERS-Rome) 2019.
12. A.M. Beckley, T.G. Brown, and M. A. Alonso, Opt. Express, **18**, 10777 (2010).
13. E.J. Galvez, S. Khadka, W.H. Schubert, and S. Nomoto, Appl. Opt. **51**, 2925 (2012).
14. J. C. Suárez-Bermejo, J.C.G. De Sande, M. Santarsiero, G. Piquero, Sensors & Transducers **247** (8), 18-24 (2020).
15. A. Ben-Israel, T.N.E. Greville, *Generalized inverses: Theory and applications (2nd ed.). New York, NY: Springer.* (2003).
16. J. C. Suárez-Bermejo, J.C.G. de Sande, M. Santarsiero, G. Piquero, *Advances in Optics: Reviews* Book Series, Vol. 5 (2021).
17. J. del Hoyo, L.M. Sanchez-Brea, J.A. Gomez-Pedrero Opt. Laser Eng., **132**, 106112 (2020).

Femtosecond laser writing of integrated photonic circuits in diamond

Giulio Coccia[1,*], *Argyro N.* Giakoumaki[1], *Vibhav* Bharadwaj[1], *Ottavia* Jedrkiewicz[2], *Roberta* Ramponi[1], and *Shane M.* Eaton[1]

[1]Istituto di Fotonica e Nanotecnologie-Consiglio Nazionale delle Ricerche (IFN-CNR) and Dipartimento di Fisica - Politecnico di Milano, Piazza Leonardo da Vinci 32, Milano 20133, Italy
[2]Istituto di Fotonica e Nanotecnologie-Consiglio Nazionale delle Ricerche (IFN-CNR), Department of Science and High Technology, Università degli Studi dell'Insubria, Via Valleggio 11, 22100 Como, Italy

Abstract. Integrated photonic circuits pave the way for next generation technologies for quantum information and sensing applications. Femtosecond laser writing has emerged as a valuable technique for fabricating such devices when combined with diamond's properties and its nitrogen vacancy color center. Such color centers are fundamental for sensing applications, being possible to excite them and read them out optically through the fabrication of optical waveguides in the bulk of diamond. We show how to integrate these building blocks in diamond, to develop proof-of-concept devices with unprecedented electric and magnetic field sensitivities.

1 Introduction

Femtosecond laser fabrication has already established itself as a powerful technique for application in the field of material processing for welding, cutting and drilling [1], and it is attracting a growing interest for its capability to directly write microstructures in the bulk of transparent materials [2]. Focused femtosecond laser pulses can be employed to induce a change in the refractive index of almost all transparent materials such as glasses, single crystals, and polymers, enabling the writing of 3D photonic structures. The locally modified refractive index profile depends on both material and laser processing parameters [3].

Typically in crystals, the irradiation of femtosecond laser pulses produces a decrease of the density at the focal volume due to the creation of lattice disorder, leading to a decrease of the refractive index. In this situation, guiding of light is usually achieved through the Type II waveguide (WG) scheme, consisting in the fabrication of two parallel lines in the bulk of the crystal, resulting in a relative refractive index increase between the two written tracks.

Synthetic diamond is emerging as a convenient platform for the fabrication of three-dimensional photonic structures, not only for its excellent transparency, hardness and chemical resistance, but because of the existence of vacancy complexes behaving as color centers. These quantum emitters can be interfaced with photonic circuits written on the surface or in the bulk of the crystal for quantum sensing or information. The most appealing of such color centers is the Nitrogen Vacancy (NV) center, whose negatively charged state (NV^-) is characterized by a spin triplet ground state, which can be polarized under green excitation [4]. One can exploit the NV^- color center also for optical read out, as the fluorescence from one state is brighter than the other due to the existence of a non-radiative decay path. Finally, the NV^-'s spin states are sensitive to electric and magnetic fields, enabling room temperature quantum sensing with unprecedented spatial resolution and sensitivity.

The implementation of sensing devices or quantum information systems based on NV^- centers and optical WGs in the bulk of diamond requires the ability to fabricate photonic circuits and color centers in a controlled manner. Both the formation of NV^- centers and optical WGs in diamond can be carried out by the same technique, namely femtosecond laser writing. We will show how this laser fabrication method is a versatile and non-invasive approach capable of retaining the optical quality of the crystal and its color centers. We will demonstrate in this paper a first proof-of-concept room temperature sensing device, based on an ensemble of color centers and accessible through laser written waveguides in the bulk of diamond, which could allow unprecedented sensitivities for the detection of magnetic and electric fields.

2 Integration of NV^- centers and WGs

The NV^- center appears intrinsically in both natural and synthetic diamond and results when two adjacent sites of the diamond carbon lattice are replaced by a nitrogen and a vacancy. Interestingly, the ground state of such a complex is sensitive to electric and magnetic fields through the Stark and Zeeman effects, respectively [5]. This property makes of NV^- centers a convenient room

* Corresponding author: giulio.coccia@polimi.it

temperature detection complex for not only electric and magnetic fields by measuring the shift of the ground state spin transition frequencies under the influence of the fields themselves, but also changes in temperature through optically detected magnetic resonance measurements (ODMR) [6]. However, the possibility to develop real-life sensing devices is bound to the ability to laser-write waveguides and color centers in diamond in a systematic way and finally to integrate them in more complex photonics circuits.

2.1 Fabrication of WGs and single NV⁻s

In 2016, the first Type II WG was successfully written in diamond by focusing a femtosecond Yb laser (300 fs pulse duration, 500 kHz repetition rate, 515 nm central wavelength, 100 nJ pulse energy) into the bulk of the single crystal synthetic diamond at different depths through an objective lens with 1.25 numerical aperture (NA)[7]. Shortly thereafter, Chen et al. [8], demonstrated how a similar femtosecond laser system could induce the creation of on-demand and high-quality single NV^- centers in the bulk of quantum grade diamond. In this case, a single low intensity femtosecond laser pulse was focused beneath the surface of diamond to form an ensemble of vacancy defects within the focal volume. After a post high temperature annealing treatment (1000 °C), the vacancies were mobilized, enabling the recombination with nitrogen impurities within diamond to form NV^- centers. The next step in the development of quantum photonic circuits is the integration of the color centers with the optical waveguides. Our group has recently demonstrated such a result [9], showing the possibility to firstly write type II WGs and then induce the creation of NV^- color centers in between the laser written tracks with a spatial accuracy of ~500 nm, fabricating a proof-of-concept device in which precisely located NV^- centers can be externally accessed, excited and read out though the laser written WGs.

2.2 High density NV^- ensembles for sensing applications

Single-NV^- centers are characterized by a nanometric spatial resolution due to the atomically small size of single isolated complexes, but by a relatively low sensitivity which can be improved by averaging over many measurements. If higher sensitivity (or speed) is required and the nanometric spatial resolution is not necessary, single NV^- centers can be replaced by high density NV^- centers (ensembles) at the expense of reduced overall spatial resolution. For this purpose, it is critical to be able to deterministically create the ensembles with a required density and to interact with them through the optical WGs. It is known that the sensitivity (η) of an ensemble of NV^- centers can be described as:

$$\eta = \frac{1}{k}\frac{1}{C\sqrt{M\Gamma}}\frac{1}{T_2^*} \quad (1)$$

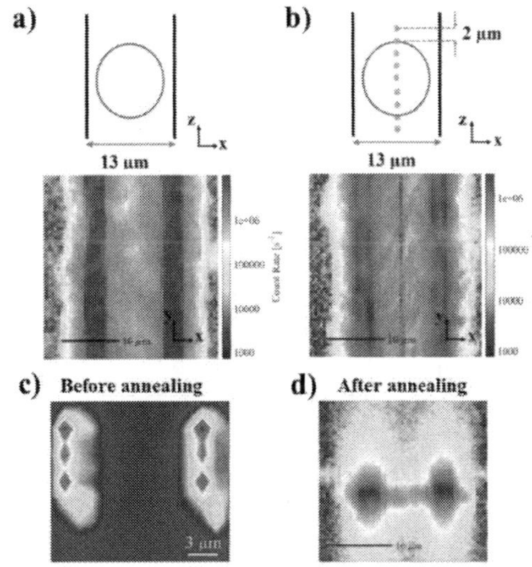

Fig. 1. Schematic cross-section of femtosecond laser written 'empty' (a, above) and 'static exposure' (b, above) waveguides, with modification lines in black and the expected mode field diameter of the waveguide mode outlined in red, with integrated static exposures marked in gray. Photoluminescence confocal maps of 'empty' (a, below) and 'static exposure' (b, below) waveguide and modification lines written in HPHT sample after annealing. PL Photoluminescence confocal cross-section of 'empty' waveguide and modification lines written in HPHT sample before (c) and after (d) annealing.

where k represents the frequency shift of the NV^- center's spin transitions under magnetic (k_m = 28 GHz T^{-1}) or electric fields (k_e = 17 Hz cm V^{-1})[10], C is the ODMR contrast, M is the number of NV^- centers in the ensemble measurement, Γ is the detected count rate for each NV^- center, and T_2^* is the inhomogeneous NV^- coherence time [11]. Following equation (1), the sensitivity of such devices based on ensembles of color centers scales as $\sqrt{M}$. This can be achieved by either increasing volume of NV^- centers or the density of NV^- centers sampled, while paying attention not to compromise their coherence time. Our group managed to write high density NV^- center ensembles in HPHT diamond through single static laser exposures followed by annealing, and their integration within a waveguide has also been studied. Arrays of 'empty' type II waveguides with 13 µm separation between the sidewalls and 10 µm estimated waveguide mode field diameter at 532 nm (Fig. 1(a), above), together with 'static exposure' ones, containing arrays of nine single 100 nJ static exposures separated by 2 µm axially (Fig. 1(b), above), were written in HPHT diamond at 18 µm depth. Fig.1(a-b, below) shows PL confocal maps in overhead view of the 'empty' and 'static exposure' waveguides, where it can be noticed how despite some brighter spots are visible in the 'static exposure' WG, a diffused bright fluorescence is observed in both cases. This is due to the fact that during the WGs fabrication, vacancies are created in the laser written sidewalls, which can migrate during the annealing treatment and combine

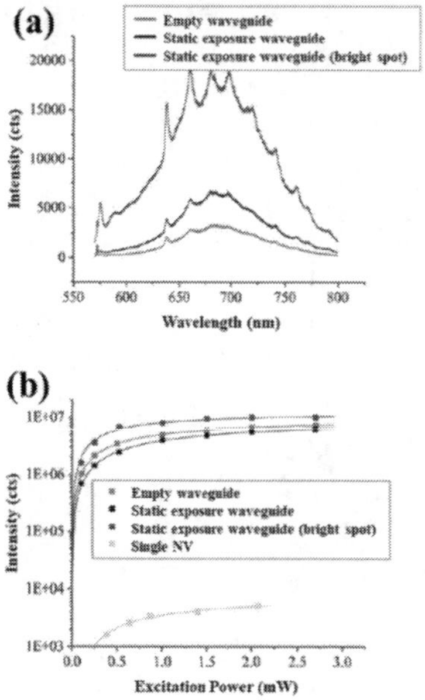

Fig. 2. PL spectra (a) and Power dependent PL (b) from 'empty' waveguide, 'static exposure' waveguide and from a bright spot (static exposure).

with the nitrogen in the diamond forming the NV^- centers, as shown from the transverse confocal microscope images of Fig.1(c-d). To validate the formation of NV^- centers after annealing, photoluminescence (PL) spectra have been taken from the 'empty', 'static exposure' WGs and from a bright spot corresponding to static exposure in the waveguide (Fig.2(a)). In all three cases, the characteristic zero phonon line (ZPL) of NV^- centers at 637 nm is observed, confirming that the created vacancies migrated in between the walls of the waveguide and created the color centers. The spectra taken from the static exposures are brighter than the surrounding area, presenting also additional structures with lines at 660, 680, 697.6, 720.7, 741.1 and 760.7 nm. A much stronger contribution from a line at 575 nm is also shown corresponding to the neutral NV^0 ZPL.

To optimize the sensitivity of the device one has to be able to evaluate the density of the created NV^- ensemble. This can be done by performing power dependent PL saturation measurements of both the "empty" and "static exposure" waveguides and comparing them to the total signal collected by a single NV^- center in another sample measured using the same conditions (Fig.2(b)). By approximating the confocal microscope's spread function through an estimation of the collection volume of the confocal microscope itself using confocal images of a single NV^-, it was possible to determine the ensemble densities. We found densities of 1.1×10^{15} cm^{-3} (or 6 ppb) in the 'empty' waveguide, and densities up to 1.4×10^{15} cm^{-3} (or 8 ppb) in the 'static exposure' ones. Although the coherence times of the generated color centers must be quantified, a first approximation of the achievable electric and magnetic sensitivities can be performed. We can

Fig. 3. Overhead microscope view of 'static exposure' waveguide showing the streak of fluorescence emission along the length of the waveguide.

consider a simple sensing device in which excitation and detection are performed through the coupling to a laser-written waveguide as shown in Fig. 3, where the sample was excited with a green laser end-fire coupled into the waveguide from the right-hand side, while the light scattering from the waveguide was filtered using a 532 nm notch filter from the top of the sample and recorded on a CCD. Assuming a 3 mm long waveguide having a 10 µm mode field diameter, and NV^- density of 1.1×10^{15} cm^{-3} we can expect to have $M = 2.6 \times 10^8$ NV^- centers contributing to the sensing signal, with a photon collection efficiency of $\Gamma = 900$ Hz per NV^- and a contrast $C = 0.05$. From this, we estimate a magnetic field sensitivity of 1.5 nT $Hz^{-1/2}$ and an electric field sensitivity of 2.4 V cm^{-1} $Hz^{-1/2}$. These results can already be compared with the state-of-the-art systems, which generally are characterized by lower sensitivities, ≈ 290 pT $Hz^{-1/2}$ for magnetic fields [12] and ≈ 1.6 V cm^{-1} $Hz^{-1/2}$ for electric fields [10]. Further optimization of laser writing parameters and nitrogen impurity levels within the diamond are anticipated to lead to record high sensitivities.

3 Conclusions

We have shown how femtosecond laser writing in diamond can be applied for the fabrication of next generation technologies, in particular for quantum optics and information. We have demonstrated how femtosecond lasers enable the formation of optical waveguides and color centers in an easy and repeatable fashion and how it is possible to integrate the two components for prototype devices exceeding the performance of the existing systems based on conventional technologies. In particular, we studied a room temperature sensing device based on ensembles of NV^- centers, which is characterized by unprecedented high sensitivities. It may be possible to improve the sensitivity even further by integrating femtosecond laser written Bragg reflectors into the waveguides to increase the effective light-NV^- center interaction length through multiple passes of the excitation laser [13]. Moreover, since the strain created between the lines confining the waveguide structure is equivalent to an electric field, the NV^- centers located in the middle of the waveguide and experiencing such a strain are affected by a constant bias electric field. Because of this bias field these NV^- centers are less sensitive to stray magnetic fields, while more sensitive to extremely weak electric fields, making them particularly attractive for the detection of such fields. Finally, further studies with the versatile 3D femtosecond laser writing method will improve the functionality and quality of the fabricated quantum devices, allowing for more complex systems capable of advanced functionalities [14].

References

1. R. Bogue, Assembly Automation, **35** **2**, 161-165, (2015).
2. K. M. Davis et al., Opt. Lett. **21**, 1729 (1996).
3. R. Gattass, E. Mazur, Nat. Photonics **2**, 219 (2008).
4. I. Aharonovich, et al., Nat. Photonics **5**, 397 (2011).
5. F. Dolde, et al., Nature Physics **7,** 459 (2011).
6. F. Jelezko, et al., Phys. Rev. Lett. **92**, 76401 (2004)
7. B. Sotillo et al., Sci Rep **6**, 35566 (2016).
8. Y. C. Chen, et al., Nat. Photonics **11**, 77 (2017).
9. J.P. Hadden, et al., Opt. Lett. **43**, 3586 (2018).
10. E. H. Chen, et al., Physical Review A **95**, 53417 (2017).
11. A. Dréau, et al., Physical Review B **84,** 195204 (2011).
12. H. Clevenson, et al., Nat. Phys. 11, 393 (2015).
13. V. Bharadwaj, et al., Opt. Lett. **42, 3451 (2017)**
14. V. Bharadwaj, et al., J. Phys. Photonics **1**, 22001 (2019).

Hybrid laser-assisted forming process for ultralight three-dimensional glass elements

Robin Hassel[1,*], *Andrea* Barz[1], *Jens* Bliedtner[1], *Jörg* Steuer[2], *Hans* Pitcha[2], and *Matthias* Eberlein[2]

[1]Ernst-Abbe University of Applied Sciences Jena, Carl-Zeiss-Promenade 2, 07745 Jena, Germany
[2]GLAMACO GmbH, Prasseweg 1, 01640 Coswig, Germany

Abstract. As part of a research project, a new hybrid forming process for ultrathin three-dimensional glass elements was developed. In a continuous process chain, the three-part procedure enables large-area gravity-assisted bending of thin glass, partial deep-drawing of individual small structures using CO_2 laser radiation and cutting of flexible inner and outer contours. Using an experimental setup with a laser scanning system, novel component designs can be generated in a highly flexible and reproducible manner. This can open up new application fields or expand existing ones, for example in the area of cover glasses for display technology.

1 Introduction

Current manufacturing processes allow large-area forming, especially for thin glasses (thickness ≤ 2.0 mm), only to a limited extent. Furthermore, the creation of filigree structures with high aspect ratios is hardly possible to reproduce for preformed glass. Due to the constantly developing technology, glass manufacturers and consumers are increasingly demanding flexibly bent thin glasses with large surface dimensions and simultaneously filigree elements. The weight of the glass should be as low as possible, while the strength and the achievable accuracies should be high. In order to realise such glass elements according to the described requirements with high optical qualities in low processing times, an innovative hybrid forming process was developed.

2 Glass forming process

2.1 Materials

The hybrid procedure can be used to process borosilicate and alumosilicate glasses with maximum dimensions of 250 x 250 mm² and thicknesses between 0.5 and 2.0 mm. To be able to bend or deep-draw the glasses, the viscosity must be reduced. This requires temperatures close to or above the material specific softening point ($T_{Softening}$). To simplify the handling of viscosity-temperature values, fixed points have been defined representing characteristic viscosities. [1]

The fixed points of the materials used are shown in Table 1.

Table 1. Viscosity fixed points of the glass materials used.

	BORO-FLOAT®33	Xensation® Cover
Annealing Point ($\eta = 10^{13.0}$dPas)	560 °C	635 °C
Softening Point ($\eta = 10^{7.6}$dPas)	820 °C	880 °C
Working Point ($\eta = 10^{4.0}$dPas)	1270 °C	1265 °C

With the help of the fixed points and the VOGEL-FULCHER-TAMMANN equation, the relationship between temperature and viscosity above the transformation range can be approximately calculated. [2]

$$lg\,\eta = A + \frac{B}{T - T_0} \quad (1)$$

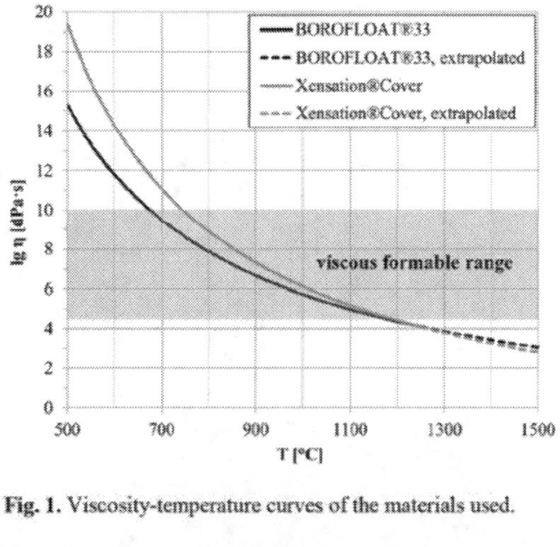

Fig. 1. Viscosity-temperature curves of the materials used.

* Corresponding author: robin.hassel@eah-jena.de

The viscosity-temperature curves shown in Fig. 1 provide information on the necessary forming temperatures.

2.2 Experimental setup and procedure

As a result of high coefficients of thermal expansion, the glasses used have a relatively low thermal shock resistance. Therefore, laser processing at room temperature is almost impossible. To avoid glass breakage, the glasses are preheated in a heating chamber (GLAMACO GmbH) before laser processing. The experimental setup, which is divided into two parts on a laboratory scale, is illustrated schematically in Fig. 2.

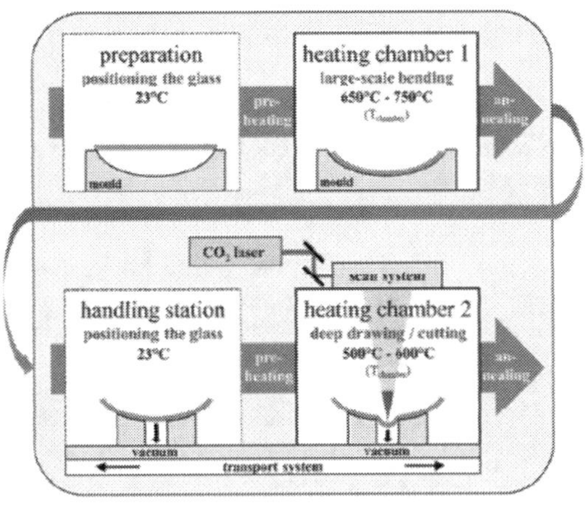

Fig. 2. Principle of the hybrid forming process of thin glass.

To bend larger geometries, the entire glass component is heated in a thermal chamber to a suitable temperature below the softening point, which allows forming but precludes flowing of the glass to prevent undesirable deformation and glass thickness changes. The mould used must be dimensionally stable and heat-resistant. It should also have a surface roughness that prevents sticking behaviour between glass and mould and at the same time ensures a high surface quality of the formed glass.

For the deep-drawing of filigree elements, defined areas of the glass are heated by CO_2 laser radiation (λ = 10,6 μm) and thermoformed with vacuum support and without an underlying forming tool. By using a laser scanning system, freely designable contours of different depths can be generated with high precision. Due to the locally limited heat input, these structures can also be realised in already preformed glass without undesirably deforming it. This results in considerable energy and time savings since not the entire glass component has to be heated to forming temperature.

In the final step of the hybrid process, the preheated glass is cut by using CO_2 laser radiation. The same beam source can be used for cutting as for deep-drawing. A focusing optic (flat field lens with a focus length of 600 mm) is used to achieve the high intensities required for the cutting process. A mechanical lifting device inside the chamber ensures the sample movement along the z-axis so that the glass can be positioned in the focal plane. The laser beam tool allows not only the individual cutting of outer contours, but also the generation of flexible inner contours with good (cutting) edge qualities. [3]

During the laser-based processes, the influence of the following parameters is investigated:

- preheating temperature [°C]
- values of laser power [W]
- values of scanning speed [$m \cdot s^{-1}$]
- laser scanning strategies
- number of scans (exposure time)

Finally, the influence on the following output parameters is examined:

- process temperature [°C]
- contour width and length
- contour depth
- surface and edge quality
- state of thermal stress

3 Results

With the described experimental setup, the sub-processes of the hybrid forming process could be successfully and reproducibly implemented on thin borosilicate and alumosilicate glasses using previously determined process parameters (such as temperature and laser power).

In numerous investigations the technological process windows for the sub-processes could be determined. For gravity-assisted bending, temperatures below the material-specific softening point are sufficient, whereas temperatures above this point are required for laser-assisted deep-drawing. The use of laser radiation in combination with the scanning system offers many advantages. In particular, the flexibility and individuality with regard to three-dimensional structures of different depths that can be generated is increased enormously. In addition, the processing times are very short due to the high scanning speeds and the fast energy input into the glass component. The hybrid process approach and the associated interaction of multiple sub-processes and energy sources can increase the performance of the glass forming procedure compared to conventional processes.

Fig. 3 shows a three-dimensionally formed borosilicate glass component with a thickness of 0.7 mm produced by the hybrid forming process.

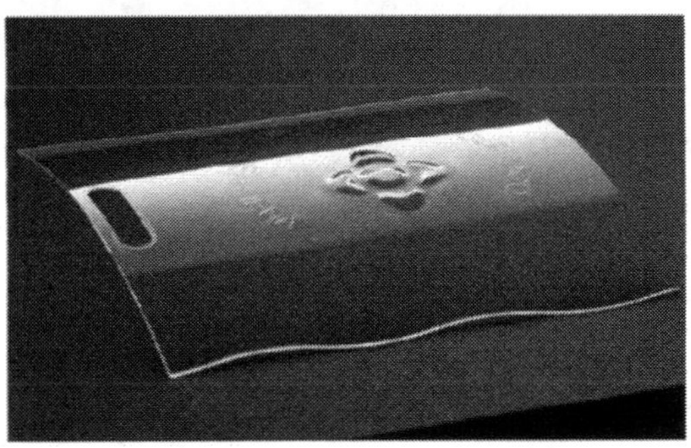

Fig. 3. Generated borosilicate glass component (0.7 mm thick).

Deep-drawing without a mould is associated with the reduction of processing time, energy and costs. Moreover, there is no (negative) mechanical impact on the glass surface. By varying the laser beam diameter, even structures with small bending radii and high aspect ratios can be generated with high accuracies. In this context, it is possible to achieve contours with a flat-edged design on the one hand and sharply defined contours on the other. The locally limited laser processing and the associated high temperature gradients result in undesirable thermal stresses in the glass. For this reason, the laser treatment is followed by an annealing process in order to reduce relieve residual internal stresses to a non-critical level. Besides annealing and chemical strengthening of glass, no time-consuming post-processing steps are necessary. [3]

Further investigations demonstrate that the process described allows not only two-dimensional bending but also three-dimensional large-area bending of thin glass. As exemplified in Fig. 4, special forming tools (a) can be used to bend thin glasses with a thickness of 1.0 mm as parabolic mirrors with a diameter of 70 mm (b). [4]

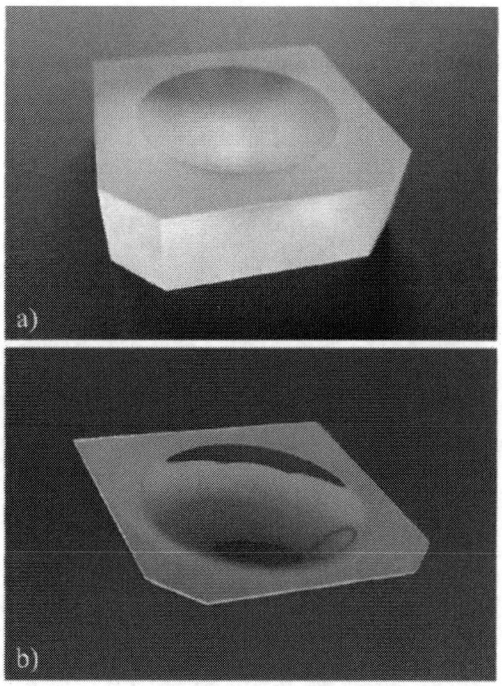

Fig. 4. Gravity-assisted bending of a parabolic mirror. a) forming tool, b) parabolic mirror made of thin glass before the coating process. [4]

The mould can preferably be made of coated steels or nickel-based alloys, ceramics or, as in this case, fused silica. It should have certain surface properties that lead to easy demoulding of the glass without any sticking behaviour. Otherwise, the glass may be damaged and, in the worst case, the mould could be destroyed. The mirror shown, made of BOROFLOAT®33 (SCHOTT AG), can subsequently be edge-machined and coated, depending on the intended application. [4, 5]

4 Conclusion and outlook

The hybrid glass forming process was successfully implemented. With the two-part experimental setup, it is possible to bend thin glasses under the influence of temperature and gravity, followed by partial deep-drawing with the help of CO_2 laser radiation and vacuum, and finally create inner and outer contours by laser cutting. As two of the sub-processes are laser-based, the procedure leads to a significant reduction in processing time compared to conventional forming processes. In the future, it is desirable to realise all sub-processes of the hybrid process in one test facility, especially with a view to industrial application. This would be accompanied by further considerable savings in terms of costs, time and energy, as the glass elements only have to be heated once, so that only one annealing process would be necessary.

Another positive aspect is that the process can be well adapted to other glass materials. If the material-specific viscosity fixed points are known, the required process temperatures, even for new types of glass, can be determined by calculation. This expands the potential fields of application of the individually formed glass elements, so that the process can be interesting for many different target groups.

In conclusion, the innovative process combination enables the production of new component designs that are not feasible with current state-of-the-art methods. This is intended to open up new areas of application, for example in the automotive industry as well as in medical technology, entertainment industry (e.g. for smartphones or tablets), building architecture or optics.

The authors gratefully acknowledge the financial support by the German Federal Ministry for Economic Affairs and Energy and the AiF-Projekt GmbH (funding reference: ZF4039612AG8).

References

1. J. Bliedtner, G. Gräfe, R. Hector: *Optical technology*. McGraw-Hill Education. (2011)
2. G. Nölle: *Technik der Glasherstellung*. Freiberg: Wiley-VCH Verlag GmbH & Co. KGaA. (1997)
3. R. Hassel, A. Barz, J. Bliedtner, J. Steuer, H. Pitcha, M. Eberlein: Neuartige Prozesskette zum hybriden laserstrahlunterstützten Thermoformen von 3D-Glaselementen. *12. Jenaer Lasertagung, DVS-Berichte* Vol. **367**, Page 252. (2021)
4. J. Bliedtner: *Optiktechnologie: Grundlagen - Verfahren - Anwendungen - Beispiele*. **3.**, überarbeitete und erweiterte Auflage. Carl Hanser Verlag GmbH & Co. KG. (to be published)
5. D. Rieser: *Untersuchungen zum Klebe- und Abriebverhalten von Formenwerkstoffen für die Heißformgebung von anorganischem Glas*. Dissertation. Universität Karlsruhe. (2004)

 https://doi.org/10.1051/epjconf/202125512008

Low-cost interrogation system for dynamic sensor applications using a resin based fiber Fabry Perot filter

João Preizal, *Lúcia* Bilro, and *Ricardo* Oliveira*

Instituto de Telecomunicações and Universidade de Aveiro, Campus Universitário de Santiago, 3810-193 Aveiro, Portugal

Abstract. In this paper it will be presented a prototype of a low-cost power interrogator for dynamic sensor applications. The proposed system is based on the edge filter technique, where the slope of an in-house resin based Fabry Perot (FP) filter is used to actively filter the power of a narrowband laser centred at the middle part of the FP slope region. The interrogator is composed of a laser source, circulator, photodetector, Arduino board and laptop. Tests have been performed in order to check the viability of the system. Finally, the measurement of a pulse wave signal has been performed by embedding the FP sensor in a 3D printed elastic membrane and reading the signal through the proposed interrogator.

1 Introduction

Fiber optic sensors are a technology that has been developed in many different areas in the last century. They present several qualities, such as durability in the most diverse environments, immunity to electromagnetic fields, compatibility, lightweight, ability to operate in long distances, they are compact, could be implemented with multiplexing capabilities, etc.

Most of the sensors reported in literature, such as fiber Bragg gratings and a variety of interferometers, rely on the wavelength detection technique. However, for such measurement schemes, it is required the use of bulk, and expensive fiber interrogators [1]. Furthermore, the acquisition frequency is normally low which is problematic when dynamic applications are required. Thus, here we propose the creation of an interrogator based on the edge filter technique [2], capable to read the data through a photodetector in an easy, simple and cost effective way. In order to keep the cost low, fiber Fabry Perot interferometers (FPIs), based on the resin splice method [3] will be used.

In this work we propose a simple and low-cost intensity detection system operating with an in-house FPI cavity. The edge filter technique will be used and the power fluctuations generated by the filter will be measured through a photodetector that is attached to the analogical input of an Arduino board. The dynamic response of the sensor will be tested for different frequencies and to finalise the system will be used to measure the pulse wave signal at the wrist region.

2 Theoretical concept

The theoretical basis of this paper is mainly based on the study of the power variation for a given wavelength, using the edge filtering technique. In this technique, a narrowband laser is actively filtered by an external filter.

The edge filter will filter the optical signal coming from the laser source and will limit the output power according to the spectral response of the filter at the laser wavelength location, as it can be seen on Fig. 1**Erro! A origem da referência não foi encontrada.**.

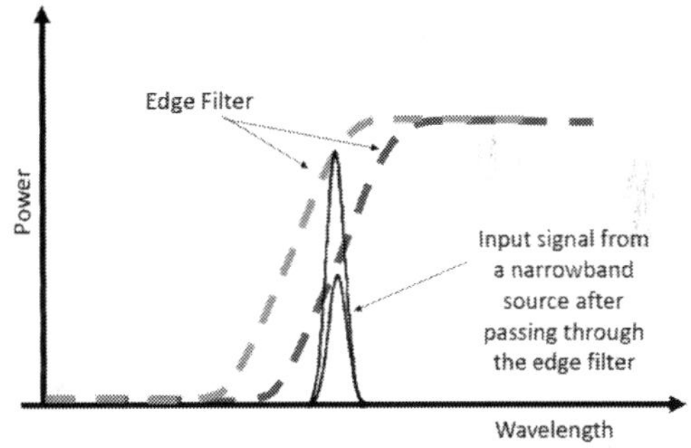

Fig. 1. Power fluctuation of a laser source as it passes through an edge filter.

In order to keep the cost of the system low, while keeping its simplicity, we decide to implement an FPI filter fabricated through the resin based method [3]. The spectral response of this filter varies between minima and maxima along the wavelength, being the transitions between minima and maxima well adapted to act as an edge filter. The wavelength range of this region is however controlled by the cavity length developed and thus, can be designed according to the desired need. When this FPI is implemented through the edge filtering technique, it is able to actively filter the incoming signal by simply adjusting the location of the filter, that is possible to be performed by adjusting the cavity length, e.g., strain, curvature, temperature, etc. The operation

*Corresponding author: oliveiraricas@av.it.pt

principle of the edge filtering technique implemented with a FPI may be seen in Fig. 2.

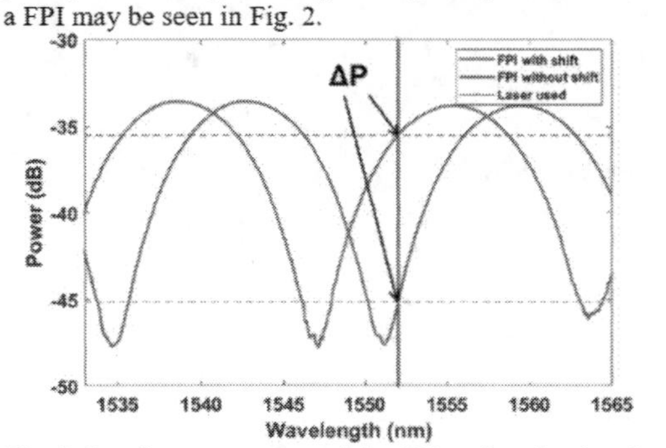

Fig. 2. Interference spectrum of the FPI cavity, showing the region where the laser operates and the correspondent power that can be filtered by the edge of the FPI, when the FPI is under strain.

As it is possibly to verify on Fig. 2, the laser power changes ΔP, when the FPI spectrum is shifted in wavelength (e.g., when subjected to an external load).

3 Methods and experimental setup

Firstly, we experimentally validate the proposed technology by measuring the movement of the membrane of a speaker that has been subjected to different sinusoidal waves. For that, it was used a resin based FPI sensor fabricated through the technique described in [3], [4], where two single mode fibers aligned transversely to each other and with a small longitudinal distance in between, are cold-spliced with a ultraviolet curable resin. To allow the FPI to read the oscillatory movement of the membrane on the speaker, one of the terminals of the optical fiber is glued perpendicular to it, while the other is used as the In/Out terminal, as is shown in Fig. 3**Erro! A origem da referência não foi encontrada.**. With this configuration, the cavity length could change as the membrane of the speaker goes up and down according the wave generated on the speaker. This will induce a shift in the spectrum of the FPI cavity that actively filters the power of the laser.

The signal detection is made in reflection using an optical circulator. For that, a narrowband laser centred at the linear region of the FPI travels up to the. The reflected power (already filtered) is fed into a photodetector through the output port of the circulator. The optical signal is converted to the electrical domain by the photodetector, and an Arduino board is used to read and to perform the digitalization of the signal. The data is transferred to a computer and finally, it is represented in real time through a dedicated graphical user interface. An FFT is also applied to the data allowing to get the frequency of the oscillation. The full schematic of the operation principle of the proposed interrogator and detection scheme may be seen in Fig. 3.

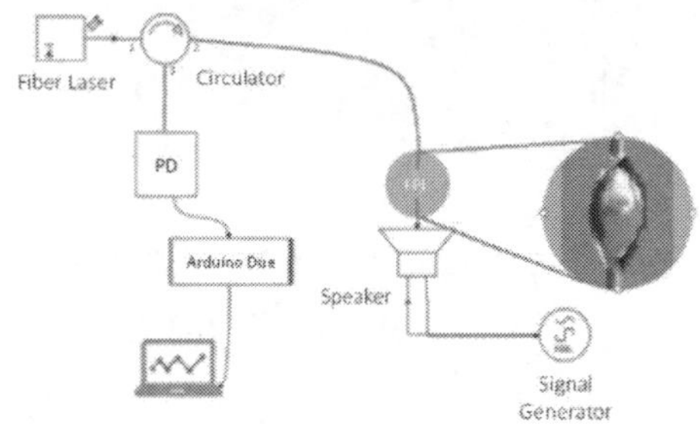

Fig. 3. Schematic of the operation principle. It is composed of a detection system on the left and by an FPI sensor attached to the membrane of a speaker on the right. The inset shows the microscope image of the resin based FPI sensor.

The data acquisition boards that compose this interrogator (i.e. circuit for the photodetector board and for the Arduino board) are assembled in a 3D printed box, allowing to get a more compact system for easy access. The final result of the interrogation system may be seen in Fig. 4**Erro! A origem da referência não foi encontrada.**.

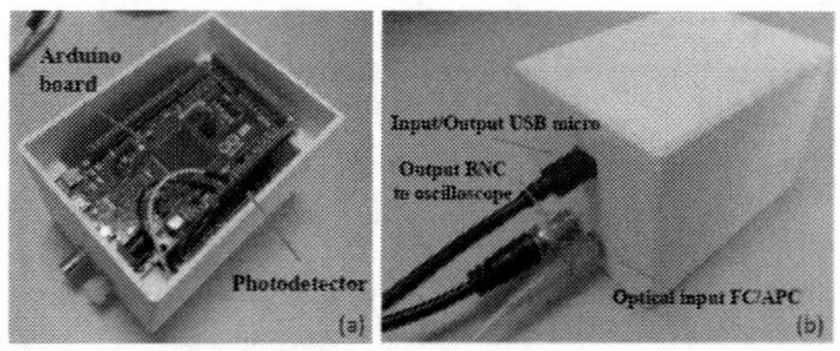

Fig. 4. (a) Inside and (b) outside picture of the data acquisition boards that compose the interrogator.

In Fig. 4, it is possible to observe that the front panel of the interrogator box contains three different connection ports, namely one FC/APC optical adaptor for the reading of the optical signal, one micro USB port to export data to a computer, and finally, a BNC port for ready visualization of the signal using an oscilloscope.

In a second stage of the work, we decided to implement the same detection scheme for the measurement of a pulse wave at the wrist region. For that, the FPI sensor was embedded on an elastic membrane that is latter attached to a fabric bracelet, allowing it to be stretched and adjusted to the wrist of different individuals. The computer aided design image of the elastic membrane may be seen in Fig. 5(a), while the fully developed sensor on Fig. 5(b).

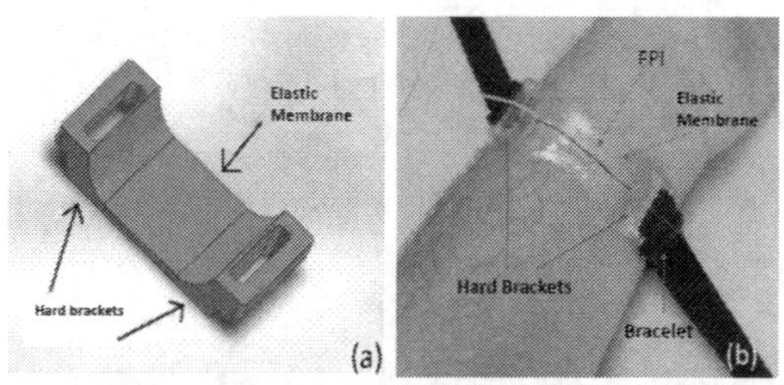

Fig. 5. (a) Drawing of the part used to measure the pulse wave and, (b) how it was implemented on the wrist region.

The principle behind the detection of the pulse wave relies on the blood pressure signal generated at the wrist region. Thus, the FPI embedded on the elastic membrane is subjected to a longitudinal strain, allowing it to change the cavity length and shift the spectrum according to the pulse wave, generating a power fluctuation during time.

5 Results and discussion

For the first part of the work, the data was taken for frequencies below and above 1 Hz. An example of the results taken for the 0.5 Hz and 2.4 kHz may be seen in Fig. 6(a) and (b), respectively.

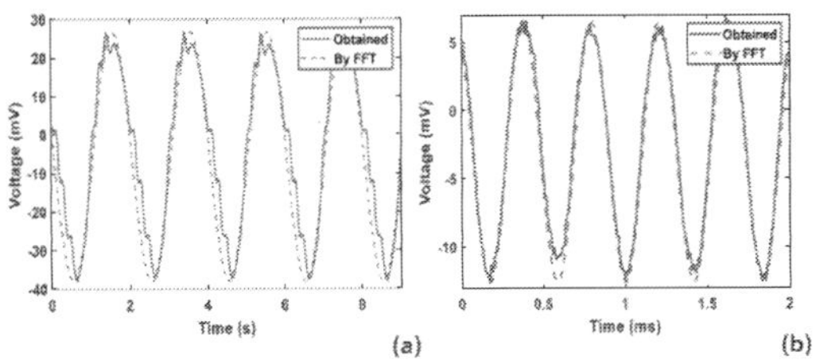

Fig. 6. Data taken for (a) 0.5 Hz and (b) 2.4 kHz

As shown in Fig. 6(a) and (b), the raw signals (blue traces), show a clear sinusoidal behaviour, that is in agreement with the movement performed by the membrane of the speaker. However, the signal also shows some noise, that is associated to the environmental conditions, such as noise from the equipment's. By performing an FFT to the data shown in Fig. 6, it is possible to calculate the frequency of the signal. Using that, we were able to reconstruct the signal as is shown by the red dotted lines.

The frequency collected from the FFT was also used to check the response of the sensor for a span of frequencies and the results may be seen in Fig. 7. From that, it is possible to observe that the sensor has a flat response up to 500 Hz. Then, the amplitude decreases possibly due to the elasticity of the membrane itself. However, the bandwidth fits quite well a variety of applications.

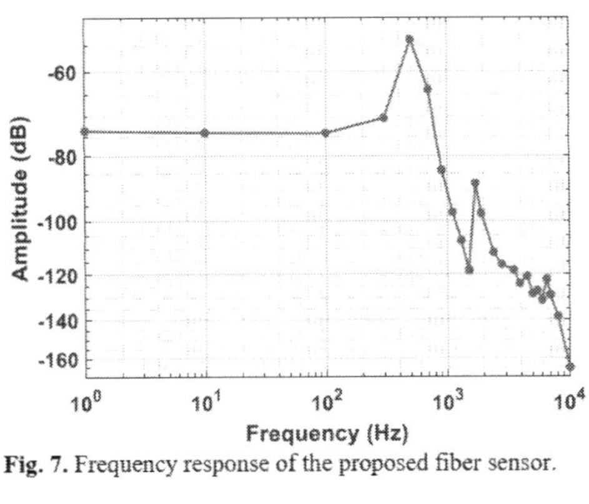

Fig. 7. Frequency response of the proposed fiber sensor.

The prototype was then experimentally tested as a wearable device for the monitoring of radial pulse wave. Despite the difficulties in acquiring the data, possibly due to the elasticity of the membrane, the results were similar to the expected ones, where an initial peak is achieved (main wave), followed by a dip (dicrotic notch), as is shown in Fig. 8.

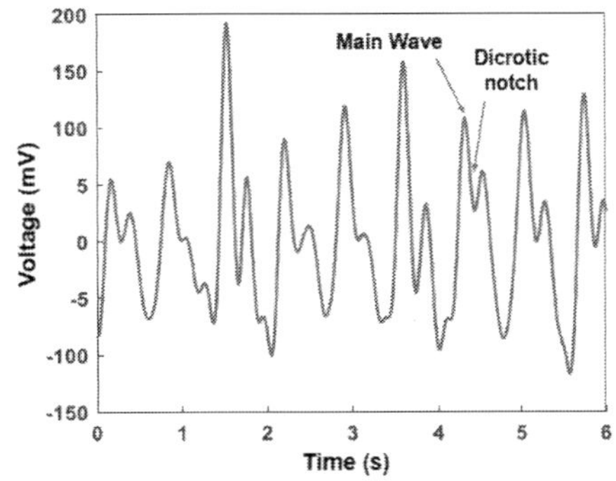

Fig. 8. Measured pulse wave signal at the wrist region.

The results revealed that the proposed fiber sensor can be used to measure the pulse wave signal in a simple and cost effective way.

In the overall we have shown that it is possible to transduce measurements commonly performed in the spectral domain, to simple power detection schemes that could be used with cost effective devices, paving the way for a simpler and cost effective use of fiber optic sensors.

6 Conclusions

In this work a low-cost fiber optic interrogation system has been implemented. The system uses an FPI sensor to actively filter the output power of a narrowband laser. By subjecting the FPI sensor to external conditions, such as load, the interference spectrum is wavelength shifted producing a power change that is then detected through a photodetector attached to an arduino board. The results showed that the proposed system has a flat response up to 500 Hz and that it could be used for instance in the detection of the pulse wave at the wrist region.

This work was funded by FCT-Fundação para a Ciência e Tecnologia and by the European Regional Development Fund (FEDER), through the Competitiveness and Internationalization Operational Programme (COMPETE 2020) of the Portugal 2020 framework and by national funds under the projects FOPEComSens (PTDC/EEI-TEL/1511/20), AQUATICsens (POCI-01-0145-FEDER-032057) and UIDB/50008/2020-UIDP/50008/2020.

4 REFERENCES

[1] A. Cusano, A. Cutolo, and J. Albert, *Fiber Bragg Grating Sensors: Research Advancements, Industrial Applications and Market Exploitation*, 1st ed. Bentham Science Publishers Ltd., 2011.

[2] Y. J. Rao, Z. L. Ran, and Y. Gong, *Fiber-optic fabry-perot sensors: An introduction*. 2017.

[3] R. Oliveira, L. Bilro, and R. Nogueira, "Fabry-

Pérot cavities based on photopolymerizable resins for sensing applications," *Opt. Mater. Express*, vol. 8, no. 8, pp. 899–902, 2018.

[4] R. Oliveira, L. Bilro, T. H. R. Marques, C. M. B. Cordeiro, and R. Nogueira, "Simultaneous detection of humidity and temperature through an adhesive based Fabry – Pérot cavity combined with polymer fiber Bragg grating," *Opt. Lasers Eng.*, vol. 114, pp. 37–43, 2019.

https://doi.org/10.1051/epjconf/202125512009

Microfluidic devices manufacturing combining stereolithography and pulsed laser ablation

Bastián Carnero[1], *Carmen* Bao-Varela[1], *Ana Isabel* Gómez-Varela[1], and *María Teresa* Flores-Arias[1,*]

[1]Photonics4Life research group, Department of Applied Physics, Facultade de Física and Facultade de Óptica e Optometría, Universidade de Santiago de Compostela, Campus Vida, E-15782 Santiago de Compostela, Spain

Abstract. 3D printing has revolutionized the field of microfluidics manufacturing by simplifying the typical processes offering a considerable accuracy and user-friendly procedures. For its part, laser ablation proves to be a versatile technology to perform detailed surface micropatterning. A hybrid technique that combines both technologies is proposed, employing them in their most suitable range of dimensions. This technique allows to manufacture accurate microfluidics devices as the one proposed: a microchannel, obtained using a stereolithographic printer, coupled with an array of microlenses, obtained by pulsed laser ablation of a 3D printed master.

1 Introduction

Optical technologies have proven their versatility in the manufacturing of microfluidic devices of different sizes, geometries, and functionalities. Historically, photolithography or reactive ion etching have been the most used manufacturing techniques in microfluidics [1], requiring the manipulation of very complex masks and producing a considerably polluting chemical waste. These techniques, although precise, show limitations by only allowing superficial structuring of various photoresists and polymers. This has recently changed thanks to advances in lasers, promoting new technologies to appear. In this regard, stereolithographic 3D printers stand out by allowing to structure many materials in volume with high structural complexity [2], that are increasingly used for manufacturing microfluidic devices for biological applications [3]. Laser ablation has also simplified the micropatterning of a vast quantity of materials since it is a self-sufficient process that does not need of posterior chemical treatments neither masks to pattern the surfaces. Recent works have even proposed strategies to perform laser ablation of 3D surfaces in real time [4].

This contribution presents a novel hybrid technique for the manufacturing of microfluidic devices that combines laser ablation with an emerging 3D printing optical technology: stereolithography (SLA). In this technique, each technology is used to manufacture structures in the dimension range that is most suitable for it. Thereby, an SLA 3D printer is used to obtain 3D printed masters in resin with microchannels of hundreds of microns. Meanwhile, laser ablation is performed on the surface of this printing resin to create fine detailed structures in the range of tens of microns, unreachable for most current 3D printers. Finally, the channels and micropatterned structures of the masters are replicated by soft lithography of polydimethylsiloxane (PDMS). The latter one of the most used polymers to manufacture biological microfluidic devices [5], given its optical transparency to visible and UV light, permeability to gases, elasticity, and biocompatibility.

2 Materials and Methods

A Form 3B SLA printer (Formlabs, Somerville, Massachusetts) was used in this work to manufacture the microchannel master as well as the substrate for ablation. SLA printers have a tank in which a liquid printing resin is deposited so a proper light source (generally a laser) performs a selectively layer-by-layer photopolymerization (Fig. 1).

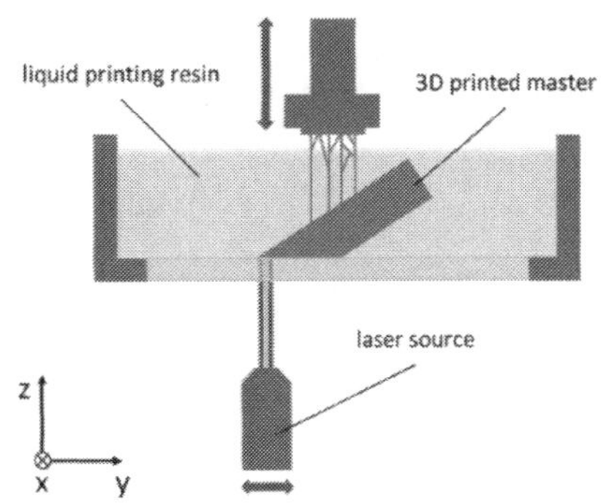

Fig. 1. SLA 3D printing principle.

The used Form 3B printer employs Low Force Stereolithography (LFS), a novel technique that improves the precision and accuracy of printed pieces, by reducing the manufacturing stresses that the pieces suffer during

* Corresponding author: maite.flores@usc.es

the printing. For this, a resin tank with a flexible sheet at its base that can be gradually deformed when the piece is pressed against it is used. For the polymerization of the resin, the laser is always addressed perpendicular to the piece thanks to a galvanometric system. Commercial Model V2 resin from Formlabs was selected as printing resin given the high precision it offers, its good performance when replicating polymers and the good response offered to laser ablation. In the postcuring process, the pieces are washed in isopropanol and introduced into a UV chamber for 60 minutes at 60°C.

Pulsed laser ablation (PLA) of the 3D printed substrate was performed in the STELA laser of the Laser Laboratory for Acceleration and Applications (L2A2), a research facility of the University of Santiago de Compostela, using a beam of 800 nm of wavelength, pulse duration of 35 fs and repetition rate of 1kHz. The laser was focused on the surface of the resin employing a Mitutoyo M Plan APO NIR 20x objective.

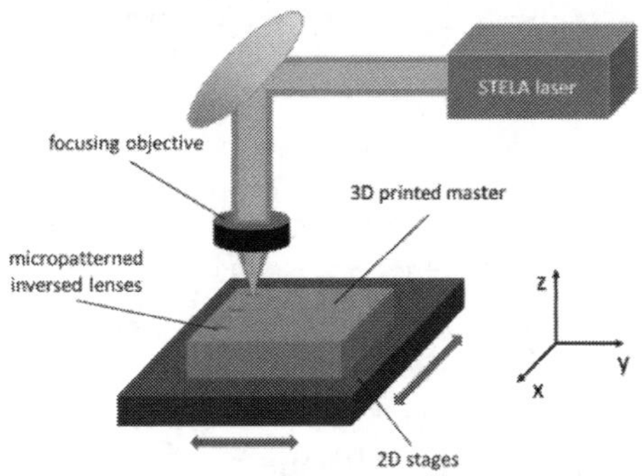

Fig. 2. PLA micropatterning principle.

PDMS was selected as soft lithography replication polymer given the previous introduced reasons. The PDMS was prepared from Sylgard 184 elastomer by mixing the monomer and the curing agent in a ratio 10:1. The mixture was then deposited over the structures fabricated by 3D printing and ablation (with walls added to avoid losses of liquid polymer) forming a thin layer, introduced in a vacuum chamber for 40 min at 400 mbar to remove bubbles produced during the mixing process and cured in an oven for 12 h at 60 °C.

For bonding, a Diener Zepto (Diener, Ebhausen, Germany) plasma cleaner was employed, exposing the PDMS surfaces to an oxygen plasma. First, vacuum was performed in the chamber until reach 0.1 mbar. Next, oxygen was introduced into the chamber during 7 minutes until a pressure of 0.4 mbar was reached. Then, the generator was activated, and the plasma initialized. Power was fixed at 50 W and exposure lasted for 20 seconds, for glass-PDMS bonding, and 10 seconds, for PDMS-PDMS bonding. Next, the plasma cleaner vented and reached ambient pressure. Finally, the pieces were put in contact and introduced into a furnace to receive a thermal treatment (to promote full bonding) of 20 minutes at 100 °C.

3 Results and Discussion

To test the possibilities offered by the proposed hybrid technique, a sandwiched microfluidic device was manufactured by bonding two structured layers of PDMS to a microscope slide. The top layer (cover) was obtained from a master superficially micropatterned by PLA, featuring an array of microlenses for focusing light on the microchannel presented in the bottom layer (substrate). The proposed device could be used to perform selective photopolymerization of a small photocurable sample that flows through the microchannel [6].

3.1 Cover manufacturing: microlenses arrays

For the manufacture of the cover, a wafer was printed in Model resin, on which surface ablation tests were performed, as shown in Fig. 2. By PLA, it was possible to create inverse microlenses in the resin, to be later replicated with PDMS and obtain various arrays of microlenses with different geometries and periods (Fig. 3).

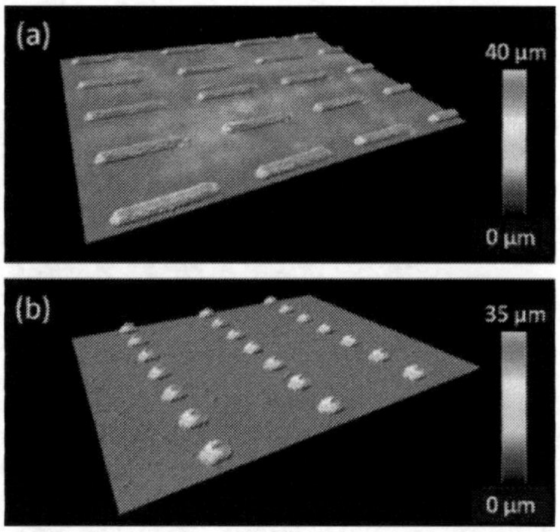

Fig. 3. Confocal images of a) cylindrical and b) hemispherical microlenses of PDMS.

Thereby, cylindrical lenses of 35 µm of radius, approximately 125 µm of focal distance and 150 µm of length were patterned with 20 mm/s of velocity and 30 mW of power on the unpolished resin surface, forming the array showed in Fig. 3a, with 240 µm of period. Moreover, hemispherical lenses of 25 µm of radius and approximately 90 µm of focal distance were patterned with 60 mW of power on the polished surface, forming the array showed in Fig. 3b, with 240 µm of period.

The different geometries of the microlenses achievable with this hybrid technique will determinethe way in which the liquid sample can be selective photopolymerized, allowing to obtain a wide range of polymerized microdroplets, with the desired size and sphericity for each specific application.

https://doi.org/10.1051/epjconf/202125512009

3.2 Substrate manufacturing: microchannel

The second manufactured element was a master to obtain a square microchannel of 200 µm in height. The outward microchannel with connectors was designed on the bottom of a 20x40 mm box, using a CAD-like software (Fig. 4a). The resin printed piece was postprocessed to optimize its mechanical properties and then easily replicated by PDMS soft lithography, forming the inward replica of the previous structure. A 300x500 µm array of hemispherical microlenses was fabricated in PDMS and bonded over the start of the channel (Fig 4b). Finally, the layers were bonded to a microscope slide, to obtain a fully functional and easily handled microfluidic device.

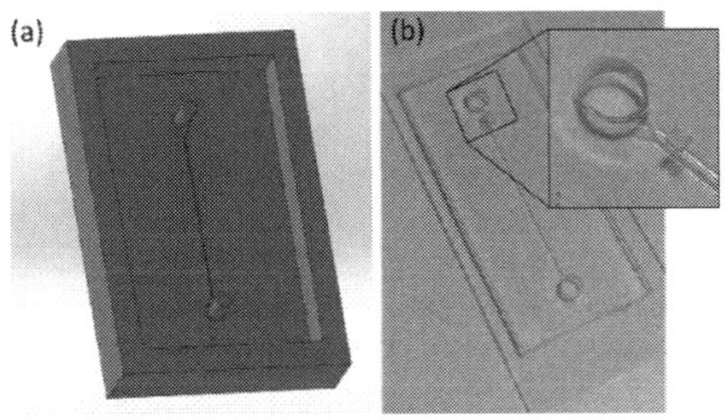

Fig. 4. a) CAD designed master for the microchannel. b) Picture of the final microfluidic device obtained after the bonding of the PDMS sheets to the microscope slide.

4 Conclusions

Microfluidics is a multidisciplinary field that requires different technologies capable of manufacturing highly accurate structures in a reliable and precise way. 3D printing has the potential to revolutionize these areas, simplifying manufacturing processes through simple procedures and very low waste production. For its part, laser ablation proves to be a versatile and much more user-friendly technique than photolithography to perform detailed surface micropatterning (unreachable for many current 3D printers).

In this work, a hybrid technique that combines both technologies is proposed, taking advantage of each of them in the range of dimensions in which they show most suitable results. For this, structures of hundreds of microns were printed using a SLA 3D printer. This structures, manufactured in printing resin, were later selective micropatterned by direct laser ablation to obtain inverted microlenses of tens of microns on its surface. Finally, these patterns can be replicated using transparent polymers (as PDMS) to obtain arrays of functional microlenses. This technique allows to manufacture accurate devices such as the one proposed: a microchannel (obtained from a 3D printed master) coupled with an array of microlenses (obtained from a 3D printed master micropatterned by PLA).

Acknowledgements: This work has been sponsored by contracts AEI RTI2018-097063-B-100, AEI/FEDER, UE; ED431B 2020/29; ED431E 2018/08 and ED481D-2021-019, Consellería de Educación Xunta de Galicia/FEDER e Estructuración Xunta de Galicia, IN607A2019-02.

References

1. R. H. Liu, M. A. Stremler, K. V. Sharp, M. G. Olsen, J. G Santiago, R. J. Adrian, H. Aref, D. J. Beebe, Journal of microelectromechanical systems, **9.2**, 190-197 (2000)
2. B. Heidt, R. Rogosic, S. Bonni, J. Passariello-Jansen, D. Dimech, J. W. Lowdon, R. Arreguin-Campos, E. Steen Redeker, K. Eersels, H. Diliën, B. van Grinsven, T. J. Cleij, Phys. Status Solidi. **217.13**, 1900935 (2020)
3. K. Piironen, M. Haapala, V. Talman, P. Järvinen, T. Sikanen, Lab Chip, **20.13**, 2372-2382 (2020)
4. A. J. Lopez, J. Lamas, J. S. Pozo-Antonio, T. Rivas, A. Ramil, Opt Lasers Eng, **126**, 105897 (2020)
5. J. C. McDonald, D. C. Duffy, J. R. Anderson, D. T. Chiu, H. Wu, O. J. A. Schueller, G. M. Whitesides, Electrophoresis, **21.1**, 27-40 (2000)
6. T. Nisisako, H. Suzuki, T. Hatsuazawa, Micromachines, **6.10**, 1435-1444 (2015).

On polarization characteristics of highly focused fields concentrated along the axis

Rosario Martínez-Herrero[1,*], *Artur* Carnicer[2,**], *Ignasi* Juvells[2,***], and *Ángel S.* Sanz[1,****]

[1]Department of Optics, Faculty of Physical Sciences, Universidad Complutense de Madrid (UCM),
Pza. Ciencias 1, Ciudad Universitaria – 28040 Madrid, Spain
[2]Universitat de Barcelona (UB), Facultat de Física, Departament de Física Aplicada, Martí i Franquès 1, 08028 Barcelona, Spain

Abstract. After having proven that an uncertainty relation holds for the on-axis power content of highly-focused fields, in this Communication we explore and discuss the consequences of such a relation concerning the polarization state characterizing the fields that satisfy it.

1 Introduction

When referring to the concept of structured light, one finds in the literature a wide variety of works concerning the generation and application of custom light fields (see, for instance, Ref. [1] and references therein). The phenomenon arises whenever a vector light beam is focused by a high numerical aperture objective. In such cases, the electric field acquires a component along the optical axis, which eventually turns into a longitudinal component [2]. Accordingly, tightly focused vector fields will be able to exhibit complex and versatile polarization topologies.

The study of the field in the focal region of an optical system with high numerical aperture has received much attention in the last two decades, because of its potential applications in microlithography, image processing, diffractive optics, adaptive optics, or holographic data storage [2–19]. Furthermore, the analysis of the on-axis intensity of highly focused optical fields has also proven to be of much practical interest, because it allows us to design structures such that their behavior along the axis can be modulated [7, 9, 11, 14–16, 20].

Within this scenario, recently [21] a novel and convenient measure for the size of the region where the axial power content mainly concentrates has been proposed on the basis of an uncertainty principle. From it, it is also possible to determine the functional form of the incident beam that warrants the equality in the uncertainty relationship. Interestingly, this incident field has not a unique polarization structure.

In this Communication, we present some results about the polarization features of the focused field generated from such paraxial incident beams, characterized by different polarization states. This analysis includes global polarization features [22]. Furthermore, we investigate and discuss the influence on the polarization and transverse structure of the focused field as the σ-parameter is changed (note that this parameter contains information about the lower bound of the uncertainty principle). To be self-contained, in next section some basic definitions are introduced, while the main finding are discussed in Sec. 3

2 Basic definitions

To better understand the point of our contribution, let us first introduce a series of basic definitions. Thus, consider a monochromatic beam at the entrance pupil of an aplanatic focusing system with a high numerical aperture. Following the theory of vector field propagation, the electric field at the focal region can be written as

$$\mathbf{E}(r,\phi,z) = A\int_0^{\theta_0}\int_0^{2\pi} \mathbf{E}_0(\theta,\varphi)e^{ikr\sin\theta\cos(\phi-\varphi)} \times e^{-ikz\cos\theta}\sin\theta d\theta d\varphi, \qquad (1)$$

in terms of the so-called Richards-Wolf integral [23]. In this expression, (r,ϕ,z) denote the cylindrical coordinates at the focal area, with θ and φ being the polar and azimuthal angles at the reference Gaussian sphere, respectively. Furthermore, in (1) the input vector angular spectrum reads as

$$\mathbf{E}_0 = \sqrt{\cos\theta}\left[(\mathbf{E}_s\cdot\mathbf{e}_1)\,\mathbf{e}_1 + (\mathbf{E}_s\cdot\mathbf{e}_2')\,\mathbf{e}_2\right], \qquad (2)$$

where $\mathbf{E}_s$ is the transverse beam distribution at the entrance pupil of the optical system, while the unit vectors $\mathbf{e}_1$ and $\mathbf{e}_2$ are defined as

$$\mathbf{e}_1 = \begin{pmatrix} -\sin\varphi \\ \cos\varphi \\ 0 \end{pmatrix}, \qquad \mathbf{e}_2 = \begin{pmatrix} \cos\theta\cos\varphi \\ \cos\theta\sin\varphi \\ \sin\theta \end{pmatrix}. \qquad (3)$$

These unit vectors point along the azimuthal and radial directions, respectively, while $\mathbf{e}_2' = (\cos\varphi, \sin\varphi, 0)$ denotes

*e-mail: r.m-h@fis.ucm.es
**e-mail: artur.carnicer@ub.edu
***e-mail: ignasi.juvells@ub.edu
****e-mail: a.s.sanz@fis.ucm.es

https://doi.org/10.1051/epjconf/202125512010

the projection of $\mathbf{e}_2$ onto the entrance pupil plane. As for the other parameters in (1), A is a constant, k is the wave number, and $\theta_0 = \max\{\theta\}$ is the semi-aperture angle, with the numerical aperture being NA $= \sin\theta_0$.

Taking into account the field given by (2), the on-axis intensity distribution reads as

$$I(0,z) = |\mathbf{E}(0,z)|^2 = \left| \int_0^{\theta_0} \tilde{\mathbf{F}}(\theta) e^{-ikz\cos\theta} \sin\theta d\theta \right|^2, \quad (4)$$

where we introduce a new vector field, defined as

$$\tilde{\mathbf{F}}(\theta) \equiv A \int_0^{2\pi} \mathbf{E}_0(\theta,\varphi) d\varphi. \quad (5)$$

After recasting this new field in terms of a new variable $\alpha \equiv \cos\theta$ (with $\alpha_0 = \cos\theta_0$) and the electric field in terms of a dimensionless variable $\bar{z} \equiv kz$, the intensity (4) can be rewritten as

$$I(0,z) = |\mathbf{E}(0,\bar{z})|^2 = \left| \int_{\alpha_0}^{1} \tilde{\mathbf{F}}(\alpha) e^{-i\bar{z}\alpha} d\alpha \right|^2. \quad (6)$$

This expression highlights a dual relationship between the electric field, $\mathbf{E}(0,\bar{z})$, and the new field $\tilde{\mathbf{F}}(\alpha)$, with $\bar{z}$ and α playing the role of canonically conjugate variables.

3 Results

As it has recently been shown [21], the dispersion in the variable α of new vector field $\tilde{\mathbf{F}}(\alpha)$ and the axial power content satisfy an uncertainty relation. Upon the basis of this uncertainty principle, it has been proposed a novel and adequate measure for the size of the region where such an axial power content mainly concentrates. The functional form for the incident beam that satisfies the equality of such an uncertainty relation is given by

$$\tilde{\mathbf{F}}(\alpha) = e^{-\frac{\sigma}{2}\frac{(\alpha-\bar{\alpha})^2}{(1-\alpha_0)^2}} \mathbf{u}, \quad (7)$$

where $\mathbf{u}$ is a unitary constant vector, σ an arbitrary real, positive parameter, and

$$\bar{\alpha} = \frac{1}{I_0} \int_{\alpha_0}^{1} \alpha |\tilde{\mathbf{F}}(\alpha)|^2 d\alpha, \quad (8)$$

with

$$I_0 = \int_{\alpha_0}^{1} |\tilde{\mathbf{F}}(\alpha)|^2 d\alpha. \quad (9)$$

By inspecting the quantity (7), it is seen that, interestingly, incident fields with different polarization states give rise to the same on-axis irradiance distribution.

In particular, here we consider two polarization states for incident paraxial field, namely, linear (l) and circular (c). That is, in Eq. (2) we insert the expression

$$\mathbf{E}_{s,j}(\theta) = g(\theta)\mathbf{u}_j, \quad (10)$$

with $j = l, c$, and where

$$\mathbf{u}_l = \begin{pmatrix} 1 \\ 0 \end{pmatrix}, \qquad \mathbf{u}_c = \begin{pmatrix} 1 \\ i \end{pmatrix}, \quad (11)$$

and

$$g(\theta) = \frac{F(\theta)}{\sqrt{\cos\theta}\,(1+\cos\theta)}. \quad (12)$$

In the latter expression, $F(\theta)$ denotes the modulus of the vector field defined in (7) – also remember that $\alpha = \cos\theta$, as defined earlier on, in Sec. 2.

If we consider the linearly polarized incident field, at the focal plane ($z = 0$) we find

$$\mathbf{E}_{0,l}(r,\phi) = \begin{pmatrix} D_0(r) + \cos 2\phi\, D_2(r) \\ \sin 2\phi\, D_2(r) \\ 2i\cos\phi\, D_1(r) \end{pmatrix} \quad (13)$$

where the following definitions have been introduced for simplicity:

$$D_0(r) = \int_0^{\theta_0} \sqrt{\cos\theta}\, g(\theta)(1+\cos\theta) J_2(kr\sin\theta) \sin\theta d\theta, \quad (14a)$$

$$D_1(r) = \int_0^{\theta_0} \sqrt{\cos\theta}\, g(\theta) \sin\theta J_1(kr\sin\theta) \sin\theta d\theta, \quad (14b)$$

$$D_2(r) = \int_0^{\theta_0} \sqrt{\cos\theta}\, g(\theta)(1-\cos\theta) J_2(kr\sin\theta) \sin\theta d\theta. \quad (14c)$$

As it can be noticed, the polarization state of the field (13) is fully nonuniform over the whole focal plane, since it depends at each point on ϕ. Therefore, the cylindrical symmetry around the optical axis is broken. We also notice that, at the focus ($r = 0$), the field is linearly polarized for any value of the σ-parameter. At any other point, the field is elliptically polarized, in general. Finally, the electric field at each point ($r \neq 0, \phi \neq \pi/2$) is contained within a plane defined by the normal

$$\mathbf{N}_l(r,\phi) = 2D_1(r)\cos\phi \begin{pmatrix} \sin 2\phi D_2(r) \\ -D_0 + \cos 2\phi D_0 \\ 0 \end{pmatrix}. \quad (15)$$

That is, the normal $\mathbf{N}_l$ is always contained within the XY-plane.

If we now consider the incident circular field, we find the following expression for the field at the focal plane:

$$\mathbf{E}_{0,c}(r,\phi) = \begin{pmatrix} D_0(r) + e^{2i\phi} D_2(r) \\ iD_0(r) - ie^{2i\phi} D_2(r) \\ 2ie^{i\phi} D_1(r) \end{pmatrix}. \quad (16)$$

where the same definitions (14) also hold. In this case, the field polarization is also totally nonuniform at the focal plane, although the polarization state is independent of ϕ, hence being the same along concentric rings around the optical axis. At the focus, $r = 0$, the field is circularly polarized for any value of the σ-parameter, while the polarization state in any other point is, in general, elliptical. Finally, it is also seen that the electric field is contained within the plane defined by the normal

$$\mathbf{N}_c(r,\phi) = \begin{bmatrix} 2D_1(r)\sin\phi\left(D_0(r) + D_2(r)\right) \\ -2D_1(r)\cos\phi\left(D_0(r) + D_2(r)\right) \\ D_0^2(r) - D_2^2(r) \end{bmatrix}, \quad (17)$$

with a z-component independent of ϕ.

4 Concluding remarks

As it is mentioned above, the purpose of this Communication is to present some results about the polarization features of the focused field generated from paraxial incident beams with different polarization states. In particular, we have considered linearly and circularly polarized incident fields, and have investigated the influence on the polarization and transverse structure of the focused field in terms of the σ-parameter, which contains information about the lower bound of the uncertainty principle introduced in Ref. [22]. In both cases, as discussed in Sec. 3, a series of common features are observed (although with some particularities proper of each polarization state). First, the field polarization at the focal plane is fully nonuniform, although circularly polarized fields preserve a rotational symmetry, which linearly polarized ones do not. Second, at the focus, the field preserves the polarization of the incident field for any value of the σ-parameter. Finally, the fields are contained within planes defined by the corresponding normal; in the case of linearly polarized incident fields this normal is perpendicular to the XY-plane, while for circularly polarized fields there is a z-component, although it is independent of ϕ, thus in compliance with the associated rotational symmetry (around the z-axis, for a given value of r).

Acknowledgments

Financial support is acknowledged to the Spanish Research Agency (AEI) and the European Regional Development Fund (ERDF) (grants Nos. PID2019-104268GB-C21 and PID2019-104268GB-C22).

References

[1] A. Forbes, M. de Oliveira, and M. R. Dennis, Nat. Photonics **15**, 253-262 (2021).

[2] R. Dorn, S. Quabis, and G. Leuchs, Phys. Rev. Lett. **91**, 233901 (2003).

[3] Y. Kozawa and S. Sato, J. Opt. Soc. Am. A **24**, 1793–1798 (2007).

[4] C. Maurer, A. Jesacher, S. Fu¨rhapter, S. Bernet, and M. Ritsch-Marte, New J. Phys. **9**, 78 (2007).

[5] Y. Zhao, J. S. Edgar, G. D. M. Jeffries, D. McGloin, and D. T. Chiu, Phys. Rev. Lett. **99**, 073901 (2007).

[6] K. Lindfors, A. Priimagi, T. Setälä, A. Shevchenko, A. T. Friberg, and M. Kaivola, Nat. Photonics **1**, 228–231 (2007).

[7] H. Wang, L. Shi, B. Lukyanchuk, C. Sheppard, and C. T. Chong, Nat. Photonics **2**, 501-505 (2008)

[8] G. M. Lerman and U. Levy, Opt. Express **16**, 4567-4581 (2008).

[9] H.T. Wang, X.L. Wang, Y. Li, J. Chen, C.S. Guo, and J. Ding, Opt. Express **18**, 10786-10795 (2010).

[10] R. Martínez-Herrero and P. Mejías, Opt. Express **18**, 7965-7971 (2010).

[11] F. Kenny, D. Lara, O. Rodríguez-Herrera, and C. Dainty, Opt. Express **20**, 14015–14029 (2012)

[12] J. A. Fan, K. Bao, J. B. Lassiter, J. Bao, N. J. Halas, P. Nordlander, and F. Capasso, Nano Lett. **12**, 2817-2821 (2012).

[13] R. Martínez-Herrero, I. Juvells, and A. Carnicer, Opt. Lett. **38**, 2065-2067 (2013).

[14] D. Maluenda, R. Martínez-Herrero, I. Juvells, and A. Carnicer, Opt. Express **22**, 6859–6867 (2014).

[15] R. Martínez-Herrero, I. Juvells, and A. Carnicer, Opt. Lett. **39**, 6025-6028 (2014)

[16] R. Martínez-Herrero, D. Maluenda, I. Juvells, and A. Carnicer, Opt. Express **22**, 32419-32428 (2014).

[17] T. Bauer, S. Orlov, U. Peschel, P. Banzer, and G. Leuchs, Nat. Photonics **8**, 23-27 (2014).

[18] T. Bauer, P. Banzer, E. Karimi, S. Orlov, A. Rubano, L. Marrucci, E. Santamato, R. W. Royd, and G. Leuchs, Science **347**, 964-966 (2015).

[19] S. Roy, K. Ushakova, Q. van den Berg, S. F. Pereira, and H. P. Urbach, Phys. Rev. Lett. **114**, 103903 (2015).

[20] L. Hang, Y. Wang, and P. Chen, Opt. Express **28**, 12028–12036 (2020).

[21] R. Martínez-Herrero, A. Carnicer, I. Juvells and A. S. Sanz, Opt. Express **28**, 29676-29690 (2020).

[22] R. Martínez-Herrero, P. M. Mejías, and G. Piquero, Opt. Commun. **265**, 6-10 (2006).

[23] B. Richards and E.Wolf, Proc. R. Soc. Lond. A **253**, 358-379 (1959).

https://doi.org/10.1051/epjconf/202125512011

Optical characterization of strain sensing cables for Brillouin optical time domain analysis

Filippo Bastianini [1], *Francesco* Falcetelli [2], *Paweł* Bocheński [3], *Leonardo* Rossi [4], *Raffaella* Di Sante [2], and *Gabriele* Bolognini [4,*]

[1]Sestosensor s.r.l., 140 via Gesso, 40069 Zola Predosa, Italy
[2]Dept. of Industrial Engineering, University of Bologna, Forlì, Italy
[3]Fibrain Sp. z.o.o., Wspólna 4A, Rzeszów 35-205, Poland
[4]Consiglio Nazionale delle Ricerche, IMM Institute, 101 via P. Gobetti, 40129 Bologna Italy

Abstract. Two innovative optical fibre cable layouts designed to improve BOTDA strain measurement accuracy through improved strain transfer efficiency are presented, discussed and tested through experiments, analytical and numerical modelling. The second improved design presents good features to minimize the mismatch between measured and actual strain.

1 Introduction

Optical fiber sensors represent one of the most interesting technologies for the monitoring of structures in many different fields. Among their many advantages compared to their electronic counterparts, optical fiber sensors are light weight, small size (which makes them suitable to be embedded in composites and other materials), immune to electromagnetic interference and capable of transmitting a large amount of information [1-7]. Among the various distributed sensing techniques, Brillouin optical time domain analysis (BOTDA) gained signficant relevanceamong the scientific and industrial community due to its capability, through the use of the Stimulated Brillouin Scattering (SBS) phenomenon, to monitor large structures by extracting temperature and strain profiles along tens of km of sensing fiber with spatial resolutions of the order of few centimetres [8-10]. While a great deal of study focuses on the interrogating system, an important factor for BOTDA sensors performance is the structure on the sensing cable. This is because in most applications, exposure to harsh environmental conditions makes it necessary for the sensing fiber to be protected by additional layers, but this can cause a mismatch between the strain at the core (the measured strain) and the strain in the outer cable sheath (the real strain) [11], which becomes even more significant. In this study, an experimental setup based on swept Swept Wavelength Interferometry (SWI) distributed strain sensing technique is used alongside analytical and numerical simulations to accurately evaluate the strain affecting sensing cables built from two new designs intended for BOTDA and is compared to the strain perceived by the structure they are attached to in order to evaluate their strain transfer efficiency.

2 Sensing cable design

The two designs are shown in Fig. 1. The first sensing cable design was developed with the following goals as standard: (i) optimized mechanical coupling with the structure; (ii) protection against harsh environmental conditions; (iii) stable attenuation and sensitivity over the range of operational conditions; (iv) capability of withstanding large deformations; (v) absence of creep and other long-term phenomena capable of degrading the measuring performance of the cable. During preliminary testing, the performance of this design raised several concerns in terms of fiber uniformity, decoupling of the temperature compensation fiber and overall mechanical coupling between the strain sensing fiber and the observed structure. The second design, compared to the first one, was introduced as a way to address these issues with the

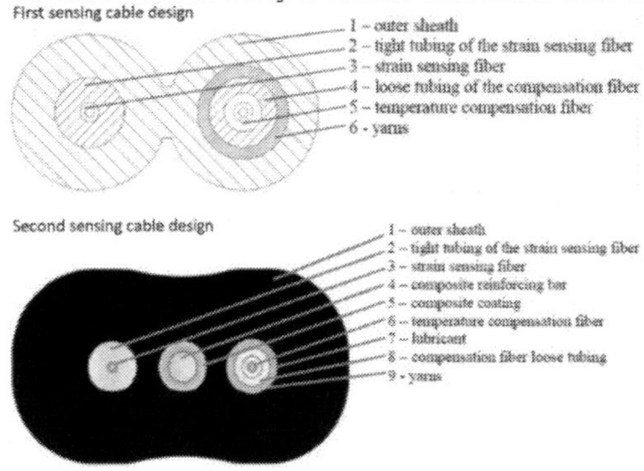

Fig. 1. Schematics of the cable prototypes.

following changes: (i) lower speed ramp-up during fabrication to ensure better uniformity of the fibre pre-

[*] Corresponding author: bolognini@bo.imm.cnr.it

https://doi.org/10.1051/epjconf/202125512011

stress; (ii) the presence of a silane-based lubricant to reduce the drag on the temperature compensation fiber; (iii) different cable jacket material to avoid slippage with the inner fibre tubing; (iv) structured surface to improve the mechanical coupling with the host structure. These changes are expected to improve overall sensing performance in terms of strain sensing accuracy, better distinction with temperature readings, and more reliable strain transfer.

3 Testing Methods

First, the performance of the devices was evaluated through numerical simulation based on the development of a Finite Element Method (FEM) model of the two cables using Abaqus/CAE, in which every single layer of the cables was taken into account. The shape of the cables was also accurately reproduced based on direct observation of the cable cross section via microscope. Between every layer was assumed to be present a perfect bonding which translated in applying a tie constraint between their surfaces. The behavior of the cables under different loads was simulated by imposing fixed displacements in the substrate structure, thus reproducing real experimental conditions.

Afterwards, the strain transfer of the two cables was experimentally assensed through the use of an Optical Backscatter Reflectometer (OBR) based on Swept-Wavelength Interferometry (SWI) [12-14]. SWI involves measuring the Rayleigh backscattered signal generated by refractive index variations in silica fibers. For each point of the fiber, the strain value can be derived from the reflected wavelength λ_R by treating the fiber as a Fiber Bragg Grating (FBG) sensor. While the change in refractive index is generally nonperiodical, it is possible to define a local average perturbation period Λ_{AVG} dependent on λ_R and the effective refractive index n_{eff} through the relation:

$$\lambda_R = 2n_{eff}\Lambda_{AVG} \tag{1}$$

By detecting the Rayleigh backscatter before the experiment, it is possible to determine a stable baseline. When a certain point of the fiber is subject to strain or temperature variations, the corresponding reflected wavelength in the trace shifts from the baseline by an amount $\Delta\lambda_R$ defined by the relation:

$$\frac{\Delta\lambda_R}{\lambda_R} = -\frac{\Delta\nu_R}{\nu_R} = K_\varepsilon\varepsilon + K_T\Delta T \tag{2}$$

Where K_ε and K_T are the strain and temperature coefficients, respectively, while ν_R and $\Delta\nu_R$ are the corresponding reference frequency and frequency shift. In order to completely exclude the contribution $K_T\Delta T$ caused by temperature variation, all experiments in this work were performed at constant ambient temperature to filter any undesired thermal evolution, and thus the temperature term can be considered negligible. For what concerns the strain term $K_\varepsilon\varepsilon$, the strain coefficient K_ε is defined as:

$$K_\varepsilon = 1 - \rho_\varepsilon \tag{3}$$

Where ρ_ε is the photoelastic coefficient, which is usually calculated with the following formula:

$$\rho_\varepsilon = \frac{n_{eff}^2}{2}[p_{12} - \nu(p_{11} + p_{12})] \tag{4}$$

Where p_{11} and p_{12} are the components of the strain optic tensor, while ν is Poisson's ratio [15]. For standard germanium doped silica fibers, like the ones used in this work, the photoelastic coefficient is known to be equal for 0.22. As a result, the strain coefficient should be equal to 0.78. Knowing these parameters and the central wavelength $\bar{\lambda}$ of the scanning process, it is possible to determine the relation between the strain and the frequency shift as a linear relation of the form:

$$\varepsilon = -\frac{\bar{\lambda}}{cK_\varepsilon}\Delta\nu_R = \alpha\Delta\nu_R \tag{5}$$

From this formula, assuming $\bar{\lambda} = 1306$ nm for this measurement, the value for α is found to be -5.59 $\mu\varepsilon$/GHz. While this value may be realistic, it is still affected by variable factors such as dopant species concentration in the fiber core and outer layer composition (including cladding and coating) which all contribute to the uncertainty of the estimation [16]. Because of this, the strain coefficient was ultimately found through an experimental calibration process, which involved inserting the fibers in a test rig employing a high-precision linear actuator with a travel range of 500 mm and a resolution of 0.02 mm, allowing for characterization of up to 3.5 m of fibers. The frequency displacement was measured while the fiber was subject to different strain values, ranging from 0 to 4000 με, with a step of 100 με. The results of the calibration for both cable design are shown in Fig. 2. From linear fitting, the α parameter, which was the inverse of the slope, was determined to be $\alpha_1 = -5.520 \pm 0.018$ $\mu\varepsilon$/GHz for the first cable design and $\alpha_2 = -5.590 \pm 0.023$ $\mu\varepsilon$/GHz for the second one.

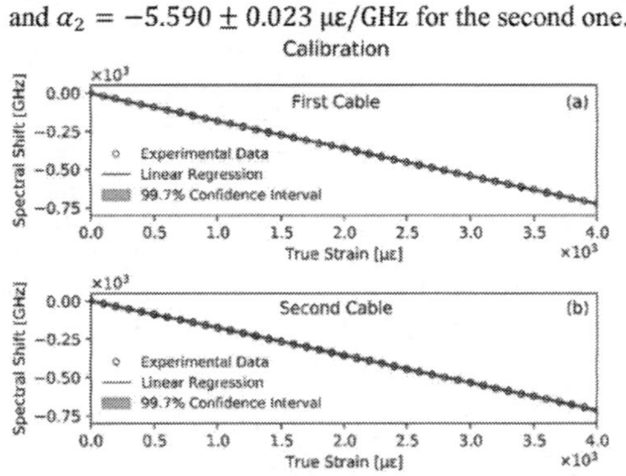

Fig. 2. Calibration data for the first (a), and the second (b), cable prototypes.

Once the calibration process was complete, SWI could be used to evaluate the strain transfer properties of the two fiber design. For the experimental evaluation, the sensing fibers were bonded through epoxy adhesive (LOCTITE®

https://doi.org/10.1051/epjconf/202125512011

EA 9466™) on the surface of an aluminium specimen, which in turn was put into a tensile strength machine, as shown in Fig. 3. Its dimensions were 300 mm in length, 8 mm in thickness and 20 mm width, a Young modulus of 71.7 GPa and a shear modulus of 26.9 GPa. Sensing cables built according to the first and second design were attached at the opposite sides of the specimen, in order not to come in contact with the clamps of the tensile strain machine, which would have added unwanted bending radii. The bonding length between the cable and the specimen required for a complete strain transfer was determined from the preliminary results of the previously showcased numerical model, which found optimal results for 270 mm.

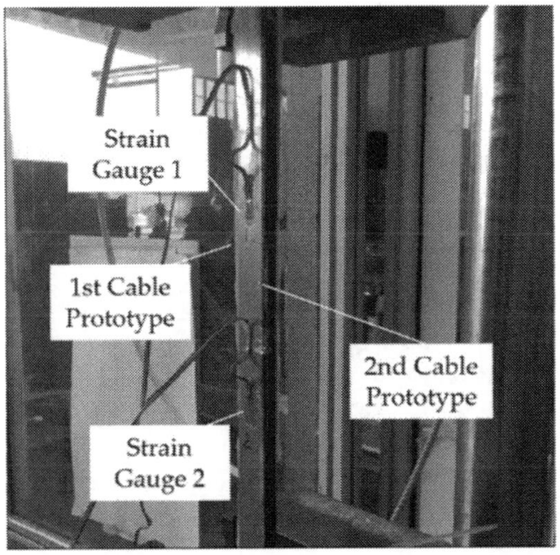

Fig. 3. Aluminium specimen in the tensile strength machine.

During the experiment, the aluminium specimen was subjected to loads ranging from 5 kN to 25 kN (5, 10, 15 20 and 25 kN). Two electrical strain gauges were fixed on the opposite ends of the specimen's length and used as a reference to measure the effective longitudinal strain value of the specimen. In particular, the average of their measurement was used as the asymptotic strain value that should be detected by the two sensing cables.

As further validation, the experimental results were compared with a known analytical model for strain transfer developed in [17], represented by the following formula:

$$\varepsilon_f(x) = \varepsilon_h\left[1 - \frac{\cosh(kx)}{\cosh(kL)}\right] \quad (6)$$

In which ε_f represents the strain profile of the fiber core, ε_h the strain profile in the host structure, L denotes the bonding length and x represents the spatial coordinate along the fiber axis. Finally, k corresponds to the shear lag parameter, and is the fundamental variable of the model, enclosing all the geometrical and mechanical aspect of the sensing cable and its interface with the structure. In particular, the higher the value, the more efficiently the strain is transferred from the structure to the sensing cable. While in principle its value can be derivable a priori from the properties of the sensing cable, it is commonly considered more feasible to tune the value based on experimental data in order to correct the initial guess for the value.

4 Results

The results are summarized in Fig. 4. Due to the symmetrical nature of the experiment, only one half of the strain profile is plotted (the first 135 mm of bonding length). The 0 point in the distance axis (the vertical line in both plots) is chosen as the first bonding point between the cable and the aluminium specimen. The horizontal solid lines correspond to the reference strain value obtained by the the two strain gauges for each load case.

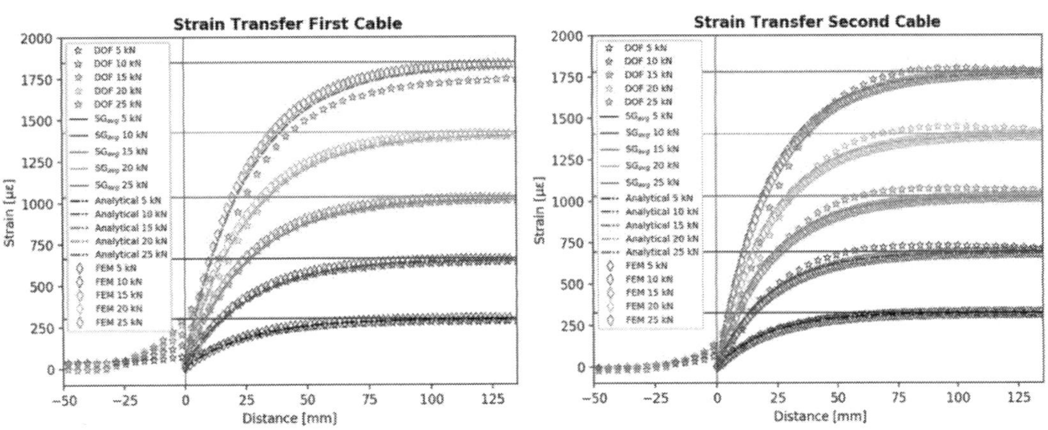

Fig. 4. Strain transfer of the first sensing cable (left) and second sensing cable (right). Horizontal lines: strains measured by the gauges. Stars: OBR measurements. Dashed lines: analytical model. Diamonds: numerical FEM model.

The diamond shaped symbols are the results obtained from the FEM numerical model. The SWI sensing measurements are represented by star symbols and are denoted in the legend with the DOF acronym. Finally, the dashed lines are the strain profiles predicted by the strain transfer model. The strain transfer improvement from the first sensing cable with respect to the second can be easily seen at the highest load levels, where even in the midline (x = 135 mm), there is a mismatch between the strain of the first sensing cable and the strain of the gauge, representing an incomplete strain transfer. On the other hand, the second developed sensing cable is always capable to reach the strain level measured by the two strain gauges, proving it as a viable alternative for BOTDA sensing cables.

In addition, for both designs it can be seen that there is a consistent overlap between the experimentally obtained strain values for the sensing cable and the one predicted with the numerical and theoretical models, showing how both these tools can allow to make reliable predictions without the need of performing experiments, although both systems still improved thanks to experimental data.

5 Conclusion

In this work, a sensing method based on swept wavelength interferometry has been used to evaluate two newly manufactured optical fibre temperature and strain sensing cables. In particular, these cables are intended to be employed in Brillouin optical time domain analysis to extract temperature and strain spatial distributions in large structures at centimetre level spatial resolution. Of the two cable designs, the first one was intended to satisfy a series of requirements for effective strain sensing performance, but a series of construction issues caused it to have non optimal strain transfer capabilities and in general issues that are expected to prevent it from operating at optimal operating conditions, leading to the realization of the second design in order to solve them. Besides swept wavelength interferometry, the two designs were also evlauated through numerical and analytical approaches. The results showed a good agreement between the experimental methods and both numerical and analytical models, showing that swept wavelength interferometry is an effective method for characterizing the strain transfer behaviour of optical sensing cables and that the numerical and analytical models can be considered reliable tools to make predictions. As expected, these results showed that the second design presented superior strain transfer capabilities, always reaching the same strain value of the actual specimen, proving it to be a viable alternative for BOTDA sensing. The first design, on the other hand, due to the issue detailed in this work, presented a mismatch between the strain detected by the cable and the one effectively applied to the specimen, especially at higher strain values.

This activity has received funding from the European Union through Horizon 2020, the Framework Programme for Research and Innovation, under project SLAM-DAST, Grant Agreement No. 971149.

References

1. R. Di Sante, Sensors, **15**, 18666–18713 (2015).
2. D. Marini, M. Iuliano, F. Bastianini, G. Bolognini, J. Light. Technol., **36**, 1131-1137 (2018)
3. M.A. Soto, P.K. Sahu, S. Faralli, G. Sacchi, G. Bolognini, F. Di Pasquale, B. Nebendahl, C. Rueck, Proceedings of SPIE, 66193B (2017)
4. F. Falcetelli, L. Rossi, R. Di Sante, G. Bolognini, Sensors. **20**, 3100 (2020).
5. D. Marini, L. Rossi, F. Bastianini, G. Bolognini, 20th Italian National Conference on Photonic Technologies (Fotonica 2018), CP748, 147646 (2018)
6. L. Rossi, D. Marini, F. Bastianini, G. Bolognini, Opt. Express, **27**, 29448-29460 (2019).
7. G. Bolognini, S. Faralli, A. Chiuchiarelli, F. Falconi, F. Di Pasquale, IEEE Photonics Technol. Lett., **18**, 1591-1593 (2006).
8. M.A. Soto, M. Taki, G. Bolognini, F. Di Pasquale, Opt. Express, **20**, 6860–6869 (2012).
9. A. H. Hartog, An introduction to distributed optical fibre sensors, 1st ed., CRC press, 2017.
10. D. Marini, L. Rossi, F. Bastianini, G. Bolognini, OSA Technical Digest, ThE71, 140688 (2018).
11. F. Bastianini, R. Di Sante, F. Falcetelli, D. Marini, G. Bolognini, Sensors, **19**, 5172 (2019).
12. E. D. Moore, R. R. McLeod, Opt. Express, **16**, 13139-13149 (2008)
13. D.K. Gifford, S.T. Kreger, A.K. Sang, M.E. Froggatt, R.G. Duncan, M.S. Wolfe, B.J. Soller, Proceedings of the Fiber Optic Sensors and Applications V, **6770**, 67700F (2007).
14. B.J. Soller, M.S. Wolfe, M.E. Froggat, Proceedings of the OFC Technical Digest; NWD3 (2005).
15. Rao, Y.J. *Fiber Bragg grating sensors: principles and applications*. In *Optical Fiber Sensor Technology*; Grattan, K.T.V., Meggitt, B.T. (Eds.; Springer US: Boston, MA, 1998), pp. 355–379.
16. LUNA Technologies Optical Backscatter Reflectometer User Guide (2009).
17. D. Li, H. Li, L. Ren, G. Song, Opt. Eng, **45**, 024402 (2006).

* Corresponding author: bolognini@bo.imm.cnr.it

https://doi.org/10.1051/epjconf/202125512012

Propagation features of a class of beams radiated from pseudo-Schell model vectorial sources

Rosario Martínez-Herrero[1,*], *Gemma* Piquero[1], *Juan Carlos* González de Sande[2], *Massimo* Santarsiero[3], and *Franco* Gori[3]

[1]Departamento de Óptica, Universidad Complutense de Madrid, Ciudad Universitaria, 28040 Madrid, Spain;
[2]Dipartimento di Ingegneria, Università Roma Tre, Via V. Volterra 62, 00146 Rome, Italy;
[3]ETSIS de Telecomunicación, Universidad Politécnica de Madrid, Campus Sur 28031 Madrid, Spain;

Abstract. Pseudo Schell-model sources were introduced and studied in the scalar domain as those partially coherent sources that present a degree of coherence depending on the difference between the radial coordinates of the two considered points. In this work we study the propagation features of a class of beams radiated from sources of this kind, but endowed with a vectorial nature. The polarization pattern and the degree of polarization are not uniform across the beam section but remain invariant upon free paraxial propagation. On the other hand, their coherence and irradiance features are also non-uniform but in this case they change at each z-plane. The field characteristics can be varied on acting on the free parameters of the cross-spectral density matrix of the source.

1 Introduction

The proposal of new types of light beams with structured coherence and polarization is a subject of increasing recent research in the literature [1–8]. The interest for such beams lies in the peculiar properties that they may present, both across the source plane and upon propagation. Just to quote some of them, we recall the self-focusing, particle trapping, or propagation through turbulent media [9–15]

A recently introduced example of structured sources in the scalar realm are the so-called pseudo-Schell model (PSM) sources [7, 8], which present coherence characteristics that only depend on the difference of the radial distance to the center of two points in the source transverse section, i.e., on the difference of their radial coordinates. Vectorial PSM (VPSM) sources have also been proposed by Hyde [16] who, in addition, presented a scheme aimed at synthesizing them.

In this work we focus our attention onto a class of vectorial VPSM sources, whose polarization characteristics only depend on the azimuthal coordinate across the source plane, and study the coherence and polarization features of the radiated beams in the paraxial regime. The proposed sources present a degree of polarization dependent on the point across the transverse section of the source, in particular on the azimuthal coordinate. The intensity at the source plane has a null at the center and the coherence characteristics only depend on the difference between the radial coordinates of the two considered points. An interesting feature of these sources is that the polarization characteristics remain invariant upon free propagation being useful, for example, in polarimetry [17].

*e-mail: r.m-h@fis.ucm.es

2 Source model

The VPSM source we are going to study is characterized by a cross-spectral density matrix (CSDM) across the source plane of the following form:

$$\widehat{W}(\mathbf{r}_1,\mathbf{r}_2) = W_{\mathrm{PS}}(r_1,r_2)\widehat{H}\left(\theta_1,\theta_2\right), \qquad (1)$$

where $\mathbf{r}_j = (r_j,\theta_j)$ (with $j = 1,2$) are the polar coordinates of two points across the source plane. $W_{PS}(r_1,r_2)$ is a scalar cross-spectral density (CSD) of the Gaussian pseudo-Schell model type [7, 8], i.e.,

$$W_{\mathrm{PS}}(r_1,r_2) = I\, r_1 r_2 \exp\left[-\frac{r_1^2+r_2^2}{w_0^2} - \frac{(r_1-r_2)^2}{\delta^2}\right], \qquad (2)$$

being I a real positive constant, and w_0 and δ two distances related the intensity width and the spatial coherence of the source, respectively. Furthemore, $\widehat{H}(\theta_1,\theta_2)$ is a 2×2 matrix having components

$$\begin{aligned} H_{xx} &= \cos(\theta_1)\cos(\theta_2)\ , \\ H_{xy} &= b\cos(\theta_1)\cos(\theta_2-\alpha)\ , \\ H_{yx} &= b^*\cos(\theta_1-\alpha)\cos(\theta_2)\ , \\ H_{yy} &= \cos(\theta_1-\alpha)\cos(\theta_2-\alpha)\ , \end{aligned} \qquad (3)$$

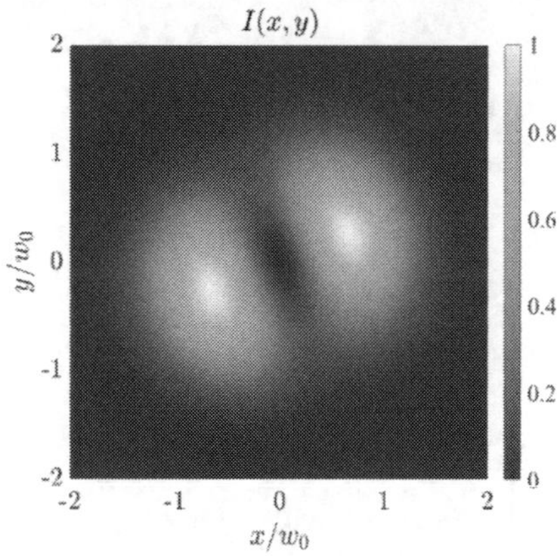

Figure 1. Intensity (normalized to its maximum) at the source plane for the source parameters $\delta = 0.25w_0$, $\alpha = \pi/4$, and $b = 0.5$.

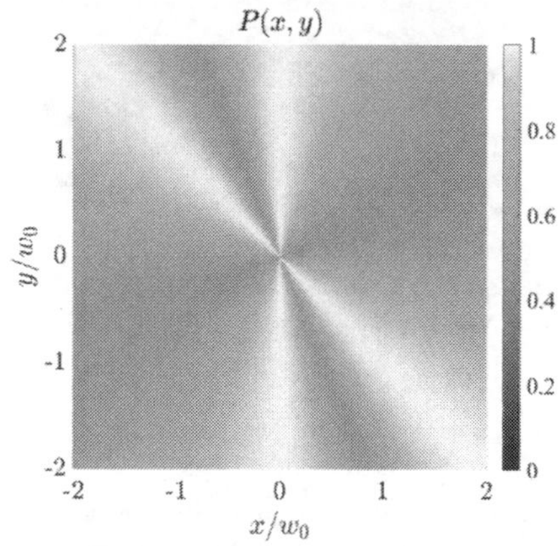

Figure 2. Degree of polarization at the source plane for the same source parameters than those in Fig. 1 ($\delta = 0.25w_0$, $\alpha = \pi/4$, and $b = 0.5$).

where b is a complex constant satisfying $|b| < 1$ and α is a real angle. It can be shown that this kind of source is well defined [18, 19].

The evaluation of the CSDM at joint points $\mathbf{r} = \mathbf{r_1} = \mathbf{r_2}$ yields the polarization matrix [20], which describes all characteristics of the source, such as its intensity and the degree of polarization (DOP) at a given position. In particular, it can be easily seen that, for the chosen CSDM, the intensity profile does not present circular symmetry, except for specific values of the source parameters, but that the intensity always vanishes at the source center. This fact can be appreciated in Fig. 1 where a particular choice of the source parameters has been made.

Furthermore, the DOP turns out to be not uniform across the source plane as it can be seen in Fig. 2. The DOP is the same along any selected radial line, but varies from a minimum value that depends on the b value to 1 when varying the angular coordinate.

Several parameters have been introduced to describe the coherence properties of a vectorial source [21–25]. Among them, we will consider the ones proposed in [22–24].

The first one, known as the electromagnetic degree of coherence, gives a measure of the correlation among all components of the fields at the two points as defined in [22]. Fig. 3 shows the behaviour with the angular coordinate of the electromagnetic degree of coherence [22] calculated between two points located on the same circle. The second point is fixed at the angular coordinate θ_2. It can be demonstrated that the electromagnetic degree of coherence depends on the angular variables through the square of cosine functions, so it shows a periodic behaviour with π period. It is worth noting that this behaviour is different from that observed for scalar pseudo Schell-model beams where the degree of coherence present a circular symmetry [7, 8].

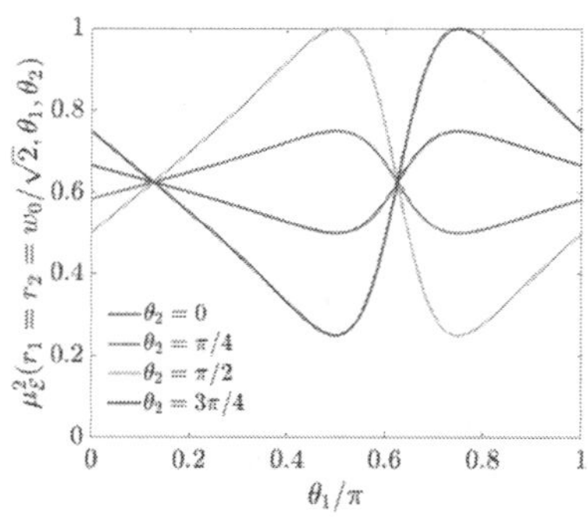

Figure 3. Angular dependence (θ_1) of the electromagnetic degree of coherence [22] relative to points located on the same circle but at different angular coordinates θ_2. The source parameters are the same as in Fig. 1 ($\delta = 0.25w_0$, $\alpha = \pi/4$, and $b = 0.5$).

The radial dependence of the electromagnetic degree of coherence [22] shows a maximum for points on the same circle but this maximum has not to be necessarily 1, even though this parameter is calculated for the field at the same point (see Fig. 4). This parameter decreases from such a maximum value in a Gaussian way when one of the two considered points apart from the other in radial direction either towards the center of the soure or to infinity. Similar behavior is observed if the two points are located on different radial lines, the only difference is that the maximum value changes.

The second analyzed parameter is related to the maximum fringe visibility obtainable in a Young experiment on modifying the interfering fields by means of local unitary transformations (i.e., by means of deterministic nonabsorbing anisotropic optical elements). The behaviour of this parameter along a cir-

 https://doi.org/10.1051/epjconf/202125512012

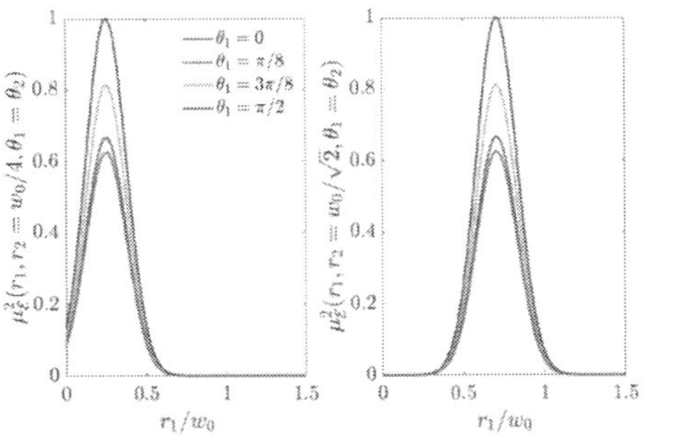

Figure 4. Radial dependence of the electromagnetic degree of coherence [22] relative to a point located at a distance from the source center $r_2 = w_0/4$ (left) or $r_2 = w_0/\sqrt{2}$ (right) and on the several radial lines forming $\theta_1 = \theta_2$ angle with x−axis. The source parameters are the same as in Fig. 1 ($\delta = 0.25w_0$, $\alpha = \pi/4$, and $b = 0.5$).

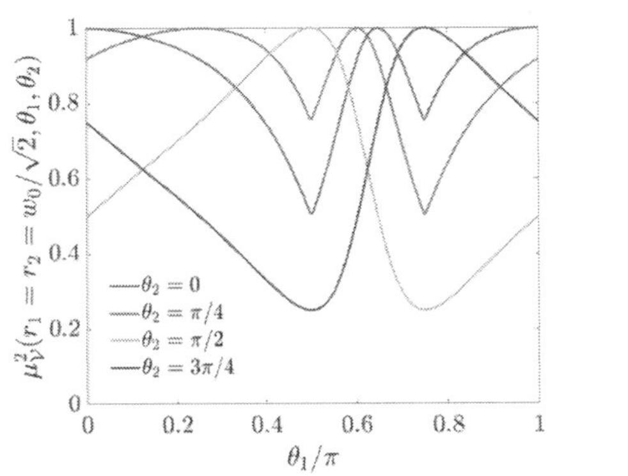

Figure 5. Angular dependence (θ_1) of the maximum fringe visibility in a Young experiment [23, 24] relative to points located on the same circle but at different angular coordinates θ_2. The source parameters are the same as in Fig. 1 ($\delta = 0.25w_0$, $\alpha = \pi/4$, and $b = 0.5$).

cle can be observed in Fig. 5. Again, due to the way the source is defined, this parameter shows a peri-dodic behaviour with the angle variation of one of the considered points with π period. Note that a maximum value of 1 (maximum visivility ain a Young experiment) is always reached when the two points coincide and when they are diametrically opposite (on the same circle).

3 Paraxial propagation

The Fresnel diffraction integral can be used to obtain the CSDM of the field radiated from the source under paraxial approximation. It can be shown that the CSDM of the propagated beam presents the same angular dependence than that at the source plane, that is, Eq. (1) holds just replacing W_{PS} . Then, the polarization properties of the beam remain invariant under propagation and the degree of polarization has the same pattern as in Fig. 2.

The intensity profile, in general changes in propagation, but it is always zero at the axis of the beam. Figure 6 shows the evolution of the intensity distribution with propagation distance. The same parameters as in Fig. 1 are chosen. It can be observed that the maximum intensity is not reached at the source plane (see the colorbar scale) but after a certain propagation distance.

4 Conclusions

In summary, a new class of partially coherent, partially and nonuniformly polarized source of the pseudo-Schell model type is proposed and analyzed in detail. The generated fields not only have a structured coherence but also exhibit a non-uniform polarization. The pattern of the degree of polarization remains invariant in propagation. Although the field radiated by these sources does not present circular symmetry, it does present a null in the center for all z=constant plane. They also have a self-focusing effect. These properties could be very useful in various applications such as particle trapping.

Acknowledgements

This work has been supported by Spanish Ministerio de Economía y Competitividad under project PID2019 104268GB-C21.

References

[1] H. Lajunen, T. Saastamoinen, Opt. Lett. **36**, 4104 (2011)
[2] Y. Cai, Y. Chen, F. Wang, J. Opt. Soc. Am. A **31**, 2083 (2014)
[3] Y. Cai, Y. Chen, J. Yu, X. Liu, L. Liu, Progress in Optics **62**, 157 (2017)
[4] M. Santarsiero, R. Martínez-Herrero, D. Maluenda, J.C.G. de Sande, G. Piquero, F. Gori, Opt. Lett. **42**, 1512 (2017)
[5] M. Santarsiero, R. Martínez-Herrero, D. Maluenda, J.C.G. de Sande, G. Piquero, F. Gori, Opt. Lett. **42**, 4115 (2017)
[6] G. Piquero, M. Santarsiero, R. Martínez-Herrero, J.C.G. de Sande, M. Alonzo, F. Gori, Opt. Lett. **43**, 2376 (2018)
[7] J.C.G. de Sande, R. Martínez-Herrero, G. Piquero, M. Santarsiero, F. Gori, Opt. Express **27**, 3963 (2019)
[8] R. Martínez-Herrero, G. Piquero, J.C.G. de Sande, M. Santarsiero, F. Gori, Applied Sciences **9** (2019)
[9] J.C. Ricklin, F.M. Davidson, J. Opt. Soc. Am. A **19**, 1794 (2002)
[10] O. Korotkova, L.C. Andrews, R.L. Phillips, Optical Engineering **43**, 330 (2004)

 https://doi.org/10.1051/epjconf/202125512012

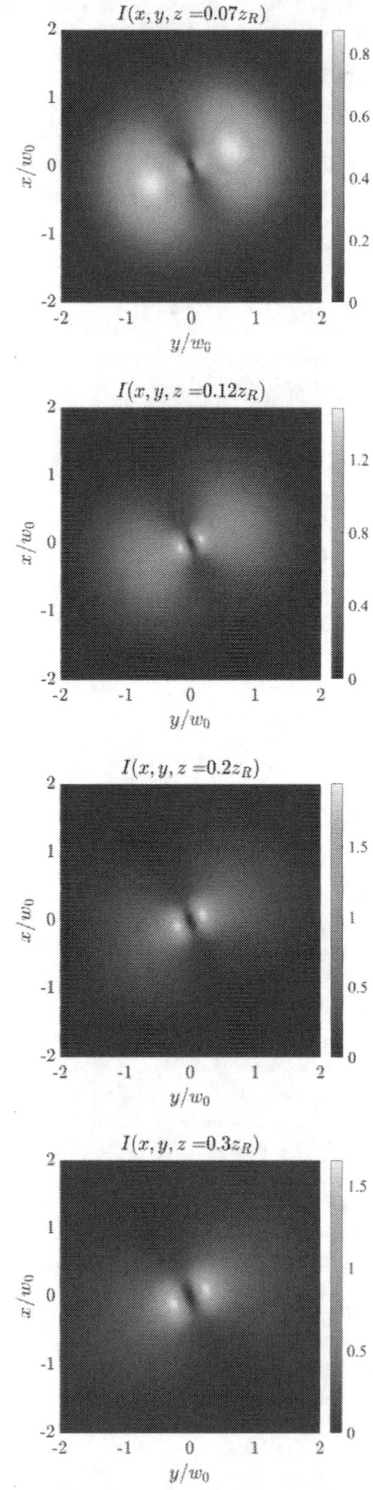

Figure 6. Intensity maps (normalized to the maximum of the intensity at the source plane) for several propagation distances, being $z_R = kw_0^2/2$ and k the wavenumber. The source parameters are the same as in Fig. 1 ($\delta = 0.25w_0$, $\alpha = \pi/4$, and $b = 0.5$)

[11] S.B. Raghunathan, T. van Dijk, E.J.G. Peterman, T.D. Visser, Opt. Lett. **35**, 4166 (2010)
[12] C. Zhao, Y. Cai, Opt. Lett. **36**, 2251 (2011)
[13] G. Wu, Y. Cai, Opt. Lett. **36**, 1939 (2011)
[14] G. Gbur, J. Opt. Soc. Am. A **31**, 2038 (2014)
[15] C. Ding, M. Koivurova, J. Turunen, L. Pan, J. Opt. Soc. Am. A **34**, 1441 (2017)
[16] M.W.H. IV, Optics **1**, 97 (2020)
[17] J.C.G. de Sande, M. Santarsiero, G. Piquero, Optics and Lasers in Engineering **91**, 97 (2017)
[18] F. Gori, V. Ramírez-Sánchez, M. Santarsiero, T. Shirai, Journal of Optics A: Pure and Applied Optics **11**, 085706 (2009)
[19] R. Martínez-Herrero, P.M. Mejías, Opt. Lett. **34**, 2303 (2009)
[20] L. Mandel, E. Wolf, *Optical Coherence and Quantum Optics* (Cambridge University Press, 1995), ISBN 9780521417112
[21] E. Wolf, *Introduction to the Theory of Coherence and Polarization of Light* (Cambridge University Press, 2007), ISBN 9780521822114
[22] J. Tervo, T. Setälä, A.T. Friberg, Opt. Express **11**, 1137 (2003)
[23] F. Gori, M. Santarsiero, R. Borghi, Opt. Lett. **32**, 588 (2007)
[24] R. Martínez-Herrero, P.M. Mejías, Opt. Lett. **32**, 1471 (2007)
[25] A. Luis, J. Opt. Soc. Am. A **24**, 1063 (2007)

https://doi.org/10.1051/epjconf/202125512013

Stabilized, short cavity Brillouin ring laser source design for fiber sensing applications

Leonardo Rossi[1], *Filippo* Bastianini[2], and *Gabriele* Bolognini[1,*]

[1]Consiglio Nazionale delle Ricerche, IMM Institute, 101 via P. Gobetti, 40129 Bologna Italy
[2]Sestosensor s.r.l., 140 via Gesso, 40069 Zola Predosa, Italy;

Abstract. A new pump-seeded, short-cavity Brillouin ring laser source layout intended for Brillouin sensing applications is showcased, showing increased high maximum output (1.5 mW), a strong linewidth narrowing effect (producing light with a linewidth of 10 kHz) and limited RIN (~-145 dB/Hz), providing an ultranarrow, highly stable BRL source that can also be employed as a pump-probe source for BOTDA applications.

1 Introduction

Brillouin Ring Lasers (BRLs) are a fiber-based light source in which a light (defined as seed pump) is injected and circulates inside a closed fiber loop, where it produces a coherent, downshifted and counterpropagating light (defined as Stokes light) through a process known as Stimulated Brillouin Scattering (SBS). Due to their capability to provide a light signal which is coherent and with a stable frequency shift from the light it originated from with limited intensity and phase noise, BRLs have been studied for applications such as fiber-optics gyroscopes [1], microwave phase-modulated generators [2] and LIDAR detection. Another potentially interesting field of application is distributed optical fiber sensing, which has gathered significant attention in recent years due to the interesting properties of optical fibers, such as resistance and flexibility [3]. A particularly interesting distributed optical fiber sensing method is Brillouin optical time-domain analysis (BOTDA) [4-6]. In this sensing scheme, which is also based on SBS, an optical pump amplifies a downshifted, counter-propagating probe along a sensing fiber. The extent of this amplification is related to the pump-probe frequency shift with a relation known as the Brillouin Gain Spectrum (BGS), and the shift at which this amplification is most efficient is known as the Brillouin Frequency Shift (BFS) which is dependent on the temperature and strain of every point in the fiber [7]. The pump and probe lights are usually provided by sources such as Phase locked loop (PLL) schemes or optical sideband generation (OSB) techniques [8], which tend to have strict requirements in terms of laser source quality and hardware components. In this context, the natural frequency shift of BRLs and the simplicity of their structure makes them an interesting alternative for low-complexity BOTDA setups. In [9], an implementation of a BRL source based on a long cavity (2 km) hybrid BRL source was developed and successfully employed in a BOTDA sensing scheme, where it allowed for measurements over up to 10 km of range with a spatial resolution of 4 m and a frequency resolution of 0.5 MHz. While these results were already viable for some BOTDA applications, the BRL cavity was affected by three issues which limited its performance: low maximum output power, high output linewidth (2 MHz) and a high relative intensity noise (RIN) for frequencies below 1 GHz, which was caused by mode hopping effects. In this work, a new BRL design is presented to overcome these limitations. Unlike the other design, this one is based on a short cavity (~4 m) Brillouin ring laser (SC-BRL) in order to suppress mode hopping and improve linewidth narrowing of the Stokes output. Through accurate tuning of the length of the cavity and self-injection locking techniques, the BRL was made resonant for both the seed pump and the Stokes output. As a result, the new source displayed significantly reduced RIN, greatly reduced linewidth of the Stokes light and higher maximum output power.

2 Short cavity BRL

The BRL cavity that is studied in this work is shown in Fig. 1. Through the exits 1 and 2 of an optical circulator (OC1), the seed pump light from a distributed feedback laser (λ=1555 nm, 350 kHz linewidth) is injected inside the ring cavity which, at its core, is composed of a loop of standard single mode optical fiber for telecommunication (SMF-28) with a length below 5m. The loop is closed by a Directional Coupler (DC1) with a 90/10 coupling ratio. While recirculating inside this cavity, the seed pump is scattered through SBS, producing a counter propagating downshifted Stokes light, a portion of which is extracted through DC1 and employed as output after going through exits 2 and 3 of OC1 to be employed in the sensing system. It is to note SBS is a process that is polarization dependant and is most efficient when the polarizations of Stokes and the seed pump are aligned. To ensure this is

* Corresponding author: bolognini@bo.imm.cnr.it

the case, two polarization controllers (PC1 and PC2) are placed before and inside the ring.
Compared to [9], the length of the ring cavity was reduced from 2 km to 5 m. As was mentioned above, one of the sources for the high relative intensity noise of the previous cavity were mode hopping effects. These effects are caused by the fact that the frequency at which a BRL emits light (the dominant frequency) is the cavity resonant mode which is closest to the BFS, and the positions of both these elements can shift due to thermal and vibrational noises. When this happens, a different resonant mode might become the one closest to the BFS, and thus the dominant mode might change. As a result, both frequency and intensity noise are increased as the BRL source "hops" from one mode to another [10]. This effect is more prevalent the lower the spacing between resonant frequencies is and conversely can be suppressed by increasing the spacing of the resonant modes enough that only one mode lies inside the BGS at any point, hence the greatly reduced cavity length presented in this work. The simplest way to achieve this is by reducing the cavity length, which is inversely proportional to the resonant mode spacing, from 2 km down to a length below 5 m, like the one we present in this work.

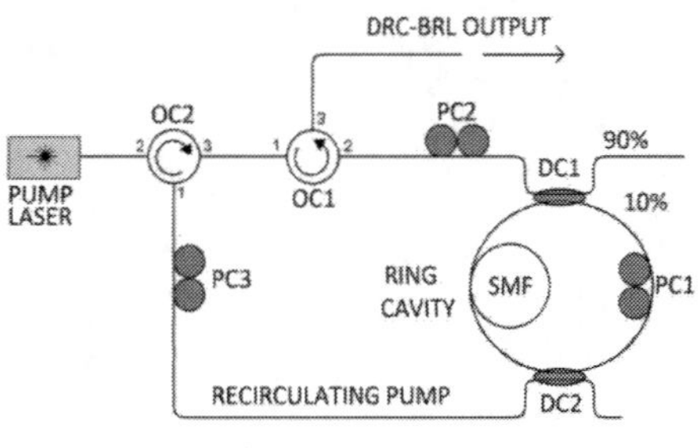

Fig. 1. Scheme of the SC-BRL with self-injection locking.

To counteract the steep reduction in gain medium compared to cavities like the one shown in [9], the cavity is made resonant with both the seed pump and the Stokes output. When this condition, known as double resonance, is satisfied, both lightwaves interfere constructively with newly injected seed pump and newly generated Stokes light at every roundtrip, thus greatly increasing the effective interaction length and overall conversion efficiency.
The resonance of the seed pump is guaranteed through the implementation of a self-injection locking technique in the BRL scheme, which is included in Fig. 1. An optical coupler with a 99/1 coupling ratio (DC2) extracts part of the seed pump inside the ring (which is at the resonant frequency) and sends it back into the DFB laser through exits 1 and 2 of a second optical circulator (OC2). As a result, the DFB is forced to emit at the frequency of the injected light, making the pump resonant with the cavity. In order to make the Stokes light resonant with the cavity, its length is accurately tuned through the use of a single cut technique developed in [11] so that the BFS of the fiber is equal to an integer multiple of the resonant mode spacing. This technique involves measuring the intensity of the Stokes output at different seed pump wavelengths using a tunable laser in place of the DFB, providing a measurement such as the example for a cavity of 5 m shown in Fig. 2. The goal of this measurement is to locate two consecutive peaks corresponding to wavelengths for which double resonance is satisfied, λ_L^k

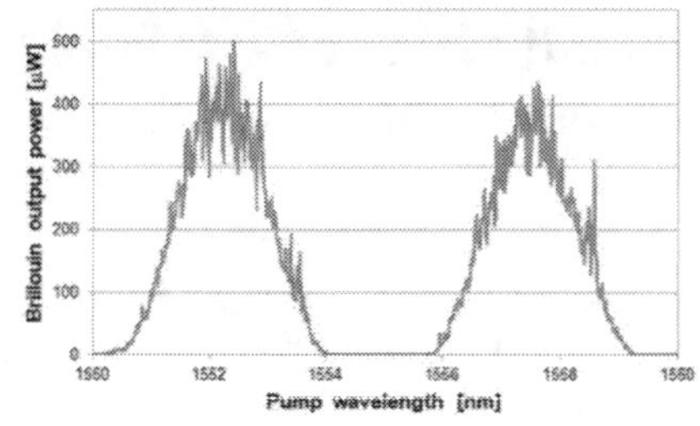

Fig. 2. Double resonance peaks for a cavity length of 5 m.

and λ_L^{k-1}. With these wavelength values, it is possible to estimate the length ΔL to be cut from the cavity to move one of those peaks to the desired wavelength λ_{LM}^k by using the following formula, where n is the refraction index and V_a is the fiber's acoustic velocity:

$$\Delta L = \frac{c\lambda_L^k}{2n^2 V_a} \frac{1}{\lambda_L^{k-1} - \lambda_L^k} (\lambda_{LM}^k - \lambda_L^k) \quad (1)$$

After the tuning, the cavity was 3.4 meters long.

3 Performance evaluation

In order to see how this new design affected the issues noted from the source developed in [6], the Stokes output for this source was evaluated in terms of maximum power, linewidth and intensity noise.

The maximum Stokes output power was found to be 1.5 mW for a pump power of 25 mW, while the design showcased in [9] was found to provide an output with a maximum power of 0.5 mW for a 17 mW pump. The lasing threshold power (the minimum pump power required to obtain Brillouin lasing from the BRL), on the other hand, was 10 mW, up from the 2 mW found in [9]. These results show that the double resonance effect allows to counteract the greatly reduced amount of gain medium, providing comparable lasing threshold power compared to a long cavity, while providing higher output power, which is a crucial parameter for improved measurement range.

Afterwards, the linewidth of the Stokes output was measured by acquiring the spectrum through the use of the self-heterodyne technique [12], performed with a Mach-Zender interferometer as the one shown in Fig. 3 employing a 150 MHz acousto-optic modulator and a delay line with a length of 12 km which, when sent into an Electrical Spectrul Analyzer, allows to measure signal spectra with minimum linewidths of 6 kHz [13].

As a result, the output spectrum, shown in Fig. 4, was found to have a full width half-maximum of 10 kHz, showing a significant narrowing compared to the output of the long cavity BRL (2.5 MHz), and also significantly narrower than the linewidth of the DFB signal used as

https://doi.org/10.1051/epjconf/202125512013

seed pump (350 kHz), showing the beneficial effect of a short cavity BRL's filtering capabilities.

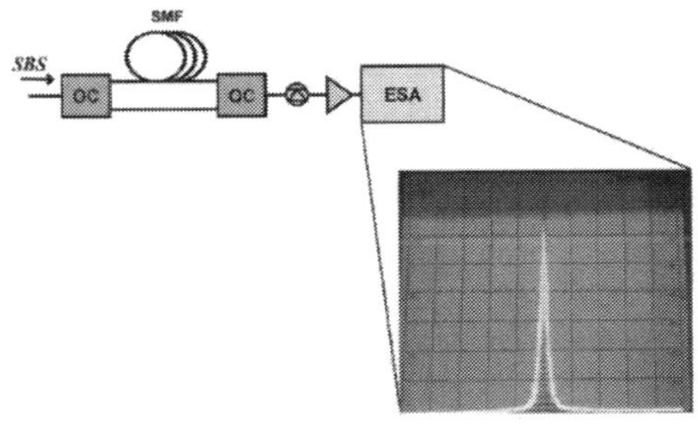

Fig. 3. Self-heterodyne method.

As a further evaluation on self injection locking's capability of suppressing noise effects in BRL, the duration of Stokes lasing intervals with and without self-injection locking were measured with a photodetector and an oscilloscope. The results are shown in Fig. 2. Without self-injection locking, the duration of Brillouin lasing was found to be between 2 and 10 ms, while with self-injection locking the duration was significantly increased with values ranging from 30 to 90 ms, which is already sufficient for BOTDA measurements. The reason for this difference is of course linked to the seed pump light remaining resonant with the cavity for a much longer time despite thermal and vibrational noise.

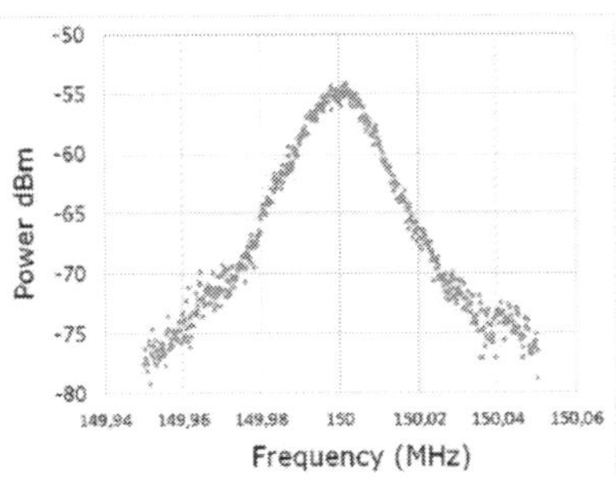

Fig. 4. Short cavity BRL output spectrum.

Finally, in order to evaluate how the employment of a short cavity suppresses noise effects such as mode hopping, RIN measurements were performed in a way similar to [9], as shown in Fig. 6. These measurements were performed by acquiring the BRL Stokes output light with a fast p-i-n photodetector (InGaAs-based, 10 GHz bandwidth) to convert a light signal into a voltage signal, whose power spectrum $\sigma^2(\omega)$ in the 1-800 MHz range was analysed through an ESA (working up to 15 GHz) after subtracting thermal and shot noise terms due to the photodetector [14]. As is evident from the image, the RIN profile of the SC-BRL ranges between -140 dB/Hz and - 150 dB/Hz across the whole 0-800 MHz range, adding a limited amount of intensity noise to the profile of the source DFB. The long-cavity BRL (> 2 km length) developed in [9], on the other hand, introduces significant intensity noise, increasing the RIN profile at frequencies below 400 MHz, with values up to -90 dB/Hz in the 10-15 MHz frequency range, and becomes similar to the DFB seed pump only beyond 500MHz, where it remains approximately constant with an average value of about - 145 dB/Hz. From these results it can be easily seen how the reduction of cavity length, mixed with the contribution of self-injection locking suppress noise sources and instability usually associated with fiber cavities that increase the RIN profile of the output compared to the source pump seed [15].

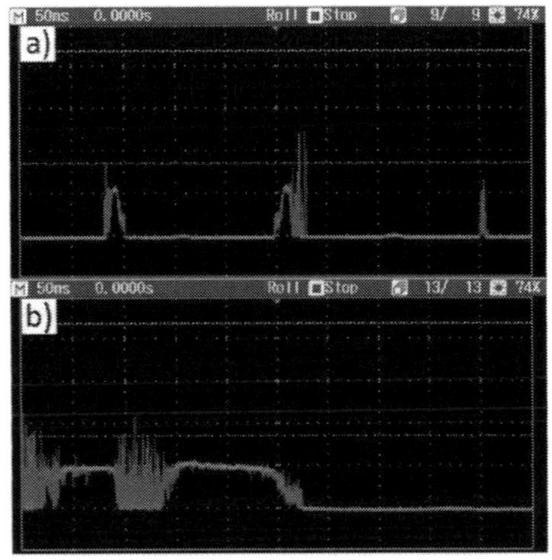

Fig. 5. Stokes output trace without self-injection (a) and with self-injection (b). Timescale: 50 ms per division.

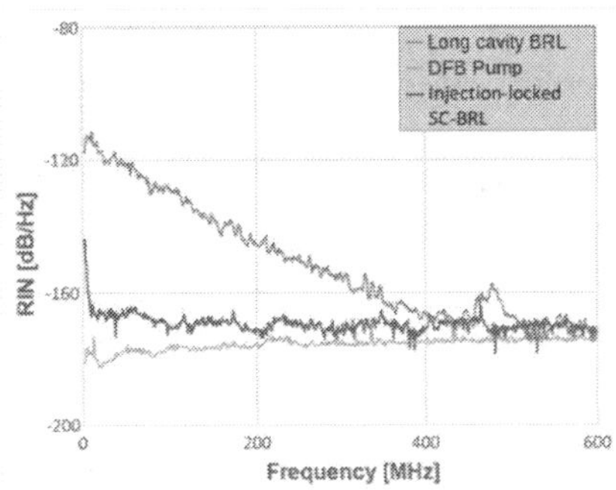

Fig. 6. Comparison of the RIN profiles of the sources.

It can be shown that the reduction in intensity noise provided by the stabilization from self-injection locking and the mode hopping suppression from the employment of the short cavity directly impact BOTDA resolution. To show this, the intensity noise limited signal-to-noise ratio (SNR) for both the SC-BRL source and the long-cavity one was calculated by integrating their RIN spectra over a frequency range equal to the bandwidth of a typical photodiode used in BOTDA (125 MHz). The resulting values were 38.7 dB for the long cavity BRL and 61.0 dB for the short cavity one, which implies a SNR improvement of 22 dB. When Lorentzian Curve fitting is used to estract the BFS in BOTDA measurements, source SNR can be translated to frequency resolution $\delta\nu_B$ with the following relation [16]:

$$\delta v_B = \frac{\Delta v_B}{\sqrt{2}\left(SNR\right)^{\frac{1}{4}}} \quad (2)$$

Where Δv_B is the linewidth of the Brillouin gain spectrum. Due to the linear dependency between the BFS and temperature and strain, their resolutions are also directly proportional to the frequency resolution, and they can be calculated as:

$$\delta\varepsilon = \frac{\delta v_B}{C_S v_B(0)} \quad \text{and} \quad \delta T = \frac{\delta v_T}{C_T v_B(t_r)} \quad (3)$$

where $v_B(0)$ and $v_B(t_r)$ are the Brillouin frequency shifts of unstrained sensing fiber and of sensing fiber at reference temperature, respectively, while C_S and C_T are the linear temperature and strain coefficients.

From the relations above is immediate to see that, for sources whose SNR is limited by intensity noise, such as BRLs, a 22 dB improvement is equivalent to a resolution improvement of 5.5 dB which, converted in linear scale, is equal to an improvement in frequency, strain and temperature resolutions by a factor of ~3.5.

4 Conclusion

In this work, the development and characterization of a Brillouin Ring Laser based on a doubly resonant short cavity to be employed as a source for Brillouin optical time domain analysis were showcased and discussed. Compared to previously developed designs, which used long (>2 km) cavities, the employment of a short cavity was expected to provide substantial improvements in terms of output linewidth and intensity noise, due to the positive correlation between mode hopping effects and cavity length. Double resonance, which was achieved through the combined effect of a self-injection locking approach and the use of a single cut technique, was expected to allow for the cavity length to be reduced down to a few meters without significant losses in threshold seed pump power and maximum output. Experimental evaluation showed that the new cavity had a threshold seed pump power of 10 mW, up from the 2 mW found in previous design, while having a maximum output power of 1.5 mW, up from 0.5 mW, showing how the single cut technique allowed for a threshold power of the same order of magnitude despite the great reduction of gain medium, while providing an even greater maximum output power. In terms of output linewidth, the short cavity and self-injection locking were found to allow for a Stokes output with an extremely narrow linewidth of 10 kHz from an original source of 350 kHz, down from the 2 MHz linewidth found for long cavity Brillouin Ring Lasers. In addition, self-injection locking was found to significantly improve lasing stability, allowing for lasing intervals of up to 90 ms, compared to the 10 ms ones obtainable without it. Finally, the intensity noise improvements provided by a short cavity, were evaluated through RIN measurements, obtaining values of ~-145 dB/Hz across the whole 0-800 MHz range, which were compared to RIN measurements performed on previous long cavity designs, which reached significantly higher values, up to -90 dB/Hz. Compared to the long cavity, the RIN improvements are shown to translate to a BOTDA strain and temperature resolution improvement by up to a factor of 3.5.

From these results it can be seen how the stabilized doubly resonant BRL design can provide substantial improvements in a variety of applications. For what concerns BOTDA, it has the potential to be successfully employed as a pump-probe source in accurate Brillouin optical time-domain sensor systems.

This activity has received funding from the European Union through Horizon 2020, the Framework Programme for Research and Innovation, under project SLAM-DAST, Grant Agreement No. 971149.

References

1. S. P. Smith, F. Zarinetchi, S. Ezekiel, Opt. Lett., **16**, 393-395 (1991)
2. P. A. Nicati, K. Toyama, S. Huang, H. J. Shaw, IEEE Photonics Technol. Lett., **6**, 801-803 (1994).
3. M.A. Soto, P.K. Sahu, S. Faralli, G. Sacchi, G. Bolognini, F. Di Pasquale, B. Nebendahl, C. Rueck, Proceedings of SPIE, 66193B (2017)
4. D. Marini, L. Rossi, F. Bastianini, G. Bolognini, 20th Italian National Conference on Photonic Technologies (Fotonica 2018), CP748, 147646 (2018)
5. D. Marini, L. Rossi, F. Bastianini, G. Bolognini, OSA Technical Digest, ThE71, 140688 (2018).
6. L. Rossi, D. Marini, F. Bastianini, G. Bolognini, Opt. Express, **27**, 29448-29460 (2019).
7. F. Falcetelli, L. Rossi, R. Di Sante, G. Bolognini, Sensors, **20**, 3100 (2020)
8. M. Nikles, L. Thévenaz, P. A. Robert, Opt. Lett. **21**, 758–760 (1996)
9. D. Marini, M. Iuliano, F. Bastianini and G. Bolognini, J. Light. Technol., **36**, 1131-1137 (2018)
10. P. Nicati, K. Toyama, H.J. Shaw, J. Light. Technol., **13**, 1445-1451 (1995).
11. V. V. Spirin, C. A. López-Mercado, S.I. Kablukov, E.A. Zlobina, I.O. Zolotovskiy, P. Mégret, A.A. Fotiadi, Opt. Lett., **38**, 2528-2530 (2013)
12. P. Horak, W.H. Loh, Opt. Expr., **14**, 3923-3928 (2006).
13. D. Derickson, *Fiber optic test and measurement* (Prentice Hall, 1998).
14. G. Bolognini, S. Faralli, A. Chiuchiarelli, F. Falconi, F. Di Pasquale, IEEE Photonics Technol. Lett., **18**, 1591-1593 (2006).
15. Z. Ou, X. Bao, Y. Li, B. Saxena and L. Chen, IEEE Photonic. Tech. L., **26**, 2058-2061 (2014).
16. M.A. Soto, G. Bolognini and F. Di Pasquale, Opt. Exp., **18**, 14878-14892 (2010).

 https://doi.org/10.1051/epjconf/202125512014

Wide band UV/Vis/NIR blazed-binary reflective gratings: two lithographic techniques investigation

Mane-Si Laure Lee[1], *Julie* Cholet[1], *Anne* Delboulbé[1], *Raphaël* Guillemet[1], *Brigitte* Loiseaux[1], *Patrick* Garabedian[1], *Thomas* Flügel-Paul[2], *Tino* Benkenstein[2], *Susann* Sadlowski[2], *Nicolas* Tetaz[3], *Roman* Windpassinger[4], *Saroj* Mahalik[4]

[1]Thales Research & Technology, 1 Av Fresnel, 91767 Palaiseau, France
[2]Fraunhofer Institute for Applied Optics and Precision Engineering, Albert-Einstein-Strasse 7, 07745 Jena, Germany
[3]Thales Alenia Space, 5 Allée des Gabians, BP 99, 06156 Cannes la Bocca Cedex, France
[4]European Space Research and Technology Centre, Postbus 299-2200 AG Noordwijk, The Netherlands

Abstract. We report on subwavelength reflective gratings for hyperspectral applications operating in the 340 nm-1040 nm spectral range. The blazed grating period is 30 µm and is composed of 2D subwavelength binary structures with sizes in-between 120 nm and 350 nm. We demonstrate the manufacturing of gratings on 3" wafers by two lithography technologies (e-beam or nanoimprint) followed by dry etching process. These subwavelength gratings enable broadband efficiency which is in average 15%-20% above the efficiency requirement for next generation of spectro-imagers for Earth observation missions and a wavefront error that is much smaller than the 100 nm requirement for space application.

1 Introduction

Next generation of spectro-imagers for future Earth observation missions requests ultra-wide band operation, very large field of view but small volume. However, the performance of these future instruments is limited by the performance of the diffraction grating, since high diffraction efficiency and reduced polarization sensitivity over wide spectral range are required, while introducing a limited wavefront error. In this paper, we report on gratings with period of 30 µm and operating at an angle of incidence of 0.5°, matching the specific requirements of a Dyson spectrometer [1]. The required minimum throughput lies only between 35% and 50% but must be achieved within 340 nm-1040 nm spectral range.

Classical continuous reflective saw-tooth profile blazed-gratings offer peak efficiencies up to ~90% (mainly limited by the reflectivity of the metallic material) but it decreases significantly when approaching the band edges in the UV and NIR regions. To overcome this limited spectral efficiency, several approaches have been proposed in the literature. The first approach consists in combining several materials to achromatize the efficiency. Originally proposed for transmission gratings [2, 3], it has been applied recently to reflection gratings [4]. Another alternative is the so-called "multi-blaze" or "multi-angle" approach [5, 6]. Different sawtooth depths or different facet angles are combined within the same grating.

A third type of approach is based on blazed-binary gratings composed of subwavelength structures [7, 8] and effective index dispersion engineering. Recently, we reported on the realization of a subwavelength grating by e-beam lithography and etching [9].

Here, we evaluate two lithography techniques: e-beam and NanoImprint Lithography (NIL) and compare their performance with next generation spectro-imager requirements, which are summarized in Table 1. NIL is a replication technology enabling cost-effective manufacturing process fitted for low/medium-volume for earth observation missions.

Table 1. Grating requirements related to UV/Vis/NIR spectro-imager.

Spectral band	340 nm-1040 nm
Diffraction efficiency in the -1st diffraction order	λ = 340 nm: > 0.5 λ = 440 nm: > 0.4 λ = 800 nm: > 0.35 λ = 1040 nm: > 0.35
Polarization sensitivity	< 10% (goal < 1%)
Spectral ghosts intensities	< 1e-4 (with respect to useful diffracted order)
Wavefront error	WFE < 100 nm RMS

* Corresponding author: mane-si-laure.lee@thalesgroup.com

2 Subwavelength gratings fabrication

The design of the blazed-binary subwavelength grating is based on artificial material engineering [10] and purposeful design of size, shape and arrangement of the sub-wavelength features for tailoring the grating's spectral characteristics. The 30 µm grating period is composed of 2D subwavelength binary structures with sizes from 120 nm to 350 nm and depth of ~ 510 nm.

For gratings realization, one main process flow, depicted in Fig.1, was followed in two locations: Institute Fraunhofer for Applied Optics and Precision Engineering (IOF) and Thales Research & Technology (TRT).

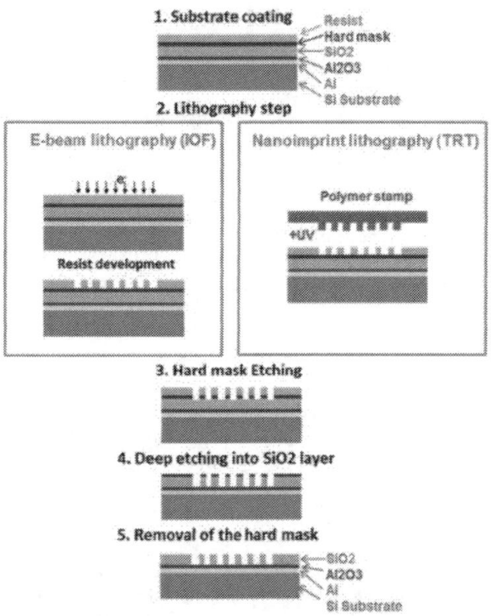

Fig. 1. Process flow for the realization of the gratings.

We use a lithography step to define the patterns into a resist layer, followed by dry etching processes to i) transfer the patterns into the metal mask and ii) to transfer the patterns into the underlying SiO_2 material and get high aspect ratio structures.
Regarding the lithography step, two techniques were investigated: e-beam lithography (IOF), which is the most common technique to tackle such small size patterns, and NIL (TRT). The NIL is a replication technology which consists in fabricating a transparent polymer stamp (polydimethylsiloxane PDMS) and applying the stamp onto a photosensitive resist layer cured under UV light. The polymer stamp which is illustrated in Fig. 2, is previously produced from a flat silicon master mold fabricated with e-beam lithography (at IOF). Thanks to a Surface Conformal Imprint Lithography module from Süss MicroTec, one stamp enables several replications with a high reproducibility and homogeneous patterning over 3" surface.

The advantage of NIL enables the same high resolution as e-beam lithography, with reduced process time (typically few minutes per wafer vs few hours). Pictures of two gratings are given in Fig. 3.

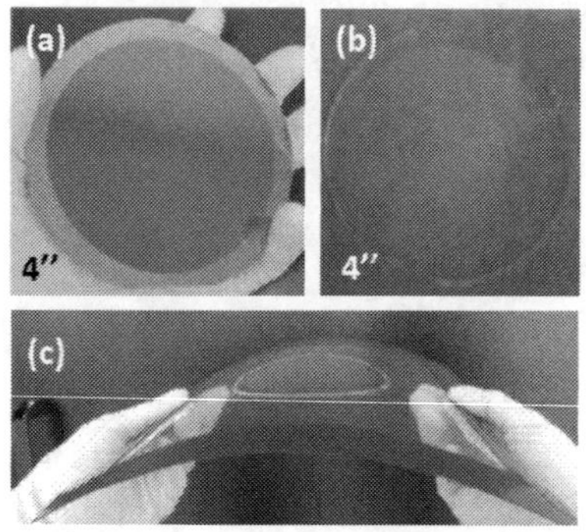

Fig. 2. Silicon master (4") (a) and manufactured PDMS mold (b) fixed on a thin glass plate (c) used for the NIL grating manufacturing.

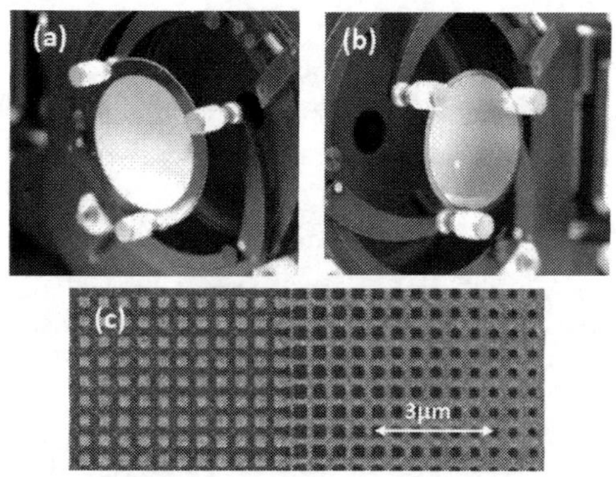

Fig. 3. Manufactured gratings on 4" wafer by e-beam lithography (a) and on 3" wafer by NIL (b). (c) SEM top view of few subwavelength structures of the NIL grating.

3 Optical measurements

The ~80 mm diameter blazed-binary subwavelength gratings were characterized optically at IOF premises.

The efficiency measurements were performed using a tunable light source. The gratings are illuminated by a monochromatic light beam (diameter: ~ 3 mm) with adaptable wavelength (FWHM < 5nm). Measurements are carried out under TE and TM polarization. The diffracted light is detected using an integrating sphere which is mounted on a goniometer setup.

Fig. 4 shows efficiency maps at 376 nm, 600 nm and 1000 nm for the subwavelength grating by e-beam lithography. The mapping sampling is 4 mm × 4 mm. It shows the large-scale uniformity of the manufactured grating.

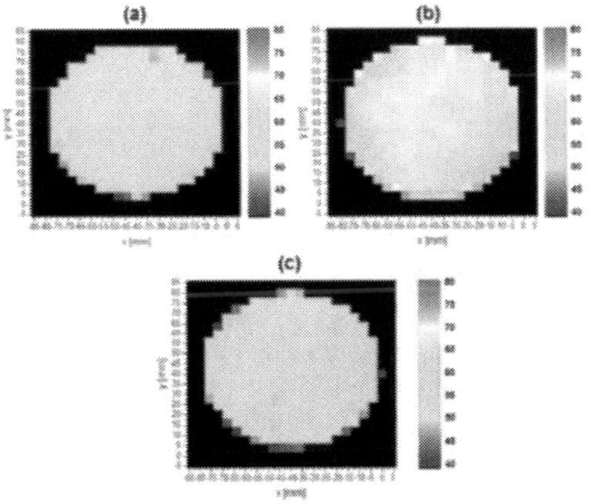

Fig. 4. Spatial mapping of the diffraction efficiency in -1st order of the prototype flat grating for three different wavelengths, i.e. 376 nm (a), 600 nm (b) and 1000 nm (c).

Fig. 5 illustrates the -1st order reflective efficiency of the gratings illuminated under unpolarized light and at 1° angle of incidence. The black line corresponds to the e-beam grating, as compared to the green line (design efficiency), and the red line (requirement). The blue line corresponds to the NIL grating with a 50 nm depth deviation with respect to the nominal depth.

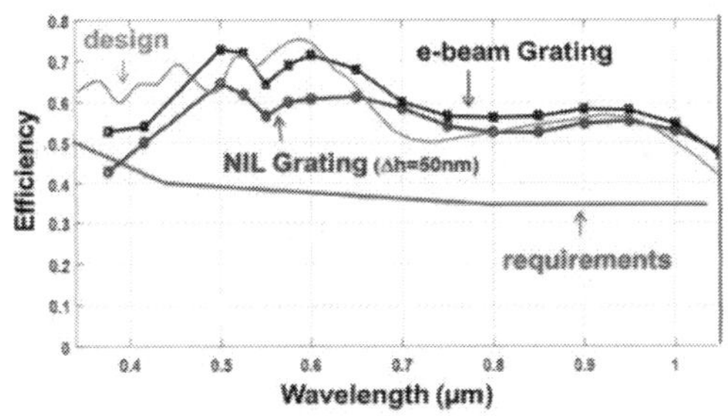

Fig. 5. Measured -1st order diffraction efficiency of blazed-binary subwavelength gratings.

From Fig. 5, two main results are straightforward.

- First, both e-beam and NIL give similar efficiency behavior which is in average 10%-20% above the requirement (red line).
- Secondly, the proposed design (green line) and sub-wavelength grating technology (black and blue line) are tolerant to height deviation of 50 nm (blue line) and efficiency remains compatible with requirements.

Fig. 6 illustrates the -1st order reflective efficiency of the gratings under TE and TM polarization. From TE and TM efficiency measurements, we calculated the polarization sensitivity, defined as

$$|\eta_{TE} - \eta_{TM}|/(\eta_{TE} + \eta_{TM}), \quad (1)$$

where η_{TE} and η_{TM} correspond to efficiencies measured under TE and TM polarized light.

It results that the polarization sensitivity is smaller than 2% for both gratings over the whole spectral band, which confirms experimentally subwavelength optics polarization property.

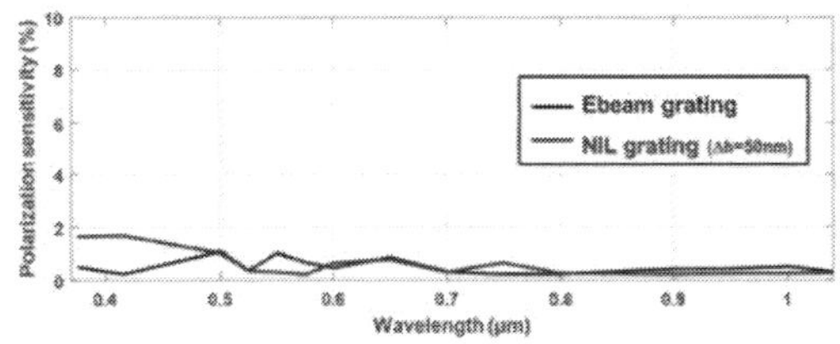

Fig. 6. Polarization sensitivity of the blazed-binary subwavelength gratings.

E-beam technology is well-known for its ability to achieve low level grating ghosts [11]. To verify this property for the NIL grating, we carried out angular scanning of many diffraction orders in the diffraction plane and plot the efficiency (power/reference) on the y-axis as a function of the detector angle θ on the x-axis. The diffracted orders are separated by Δθ ≈ 1.3° and the -1st order is located at θ = - 6.3°. It results that no additional peaks visible, meaning that no grating ghosts were measured. Note that this result is not surprising as the NIL technology uses a master based on e-beam lithography. However, it shows that the NIL technology doesn't introduce additional ghosts.

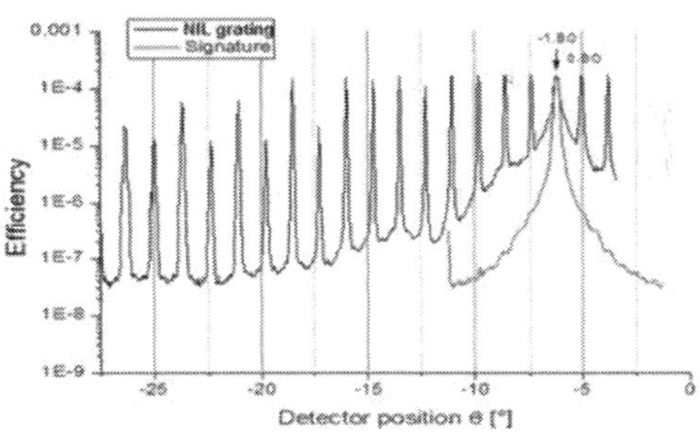

Fig. 7. Angular scanning of the efficiency in the diffractive plane at 633 nm for the NIL grating.

The wavefront error (WFE) introduced by the fabrication process is evaluated by measuring the reflected WFE of the gratings, using a plane-wave Zygo Verifire AT 12" interferometer at a wavelength of 633 nm.

The wavefront deformation introduced by the fabrication process is evaluated as follows: 1) The retro-reflected wavefront of the grating is measured for collimated illumination obeying Littrow condition. A first measurement is performed in +1st diffraction order and a second measurement is independently performed in -1st diffraction order. 2) Both wavefront maps are then subtracted. This allows removing the influence of the grating substrate's deformation. The result only covers the inaccuracy introduced by the respective patterning process, see Fig. 8.

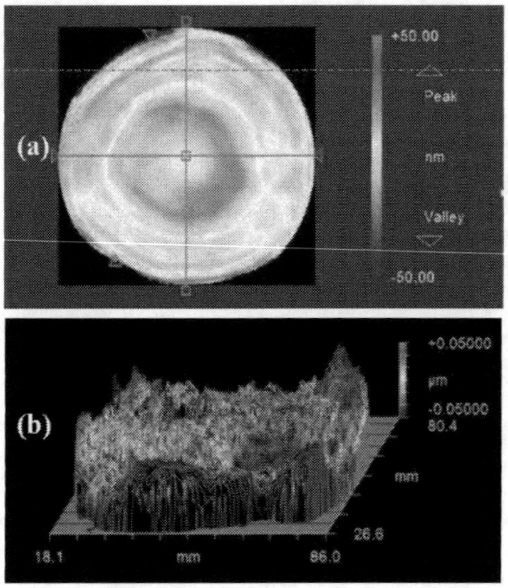

Fig. 8. Reflected wavefront introduced by the grating fabrication process (excluding substrate deformation) for e-beam grating (a) and for NIL grating (b).

For both technologies, the WFE is smaller than 20 nm RMS which is 5 times smaller than the 100 nm RMS space application requirement. Such results have already been demonstrated for e-beam technology [12], but it is striking that although NIL technology is less mature, requests additional process steps and is based on soft PDMS mold, its WFE over 3" size component is also excellent.

4 Conclusion

Starting from one single design, we have manufactured large area 3" subwavelength reflective gratings by two technologies (e-beam lithography and NIL). They are composed of high aspect ratio (~5), high density and one level subwavelength binary structures.

Optical measurements show that the subwavelength blazed-binary gratings technologies based on e-beam lithography or NIL are promising for space application: they enable superior performance, in term of spectral band (efficiency margin of 10-20% in average), polarization sensitivity (smaller than 2%), very low ghost level (not detectable) and WFE level (< 20 nm RMS) with respect to requirements for observation imaging spectrometer operating in UV/Vis/NIR spectrometer requirements.

Part of this work was carried out within the framework of the ESA TRP project "Wide band and high efficiency reflective grating", No 4000118037/16/NL/PS.

References

1. N. Tetaz, C. Ruilier, M. Taccola, T. Viard, M.-S.L. Lee Bouhours, B. Loiseaux, G. Lehoucq, "Advanced large FOV UV/VIS/NIR/SWIR spectrometers for future earth observation instruments", *Proc. SPIE* **10563**, 1056333 (2017).
2. S. M. Ebstein, " Achromatic diffractive optical elements", *SPIE* Vol. **2404**, p211-216 (2001).
3. B. H. Kleemann, M. Seesselberg, and J. Ruoff, "Design concepts for broadband high-efficiency DOEs", J. Eur. Opt. Soc. **3**, 08015 (2008).
4. O. Sandfuchs, M. Kraus, R. Brunner, "Structured metal double-blazed dispersion grating for broadband spectral efficiency achromatization", J. Opt. Soc. Am. A **37**, 1369-1380 (2020).
5. P. Mouroulis, S.W. Wilson, P.D. Maker , R.E. Muller, "Convex grating types for concentric imaging spectrometers", Appl. Opt. **37**(31), p7200-7208 (1998).
6. B. Van Gorp, P. Mouroulis, D. W. Wilson, R. O. Green, "Design of the Compact Wide Swath Imaging Spectrometer (CWIS) ", *Proc. SPIE* **9222**, *Imaging Spectrometry XIX*, 92220C (2014).
7. C. Sauvan, P. Lalanne, M-S. L. Lee, "Broadband blazing with artificial dielectrics", Opt. Lett. **29**, 1593-1595 (2004).
8. C. Ribot, M.-S.L. Lee, S. Collin, S. Bansropun, P. Plouhinec, D. Thenot, S. Cassette, B. Loiseaux, P. Lalanne, "Broadband and Efficient Diffraction", Adv. Opt. Mat. **1**(7), p489-493 (2013).
9. F. Burmeister, T. Flügel-Paul, U.D. Zeitner, M.-S. L. Lee-Bouhours, G. Lehoucq, J. Cholet, B. Loiseaux, N. Tetaz, R. Windpassinger, M. Taccola, "Binary blazed reflection grating for UV/VIS/NIR/SWIR spectral range", *Proc. SPIE* **11180**, *ICSO 2018*, 111801J (2019).
10. W. Stork, N. Streibl, H. Haidner, and P. Kipfer, "Artificial distributed-index media fabricated by zero-order gratings", Opt. Let. **16**(24), pp. 1921-1923 (1991).
11. M. Heusinger, M. Banasch, T. Flügel-Paul, U. D. Zeitner, "Investigation and optimization of Rowland ghosts in high efficiency spectrometer gratings fabricated by e-beam lithography", *Proc. SPIE* **9759**, *Advanced Fabrication Technologies for Micro/Nano Optics and Photonics IX*, 97590A (2016).
12. M. Erdmann, E.-B. Kley, U. Zeitner, "Development of a large blazed transmission grating by effective binary index modulation for the GAIA radial velocity spectrometer", *Proc. SPIE* **105650**, *ICSO 2010*, 105651N (2017).

Infrared nanospectroscopy study of the light-induced conformational changes of Channelrhodopsin

Maria Eleonora Temperini[1,2,*], *Raffaella* Polito[2], *Antonia* Intze[1], *Ulrich* Schade[3], *Ljljana* Puskar[3], *Eglof* Ritter[3], *Leonetta* Baldassarre[2], *Michele* Ortolani[2], *Valeria* Giliberti[1]

[1] Istituto Italiano di Tecnologia, Center for Life NanoScience, Viale Regina Elena 291, I-00161 Roma, Italy
[2] Department of Physics, Sapienza University of Rome, Piazzale Aldo Moro 2, I00185 Roma, Italy
[3] Helmholtz-Zentrum Berlin für Materialien und Energie GmbH, Albert-Einstein-Str. 15, 12489 Berlin, Germany

Abstract. The channelrhodopsin-ChR2 is a light-sensitive transmembrane protein that acts as a selective ion channel between the intra- and the extra-cellular environments. In the last decade, ChR2 has proven to be essential for optogenetics, because, if expressed in mammalian neural cells, it enables the control of neuronal activity in response to visible light. Mid-infrared difference spectroscopy can probe the functional conformational changes of light-sensitive proteins, however intrinsic limitations of standard IR spectroscopy in terms of diffraction, and therefore number of probed proteins, require that the mid-IR experiments be performed on huge numbers of lipid membrane patches with overexpressed proteins. In this work, we apply for the first time IR difference nanospectroscopy, based on the use of mid-IR lasers and an atomic force microscope (AFM), to single membrane patches containing ChR2, obtaining relevant spectroscopy results for optogenetic applications and, more generally, for future experimental studies of light-sensitive proteins at the nanoscale.

1 Introduction

Light-sensitive microbial rhodopsins are proteins that play a crucial role in the cell functions acting as ion pumps or channels across the cell membrane in response to absorption of visible light. Although their native environment belongs to the world of archaea and bacteria, in the last decades the deep understanding of the molecular mechanisms that allow their functions has led to their application in diverse fields such as bioelectronics and neuroscience, depending on the specific protein and corresponding mutants.

The activity of light-sensitive proteins is triggered by photons in the visible range at a specific wavelength, which are absorbed by a retinal cofactor, covalently bounded to the protein structure, typically triggering a trans-to-cis isomerization. Starting from this event the protein undergoes a sequence of states conformationally and spectroscopically distinguishable that constitute the photocycle, during which the pump or the channel is active, until the initial state (i.e. the dark-adapted state) is recovered.

In this context, infrared (IR) spectroscopy in difference-absorption mode permits to investigate the functional conformational changes of the proteins relying on the vibrational excitation of the chemical bonds [1]. Conventional IR spectroscopy is limited in terms of sensitivity and it requires large amounts of proteins to be probed, preventing its application to the study of the single cell membrane patch.

We present an innovative IR difference spectroscopy approach at the nanoscale, based on the photothermal-induced resonance (PTIR) method, also called AFM-IR [2], to overcome the IR microscopy limitations in studying the light-sensitive conformational changes.

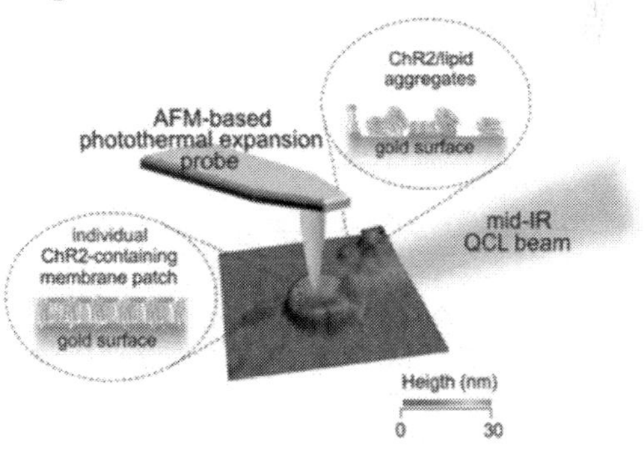

Fig. 1. Sketch of the AFM-IR setup based on a gold-coated tip probe and a mid-IR QCL laser beam. The AFM tip allows the topographic identification of a single membrane patch embedding ChR2 molecules, featuring a flat top surface and a height of ≈ 7 nm.

We are interested in the specific case of the light-activated conformational changes of the channelrhodopsin-ChR2 [3], an ion-gate protein used in the field of optogenetic to control neuronal stimuli with light in genetically modified neurons. Although

* Corresponding author: maria.eleonora.temperini@roma1.infn.it

optogenetics is now widely employed [4], a method of analysis of the detailed conformational changes of light-sensitive proteins at the single membrane level is obviously required to understand the opening/closing of the optogenetic gate in neurons. We then propose IR nanospectroscopy as an innovative method to study the protein activity at the level of a single cell membrane patch immobilized on a solid substrate. This configuration, in a broad sense, is more similar to that of neuronal cells with respect to the more widely studied sets of proteins in an aqueous suspension, and therefore it may provide meaningful information for optogenetic applications that sometimes require direct contact between the neuronal cells and metal probes.

2 Experimental

The sample investigated is prepared from a liquid suspension containing membrane patches loaded with ChR2 proteins. A microliter-size drop is cast on an ultra-flat gold surface and it is left to dry in a humidity-controlled atmosphere.

The AFM-IR vibrational imaging and nanospectroscopy are performed using a NanoIR2 (Anasys Instruments / Bruker Nano Surfaces) based on the photothermal expansion effect (see Fig.1 for a sketch of the experiment). The mid-IR laser source is an external-cavity broadly tunable quantum cascade laser (QCL) with a spectral range 900-1800 cm^{-1} (MIRcat by Daylight Solutions / Leonardo DRS). The AFM-IR technique exploits the photothermal expansion measured by the AFM readout circuit: when the QCL emits at a wavelength that matches one of the vibrational bands of the sample, the latter heats up and expands, exerting an extra force on the AFM tip that triggers the excitation of the natural cantilever bending modes. The amplitude of the cantilever oscillation is proportional to the absorption coefficient of the sample, and one can therefore obtain its IR vibrational spectrum by monitoring the deflection of the AFM probe [5]. The use of a gold-coated tip induces a strong plasmonic field enhancement at the apex, which allows one to collect IR response from a sample volume smaller than the diffraction limit by many orders of magnitude. The reason for that is the confinement of the temperature increase to the very small volume below the tip apex (thermoplasmonic effect). Furthermore, sensitivity down to the molecular monolayer can be reached by exploiting the plasmonic field enhancement in the nanogap between the gold-coated tip and a metallic surface used as support for the monolayer sample, in our case the single membrane patch [5,6]. In our experiment, AFM-IR difference nanospectroscopy is performed by implementing an external system of visible illumination to control protein photocycle based on two LEDs: a blue one (470 nm) to excite the ChR2 photocycle and a green one (505 nm) in order to recover the rest state of the proteins. The AFM-IR difference-spectra ΔA are obtained by averaging over 15 differences $A_{blue}(\omega) - A_{green}(\omega)$, where $A_{blue}(\omega)$ and $A_{green}(\omega)$ are the AFM-IR spectra acquired under blue and green light, respectively.

We perform micro-FTIR measurements on the same sample cast on an ultraflat gold surface to validate the difference-absorbance results obtained at the nanoscale and to highlight some differences between the nanospectroscopy experiment and the one realized probing a micrometer area of sample, i.e. a huge number of proteins (more than 10^7) embedded in stacked lipid membranes. The experiment has been realized at the synchrotron radiation source BESSY II of Helmoltz-Zentrum Berlin using an internal IR source of a Nicolet Continuum Infrared Microscope with a knife-edge aperture of 100μm x 200μm. The micro-FTIR difference-spectra ΔA were calculated as $\Delta A = A_{blue} - A_{green} = log(R_{green} / R_{blue})$, where R_{blue} and R_{green} are the reflection spectra obtained under illumination at 470 nm and 505 nm, respectively.

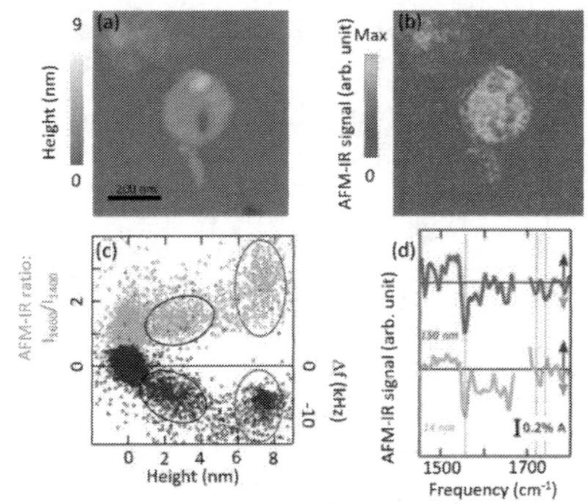

Fig. 2. (a) AFM topography of a single ChR2 membrane patch. (b) AFM-IR map acquired at 1660 cm^{-1} normalized by the map acquired at 1400 cm^{-1}. (c) Scatter plot of the ratio I_{1660}/I_{1400} (green dots) and of the variation of the resonance frequency $\Delta f = f - f_{Au}$ of the tip (blue dots) as a function of topographic height. (d) AFM-IR difference-absorbance spectra obtained on 150 nm-thick (red curve) and 14 nm-thick (yellow curve) areas of the ChR2 sample.

3 Results

The use of an AFM-based platform enables the identification of sample areas with membrane patches adhering to the ultraflat metal surface, in which the light-induced activity of proteins can then be properly investigated. In Fig. 2a we report the AFM topography map of a representative lipid membrane flake containing ChR2, whose thickness is less than 7 nm, confirming the adhesion of a single membrane patch to the substrate. IR nanoimaging may also provide information on the homogeneity of mechanical and spectroscopic properties of the assemblies. We indeed report an AFM-IR map acquired at 1660 cm^{-1} (in Fig. 2b) normalized by that acquired at a frequency that does not correspond to any IR absorption of the membrane (1400 cm^{-1}), so as to remove the background originating from thermal expansion of the gold-coated tip and of the substrate. Moreover, in Fig.2c we show the scatter plot as a function of the thickness of the sample of (i) the ratio I_{1660}/I_{1400} of the AFM-IR map (green dots), where I is the photothermal expansion signal intensity, and of (ii) the stiffness variation (blue dots) in the investigated area. For the latter one we report in particular the variation of the resonance frequency $\Delta f = f - f_{Au}$ of the tip in contact with the sample respect to the frequency f_{Au} in contact with the bare ultraflat gold substrate. Each plot has one cluster centred at a null value of AFM height that represents the bare ultraflat gold surface and a second cluster centred at height value of 7 nm for the ChR2 samples (red ellipses in Fig.2c). There is also a third cluster (black ellipses in Fig.2c), whose pixels correspond to the top-left corner of Fig. 2b: the lower value of both the mean height and the absolute value of Δf of this cluster compared to the others, suggests the simultaneous presence on the gold surface of unoriented proteins and lipid aggregates (see also Fig. 1). The capability of discarding non-membrane lipid-protein aggregates is a characteristic of the AFM-IR nanoimaging method that cannot be matched by standard FTIR spectroscopy, even if employed in microscopy mode at a synchrotron.

After the identification of sample areas with single flat membrane patches on the ultraflat gold surface, we also use the AFM-IR platform to perform a functional study of ChR2 by probing its light-activated conformational changes connected to the ion channel function, measuring the difference-absorbance ($\Delta A = A_{blue} - A_{green}$) of the sample between the blue and the green illumination state. We perform nano-IR spectroscopy on both a 150 nm-thick area and a double membrane structure (14 nm-thick). As shown in Fig. 2d, the ΔA curves display common features known to be related to the vibrational modes involved in the protein activity, including the sharp C=C retinal stretching vibration peak (around 1555 cm^{-1}) and the two negative peaks in the carbonyl region around 1720 cm^{-1} and 1736 cm^{-1}. The comparison with the micro-FTIR spectroscopy measurements also confirms the validity of the nanoscale approach. Beyond these common features, there are slight differences between the two spectra in Fig. 2d that can be attributed to the direct adhesion of the membrane patch to the gold surfaces in the case of the 14-nm thick sample (yellow curve), which should have an effect on the ChR2 conformations during the photocycle [7].

References

1. V.A. Lorenz-Fonfria, Chem. Rev. **120**, 3466 (2020).
2. A. Dazzi and C. B. Prater, Chem. Rev. **117**, 5146-5173 (2017).
3. F. Schneider, C. Grimm and P. Hegemann, Annu. Rev. Biophys **44**, 167-186 (2015).
4. A. M. Packer, B. Roska, and M. Häusser, Nat. Neurosci. **16**, 805 (2013).
5. F. Lu, M. Jin and M. A. Belkin, Nat. Photonics, **8**, 307 (2014).
6. V. Giliberti et al. Nano Lett. **19**, 3104 (2019).
7. R. Polito, M.E. Temperini et al. Phys. Rev. Applied **16**, 014048 (2021).

Intracellular SERS monitoring of drug release from plasmonic-assisted biosilica nanoparticles

Chiara Tramontano[1,2], *Stefano* Managò[3], *Donatella* Delle Cave[4], *Giovanna* Chianese[1], *Enza* Lonardo[4], *Luca* De Stefano[1], *Anna Chiara* De Luca[3], and *Ilaria* Rea[1]

[1]Institute of Applied Science and Intelligent Systems (ISASI), National Research Council of Naples, Via Pietro Castellino 111, Italy.
[2]Department of Pharmacy, University of Naples Federico II, Via Domenico Montesano 49, 80131, Italy
[3]Institute of Biochemistry and Cell Biology (IBBC), National Research Council of Naples, Via Pietro Castellino 111, Italy
[4]Institute of Genetics and Biophysics (IGB), National Research Council of Naples, Via Pietro Castellino 111, Italy

Abstract. Nanoscale delivery systems have been investigated for therapy due to their advantages, including the sustained delivery of drugs to cells and reduction of systemic toxicity compared to conventional treatments. However, their application is still hampered by experimental challenges, such as the investigation of the drug release in cells rather than *in vitro*. Here, we describe a hybrid nanoplatform for monitoring the drug release in living colorectal cancer (CRC) cells by Surface-Enhanced Raman Scattering (SERS). Specifically, the anticancer drug Galunisertib is encapsulated in diatomite nanoparticles (DNPs) decorated by gold nanoparticles (AuNPs) and capped by gelatin. The combination of DNP loading capacities with the Raman enhancement of Galunisertib provided by AuNPs enables bio-imaging and drug delivery without using fluorophores or markers, avoiding fluorescence-quenching issues. Thanks to the Raman enhancement of Galunisertib, the drug release profile is monitored and quantified in living cells by SERS with a femtogram scale resolution. When the gelatin shell is digested by proteases, Galunisertib is released and its SERS spectrum decreases, allowing real-time quantification in CRC cells. The therapeutic efficiency of the Galunisertib delivery platform offers an alternative route for lowering drug dose and toxicity.

1 Introduction

The research for nanoscale Drug Delivery Systems (DDS) is driven by their potentiality to reduce the systemic toxicity associated with small-molecule administration. Although many advantages have been reported in the medical application of nanoparticles (NPs), their use in cancer treatment is hampered by diverse experimental challenges, such as investigating the drug release behavior in cells [1]. The quantification of the intracellular released drug is crucial to understand the effective amount of the molecule that reaches the site of action for disease treatment. Sophisticated analytical tools (*e.g.*, High-Performance Liquid Chromatography HPLC, fluorescence or absorbance spectroscopy) allows studying the drug release profile of nanoscale DDS *in vitro*, but not in living cells, and the time required for sample preparation and running hinders tracing the intracellular release in real-time modalities. To avoid these limitations, it is necessary to develop sensitive platforms enabling the real-time intracellular monitoring of chemotherapeutics to improve the therapeutic efficacy of DDS and get a deeper understanding of how they react to the cell environment. Recently, Surface-Enhanced Raman Spectroscopy (SERS) has offered advancements in label-free biosensing thanks to the near-field optical amplification of molecules provided by metal nanostructures, such as metal NPs [2]. The plasmonic and electromagnetic properties of gold NPs (AuNPs) offer high-sensitivity and Raman enhancement factors up to 10^{12} for absorbed or attached molecules on metal surfaces or NPs. Therefore, the combination of the strong Raman enhancement of molecules provided by AuNPs with the loading capacity of porous NPs can be used to develop hybrid platforms capable of both delivering and real-time quantifying the released drug in cells [3]. In the last years, inorganic diatomite nanoparticles (DNPs) made of diatom biosilica have been explored as DDS thanks to their intriguing properties for biomedical applications, such as the porous structure, long half-life, chemical stability, and ease-to-tune surface properties [4,5]. The tunability of the DNP surface allows to modify NPs through several chemical

 https://doi.org/10.1051/epjconf/202125513002

approaches and provides them with novel captivating properties. Among all, the decoration of the DNP surface with AuNPs confers to DNPs optical properties suitable for bioimaging applications via SERS analysis [6]. In this study, the small molecule Galunisertib (LY), currently used for colorectal cancer (CRC) treatment, is encapsulated in a hybrid nanoplatform made of DNPs decorated by AuNPs and capped by gelatin (DNP-AuNPs-LY@Gel). The monitoring of DNP internalization in cells and quantification of the LY release is made possible by the plasmonic resonance of AuNPs providing a highly sensitive SERS readout of LY. Therefore, the intracellular release of LY is measured in CRC cells by SERS and correlatively quantified to provide the LY sensogram evolution with a sub-femtogram scale resolution [7] (Fig.1).

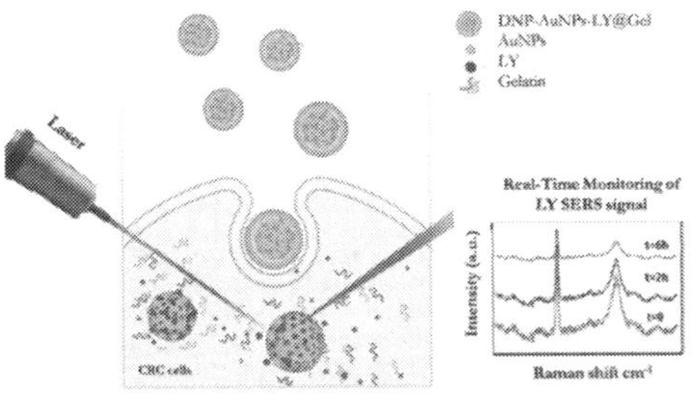

Fig.1. Scheme of the hybrid nanoplatform internalization and real-time monitoring of the released drug. The gelatin shell around the hybrid system acts as a pH-sensitive drug gatekeeper. It is disrupted in the acidic microenvironment of the CRC cells, enabling the controlled release of LY in malignant cells. The monitoring and quantification of the released drug are performed by SERS technique.

2 Results and Discussion

The hybrid nanoplatform was developed through an *in-situ* synthesis of pegylated AuNPs on the amino-modified surface of DNPs. Before this, the diatom powder, mainly constituted by diatom skeletons having different sizes and shapes, was ultrasonicated, purified, and filtered to obtain DNPs with a mean diameter of 400 (50) nm. The hydroxyl groups (-OH) exposed on the surface of the nanostructured silica were then activated by a Piranha treatment and an amino-silanization process was carried out according to methods described earlier [8]. The amino groups (-NH_3^+) promoted the nucleation of pegylated AuNPs on the surface of DNPs thanks to the electrostatic attraction to the negatively charged groups of the dicarboxylic polyethylene glycol (PEG) used as a stabilizing agent. The presence of AuNPs was confirmed by Transmission Electron Microscopy (TEM) investigation (Fig. 2A), showing that the surface of DNPs was covered by a homogeneous carpet of AuNPs after the synthesis process. According to the TEM investigation, the mean size of the spherical AuNPs was about 25 nm (calculated by Image J software).

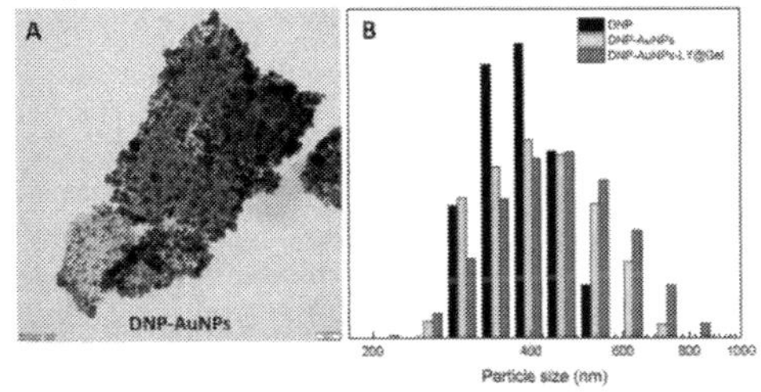

Fig.2. A) TEM image of a DNP decorated by 25 nm- in size spherical AuNPs. The scale bar is 50 nm. **B)** Size distribution (nm) of DNP before surface functionalization (black), after the decoration with AuNPs (light green), and gelatin capping (dark green).

Although DNPs show an irregular shape due to the process of ultrasonication performed to reduce the diatom powder at the nanoscale level, their biocompatibility and safety have been extensively demonstrated *in vitro* and animal models [9]. Moreover, the presence of metal NPs on the silica structure conferred to the DNPs a localized surface plasmon resonance (LSPR) at 576 nm wavelength, which enabled monitoring the functionalization *via* absorbance analysis (UV-Visible spectroscopy). The mean diameter of the hybrid system (DNP-AuNPs) calculated by Dynamic Light Scattering (DLS) was equal to the bare DNPs (400 ± 50 nm), revealing that the process did not influence the size and stability of the nanoplatform (Fig. 2B) Then, the hybrid system DNP-AuNPs was loaded with the anticancer small molecule Galunisertib (LY2157299), a Transforming Growth Factor β (TGF β) Receptor Inhibitor able to block the metastatic process in diverse types of cancer, including CRC. The loading of LY in the nanoplatform was performed by both physisorption (through the porous structure) and electrostatic attraction forces between the drug molecules and the surface of DNP-AuNPs. Finally, the loaded nanoplatform was covered by a shell of cross-linked gelatin to avoid burst release phenomena and promote a pH-controlled release of the drug *in vitro*. The encapsulation of LY in the nanocomplex did not cause changes in the LSPR of the hybrid suspension, due to the small size of LY. Rather, the LSPR peak of the gelatin-functionalized hybrid system (DNP-AuNPs-LY@Gel) red-shifted by 10 nm after the gelatin shell formation, due to a change of the local refractive index at AuNPs surroundings. The gelatin shell caused an overall size increment of 50 nm in the final nanoplatform that showed a mean diameter of 450 (50) nm. The generation of a gelatin shell surrounding the NPs was effectively confirmed by the analysis of the LY release from the nanoplatform, which followed a sigmoidal and a pH-responsive behavior (Fig. 3A). The loading capacity (LC) of the DNP-AuNPs-LY@Gel system was investigated by dispersing the hybrid nanoplatform in Phosphate Buffer Saline solution (PBS) for 60 hours and

* Corresponding author: chiara.tramontano@na.isasi.cnr.it

quantifying the amount of released drug by HPLC. The LC was estimated to be 20 (4) μg/mg of DNPs. Figure 3a shows the drug release profiles performed both in acidic and physiological conditions. The gelatin shell around the DNP-AuNPs-LY@Gel system was responsible for the controlled and pH-responsive LY release. The acidic microenvironment, as cancer cell microenvironment as well, promoted the relaxation of the gelatin chains and the release of the drug in the medium. The release profile reached a plateau corresponding to 100% of the total encapsulated drug after 60 hours, ensuring that no drug dispersion phenomena occurred in the system. In physiological conditions, instead, the gelatin shell retained the drug and hampered its release from the nanoplatform. Therefore, the pH-responsive profile exhibited by the developed DDS can address the release of LY in malignant cells, in which the acidic microenvironment is caused by high levels of lactate generated by cell activity [10].

The investigation of the drug release *in vitro* is generally considered mandatory to understand the nanoplatform behavior in different pH environments. The *in vitro* release tests offer many advantages in the comprehension of the DDS stability and performance, but they cannot replicate the cell environment at all, especially the malignant one. To overcome this limitation, we developed a SERS hybrid nanoplatform that allowed us to investigate, real-time monitor, and quantify the LY release in living CRC cells without the need of fluorescence payloads or markers, avoiding fluorescence-quenching issues or modification of the nanoprobe surface [11]. To this aim, we studied the Raman fingerprint of LY alone between 1300-1650 cm^{-1} and the SERS fingerprint of the LY from the DNP-AuNPs-LY@Gel nanoplatform dried on a coverslip. The enhancement factor (EF) analysis confirmed that drug molecules strongly interacted with the AuNPs decorating the nanoplatform, providing an EF of the LY SERS signal of 4.5×10^5. The most intense SERS vibration, ascribed to a combined action of ring C–N stretching and bending in the drug structure, was found at 1360 cm^{-1} and used for monitoring the LY intracellular release *in vitro* and cell. The release behavior of the DNP-AuNPs-LY@Gel platform was first investigated by SERS in PBS solutions 5.5 and 7.4 and then in the CRC cell. To this aim, 1.5 μg of the nanoplatform (in which 30 ng of LY was loaded according to the reported LC) was deposited on a CaF_2 slide. The release profile investigated by SERS measurements was compared with the HPLC investigations reported in Figure 3a, confirming that the acidic microenvironment triggered the LY release from the nanoplatform in a sigmoidal and pH-responsive manner. The nanoplatform released 50% of the total encapsulated drug after 24 hours in PBS solution 5.5, as already observed by HPLC and reported in Fig 3a. Furthermore, thanks to the strong SERS enhancement of the LY signal provided by the hybrid nanoplatform, we managed to quantify the amount of drug released from a single nanovector unit in the medium at each time interval. Each DNP unit contained 0.25 fg of LY and released 0.125 fg after 24 hours of immersion in the PBS solution pH 5.5, corresponding to 50% of the total loaded mass. Conversely, the gelatin shell was tightly folded in PBS solution 7.4 mimicking the physiological conditions and the LY release was prevented (data shown in [7]).

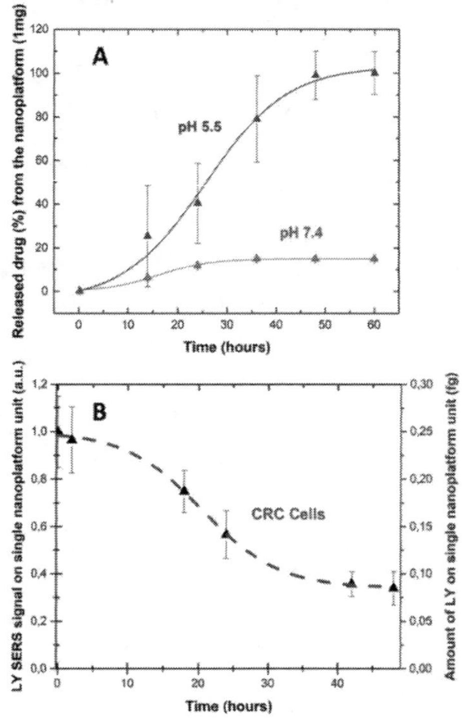

Fig.3. A) Cumulative release (%) of LY from DNP-AuNPs-LY@Gel (1mg) in PBS pH 5.5 (blue line) and 7.4 (red line). B) Sensogram of the LY SERS signal evolution after release in CRC cells. The SERS signals are correlated to the HPLC analysis to quantify the amount of drug-loaded in /released from a single nanovector unit. Left Y reports the normalized LY SERS intensity counts monitored on the nanoplatform. Right Y reports the amount of LY measured in a single nanovector unit.

The great advantage offered by this hybrid nanosensor was the possibility to real-time monitor the drug release in the LS-174T cell line and quantify the effective internalized dose. The LS-174T cell line was chosen due to its metastatic potential and mesenchymal phenotype related to an upregulation of the TGF βI receptor pathway [12]. The release of LY from the DNP-AuNPs-LY@Gel nanoplatform was expected to block the TGF β-induced metastatic process, promoting the transformation from epithelial to mesenchymal phenotype.

For the intracellular release tests, the DNP-AuNPs-LY@Gel was incubated with cells at a concentration of 50 μg/mL and the LY SERS signal was monitored in a cluster of cells at different times (Fig 3B). The intracellular release followed a sigmoidal trend

* Corresponding author: chiara.tramontano@na.isasi.cnr.it

similar to one investigated by HPLC in PBS 5.5 since the acidic tumor microenvironment favored the LY release from the gelatin shell. The delivered LY in cells was about 30% of the total drug after 18 hours. Considering that each DNP unit contained 0.25 fg of LY, the LY delivered in the cell by a single DNP corresponded to 0.075 fg after 18 hours. The released LY was 50% (0.125 fg for DNP unit) after 24 h and the remaining mass of drug (0.16 fg) was released after 48 h. So, the combination of metal NPs with porous DNPs enabled an efficient SERS intracellular drug tracing, sensing, and quantification in living CRC with a sensing resolution down to 7.5×10^{-18} g. The therapeutic outcomes of the LY DDS were finally tested in the LS-174T cell line to analyze whether the therapeutic efficacy of the drug could be enhanced by the nanoplatform or not. We assessed the biocompatibility of the nanoplatform in the CRC cell line and indicated as safe for the treatment a concentration of the DNP-AuNPs-LY@Gel suspension of 50 μg/mL (in which 2.5 μM of LY was loaded according to the reported LC). Since the chosen drug was capable of binding to the TGF β receptor and affecting the downstream genetic pathway, we evaluated the expression levels of E-Cadherin (involved in the epithelial process) and Snail-1 and Twist-1 (involved in the mesenchymal one) genes upon treatment with 50 μg/mL of the hybrid system. The encapsulation of LY in the nanoplatform enhanced its ability to block metastases by downregulating the pro-metastatic genes and enhancing cell epithelialization. As a consequence, the treatment with the hybrid nanoplatform induced a strong morphological cell transformation from elongated and spindle-shaped (aggressive and metastatic) to rounded (less aggressive and wild-type) cells within 48 hours.

Considering these results, it was clear that the encapsulation of LY in the hybrid nanoplatform made the LY therapeutic effect more than three times stronger than the conventional administration of the drug.

3 Conclusions

In this work, a hybrid nanoplatform is developed to trace and quantify the Galunisertib intracellular release with a sub-femtogram scale resolution. Galunisertib is loaded in the DNP with a loading capacity of 20 $\mu g/mg^{-1}$ of DNPs and its release is label-free monitored with extreme sensitivity in living cells thanks to the enhancement of the Galunisertib SERS signal provided by gold nanoparticles. The gelatin shell prevents burst release effects and addresses the drug release in malignant cells, where the acidic microenvironment promotes gelatin degradation. The modulation of the expression levels of E-Cadherin Snail-1, and Twist-1 genes is significantly higher upon treatment with the loaded nanoplatform than the free Galunisertib. The delivered drug promotes the transformation of the LS-174T cancer cells from mesenchymal to epithelial-like phenotype efficiently. Thereby, it is possible to lower the administered drug dose through the encapsulation of Galunisertib in the developed system to reduce the reported systemic effects.

References

[1] J. K. Patra *et al.*, Nano based drug delivery systems: recent developments and future prospects, J. of Nanobiotechnology, vol. **16**, n. 1, pag. 71 (2018), doi: 10.1186/s12951-018-0392-8.

[2] S. Managò *et al.*, Bioderived Three-Dimensional Hierarchical Nanostructures as Efficient Surface-Enhanced Raman Scattering Substrates for Cell Membrane Probing, ACS Appl. Mater. Interfaces, vol. **10**, n. 15, pagg. 12406–12416, (2018), doi: 10.1021/acsami.7b19285.

[3] I. Rea *et al.*, Hybrid Organic/Inorganic Nanomaterials for Biochemical Sensing, in Sensors and Microsystems, vol. **753**, G. Di Francia e C. Di Natale, A c. di Cham: Springer International Publishing (2021), pagg. 93–99. doi: 10.1007/978-3-030-69551-4_13.

[4] S. Managò *et al.*, Internalization kinetics and cytoplasmic localization of functionalized diatomite nanoparticles in cancer cells by Raman imaging, J. Biophotonics, vol. **11**, n. 4, pag. e201700207 (2018), doi: 10.1002/jbio.201700207.

[5] C. Tramontano, G. Chianese, M. Terracciano, L. de Stefano, e I. Rea, Nanostructured Biosilica of Diatoms: From Water World to Biomedical Applications», Appl. Sci., vol. **10**, n. 19, pag. 6811, (2020), doi: 10.3390/app10196811.

[6] M. Terracciano, M. Napolitano, L. D. Stefano, A. C. D. Luca, e I. Rea, Gold decorated porous biosilica nanodevices for advanced medicine, Nanotechnology, vol. **29**, n. 23, pag. 235601, (2018), doi: 10.1088/1361-6528/aab7c4.

[7] S. Managò *et al.*, SERS Quantification of Galunisertib Delivery in Colorectal Cancer Cells by Plasmonic-Assisted Diatomite Nanoparticles», *Small*, vol. **17**, n. 34, pag. 2101711 (2021), doi: 10.1002/smll.202101711.

[8] I. Rea *et al.*, Diatomite biosilica nanocarriers for siRNA transport inside cancer cells, *Biochim., Biophys. Acta (BBA) - General Subjects*, vol. **1840**, n. 12, pagg. 3393–3403,(2014),,doi:/10.1016/j.bbagen.2014.09.009.

[9] M. Terracciano, L. Stefano, C. Tortiglione, A. Tino, e I. Rea, In Vivo Toxicity Assessment of Hybrid Diatomite Nanovectors Using *Hydra vulgaris* as a Model System, Adv. Biosys., vol. **3**, n. 4, pag. 1800247 (2019), doi: 10.1002/adbi.201800247.

[10] K. G. de la Cruz-López, L. J. Castro-Muñoz, D. O. Reyes-Hernández, A. García-Carrancá, e J. Manzo-Merino, Lactate in the Regulation of Tumor Microenvironment and Therapeutic Approaches, *Front. Oncol.*, vol. **9**, pag. 1143, 2019, doi: 10.3389/fonc.2019.01143.

[11] D. Poplinger *et al.*, Ratiometric Fluorescence Monitoring of Antibody-Guided Drug Delivery to Cancer Cells, *Bioconjugate Chem.*, vol. **32**, n. 8, pagg. 1641–1651 (2021), doi: 10.1021/acs.bioconjchem.1c00205.

[12] D. V. F. Tauriello, A. Calon, E. Lonardo, e E. Batlle, Determinants of metastatic competency in colorectal cancer, Mol Oncol, vol. **11**, n. 1, pagg. 97–119 (2017), doi: 10.1002/1878-0261.12018.

* Corresponding author: chiara.tramontano@na.isasi.cnr.it

Optical multiplexed bioassays on photonic crystals for breast cancer biomarker detection

Tommaso Pileri[1*], *Alberto* Sinibaldi[1], *Agostino* Occhicone[1], *Elena* Giordani[2], *Matteo* Allegretti[2], *Peter* Munzert[3], *Frank* Sonntag[4], *Norbert* Danz[3], *Patrizio* Giacomini[2] and *Francesco* Michelotti[1]

[1]Department of Basic and Applied Science for Engineering, Sapienza University of Rome, Via A. Scarpa 16, 00161, Rome, Italy.
[2]IRCCS Regina Elena National Cancer Institute, Via Elio Chianesi 53, 00144 Rome, Italy.
[3]Fraunhofer Institute for Applied Optics and Engineering IOF, A. Einstein-Str. 7, 07745 Jena, Germany.
[4]Fraunhofer Institute for Material and Beam Technology IWS, Winterberg-Str. 28, 01277 Dresden, Germany.

Abstract. An optical biosensor for proteomic breast cancer biomarker detection in complex media is presented. Bloch Surface Waves (BSW) excited onto one dimensional photonic crystal (1DPC) were used to probe the interaction of HER2 with three antibody species and an inert protein (Bovine Serum Albumin - BSA). The optical system combines Label-Free readings to track the bioassay real-time development and Fluorescence emission quantification to evaluate the level of specific interaction between the antigen and the antibodies. The results confirm a distinguishable level of affinity between the antibodies and the analyte according to their specificity even at low antibody surface density (about 1173 pg/mm^2).

1 Introduction

Non-invasive diagnostics via cancer biomarker detection is a constantly increasing focus of research [1]. Among other techniques, optical biosensors based on electromagnetic surface waves represent an opportunity to improve ease-to-use and multiplexing capability while lowering costs and processing time [2]. The first known approach exploiting a spatially confined interaction between a highly enhanced electromagnetic field at the optical interface and the molecular recognition layer, are the surface plasmon polaritons (SPR). However, further significant enhancement of resolution can be obtained by using low lossy all-dielectric multi-layered structures, typically used to excite BSW [3]. This is achieved by combining both Label-Free and Fluorescence investigation of the bioassay occurring over the sensor surface [4,5].

The main purpose of this work is the study of a multiplexed optical biosensor based on BSW able to reveal the presence of the HER2 receptor tyrosine-protein kinase. Altered signalling by epidermal growth factor receptors, like HER2 and other family members, is involved in 20-30% of all breast cancers [6], representing an appealing subject of investigation in neoplastic transformation and progression. Indeed, these receptors play a critical role because of their involvement in the three main regulation pathways of a cell such as differentiation, growth and proliferation [7].

2 Methods

2.1 1DPC sensors and optical readout system

The 1DPC biochips is designed to excite BSW in the visible range when in contact with aqueous solutions. It is composed by two periods of SiO_2 (275 nm, n = 1.474 + j5 × 10^{-6} @ 670 nm) and Ta_2O_5 (120 nm, n = 2.16 + j5 × 10^{-5} @ 670 nm) terminated by a topping layer of TiO_2 (20 nm, n = 2.28 + j1.8 × 10^{-3} @ 670 nm) and SiO_2 (20 nm). The final silica layer is deposited to allow for the use of silane based chemical surface functionalization (see next section). For further details on the 1DPC design see ref [8]. The 1DPC deposition process is performed by plasma ion assisted deposition technology (PIAD) onto a prism-shaped plastic (TOPAS) substrate (Fig. 1a-b - insets). This allows to operate in the attenuated total internal reflection configuration (TIR) (Kretschmann-Raether configuration). By using cylindrical lenses, for both modes, the laser beam is focalized on a line along the longitudinal biochip axis [4]. The optical readout system is able to operate in two independent modes.

The first is the Label-Free mode. Using a s-polarized light beam (λ_{LF} = 670 nm), the electromagnetic properties evolution due to biorecognition binding events over the surface can be real-time detected. This is performed by tracking the minima position of the reflectivity angular spectra recorded over the entire investigated photonic crystal midline (Fig. 1a).

The second operation mode is the Fluorescence mode. This uses a different light source (λ_{FLUO} = 635 nm) to excite the same surface mode. Sacrificing the excitation efficiency (3% at this wavelength), the present

* Corresponding author: tommaso.pileri@uniroma1.it

design satisfies the BSW coupling condition with the emission peak of the Qdot™ 655 used as fluorescent label in the biorecognition protocol (described in the paragraph 2.2). The dye emitted fluorescence, after being stimulated matching the resonance condition of the optical surface states, is channelled back through the same BSW mode and through the dielectric structure, radiated into a narrow angular range through the TOPAS, and imaged by the CCD (Fig. 1b inset). A dichroic beam splitter separates Label-Free and Fluorescence emission from Fluorescence excitation beam (Fig. 1b inset). For both operation modes, rows and columns of each image recorded by the CCD refers to spectral angular information and lateral position along the biochip midline respectively.

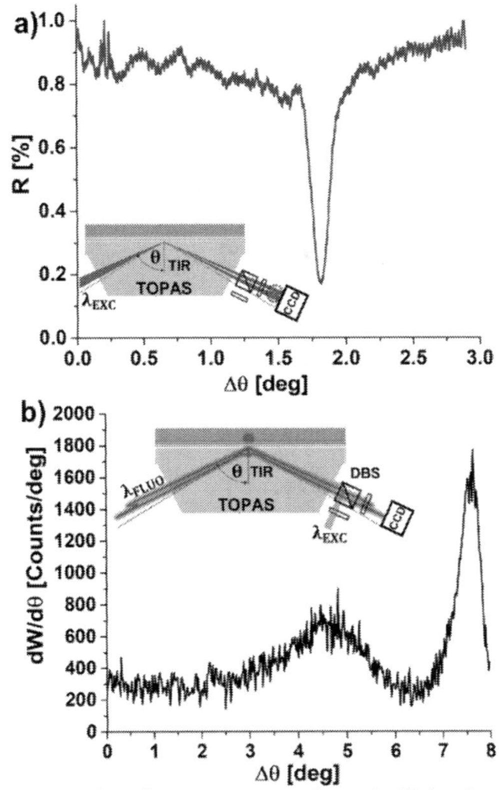

Fig. 1 – The reflectance spectrum shown in (a) has been recorded with Phosphate Buffered Saline 1x inside the microfluidics (represented by the dark blue layer in the inset). The inset of (a) depicts the light path through the biochip cross-section and light recording system in Label-Free configuration. (b) shows the angular fluorescence intensity emitted by Qdot™ 655 at the surface of the 1DPC. The inset of (b) shows the fluorescence excitation scheme where the fluorescence emitted is collected with a remarkable directionality.

2.2 Surface chemistry and bioassay

In order to covalently bind antibody and proteins onto the 1DPC surface, we silanize the previously piranha cleaned SiO_2 thin 1DPC top layer with the (3-aminopropyl)triethoxysilane (APTES) and, subsequently, surface activation with glutaraldehyde is performed [9]. The bio-functionalization step is carried-out by means of a GeSiM Nano-Plotter. The multiplexing characteristic is conferred to the 1DPC biosensor surface by printing three solutions at a concentration of 100 µg/mL in PBS 1x of three different species of gamma-immunoglobulin (Capture-IgG) over three isolated regions of interest (ROI, Fig. 2a). In particular, onto the three ROIs, we immobilized two antibodies to HER2, e.g. W6/300G9 (ROI 1) and Herceptin (ROI 2). A third antibody to Human Leukocyte antigens (HLA) named L31 (ROI 3) was used as a control of irrelevant specificity. To avoid coalescence between the printed solutions, drops containing a total volume of 10 nL of solution are printed onto each ROI. The ROIs are separated by a pitch of 1.25 mm. After the Capture-IgG incubation, 45 min at +4°C, the surface is rinsed pipetting Phosphate Buffered Saline solution (PBS 1x) and the microfluidics is mechanically coupled to the biochip.

The first solution injected (Fig.2b I) is BSA [1 mg/mL] to passivate the 1DPC surface. The concentrated solution saturates the remaining free glutaraldehyde binding sites preventing undesired non-specific signals over the ROIs. Whereas, in the formerly dry areas, the BSA binds granting the possibility of setting control regions to track the potential non-specific molecular interactions. The immunoassay performed in this work is a sandwich type. Then, after the injection of a solution of HER2 [1 ng/mL] in skim milk (SM) with a final matrix complexity of 1 mg/mL (Fig.2b II), fluorescence background is recorded (Fig.2b III). The last step is the injection of a Detection-IgG* labelled with Qdot™ 655 [10 µg/mL] in PBS 1x. Being affine for a different epitope of the HER2, the Detection-IgG* brings the fluorophore selectively in those regions of the surface where the analyte has been recognized and captured.

3 Results

The focus of this work is to set-up a multiplexed bioassay combining Label-Free and Fluorescence investigation. The Label-Free analysis confirms the successful binding of the Capture-IgG onto each ROI. It can be seen in Fig.2(a) inside the boxed regions where the resonance angle of excitation of the BSW is blue shifted compared to the neighbouring areas. This resonance shift can be translated in surface molecular density by the De Feijter's formula [10] obtaining a mean value of 1173 ± 77 pg/mm^2 of Capture-IgGs.

The recorded sensorgram starts with BSA injection (Fig. 2b I) and clearly shows the difference between the control areas (black, blue, green and purple lines) and the ROIs (red, pink and dark blue lines). This discrepancy is entirely due to the occupancy rate of the glutaraldehyde binding sites. They are mostly saturated by the Capture-IgGs onto the ROIs and completely receptive onto the regions designed to work as reference.

The second resonance shift (Fig. 2b II) is relative to the injection of the skim milk solution containing HER2 [1 ng/mL]. All the ROIs and the control regions are characterized by the same residual resonance shift, that can be read as the difference between the angular resonance position before the biomarker solution injection and after the following PBS 1x washing step.

The last section of the sensorgrams shows the response of the biochip to the injection of a solution of Detection-IgG* labelled with Qdot™ 655 (Fig. 2b IV) and the subsequent washing step with PBS 1x. As in the previous step, the same signal is recorded from all the differently bio-functionalized areas. Therefore, the Label-Free signal analysis gives no information about the interaction between the biosensor surface and the solutions of the biomarker in complex matrix and the Detection-IgG* in PBS 1x.

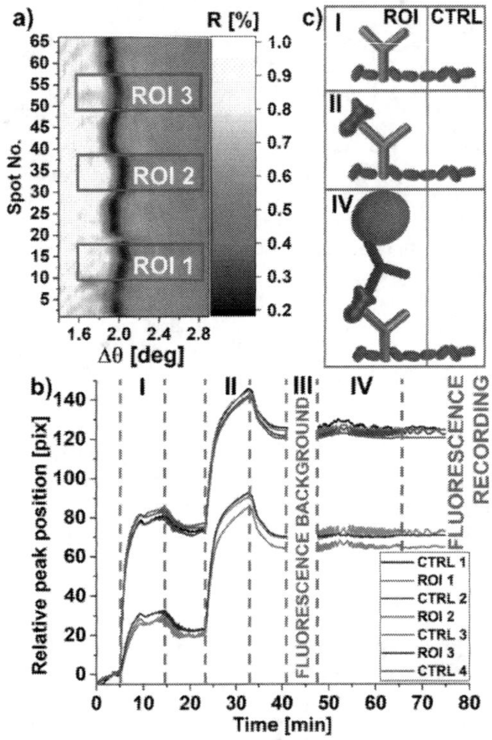

Fig. 2 – In (a) the reflectance angular spectral map along the biochip investigated area is reported. The regions where the three species of antibodies are immobilized are clearly visible. In (b) the real time sensorgrams recorded in Label-Free mode show how the refractive index of the bio-functionalized surface changes when each solution is injected: BSA (I), HER2 in skim milk (II), Detection-IgG* (IV). A surface washing step with PBS 1x follows each injection. Red, pink and dark blue lines refer to ROIs (1,2 and 3 respectively), while black, blue, green and purple lines refer to control regions (listed as CTRL in (c)).

More information about the last step can be extracted from the Fluorescence signal analysis. To take into account signal aberration due to either stray light from the excitation or fluorescence of the 1DPC, the fluorescence background signal is collected and quantified before the labelled Detection-IgG* (Fig. 2a III) injection. Subtracting this background to the total fluorescence signal collected at the end of the biorecognition protocol, the emitted fluorescence spectral map reported in Fig. 3a is obtained.

In this map the three ROIs are clearly visible where the bioassay ended characterized by different yields (black, blue and green in Fig. 3a-c) and the neighbouring areas used as reference (red in Fig. 3a-c).

From the quantitative analysis of the collected fluorescence (Fig. 3c), one can discriminate the specific (ROI No. 1 and 2) from the non-specific (ROI No. 3) immunoglobulin type. Moreover, the affinity towards the biomarker of the antibody Anti-HER2 (W6/300G9), immobilized onto the ROI No. 1 (black in Fig. 3a-c), is comparable to the affinity shown by the Herceptin (blue in Fig. 3a-c), immobilized onto ROI No. 2.

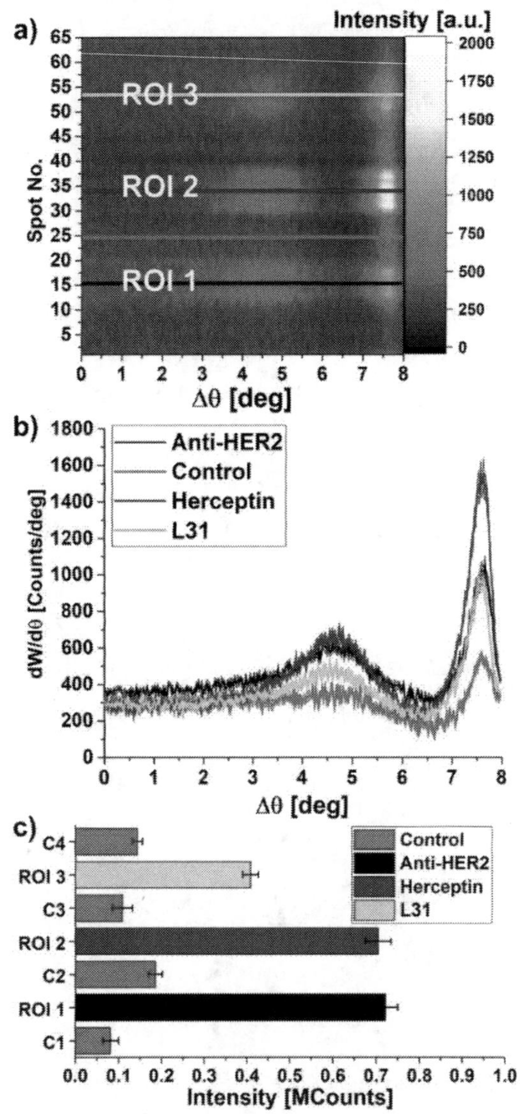

Fig. 3 – In the fluorescence spectral map (a) recorded over the investigated area the three brighter regions correspond to the three ROIs. (b) shows the Fluorescence angular spectra extracted from map (a): black, blue and green spectra correspond to ROI No. 1,2 and 3 respectively, while the red spectrum is extracted from a point of control region No. 2 (listed as C2

https://doi.org/10.1051/epjconf/202125513003

in (c)). The bar plot in (c) illustrates the results including the control signal recorded from the areas between the ROIs.

4 Conclusions

In this work we presented a multiplexed optical biosensor based on 1DPC to detect the HER2 biomarker in complex biological matrix. Three different species of immunoglobulins were successfully immobilized onto the biochip surface, achieving a molecular probe density of 1173 ± 77 pg/mm^2 onto each region of interest. The direct comparison of the affinity of the three immunoglobulins for the same antigen was performed in parallel. To take into account of totally non-specific binding events, control areas in which bovine serum albumin was immobilized, were designed. This resulted in a comparable affinity of the two HER2-specific immunoglobulins (W6/300G9 and Herceptin, that is a widely used, well-known therapeutic agent). Furthermore, subtracting the fluorescence signal recorded from the control areas, it turns out that the affinity of the specific immunoglobulins is twice the affinity of the non-specific antibody (L31) towards the same biomarker (Fig. 3c) even at low concentration (1 ng/mL).

As described herein, combining the two reading modes (Label-Free and Fluorescent), it is possible to discriminate the affinity level of two distinct antibodies to HER2 from the affinity of a non-specific antibody. Furthermore, we proved that the previously demonstrated biosensor performances are not undermined by the integration of three different antibodies and an inert protein onto the same biochip surface. This represents a technical improvement of our previous works opening an integration route and expanding the applicability of an already robust optical biosensor [5].

References

1. Delfino C.S.C., Perez Cordovés A.I., Souza Jr F.G., *Res. Dev. Material. Sci.*, **7** (2018)
2. Vollmer F., Arnold S., *Nat. Methods*, **5** (2008)
3. Danz N., Sinibaldi A., Michelotti F., Descrovi E., Munzert P., Shultz U., Sonntag F., *Biomed. Tech.*, **57** (2012)
4. Rizzo R, Alvaro M, Danz N, Napione L, Descrovi E, Schmieder S, et al. Biomed Opt. Express., **9** (2018)
5. Sinibaldi A., Doricchi A., Pileri T., Allegretti M., Danz M., Munzert P., Giordani E., Giacomini P., Michelotti F., *Anal. Bioanal. Chem.*, **412** (2020).
6. Dawood S., Broglio K., Buzdar A.U., Hortobagyi G.N., Giordano S.H., *J. Clin. Oncol.*, **28** (2010)
7. Meric-Bernstam F., Hung M.C., *Clinical Cancer Research*, **12** (2006)
8. Munzert P, Danz N, Sinibaldi A, Michelotti F., *Surf. Coat. Technol.*, **314** (2017)
9. Gunda N. S. K., Singh M., Norman L., Kaur K., Mitra S.K., *Appl. Surf. Sci.*, **305** (2014)
10. DeFeijter J.A., Benjamins J. and Veer F.A., *Biopolymers*, **17**,1759-1772 (1978)

Pump-Probe Delay Controlled by Laser-dressed Ionization with Isolated Attosecond Pulses

Martin Luttmann[1*], *David* Bresteau[1], and *Thierry* Ruchon[1**]

[1]Université Paris-Saclay, CEA, CNRS, LIDYL, 91191 Gif-sur-Yvette, France

Abstract. In a recent work [1], we demonstrated how laser-dressed ionization can be harnessed to control with attosecond accuracy the time delay between an extreme-ultraviolet (XUV) attosecond pulse train and an infrared (IR) femtosecond pulse. In this case, the comb-like photoelectron spectrum obtained by ionizing a gas target with the two superimposed beams exhibits peaks oscillating with the delay. Two of them can be found to oscillate in phase quadrature, allowing an optimal measurement and stabilization of the delay over a large range. Here we expand this technique to isolated attosecond pulses, by taking advantage of the delay-modulation of attosecond streaking traces. Although the photoelectron spectrum contains no peaks in that case, it is possible to reconstruct the pump-probe delay by simply monitoring the mean energy of the spectrum and the amplitude at this energy. In general, we find that active delay stabilization based on laser-dressed ionization is possible as long as the XUV pulses are chirped.

1 Introduction

Attosecond pump-probe experiments require to control, with sub-femtosecond precision, the delay between two ultrashort light pulses. One of them is usually intense and in the Vis-infrared (IR) spectral range, while the second is weak and in the extreme-ultraviolet (XUV) spectral range. In order to perform experiments asking for high statistics, such a stability should be maintained over hours, while scanning the delay. This can only be achieved through active stabilization schemes, usually relying on the live measurement of interferences from an additional continuous laser beam co-propagating in the interferometer [2]. Such techniques have drawbacks: they require many additional optics, and most importantly do not provide an *in situ* control of the delay. Indeed, what is actually stabilized is the optical path difference accumulated by the continuous laser, which may differ from the XUV/IR optical path difference. In a previous work [1], we introduce a new technique, named LIZARD (Laser-dressed IoniZation for the Adjustment of the pump-pRobe Delay), which does not require any such additional laser beam and allows an in-situ delay control. We implemented LIZARD on an attosecond beamline delivering attosecond pulses trains (APT), reaching an excellent delay stability. The goal of this paper is to show how LIZARD can also be applied to the stabilization of interferometers that use isolated attosecond pulses (IAP, i.e. attosecond pulses that do not come in trains). Such light sources are now routinely used in experiments [3] and would draw great benefits from the stability provided by LIZARD. We first recall the basic principle of the technique, then expand it to IAP.

*e-mail: martin.luttmann@cea.fr
**e-mail: thierry.ruchon@cea.fr

2 Principle of LIZARD

When performing scans of the delay, one has to be able to continuously lock the pump-probe phase at arbitrary set-points, with a uniform stability. A solution consists in using two sinusoidal signals oscillating with the delay, which are in phase quadrature [5]. The principle of LIZARD is to extract those signals from two-photon photoelectron spectra obtained by photoionizing a gas target in an electronic spectrometer, with the superimposed XUV and IR beams. In the case of laser-dressed ionization by APT, the spectra typically show a series of peaks corresponding to the XUV harmonic energies; and sidebands in-between two harmonic peaks, oscillating with the delay [4]. Interestingly, the natural frequency chirp of attosecond pulses trains obtained via high harmonic generation (better known as "atto-chirp") garantees that, in most practical cases, two sidebands close to phase quadrature can be found, regardless of the gas chosen or the intensity of the driving laser. The signal of such sidebands are used in the active stabilization loop. Small delay fluctuations are measurable, since at least one of the signals has a significant derivative, for any given delay. The phase quadrature hence garantees that LIZARD provides an optimal and uniform stability over the whole optical cycle. The sidebands signals amplitude and offset first have to be calibrated. To do so, a fast delay scan is performed (during which we consider that unintended delay fluctuations do not occur), and the sidebands modulation is fitted with sines.

This technique was successfully implemented on the FAB10 beamline at the ATTOLab facility (CEA Saclay) [1]. The signal of two sidebands, separated in energy by 12 eV, and in phase by 0.46 π rad, was used to perform

https://doi.org/10.1051/epjconf/202125513004

the active stabilization. Despite an interferometer length of several meters, a long term delay stability of 28 as RMS was achieved.

3 Expansion of the technique to isolated attosecond pulses

Isolated attosecond pulses can be generated through various schemes. HHG can be driven by few or single cycle IR pulse. The electric field of such pulses peaking only once, XUV light is emitted in a single electron recombination event, which results in IAP. Other techniques use polarization gating [6] or double-optical gating [7]. In all cases, the supression of the $T_0/2$ periodicity of the emitted light (with T_0 the IR period) yields an XUV supercontinuum in the frequency domain. It is thus not possible to obtain delay-oscillating sidebands. A variant of LIZARD may still be implemented, that we refer to as LIZARD-IAP. It also relies on laser-dressed photo-ionization of a gas target by the XUV pulse (Fig. 1), but in the "attosecond streaking" [8] or FROG-CRAB [9] regime. It corresponds to an intensity of the dressing laser field of about 10^{13} W/cm^2 instead of about 10^{11} W/cm^2 in the previously used RABBITT regime.

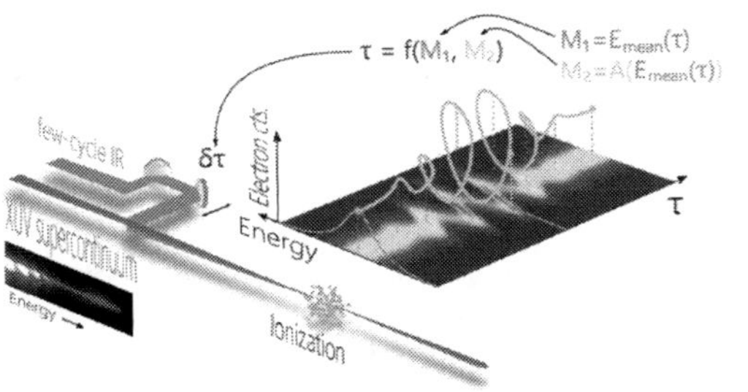

Figure 1. Principle of LIZARD-IAP. A feedback loop is performed in order to stabilize the XUV/IR delay τ, whose inputs are two quantities extracted from the streaked photoelectron spectrum: the mean energy ($M_1(\tau)$, vivid pink arrow) and the amplitude at this energy ($M_2(\tau)$, yellow arrow). If the XUV pulse is chirped, the correlation of these two quantities draws an ellipse (cyan curve). The delay τ can be reconstructed unambiguously, and any variation from the set-point value is compensated using two mirrors mounted on a piezo-electric stage. The streaking trace used in the figure is extracted from [10].

We consider an electron suddenly freed in a continuum state by absorption of an XUV photon, in the presence of a strong IR streaking field. The XUV field is denoted $\mathbf{E}_{XUV}$, the IR vector potential is $\mathbf{A}$, they are delayed in time by τ. The atom ionization energy is I_p. In the frame of first order perturbation theory, the probability amplitude for a photoelectron to have the momentum $\mathbf{p}$ in the continuum, and the energy $W = \mathbf{p}^2/2$, writes [11]

$$a_{\mathbf{p}}(\tau) = -i \int_{-\infty}^{+\infty} \mathrm{d}t\, e^{i\phi(t)}\, \mathbf{d}_{\mathbf{p}(t)} \cdot \mathbf{E}_{XUV}(t-\tau) e^{i(W+I_p)t}, \quad (1)$$

with

$$\phi(t) = -\int_t^{+\infty} \mathrm{d}t'(\mathbf{p}\cdot\mathbf{A}(t') + \mathbf{A}^2(t')/2). \quad (2)$$

$\mathbf{d}_{\mathbf{p}(t)}$ is the dipole transition matrix element from the ground state to the continuum state. $\phi(t)$ is the phase accumulated by the electron freed at time t, under the influence of an IR field. The effect of this streaking field is therefore to modulate the phase of the electron wave-packet (EWP), which in general results in distortion and shift of the photoelectron spectrum.

We propose here an intuitive explanation of the effect of the streaking field. In a semi-classical model, the atom is ionized at time t_i by the XUV field, an electron being ejected with an initial momentum $\mathbf{p}_0$. Discarding effects of the atomic potential, the electron accelerates classically in the continuum under the influence of the streaking IR field. Since we have $\partial\mathbf{p}/\partial t = -E_{IR} = \partial A/\partial t$, the quantity $\mathbf{p}(t) - A(t)$ is constant during this acceleration. Its final value equals the final momentum $\mathbf{p}_f$ of the electron, as the IR field vanishes for long times. Therefore, the momentum of the electron is shifted by the IR potential at the time of ionization: $\mathbf{p}_f = \mathbf{p}_0 - A(t_i)$, which means that the photoelectron average energy will oscillate with the XUV/IR delay (see Fig.2.a).

To get more insights on the effect of the EWP phase modulation, we consider the specific case of a sinusoidal IR field with a slowly varying enveloppe $\mathbf{E}_{IR}(t) = -\partial\mathbf{A}/\partial t = \mathbf{E}_0(t)\cos(\omega_0 t)$. While this approximation is clearly not valid for few-cycle pulses, it will still help deepening our general understanding. The phase modulation is given by [11]

$$\begin{aligned}
\phi(t) &= \phi_1(t) + \phi_2(t) + \phi_3(t)\\
\text{with} \quad \phi_1(t) &= -\int_t^{+\infty} \mathrm{d}t U_p(t),\\
\phi_2(t) &= \frac{\sqrt{8WU_p(t)}}{\omega_0}\cos\theta\cos(\omega_0 t),\\
\phi_3(t) &= -\frac{U_p(t)}{2\omega_0}\sin(2\omega_0 t).
\end{aligned} \quad (3)$$

In the above equations, U_p and θ are respectively the ponderomotive energy of the field and the angle between the momentum $\mathbf{p}$ and the laser polarization direction. We restrain our analysis to $\theta = 0$ and $U_p \ll W$, hence ϕ_2 dominates over the two other terms.

We simulate photoelectron spectra of Ar gas ionized by a single attosecond pulse in the presence of a time-delayed low-frequency few-cycle IR field (Fig. 2). The duration of the gaussian XUV pulse is set at 300 as, the atto-chirp can be adjusted. We consider a gaussian, 4 fs, 800 nm central wavelength, 10^{13} W/cm^2 peak intensity dressing pulse with a null CEP. The temporal phase modulation of the EWP is calculated from Eq. 2. The energy *vs* delay trace is generated by computing the outer product of the two fields, and Fourier transform it along one dimension.

For a given delay τ between the two pulses, the photoelectron spectra in Fig. 2 are continuous, which is characteristic of ionization by IAP. The mean energy and peak amplitude vary with the delay, which can be understood by

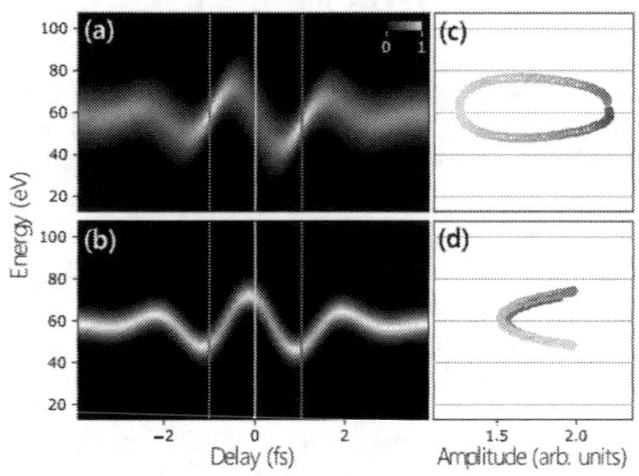

Figure 2. (a,b) Simulated attosecond streaking traces, with a 3×10^2 fs^{-2} atto-chirp (a), and with a null atto-chirp (b). (c,d) Corresponding correlation graphs between the mean energy (vertical axis) and the amplitude at this energy (horizontal axis), for delays varying between the red and blue vertical lines in (a,b). The delay is increasing from red to blue, and the horizontal axis of (c) and (d) have the same scale.

considering Eq. 3. When the attosecond pulse is synchronized with a zero of the IR field, the phase modulation is quasi-linear in the neighborhood of the pulse. This corresponds to a frequency shift of the whole spectrum (without distortion) towards high or low energies depending on whether the field amplitude is decreasing or increasing. In the case where the attosecond pulse is synchronized with a maximum of the IR field, the phase modulation is quadratic, and its effect adds up to the atto-chirp, yielding a spectral broadening and a decrease of the peak amplitude of the photoelectron spectrum. This is the case, for instance, along the green vertical line in Fig. 2.a. Conversely, when the attosecond pulse is synchronized with a minimum of the streaking field, the two effects partially compensate each other, yielding a narrower spectrum and a higher peak amplitude (red and purple lines in Fig. 2.a). Therefore, the photoelectron mean energy $E_{\mathbf{mean}}(\tau)$ and the amplitude at this energy $A(E_{\mathbf{mean}}(\tau))$ are in phase quadrature (Fig. 1) ,which provides an optimal stability. Such an asymmetry in the streaking trace is a signature of frequency chirp, and is key to LIZARD: the XUV frequency chirp allows to discriminate between a maximum and a minimum of the IR amplitude, hence to measure the pump-probe delay over the whole IR period. This conclusion applies both for LIZARD with APT and LIZARD-IAP.

We test these predictions with the simulated streaking traces. In Fig. 2.c, we display the mean energy of the spectrum *vs* the amplitude at this energy, as the delay is varied over one IR period. The correlation graph takes the form of an ellipsoid, i.e. the time delay is encoded azimutally and can be reconstructed. When the attosecond pulse has no chirp (Fig. 2.c), the graph collapses, several points ending up at the same angle, preventing delay retrieval over the whole optical cycle. Just like regular LIZARD, LIZARD-IAP needs a preliminary calibration of the ellipse, consisting in a fast delay scan.

We must highlight that LIZARD-IAP requires carrier-phase-envelope (CEP) stabilization, since changes of the CEP affect dramatically the streaking spectrum. Such stabilization loops are routinely used in beamlines delivering IAP along with single-cycle IR pulses. Conversely, if the interferometer is stabilized by another mean, monitoring the mean energy and peak amplitude of the photoelectron spectrum could be a way to access the CEP.

4 Conclusion

We have shown that the LIZARD technique, allowing active stabilization of the time delay between an attosecond and a femtosecond light pulse, can be applied to IAP. LIZARD requires two phase quadrature signals that are modulated with the delay; such signals can be extracted from the spectrum obtained by photo-ionizing a gas target with the two superimposed beams, in the FROG-CRAB regime. Provided that the IAP are chirped, the mean energy of the photoelectron spectrum and the amplitude at this energy vary with the delay and are $\pi/2$ out of phase.

The LIZARD-IAP technique is robust with respect to the experimental conditions. It is indeed applicable with any IR intensity (provided that it remains in the FROG-CRAB regime, i.e. about 10^{13} W/cm^2), any value of the atto-chirp (as long as it is significant), and regardless of the gas chosen.

The concepts developped here could be applied to the spectrograms obtained with free-electron-lasers (FEL). Such instruments are currently reaching attosecond durations, delivering sub-femtosecond XUV pulses. FEL, however, cannot provide sub-cycle synchronization between the XUV and IR field. It was shown in the case of multiple harmonics that an analysis of the correlation between the different sidebands fluctuations provides a measurement of the temporal jitter [12, 13]. We believe that a similar post-treatment can be made when the X-ray field consists of IAP. The correlation between the mean energy of the photoelectron spectrum and the peak amplitude would take the shape of an ellipse, allowing the retrieval of the pump-probe delay.

References

[1] M. Luttmann *et al*, Phys. Rev. Applied **15**, 034036 (2021)
[2] F. Böttcher *et al*, Applied Physics B **91**, 287–293 (2008)
[3] M. Chini *et al*, Nature Photon **8**, 178-186 (2014)
[4] V. Véniard *et al*, Phys. Rev. A **54**, 721 (1996)
[5] V. V. Krishnamachari *et al*, Opt. Express **14**, 5210 (2006)
[6] P. B. Corkum *et al*, Opt. Lett. **19**, 1870-1872 (1994)
[7] H. Mashiko *et al*, Phys. Rev. Lett. **10**, 103906 (2008)
[8] J. Itatani *et al*, Phys. Rev. Lett. **88**, 173903 (2002)
[9] Y. Mairesse *et al*, Phys. Rev. A. **71**, 011401 (2005)

[10] L. Cattaneo *et al*, Opt. Express **25**, 29060-29076 (2016)
[11] F. Quéré *et al*, Journal of Modern Optics **52**, 339-360 (2005)
[12] P.K. Maroju *et al*, Nature **578**, 386–391 (2020)
[13] C. Krüger *et al*, Opt. Express **28**, 12862-12871 (2020)

Ultrafast exciton dynamics reconstruction with a ptychographic approach

Gian Luca Dolso[1,*], *Bruno* Moio[1,2], *Giacomo* Inzani[1], *Nicola* Di Palo[1,2], *Rocío* Borrego-Varillas[2], *Mauro* Nisoli[1,2], and *Matteo* Lucchini[1,2]

[1]Department of Physics, Politecnico di Milano, 20133 Milano, Italy
[2]Institute for Photonics and Nanotechnologies, IFN-CNR, 20133 Milano, Italy

Abstract. Excitons characterize the ultrafast response of many materials of technological interest. While the development of attosecond science has unlocked the possibility of performing experiments with a suitable time-resolution, the access to the exciton properties remains a non-trivial step. We propose therefore a novel approach to disclose the physical properties behind the ultrafast exciton dynamics based on a phase-retrieval method.

1 Introduction

Excitons are at the basis of several technological applications in the field of electronics and optoelectronics [1], and their relevance for future devices is expected to grow. The thorough study of their physical properties and optical response is, therefore, a necessary endeavor to achieve their active (optical) control. In this context, attosecond transient absorption and reflection spectroscopies have allowed for the study of ultrafast exciton and electron dynamics in solid-state materials [2–4]. However, the direct extraction of the relevant parameters governing the excitonic response, such as the lifetime and the phonon coupling strength, is a non-trivial procedure and often requires *a priori* assumptions. Here, we propose a novel method to achieve this goal: ePIX (the extended Ptychographic Iterative engine for eXcitons), based on an iterative ptychographic approach [5].

2 Methods

To describe the ultrafast exciton dynamics, we employ the model reported in [6] for atomic-like quasi-particles in solids. The attosecond extreme-ultraviolet (XUV) pulse induces an electronic dipole in the material

$$O(t) \propto \theta(t)\, e^{-\frac{t}{T_e}}\, e^{-i\omega_e t}\, e^{\Phi(t)} \tag{1}$$

where T_e is the Auger lifetime of the exciton, $e^{\Phi(t)}$ describes phonon coupling, as discussed in [7], and $\hbar\omega_e$ is the transition energy associated to the XUV-induced excitation. $\theta(t)$ is the Heaviside function, that physically describes the instantaneous creation at time $t = 0$. The complex dynamic permittivity $\epsilon(\omega, \tau)$ resulting from the interaction of the dipole with an intense infrared (IR) field can be expressed as

$$\epsilon(\omega, \tau) = \epsilon_0 + k \int_{-\infty}^{+\infty} O(t)\, P(t, \tau)\, e^{-i\omega t}\, dt \tag{2}$$

where ϵ_0 is the permittivity response of the crystal structure (for simplicity supposed to be constant), k is a constant and $P(t, \tau)$ is defined as [8]

$$P(t, \tau) = e^{-(\gamma + i\frac{\alpha}{2}) \int_0^t E^2(t' - \tau)\, dt'} . \tag{3}$$

The quantity $P(t, \tau)$ models the interaction of the dipole $O(t)$ with the IR field $E(t)$ through the dynamical Stark shift, and the quantity $\gamma + i\frac{\alpha}{2}$ is the complex polarizability associated to the exciton state [9]. The result of a transient reflection experiment, where the XUV induces the formation of an exciton, can be modelled by applying the Fresnel equations to the permittivity $\epsilon(\omega, \tau)$. The IR vector potential $A(t)$ defines the IR field $E(t)$ and is assumed to be Gaussian within the reconstructing algorithm

$$A(t) = A_0 \cos\left\{\omega_{IR}\, t + \Phi_{IR} + CR\, t^2\right\} e^{-4\log 2\, t^2 / FWMH^2} \tag{4}$$

where A_0 is the amplitude, ω_{IR} is the frequency, Φ_{IR} is the phase, CR is the chirp rate and FWMH is the amplitude full width at half maximum. Our reconstruction method employs the theoretical model of Eq. (2) to extract $O(t)$ and $E(t)$ from a transient reflection trace. The algorithm is a suitably-modified version of the extended ptychographic iterative engine (ePIE), already employed in the time domain to perform attosecond pulse reconstruction. ePIX, however, significantly differs from ePIE: the constraints set on the object $O(t)$ and probe $P(t)$ must be appropriately recast, in order to apply the algorithm to the new physical context. The reconstruction takes place in two distinct steps within the algorithm, to enhance robustness of the convergence. The first stage (Ptychographic Iterative Engine, PIE) updates the the complex polarizability $\gamma + i\frac{\alpha}{2}$, while keeping the IR field $E(t)$ fixed. The second step (extended Ptychographic Iterative Engine, ePIE) updates the field, within the assumption of a Gaussian temporal profile. If convergence is not reached, it is always possible to iterate this procedure. Figure 1 summarizes the reconstruction process.

*e-mail: gianluca.dolso@polimi.it

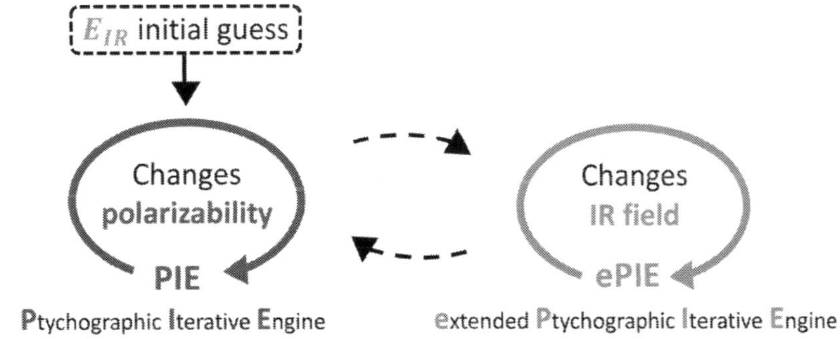

Figure 1. Scheme of the reconstruction algorithm. The reconstruction takes place in two steps: in the first stage (PIE), the algorithm updates the complex polarizability $\gamma + i\frac{\alpha}{2}$; in the second stage (ePIE), the polarizability is kept constant, and the IR field is modified. Both PIE and ePIE update the dipole $O(t)$, associated to the exciton dynamics.

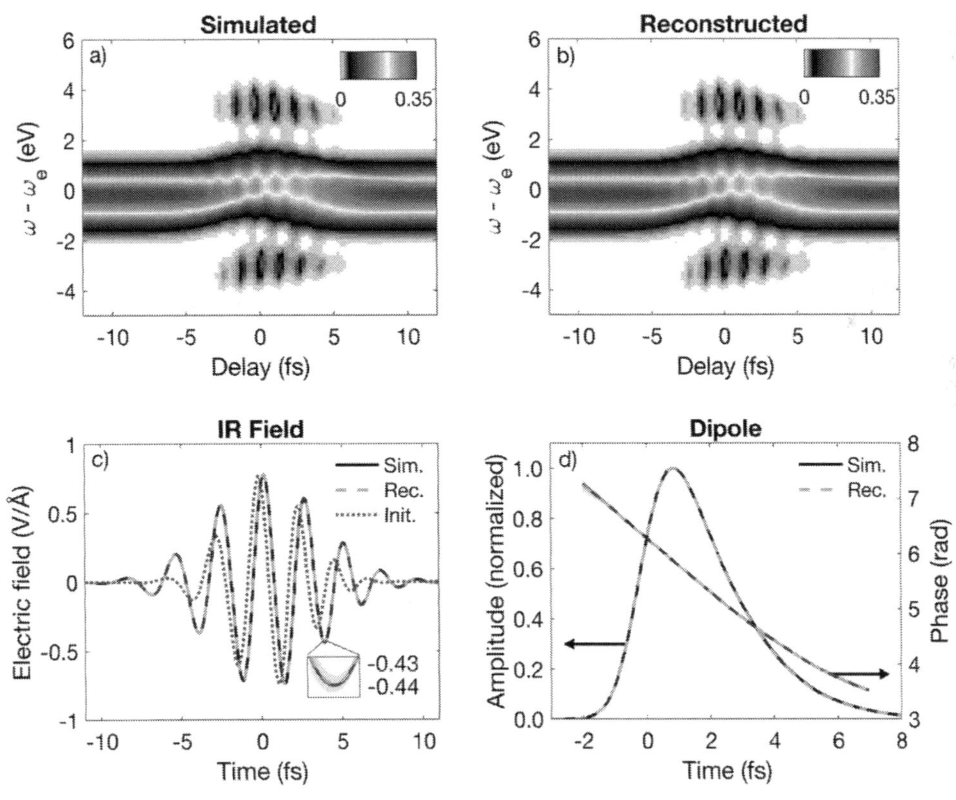

Figure 2. Simulated, a), and reconstructed, b), transient reflectivity trace associated to an excitonic transition, described by the parameters of Table 1. c) Simulated, black solid, and reconstructed, red dashed, dipole moments. d) IR field: simulated, black solid, reconstructed, red dashed, and an example of initial guess, gray dotted. Both amplitude and phase of the dipole moment $O(t)$ are correctly reconstructed. Notice, also, that the ptychographic approach also allows for the retrieval of the IR field.

https://doi.org/10.1051/epjconf/202125513005

Table 1. Simulated and reconstructed parameters, defining the exciton, its interaction with the phonons and with the IR field.

Parameter	Sim.	Rec.
Amplitude k (fs^{-1})	30.0	30.0 ± 0.22
Lifetime T_e (fs)	4.0	3.99 ± 0.076
Exciton freq. ω_e (rad fs^{-1})	84	84 ± 0.0083
Phonon coupl. M_0 (fs^{-1})	0.3	0.3 ± 0.0014
Phonon freq. ω_0 (rad fs^{-1})	0.15	0.15 ± 0.0057
IR FWHM (fs)	8.0	7.99 ± 0.045
IR phase Φ_{IR} (rad)	1.26	1.25 ± 0.013
IR chirp CR (fs^{-2})	0.05	0.051 ± 0.0035
α (rad fs^{-1} Å^2 V^2)	−5.0	−5.02 ± 0.037
γ (rad fs^{-1} Å^2 V^2)	2.0	2.01 ± 0.022

3 Results

To verify the validity of our approach, we tested ePIX on a set of simulated traces, where we include Auger decay and a finite phonon relaxation. The values used for the simulations are summarized in Table 1. Figure 2a shows the transient reflectance trace, described by the parameters of Table 1 and simulated with Eq. (2). The reconstructed trace obtained with ePIX is shown in Fig. 2b. All features in the transient reflectivity trace are correctly recovered. The comparison between the simulated and reconstructed dipole moments, in Fig. 2c, confirms the accuracy of the reconstruction. Notice that the phase of the dipole is correctly recovered, as well as the modulus. The exciton and phonon parameters can then be determined by fitting the output dipole $O(t)$. The results are reported in Table 1: the indicated errors are standard deviations, obtained by performing 4 reconstructions with different initial guesses. We found the final reconstructed values of all parameters to agree with the simulated with high degree of accuracy. Note that our method also allows for the recovery of the IR field, as shown in Fig. 2d. We have also verified that our method is more accurate when compared to an ordinary non-linear fitting procedure, and requires fewer *a priori* information on the exciton dynamics, since ePIX makes no assumption regarding the dipole $O(t)$. To further demonstrate that ePIX is a valuable tool for the investigation of exciton dynamics in transient reflectance measurement, we applied the algorithm to different transient reflectance traces, obtained with different choices of the physical parameters: in all cases, the reconstruction accuracy was the same as discussed here.

4 Conclusions

We have proved that our novel approach to reconstruct exciton dynamics, ePIX, is effective and precise, and allows to determine the characteristics of the exciton interaction with the phonon population and the external IR field. The method fully exploits the potential and versatility of ptychographic algorithms: the reconstruction of the exciton time-dependent dipole is robust and independent of the initial guess. In addition, the IR field can be characterized within the same reconstruction procedure. ePIX can hence be a useful tool in understanding ultrafast exciton dynamics, paving the way towards a more complete comprehension and control of their optoelectronic properties.

References

[1] L. V. Butov, Superlattice Microst. **108**, 2-26 (2017)
[2] R. Geneaux et al., Philos. T. R. Soc. A **377**, 20170463 (2019)
[3] M. Lucchini et al., Nat. Commun. **12**, 1021 (2021)
[4] G. D. Lucarelli et al., Rev. Sci. Instrum. **91**, 053002 (2020)
[5] M. Lucchini et al., Opt. Express **23**, 29502 (2015)
[6] A. Moulet et al., Science **357**, 1134-1138 (2017)
[7] G. D. Mahan, Phys. Rev. B **15**, 4587 (1977)
[8] A. Wirth et al., Science **334**, 195-200 (2011)
[9] R. Géneaux et al., Phys. Rev. Lett. **124**, 207401 (2020)

AUTHOR INDEX

AUTHOR INDEX

AUTHOR INDEX

EDP Sciences – Web of Conferences
17, Avenue du Hoggar
Parc d'Activité de Courtabouf
BP 112
F-91944 Les Ulis Cedex A
France

ISBN 978-1-7138-3916-3

CURRAN ASSOCIATES INC.
proceedings
.com